Texts and Monographs in Physics

Series Editors: R. Balian W. Beiglböck H. Grosse E. H. Lieb
N. Reshetikhin H. Spohn W. Thirring

Springer
Berlin
Heidelberg
New York
Barcelona
Budapest
Hong Kong
London
Milan
Paris
Santa Clara
Singapore
Tokyo

Texts and Monographs in Physics

Series Editors: R. Balian W. Beiglböck H. Grosse E. H. Lieb
N. Reshetikhin H. Spohn W. Thirring

**From Microphysics to Macrophysics
I + II** Methods and Applications
of Statistical Physics By R. Balian

**Variational Methods
in Mathematical Physics**
A Unified Approach
By P. Blanchard and E. Brüning

**Quantum Mechanics:
Foundations and Applications**
3rd enlarged edition By A. Böhm

The Early Universe
Facts and Fiction 3rd corrected
and enlarged edition By G. Börner

**Geometry of the Standard Model
of Elementary Particles**
By A. Derdzinski

**Random Walks, Critical Phenomena,
and Triviality in Quantum Field
Theory** By R. Fernández, J. Fröhlich
and A. D. Sokal

Quantum Relativity
A Synthesis of the Ideas of Einstein
and Heisenberg
By D. Finkelstein

Quantum Mechanics I + II
By A. Galindo and P. Pascual

The Elements of Mechanics
By G. Gallavotti

Local Quantum Physics
Fields, Particles, Algebras
2nd revised and enlarged edition
By R. Haag

Elementary Particle Physics
Concepts and Phenomena
By O. Nachtmann

**Inverse Schrödinger Scattering
in Three Dimensions**
By R. G. Newton

**Scattering Theory of Waves
and Particles** 2nd edition
By R. G. Newton

Quantum Entropy and Its Use
By M. Ohya and D. Petz

**Generalized Coherent States
and Their Applications**
By A. Perelomov

Essential Relativity Special, General,
and Cosmological 2nd revised edition
By W. Rindler

**Path Integral Approach
to Quantum Physics** An Introduction
2nd printing By G. Roepstorff

**Advanced Quantum Theory
and Its Applications Through Feynman
Diagrams** 2nd edition
By M. D. Scadron

Finite Quantum Electrodynamics
The Causal Approach 2nd edition
By G. Scharf

From Electrostatics to Optics
A Concise Electrodynamics Course
By G. Scharf

**Large Scale Dynamics of Interacting
Particles** By H. Spohn

**General Relativity and Relativistic
Astrophysics** By N. Straumann

The Dirac Equation By B. Thaller

**The Theory of Quark and Gluon
Interactions** 2nd completely revised
and enlarged edition By F. J. Ynduráin

Gert Roepstorff

Path Integral Approach to Quantum Physics

An Introduction

With 26 Figures

 Springer

Professor Dr. Gert Roepstorff

Institut für Theoretische Physik, Rheinisch-Westfälische Technische Hochschule Aachen
RWTH-Physikzentrum, D-52062 Aachen, Germany

Editors

Roger Balian

CEA
Service de Physique Théorique de Saclay
F-91191 Gif-sur-Yvette, France

Wolf Beiglböck

Institut für Angewandte Mathematik
Universität Heidelberg
Im Neuenheimer Feld 294
D-69120 Heidelberg, Germany

Harald Grosse

Institut für Theoretische Physik
Universität Wien
Boltzmanngasse 5
A-1090 Wien, Austria

Elliott H. Lieb

Jadwin Hall
Princeton University, P. O. Box 708
Princeton, NJ 08544-0708, USA

Nicolai Reshetikhin

Department of Mathematics
University of California
Berkeley, CA 94720-3840, USA

Herbert Spohn

Theoretische Physik
Ludwig-Maximilians-Universität München
Theresienstraße 37
D-80333 München, Germany

Walter Thirring

Institut für Theoretische Physik
Universität Wien
Boltzmanngasse 5
A-1090 Wien, Austria

Library of Congress Cataloging-in-Publication Data. Roepstorff, Gert. [Pfadintegrale in der Quantenphysik. English] Path integral approach to quantum physics: an introduction / Gert Roepstorff. p. cm. — (Texts and monographs in physics) Includes bibliographical references and index. ISBN 3-540-55213-8 (Berlin: alk. paper). — ISBN 0-387-55213-8 (New York: alk. Paper) 1. Integrals, Path. 2. Quantum theory. I. Title. II. Series. QC174.17.P27R6413 1994 530.1'2—dc20 93-43689

Title of the original German edition: Gert Roepstorff – *Pfadintegrale in der Quantenphysik*
© Friedr. Vieweg & Sohn Verlagsgesellschaft mbH, Braunschweig 1991

1st Edition 1994 · 2nd Printing 1996

ISSN 0172-5998
ISBN 3-540-61106-1 Springer-Verlag Berlin Heidelberg New York

Typesetting: Camera ready copy from the author using a Springer TeX macro package
SPIN: 10535691   55/3144-543210 - Printed on acid-free paper

To Ingrid
who made it all worthwhile

Preface

This book has been written twice. After having written and published it in German in 1990, I started all over again and rewrote the whole story for an English speaking audience. During the first round I received encouraging words and critical remarks from students and colleagues alike which have helped to sustain me the second time around.

In the preface the author usually states that his or her book resulted from a course that he or she gave at some university. I cannot claim that the present book is any exception to the rule. But I expanded and remodelled the original material which circulated as a manuscript so that the printed version would follow a more stringent and coherent architectural plan. In doing so I have concentrated on the conceptual problems inherent in the path integral formalism rather than on certain highly specialized techniques used in applications. Nevertheless, I have also included those methods that are of fundamental interest and have treated specific problems mainly to illustrate them.

The text is designed to introduce graduate students to the functional integration method in contemporary physics as painlessly as possible without their being forced to spend too much time solely in getting oriented in the mathematical intricacies of measure theory. In the development of the method, there is a striking interplay between stochastic processes, statistical physics, and quantum mechanics. This aspect, I felt, should be stressed in a text on path integration. As for the prerequisites, the student is assumed to be familiar with quantum mechanics and, on the mathematical side, with probability theory. Moreover, it is hoped that he or she has grasped the essentials of quantum field theory and elementary particle physics and is open minded.

This expository work is certainly not meant to be a substitute for the conventional operator version of quantum physics: in fact, no attempt has been made to rewrite parts of quantum mechanics or field theory using the rather sophisticated language of path integration. Nor is this work a formal account of all the activities in the field but largely a personal book whose factual details are organized and dominated by my views.

Since I began thinking about path integration, I owe thanks for their guidance to more colleagues than I could possibly name here. My special thanks are due to L. S. Schulman; without him the book could not have come into existence. I owe a great deal to E. H. Lieb who stopped me from making foolish mistakes. I would like to thank R. Haag, J. E. Roberts, and J. Challifour for their sharpening my understanding of quantum field theory. I received advice from W. Thirring, W. Beiglböck, A. Uhlmann, L. Streit, A. Martin, W. Bietenholz, and H. Siedentop. Special thanks go to H. Spohn and M. Demuth for their wisdom and encouragement. I should also like to thank my students C. Beck and M. Ringe for valuable discussions during the preparation of the early draft. I also thank M. Seymour who went carefully through the final draft and pointed out a number of misprints and suggested improvements with regard to style and clarity. Last but not least I am grateful to the Institute for Advanced Study, Princeton, for hospitality.

September 1993   Gert Roepstorff

Contents

1. Brownian Motion 1
  1.1 The One-Dimensional Random Walk 2
  1.2 Multidimensional Random Walk 6
  1.3 Generating Functions 10
    1.3.1 Return or Escape? 11
  1.4 The Continuum Limit 13
  1.5 Imaginary Time 15
  1.6 The Wiener Process 20
    1.6.1 The Analysis of Random Paths 20
    1.6.2 Multidimensional Gaussian Measures 24
    1.6.3 Increments 27
  1.7 Expectation Values 29
  1.8 The Ornstein–Uhlenbeck Process 33
    1.8.1 The Oscillator Process 36

2. The Feynman–Kac Formula 39
  2.1 The Conditional Wiener Measure 40
    2.1.1 The Path Integral 41
    2.1.2 The Stochastic Formulation 45
  2.2 The Integral Equation Method 48
    2.2.1 Stochastic Representation of Operator Norms ... 51
    2.2.2 Stochastic Representation of Green's Functions . 53
  2.3 The Lie–Trotter Product Method 54
    2.3.1 The Lie–Trotter Product Formula 55
    2.3.2 Miscellaneous Remarks and Results 57
    2.3.3 Several Particles with Different Masses 61
  2.4 The Brownian Tube 62
  2.5 The Golden–Thompson–Symanzik Bound 66
  2.6 Hamiltonians and Their Associated Processes 74
    2.6.1 Correlation Functions 75
    2.6.2 The Oscillator Process Revisited 76
    2.6.3 Nonlinear Transformations of Time 78
    2.6.4 The Perturbed Harmonic Oscillator 79
  2.7 The Thermodynamical Formalism 81
  2.8 A Case Study: the Harmonic Spin Chain 85
    2.8.1 The Inverted Harmonic Oscillator 88

2.9   The Reflection Principle 90
       2.9.1   Reflection Groups of Order Two 91
       2.9.2   Reflection Groups of Infinite Order 94
2.10   Feynman Versus Wiener Integrals 101
       2.10.1   Summing over Histories in Configuration Space . 102
       2.10.2   The Method of Stationary Phase 104
       2.10.3   Summing over Histories in Phase Space 105
       2.10.4   The Feynman Integrand as a Hida Distribution . 107

3. **The Brownian Bridge** 109
    3.1   The Canonical Scaling of Brownian Paths 109
           3.1.1   The Process $\bar{X}_\tau$ 112
           3.1.2   Rescaling of Path Integrals 113
           3.1.3   The Stochastic Integral with Respect
                    to the Brownian Bridge 114
    3.2   Bounds on the Transition Amplitude 115
           3.2.1   Defining a Subset of Paths 115
           3.2.2   The Semiclassical Approximation 117
           3.2.3   Bounds on the Functional $\Phi(V)$ 118
           3.2.4   Convexity of the Functional $\Phi(V)$ 120
    3.3   Variational Principles 122
           3.3.1   The Mean Position of a Path 125
    3.4   Bound States 126
           3.4.1   Moment Inequalities for Eigenvalues 135
    3.5   Monte Carlo Calculation of Path Integrals 142

4. **Fourier Decomposition** 150
    4.1   Random Fourier Coefficients 150
           4.1.1   Fourier Analysis of Time Integrals 151
    4.2   The Wigner–Kirkwood Expansion
           of the Effective Potential 154
    4.3   Coupled Systems 157
           4.3.1   Open Systems 159
    4.4   The Driven Harmonic Oscillator 161
           4.4.1   From Time Integrals to Sums 162
           4.4.2   From Sums Back to Time Integrals 162
    4.5   Oscillating Electric Fields 166
           4.5.1   Poisson Statistics 167

5. **The Linear-Coupling Theory of Bosons** 170
    5.1   Path Integrals for Bosons 170
           5.1.1   The Partial Trace and Its Evaluation 172
    5.2   A Random Potential for the Electron 175
    5.3   The Polaron Problem 177
           5.3.1   The Limit $L \to \infty$, $b \to 0$ 179

5.3.2   The Free Energy of the Polaron 182

5.3.3   Bounds on the Polaron Free Energy 183

5.3.4   Pekar's Large-Coupling Result 185

5.4   The Field Theory of the Polaron Model 186

6. Magnetic Fields 192

6.1   Heuristic Considerations 192

6.2   Itô Integrals 195

6.2.1   The Feynman–Kac–Itô Formula 197

6.2.2   The Semiclassical Approximation 199

6.3   The Constant Magnetic Field 201

6.3.1   A Brief Discussion of the Result 204

6.4   Diamagnetism of Electrons in a Solid 205

6.5   Magnetic Flux Lines 208

6.5.1   Winding Numbers 209

6.5.2   Spectral Decomposition 210

6.5.3   Imaginary Times 212

7. Euclidean Field Theory 215

7.1   What Is a Euclidean Field? 216

7.2   The Euclidean Two-Point Function 218

7.3   The Euclidean Free Field 222

7.3.1   The n-Point Functions 222

7.3.2   The Stochastic Interpretation 225

7.4   Gaussian Functional Integrals 227

7.5   Basic Postulates 233

7.5.1   The Hamiltonian 236

7.5.2   The Free Field Revisited 238

8. Field Theory on a Lattice 242

8.1   The Lattice Version of the Scalar Field 242

8.2   The Euclidean Propagator on the Lattice 245

8.2.1   The Fourier Representation 245

8.2.2   Random Paths on a Lattice 250

8.3   The Variational Principle 252

8.3.1   The Case of a Discrete Configuration Space 252

8.3.2   The Deterministic Limit 254

8.3.3   Continuous Configuration Space 255

8.3.4   The Classical Limit 257

8.3.5   Fluctuations Around the Classical Solution 258

8.4   The Effective Action 260

8.5   The Effective Potential 265

8.5.1   Spontaneous Breakdown of Symmetry 266

8.5.2   Order Parameters 267

8.6   The Ginzburg–Landau Equations 268

8.7   The Mean-Field Approximation 272
        8.7.1   The Curie–Weiss Approximation
                of the Ising Model 274
        8.7.2   The Ising Spin Limit of the Neutral Scalar Field 277
8.8   The Gaussian Approximation 278
        8.8.1   A Case Study 278

9.   The Quantization of Gauge Theories 281
9.1   The Euclidean Version of Maxwell Theory 281
        9.1.1   The Classical Situation ($\hbar = 0$) 282
        9.1.2   Gauge Fixing 285
        9.1.3   The Quantized Situation ($\hbar > 0$) 287
9.2   Non-Abelian Gauge Theories: Preliminaries 289
9.3   The Faddeev–Popov Quantization 292
        9.3.1   Division by $|\mathcal{G}|$ 295
        9.3.2   Faddeev-Popov Ghosts 297
9.4   Gauge Theories on a Lattice 300
9.5   Wegner–Wilson Loops 306
        9.5.1   The Static Approximation in Minkowskian
                Field Theory 306
        9.5.2   Loop Variables in Euclidean QED 308
        9.5.3   Area Law or Perimeter Law? 310
9.6   The $SU(n)$ Higgs Model 312

10. Fermions 316
10.1   The Dirac Field in Minkowski Space 316
10.2   The Euclidean Dirac Field 319
        10.2.1 External Vector Potentials 324
10.3   Grassmann Algebras 326
        10.3.1 When E Is a Function Space 329
10.4   Formal Derivatives 331
10.5   Formal Integration 334
        10.5.1 Integrals in $A(E)$ 334
        10.5.2 Integrals in $A(E \oplus F)$ 336
        10.5.3 Integrals of the Exponential Type 337
        10.5.4 The Fourier–Laplace Transformation 339
10.6   Functional Integrals of QED 342
10.7   The $SU(n)$ Gauge Theory with Fermions 346

Appendices
A   List of Symbols and Glossary 349
B   Frequently Used Gaussian Processes 357
C   Jensen's Inequality 360
D   A Table of Path Integrals 362

References 369

Index 383

1 Brownian Motion

The main advantages of a discrete approach are pedagogical, inasmuch as one is able to circumvent various conceptual difficulties inherent to the continuous approach. It is also not without a purely scientific interest [...].

Marc Kac

The shortest path between two truths in the real domain passes through the complex domain.

J. Hadamard

Physics is often seen as being rooted in, and to the present day deals with, the study of moving bodies. We have every reason to believe that, historically, phenomenological attempts at describing the observed preceded speculations about the underlying dynamical law. This chapter focuses on random motion, first described in 1828 by the British botanist R. Brown, who investigated the pollen of different plants dispersed in water. Years after the discovery scientists began to realize that any kind of inorganic substance, not just "living matter", presents, in principle, the same phenomenon, and thus looked for an explanation. In fact a respectable *theory of Brownian motion* emerged much later (not before 1905) as a result of an interplay between physics and mathematics. At present the prospects for possible applications[1] in the exact sciences seem unlimited: the concepts of Brownian motion are now being used in fields as different as astronomy (stellar dynamics), diffusion, colloid chemistry, polymer physics, quantum mechanics, and elementary-particle physics. The surprise is that the scale of length does not matter at all. Instead what strikes the eye, as a common characteristic, is some kind of universal mathematical structure. Nevertheless, the concept of Brownian motion as a whole together with its highly specialized tools does not seem to fit into the traditional framework of classical mechanics as a deterministic theory that, according to a dictum of A. Sommerfeld, represents the "backbone of mathematical physics".

The apparent irregular motion that we shall describe, nondeterministic as it may be, does not take place without obeying certain rules, and it took the genius of A. Einstein to notice and apply these rules successfully. Through his pioneering work [1.2] the theory of Brownian motion acquired a firm position within the fabric of physics.

---

[1] A collection of early significant contributions is presented in [1.1].

1.1 The One-Dimensional Random Walk

The principal features of the problem of the random walk can be elucidated by analyzing the simplest of all cases: the erratic motion of a single point particle in one dimension [1.3]. Imagine the particle suffers displacements along the x-axis in the form of a series of steps of the same length h, each step being taken in either direction within a certain period of time, say of length τ. In essence, one may think of both space and time as being replaced by sequences of equidistant marks: from now on we shall call such models *discrete*.

In addition, supposing that there is no physical factor preferring *right* over *left*, we may postulate that forward and backward steps occur with equal probability $\frac{1}{2}$. Successive steps are assumed to be statistically independent. Hence, the probability is

$$P(ih - jh, \tau) = \begin{cases} \frac{1}{2} & \text{if } |i - j| = 1 \\ 0 & \text{otherwise} \end{cases} \qquad (i, j \in \mathbb{Z}) \qquad (1.1.1)$$

for a transition from $x = jh$ to the new position $x = ih$ during the time τ.

In still other words and at a higher level, what we have before us is an example of a stochastic process, more precisely, of a *Markov chain* with a denumerable set of states [1.4]. The process has two obvious properties. It is

homogeneous: the transition probability P is merely a function of the difference $i - j$; and

isotropic: the transition probability does not depend on the direction in space, i.e., P is left unchanged if we replace (i, j) by $(-i, -j)$.

Quite generally, a Markov chain may be characterized by a pair (P, p), where $P = (P_{ij})$ stands for what is called a *transition matrix* and $p = (p_i)$ is the *initial probability distribution*. In simpler terms: p_i is the probability of the event i occurring at starting time $t = 0$. One always has $0 \leq p_i \leq 1$, $\sum_i p_i = 1$, $0 \leq P_{ij} \leq 1$, and $\sum_i P_{ij} = 1$. As for our example, the event i is identified with the particle's position $x = ih$ and the matrix P has components

$$P_{ij} = P(ih - jh, \tau). \qquad (1.1.2)$$

Caution: this matrix is doubly infinite, $-\infty < i, j < \infty$, and from another point of view it seems more appropriate to call it an *operator*.

After the elapse of time $n\tau$ $(n \in \mathbb{N})$ the accumulated transition probabilities are

$$P(ih - jh, n\tau) = (P^n)_{ij}, \qquad (1.1.3)$$

where $P^n = P \cdot P \cdots P$ (n factors) stands for the n-fold matrix product. What about the initial distribution? If, at time $t = 0$ the position of the particle is known with certainty, say $x = 0$, we have $p_i = 0$ for $i \neq 0$ and $p_0 = 1$. After time $n\tau \geq 0$, the system has evolved and produced a new distribution which is $P^n p$. As a matter of convenience, p and $P^n p$ are treated here as vectors. Phrased differently, as a function of n, P^n is the operator of evolution for the system, where we regard n as the relevant time variable. Within this setting, time never assumes negative values.

The operators

$$
R = \begin{pmatrix} \ddots & & & & 0 \\ 1 & 0 & & & \\ & \ddots & \ddots & & \\ & & 1 & 0 & \\ 0 & & & \ddots & \ddots \end{pmatrix} \qquad L = \begin{pmatrix} \ddots & \ddots & & & 0 \\ & 0 & 1 & & \\ & & \ddots & \ddots & \\ & & & 0 & 1 \\ 0 & & & & \ddots \end{pmatrix} \tag{1.1.4}
$$

shift the particle's position to the right and left respectively by the amount h. Obviously, $L = R^{-1}$ and thus $RL = LR$. This clarifies the structure of the operator P and its powers: first notice that

$$
P = \tfrac{1}{2}(R + L) \tag{1.1.5}
$$

and then write

$$
P^n = \frac{1}{2^n} \sum_{k=0}^{n} \binom{n}{k} R^k L^{n-k} \tag{1.1.6}
$$

to obtain the transition probabilities after n time steps:

$$
P(ih - jh, n\tau) = \frac{1}{2^n} \binom{n}{k}, \qquad i - j = k - (n - k). \tag{1.1.7}
$$

By appeal to the recursion formula for the binomial coefficients,

$$
\binom{n+1}{k} = \binom{n}{k} + \binom{n}{k-1}, \tag{1.1.8}
$$

one gets the following remarkable difference equation:

$$
P(x, t + \tau) = \tfrac{1}{2} P(x + h, t) + \tfrac{1}{2} P(x - h, t) \tag{1.1.9}
$$

with $x = (i - j)h$ and $t = n\tau$. Equation (1.1.9) may be rewritten as

$$
\frac{P(x, t + \tau) - P(x, t)}{\tau} = \frac{h^2}{2\tau} \frac{P(x + h, t) - 2P(x, t) + P(x - h, t)}{h^2}. \tag{1.1.10}
$$

The point is, the difference equation has got pretty close to some differential equation. Now, think of h and τ as *microscopic* quantities and pass to a macroscopic (large scale) description of a random walk by a limiting process $h \to 0$, $\tau \to 0$ with

$$D = \frac{h^2}{2\tau}\,, \qquad\qquad (1.1.11)$$

the *diffusion constant*, held fixed. This process turns x und t into continuous variables: $x \in \mathbb{R}\,,\ t \in \mathbb{R}_+$ which conforms much better with our normal view of space and time. As a highly satisfactory result, we obtain from (1.1.10) the one-dimensional diffusion equation[2]:

$$\frac{\partial}{\partial t} P(x,t) = D \frac{\partial^2}{\partial x^2} P(x,t). \qquad\qquad (1.1.12)$$

The results of a computer simulation of the one-dimensional diffusive motion using two different diffusion constants are shown in Fig.1.1.

Equation (1.1.12) and its multidimensional variants form the basis of Einstein's theory of Brownian motion. As a result of the limiting procedure, no meaning can be attributed to the velocity of the Brownian particle. This is clearly indicated by $h/\tau \to \infty$. Phrased in more mathematical terms: though continuous, a typical Brownian path is nowhere differentiable as a function of time.

Einstein reasoned that the diffusion constant should be of the form $D = k_{\mathrm{B}}T/f$, where k_{B}, T, and f are Boltzmann's constant, the temperature, and the friction constant respectively. For f he used Stokes's law $f = 6\pi a\eta$ for a single, rigid, spherical particle of radius a inserted into a fluid of viscosity η. The radius of the Brownian particle is to be taken large compared to both the radius of the bombarding solvent molecules and their mean free path. Next, Einstein suggested using the mean square displacement for a Brownian particle starting at the origin, $\langle x^2 \rangle = 2Dt$, to determine the diffusion constant D, which, when we know a and η, ultimately yields a value for Avogadro's constant N since $k_{\mathrm{B}} = R/N$. Consistency with other ways of obtaining N showed once more the validity of the molecular kinetic theory and thus the reality of atoms.

Einstein's relation between D and f represents the first instance of the fluctuation–dissipation connection of statistical physics: a fluctuation (the mean square displacement per unit time) is connected with a dissipative quantity (the friction constant).

The diffusion equation and the heat equation formally look the same. However, there is a distinction in the interpretation of the function $P(x,t)$ and the constant D. The reader familiar with the theory of heat conduction knows that the solution of the initial value problem $P_0(x,0) = \delta(x)$ is the Gauss function

$$P_0(x,t) = \frac{1}{2\sqrt{\pi Dt}} \exp\left\{ -\frac{x^2}{4Dt} \right\} \qquad (t > 0). \qquad\qquad (1.1.13)$$

In terms of Brownian motion this choice of initial data means that the particle starts at the origin. By the classical theorem of Laplace and De Moivre, i.e., convergence of the binomial distribution towards the normal distribution, the transition probability for the discrete random walk, P,

---

[2] Notice: it is $h^{-1}P_{ij}$ that approaches $P(x,t)$. The extra factor h^{-1} takes care of the fact that the normalization $\sum_i P_{ij} = 1$ changes to $\int dx\, P(x,t) = 1$.

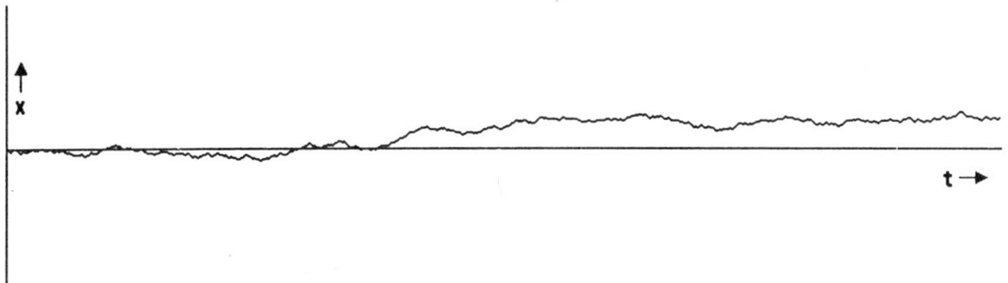

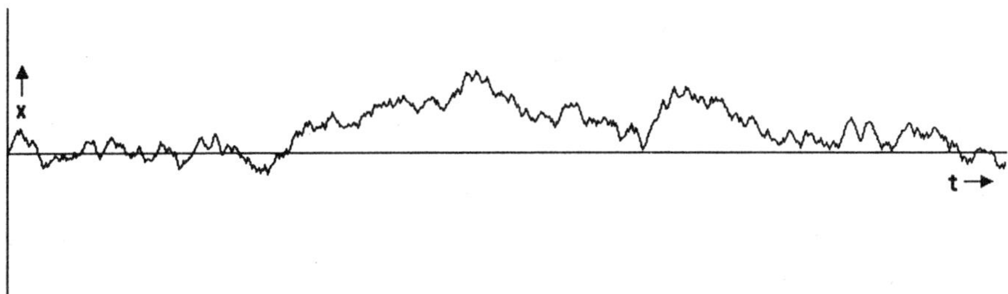

Fig. 1.1. Diffusive motion in one dimension for $D = \frac{1}{2}$ (*upper curve*) and $D = 10$ (*lower curve*). For the simulation on the computer, space and time have been discretized; 500 time steps are depicted.

will be replaced by the appropriate Gaussian in the continuum limit (i.e., $h, \tau \to 0$, $n \to \infty$, $h^2/(2\tau) \to D$, $n\tau \to t$):

$$\lim_{\substack{i \\ x_1 < ih < x_2}} \sum P(ih, n\tau) = \frac{1}{2\sqrt{\pi Dt}} \int_{x_1}^{x_2} dx \, \exp\left\{-\frac{x^2}{4Dt}\right\}. \qquad (1.1.14)$$

It should be emphasized that, on the left, i assumes only integer values, whereas, on the right, x is a continuous variable. If, at time $t = 0$, the particle starts at the origin, the probability of finding it at some later time t within the region $A \subset \mathbb{R}$ is given by the integral

$$W(A, t) = \int_A dx \, P_0(x, t). \qquad (1.1.15)$$

Integration over all space yields $\int dx \, P_0(x, t) = 1$ for all $t \geq 0$, thus guaranteeing $0 \leq W(A, t) \leq 1$.

As we pass to the continuum, the almost trivial matrix identity $P^n P^m = P^{n+m}$ changes to something more profound, i.e., we obtain the *Chapman–Kolmogorov equation*:

$$\int dx' \, P_0(x - x', t) P_0(x' - x'', t') = P_0(x - x'', t + t'). \qquad (1.1.16)$$

This equation may be looked at in different ways. First of all, it expresses the semigroup property of the integral kernel P_0, a general property which is automatic for any random motion without memory. It also expresses the temporal homogeneity of the process [1.5].

Throughout the discussion we have assumed the probabilities of left and right moves to be equal, which is characteristic of what is called the *simple random walk*. In a more general situation termed the *Bernoullian random walk* one has probabilties p and $q = 1-p$ so that in place of (1.1.7) one writes

$$P(x, t + \tau) = pP(x + h, t) + qP(x - h, t). \tag{1.1.17}$$

No great intrinsic interest is claimed for such a system: it is easy. The continuum limit may be applied as before, i.e., $h \to 0, \tau \to 0, h^2(2\tau)^{-1} \to D$. However, to avoid misbehavior we require the difference $p - q$ to approach zero so as to make

$$v = \lim \frac{h}{\tau}(p - q) \tag{1.1.18}$$

a finite quantity. Ultimately, what happens is that the ordinary diffusion equation (1.1.9) is replaced by a more general equation:

$$\frac{\partial}{\partial t}P(x, t) = \left(D\frac{\partial^2}{\partial x^2} + v\frac{\partial}{\partial x} \right) P(x, t). \tag{1.1.19}$$

The parameter v plays the role of a mean *drift velocity*. Compare $P_v(x, t)$, the solution of (1.1.19) using the initial data $P_v(x, 0) = \delta(x)$, with $P_0(x, t)$. As can be easily shown, one has

$$P_v(x, t) = P_0(x - vt, t) \tag{1.1.20}$$

which means that the general situation $v \neq 0$ may always be reduced to the simpler situation $v = 0$ by a Galileian transformation $x' = x - vt$.

1.2 Multidimensional Random Walk

We want to extend our preliminary discussion of the random walk and to introduce new tools so as to be able to treat cases where the particle performs an erratic motion on a d-dimensional hypercubic lattice $(\mathbb{Z}h)^d$ with lattice constant h and equal probabilities in all directions. Figure 1.2 shows a typical path on a two-dimensional lattice.

Obviously, there is more freedom of choice for the particle in higher dimensions. Within a period of time of length τ, the particle may proceed along any of the $2d$ directions of the lattice, the probability being $(2d)^{-1}$ in each direction. Of course, the step size is h as before. Viewing the lattice from a large distance and waiting a sufficient length of time we get a new impression of the motion which loses all reference to the lattice structure. The smaller the lattice spacing, the more chaotic the path. In some sense,

Fig. 1.2. Random motion on a two-dimensional square lattice, from a computer simulation

the limit describes what is called d-dimensional Brownian motion. For a demonstration see Fig. 1.3.

When the lattice spacing is no longer visible, formally, when $h \to 0$, we may treat space as if it where the d-dimensional continuum. Plotting the trail of a Brownian path[3] in a continuum – as has been done in Fig. 1.3 – does not allow us to determine its length[4] or how the system evolved in time.

Many details are known about Brownian paths and their trails. By construction, they are continuous, i.e., they may be drawn without lifting the pencil. More subtle properties are (here we restrict ourselves to $d = 2$): (1) The trail of a typical Brownian path returns infinitely often to a given open set, say the vicinity of a point. Such behavior is not unfamiliar to the physicist: in statistical mechanics, it is called *recurrence*. (2) Also, the trail fills more space than is filled by any smooth curve. Actually, it is a fractal set [1.6] with fractal dimension $d_{fr} = 2$. (3) The Brownian path is another fractal set (in spacetime) with $d_{fr} = 3/2$. (4) Level sets $M_x = \{t : \omega(t) = x\}$ have $d_{fr} = 1/2$. This is to be contrasted with $d_{fr} = 0$ for most level sets of a smooth function. For further properties of Brownian curves as fractal sets see [1.7].

We focus once more on the lattice version of Brownian motion and the description of a transition from one lattice point to the next. In the one-

---

[3] We need to distinguish carefully between the trail of a path as a geometric object in space and the associated graph in spacetime, the path itself, carrying the full information.

[4] The length of the trail of a typical Brownian path between any two of its points is infinite as a consequence of the velocity of the path being infinite.

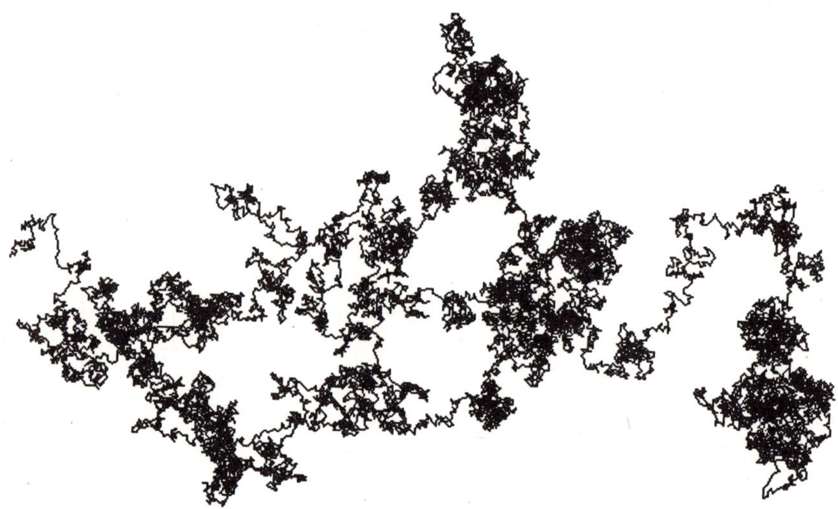

Fig. 1.3. The trail of a Brownian path in the plane, from a computer simulation

dimensional case, transition probabities were thought of as entries of a matrix P (with index set the lattice points) and most problems can be solved with the aid of matrices: they reduce probabilistic statements to explicit algebraic computations. In higher dimensions, the index sets occurring have more structure (they form a lattice). It seems more appropriate to regard matrices with large index sets as operators on some infinite-dimensional vector space, e.g., a function space. The correspondence from operators to matrices is customarily induced by fixing a basis.

For the moment let us be vague about the precise definition of the space of functions f on which the transition operator P is supposed to act:

$$[Pf](x) = \frac{1}{2d} \sum_{k=1}^{d} \{f(x + he_k) + f(x - he_k)\}. \tag{1.2.1}$$

Ordinary space is now represented by the discrete variable $x \in (\mathbb{Z}h)^d$ and e_k is the unit vector pointing in the direction of the kth axis: $(e_k)^i = \delta_k^i$. There is a particular instance where (1.2.1) has an obvious interpretation. If $0 \le f(x) \le 1$ and $\sum_x f(x) = 1$, f may be considered a distribution of the particle's position at some time t. In this case, Pf represents the transformed distribution at time $t + \tau$.

It follows from the general theory of spectral analysis of linear operators that it is advisable to pass to a space of complex functions f for which we choose the Hilbert space

$$\mathcal{H} = \{ f \mid \sum_x |f(x)|^2 < \infty \}. \tag{1.2.2}$$

As an operator on $\mathcal{H}$, P turns out to be self-adjoint and bounded. The quickest way to calculate powers such as P^n and even functions of P is to spectrally decompose P. A Fourier transformation

$$f(x) = \int_B dp\, e^{ipx}\, \tilde{f}(p), \qquad\qquad dp = dp_1 \cdots dp_d \qquad (1.2.3)$$

$$\tilde{f}(p) = \left(\frac{h}{2\pi}\right)^d \sum_x e^{-ipx} f(x), \qquad px = p_1 x^1 + \cdots + p_d x^d \qquad (1.2.4)$$

will do the job for us. Such formulas show a striking similarity to the decomposition of an arbitrary wave traveling through a crystal into plane waves with momentum p. Within the theory of the random walk, however, we may refer to p as some kind of *momentum* only by analogy.

The domain of integration for the momentum variable is known as the *Brillouin zone*

$$B = \{\, p \in \mathbb{R}^d \mid -\pi/h \le p_i \le \pi/h,\ i = 1, \ldots, d \,\} \qquad (1.2.5)$$

with volume

$$|B| = \left(\frac{2\pi}{h}\right)^d. \qquad (1.2.6)$$

Conclusion: Fourier transformation brings (1.2.1) into a form from which the spectrum of the operator P can be read off immediately:

$$[\widetilde{Pf}](p) = \lambda(p)\tilde{f}(p) \,,$$

$$\lambda(p) = \frac{1}{d}\sum_{i=1}^d \cos(p_i h) \,, \qquad (1.2.7)$$

$$\operatorname{spec} P = \{\, \lambda(p) \mid p \in B \,\}.$$

As this reasoning shows, the spectrum of P is continuous.

Writing powers of P in terms of their kernels,

$$[P^n f](x) = \sum_{x'} P(x - x', n\tau) f(x') \qquad (n \in \mathbb{N}). \qquad (1.2.8)$$

makes $P(x - x', n\tau)$ the probability of a transition $x' \to x$ during a period of time of length $n\tau$. From the spectral decomposition of P one has the analytic result

$$P(x, n\tau) = \frac{1}{|B|} \int_B dp\, e^{ipx} \lambda(p)^n \qquad (1.2.9)$$

for the probability in terms of a d-dimensional integral.

1.3 Generating Functions

There is a more powerful approach [1.8,9] to effectively calculating $P(x, n\tau)$ in the multidimensional case, say if $d \geq 3$. This is so since experience with generating functions properly chosen tells us that they are simpler in structure and easier to handle. Let z be some complex variable, and consider the power series

$$P(x|z) = \sum_{n=0}^{\infty} P(x, n\tau) \frac{z^n}{n!}. \tag{1.3.1}$$

Now use (1.2.9) and (1.2.7). Interchanging the summation and the integration leads to a new formula for $P(x|z)$ in terms of a product of one-dimensional integrals. These integrals are all of the same type and are related to the familiar integral representation of the modified Bessel functions:

$$I_n(z) = \frac{1}{2\pi} \int_{-\pi}^{\pi} d\theta \, \exp(in\theta + z\cos\theta) \qquad (n \in \mathbb{Z}). \tag{1.3.2}$$

With the help of these functions we may write

$$P(x|z) = \prod_{i=1}^{d} I_{x^i/h}(z/d). \tag{1.3.3}$$

To extract expressions for the probabilities $P(x, n\tau)$ we would simply expand both sides into power series with respect to the variable z using the well known series from the theory of Bessel functions.

The case $x = 0$ is easiest with $P(0, n\tau)$ characterizing the recurrent events, i.e., the return of a path to its starting point after n time steps. Obviously, $P(0, n\tau) = 0$ if n is odd. For an even number of time steps, one finds (see [1.9], for instance):

$$P(0, 2n\tau) = \begin{cases} 2^{-2n} \binom{2n}{n} & \text{if } d = 1 \\ 4^{-2n} \binom{2n}{n}^2 & \text{if } d = 2 \\ 6^{-2n} \binom{2n}{n} \sum_{k=0}^{n} \binom{2k}{k} \binom{n}{k}^2 & \text{if } d = 3 \end{cases} . \tag{1.3.4}$$

These formulas, though obtained from analysis, suggest to us an entirely different setting, i.e., the application of combinatorial methods: $P(0, 2n\tau)$ appears as a quotient of two natural numbers, Z/N, where N is the number of *all* paths of length $2n$ starting at a given point of the lattice; thus $N = (2d)^{2n}$, and Z is the number of closed paths of the same length. However, as is often the case, the analytic route is more convenient and serves to answer difficult questions in combinatorics. For those who don't know how to count cyclic paths on a lattice, the above analysis comes to the rescue[5].

If after $2n$ time steps some random path returns to the starting point, it may very well be that it has already visited that point earlier, say after $2m$ steps, where $1 \leq m \leq n$. Suppose the *first visit* of the initial point occurs after $2m$ steps. Then events with different m but the same n are statistically independent with probability $P(0, 2(n-m)\tau)Q(0, 2m\tau)$, where $Q(0, 2m\tau)$ stands for the (unknown) first visit probability. Conclusion: $P(0, 2n\tau)$ may be written as a sum

$$P(0, 2n\tau) = \sum_{m=1}^{n} P(0, 2(n-m)\tau)Q(0, 2m\tau). \qquad (1.3.5)$$

The set of equations (1.3.5) implicitly defines the numbers $Q(0, 2m\tau)$ which follow from it by recursion. Standard reasoning shows that it is more profitable to pass to suitably defined generating functions:

$$P(z) = P(-z) = 1 + \sum_{n=2}^{\infty} P(0, n\tau)z^n = \frac{1}{|B|} \int_B \frac{dp}{1 - z\lambda(p)} \qquad (1.3.6)$$

$$Q(z) = Q(-z) = \sum_{n=2}^{\infty} Q(0, n\tau)z^n. \qquad (1.3.7)$$

Here we regard $P(z)$ and $Q(z)$ as formal power series without specifying the radius of convergence. The set of equations (1.3.5) is immediately seen to be equivalent to the following identity:

$$P(z) - 1 = P(z)Q(z). \qquad (1.3.8)$$

Note that one way to determine the unknown quantities $Q(0, n\tau)$ is to expand $1 - P(z)^{-1}$ into a power series with respect to z.

1.3.1 Return or Escape?

All events n (first return after n steps) occurring with probability $Q(0, n\tau)$ are statistically independent. Therefore, the probability for a random path to return at all is represented by a sum over all possible cases:

$$Q(1) = \sum_{n=2}^{\infty} Q(0, n\tau) = 1 - P(1)^{-1}. \qquad (1.3.9)$$

---

[5] There are situations in statistical mechanics where it seems desirable to estimate the number of random cyclic paths or, more generally, the number of closed surfaces on a lattice. The idea behind such applications is that nonintersecting cyclic paths (which form a subset of all cyclic paths) constitute phase boundaries separating regions of different phases of a two-dimensional spin lattice. For an account of the so-called droplet model and *Peierls's argument* see [1.10, Sect.5.4].

The answer involves the special number $P(1)$ which is the key to the problem but as it stands cannot be interpreted as a probability. How should we proceed to calculate this number? The trick is to use the identity

$$\int_0^\infty dz\, e^{-z} \frac{z^n}{n!} = 1 \tag{1.3.10}$$

together with (1.3.1) and (1.3.3). We get the result in closed form:

$$P(1) = \int_0^\infty dz\, e^{-z} \Big(I_0(z/d)\Big)^d. \tag{1.3.11}$$

Or else we may write it as an integral over the Brillouin zone:

$$P(1) = \frac{1}{|B|} \int_B \frac{dp}{1 - \lambda(p)}. \tag{1.3.12}$$

Both integral representations have their merits: (1.3.11) is best suited for numerical calculations, whereas (1.3.12) shows that the integral is *infrared divergent* for $d = 1$ and $d = 2$. We argue as follows: in a neighborhood of $p = 0$

$$1 - \lambda(p) \approx \frac{1}{2d} \sum_{i=1}^d p_i^2. \tag{1.3.13}$$

Let this neighborhood be a ball of radius ϵ and let us introduce spherical coordinates within that ball. Once the integration with respect to the angles is carried out, we are left with some integral of the kind

$$\int_0^\epsilon dr\, r^{d-3} \quad \begin{cases} = \infty & \text{if } d = 1, 2 \\ < \infty & \text{if } d \geq 3. \end{cases} \tag{1.3.14}$$

Thus we have obtained an important result about the random walk due to Polya [1.11]:

$$d = 1, 2, \quad P(1) = \infty, \quad Q(1) = 1 \ : \quad \text{return with certainty} \\ \text{and hence no escape,}$$

$$d \geq 3, \quad P(1) < \infty, \quad Q(1) < 1 \ : \quad \text{return and escape with} \\ \text{finite probability.}$$

Also, the random walk is called *recurrent* for $d = 1$ or 2 and *transient* for $d \geq 3$. Since assertions of this type concern the large-time behavior of a typical path, the lattice spacing does not matter at all in the argument. It is the space dimension that matters. Therefore, we may safely state that Brownian motion is recurrent for $d = 1$ or 2 and transient for $d \geq 3$.

With dimension d increasing it becomes less probable that a random path will ever return to its origin. This may be learned from Table 1.1.

Table 1.1. The numbers $P(1)$ and $Q(1)$ for dimensions $d = 3, \ldots, 6$, where $Q(1)$ represents the return probability. The asymptotic values $(2d - 2)^{-1}$ are shown for comparison.

d	$P(1)$	$Q(1)$	$(2d - 2)^{-1}$
3	1.516 386 059	0.340 537 329 6	0.250
4	1.239 467 122	0.193 201 673 3	$0.16\bar{6}$
5	1.156 308 125	0.135 178 609 5	0.125
6	1.116 963 373	0.104 715 495 5	0.100

Asymptotically, $Q(1) \to (2d - 2)^{-1}$ $(d \to \infty)$. It it wise to remember the random walk when working in other areas of physics: the numbers $P(1)$ often come up in the context of phase transitions (see, for instance, [1.12] and [1.10, Sect.16.4].

Especially for $d = 1, 2$ it appears reasonable to ask: what is the mean time a path needs to return to its origin? The answer to this question simply is that the *mean return time* (in units of τ) coincides with the derivative of $Q(z)$ at $z = 1$:

$$Q'(1) = \sum_{n=2}^{\infty} nQ(0, n\tau) \tag{1.3.15}$$

One may also look upon this quantity as the mean length (in units of h) of such a path. We leave it to the curious reader to decide whether $Q'(1)$ is finite.

1.4 The Continuum Limit

We have indicated in Sect. 1.1 how the continuum limit, i.e., the passage to continuous spacetime, can be performed in the one-dimensional case. As for a general dimension d, we have postponed this task which will now be taken up anew. So we shall try to see what happens if both the lattice spacing h, and the unit of time, τ, approach zero in such a way that the diffusion constant

$$D = \frac{h^2}{2\tau d} \tag{1.4.1}$$

is kept fixed. As h decreases, the Brillouin zone becomes larger until, for $h = 0$, it fills the entire space $\mathbb{R}^d$.

For small h (large $n = t/\tau$, and hence $t > 0$),

$$\lambda(p)^n = \exp\left\{ n \log\left(\frac{1}{d} \sum_{i=1}^{d} \cos p_i h \right) \right\}$$

$$= \exp\left\{ \frac{t}{\tau} \log\left(1 - \frac{h^2}{2d}p^2 + O(h^4) \right) \right\}$$

$$= \exp\left\{ -Dtp^2 + O(h^2) \right\}, \tag{1.4.2}$$

where $p^2 = \sum_i p_i^2$. It then follows from (1.2.9) that the probability per volume, $h^{-d}P(x,t)$, approaches a Gaussian density:

$$P_0(x,t) = (2\pi)^{-d} \int dp\, e^{ipx} e^{-Dtp^2}$$

$$= (4\pi Dt)^{-d/2} \exp\left\{ -\frac{x^2}{4Dt} \right\}. \tag{1.4.3}$$

This describes the distribution of the position of a Brownian particle, given the information that it started at $x = 0$ at time zero. Phrased differently: the initial distribution is a δ-function. The distribution varies with time and its main characteristic is that the width broadens according to a $\sqrt{t}$ law. More precisely, the *mean square displacement* comes out to be

$$\langle x^2 \rangle = \int dx\, x^2 P_0(x,t) = 2dDt. \tag{1.4.4}$$

Again, this is the basic relation used in laboratory experiments in order to determine the diffusion constant (in principle, a stop-watch and a microscope will do).

Another simple but important observation deserves special attention: while the lattice is cubic, admitting only a limited set of symmetry operations, the continuum limit restores the full rotational symmetry of the Euclidean space. This is so because the density $P_0(x,t)$ miraculously comes out as a function of the Euclidean distance $r = \sqrt{x^2}$. We ought to be aware that such behavior of the limit is not automatic. Nevertheless, it is gratefully acknowledged.

Ad hoc examples show that full rotational symmetry may not always be restored. Take, for example, the following fictitious spectral function on a cubic lattice:

$$\lambda(p) = 1 - \{d^{-1} \textstyle\sum_{i=1}^{d} |\sin p_i h|\}^2. \tag{1.4.5}$$

Obviously, this function is invariant under reflections and 90° rotations of the lattice. In the continuum limit, the following awkward expression appears:

$$\lim \lambda(p)^n = \exp\{-D't|p|^2\}, \qquad |p| = \textstyle\sum_i |p_i| \tag{1.4.6}$$

($D' = \lim h^2/(\tau d^2) = 2D/d$). A strange metric has been created such that the length of a vector, $|p|$, is not invariant under arbitrary rotations.

Turning to our main result (1.2.4) for the spreading of an initial δ-distribution of the Brownian particle, the next thing we encounter is a differential equation for $P_0(x,t)$, called the *d-dimensional diffusion equation*:

$$\frac{\partial}{\partial t} P_0(x,t) = D\Delta P_0(x,t). \tag{1.4.7}$$

Here, Δ denotes the d-dimensional Laplace operator. And once we have gotten to that level, there is only a bit more to learn before we really understand why statistical mechanics and quantum mechanics are so closely related. Formally, the Schrdinger equation of a free particle can be obtained from the diffusion equation by introducing an imaginary time variable.

1.5 Imaginary Time

In this section we briefly review the basic concepts from quantum mechanics pertinent to the description of a free spinless particle of mass m in terms of a wave function ψ. The essential new ingredient is Planck's constant $\hbar$. By a proper choice of physical units we may take $m = \hbar = 1$. Generally, the space dimension is $d = 3$. In some applications, however, one assumes $d = 1$ or 2. The Schrödinger equation for the wave function ψ may be written in such a way that, if compared with the diffusion equation, the imaginary variable it appears in place of the real time t:

$$\tfrac{1}{2}\Delta\psi = \frac{\partial}{\partial(it)}\psi. \tag{1.5.1}$$

To be consistent let us pretend that time is imaginary and let us continue writing it for this variable whenever some quantum mechanical quantity depends on time. For instance, the solution of the initial value problem may then be presented as[6]

$$\psi(x,it) = \int dx'\, K(x - x', it)\psi(x',0), \tag{1.5.2}$$

where the complex *transition function*

$$K(x,it) = \begin{cases} (2\pi it)^{-d/2} \exp(-(2it)^{-1}x^2) & \text{if } t \neq 0 \\ \delta(x) & \text{if } t = 0 \end{cases} \tag{1.5.3}$$

results from an analytic continuation (with respect to time) of the previously introduced density $P_0(x,t)$ with diffusion constant $D = \tfrac{1}{2}$. Another way to look at the function $K(x,it)$ is to interpret it as the kernel of the unitary operator $e^{it\Delta/2}$, i.e., of the operator of time evolution:

---

[6] Of course, all authors of textbooks on quantum mechanics prefer to write $\psi(x,t)$ instead of $\psi(x,it)$, which is reasonable from their point of view.

$$\psi(x, it) = [e^{it\Delta/2}\phi](x) \qquad (t \in \mathbb{R}),$$
$$\psi(x, 0) = \phi(x). \tag{1.5.4}$$

Our short list of quantum mechanical formulas is ample evidence for the fact that the imaginary variable it appears naturally in the present context.

As we vary t, the operators $e^{it\Delta/2}$ form what is commonly called a *one-parameter unitary group*. In terms of the kernel, this amounts to an equation of the form

$$\int dx' \, K(x - x', it) K(x' - x'', it') = K(x - x'', it + it'), \tag{1.5.5}$$

quite analogous to the Chapman–Kolmogorov equation, though there are marked differences:

1. The variable t is no longer restricted to the half-axis $\mathbb{R}_+$ but assumes arbitrary real values. Hence, the arrow of time may be reversed with the consequence that a preferred direction (something we might call *future*) cannot be derived from formalism as it stands.
2. The kernel $K(x, it)$ is no longer positive but complex and of oscillatory character. As opposed to the diffusive situation, the Schrödinger equation admits wave solutions.

By the oscillatory character of the kernel $K(x, it)$, wave functions, during the course of time, do not decay exponentially (as they would do by diffusion) but rather follow a power law, giving rise to the *spreading of the wave packet*. Clearly, the effect results from

$$|K(x, it)| = |2\pi t|^{-d/2} \qquad (t \neq 0), \tag{1.5.6}$$

since then

$$|\psi(x, it)| \leq \int dx' \, |K(x - x', it)||\psi(x', 0)|$$
$$= |2\pi t|^{-d/2} \int dx' \, |\psi(x', 0)| \,, \tag{1.5.7}$$

granted absolute integrability of the initial values.

There is still another conceptual distinction between diffusive theories and quantum mechanics. In diffusion, density functions may be prepared and observed without ambiguity (in principle at least). By contrast, wave functions are not directly observed and the way in which they characterize the *state* is ambiguous. To what is called a density in a classical context there corresponds the quantity $|\psi|^2$ in quantum mechanics, a nonnegative function of space and time which, however, discloses only part of the information coded in the wave function ψ: the distribution of position. This particular facet of the statistical interpretation of quantum theory gives rise to the so-called interference phenomena which also rely on the superposition principle: if ϕ_1 and ϕ_2 are solutions of the Schrödinger equatation, so is $\phi_1 + \phi_2$. From a logical point of view, this statement may be resolved into a definition, an axiom, and an assertion. It defines "superposition" as a mathematical term, contains the implied axiom that, if ψ_1 and ψ_2 taken at time t each represent a state, then so does $\psi_1 + \psi_2$, and asserts that the time translation preserves superpositions[7]. The assertion follows from the linearity of the Schrödinger equation.

Though no reasonable physicist seriously questions the validity of the superposition principle, some scientists feel uneasy about beeing confined to measuring the square of the amplitudes rather than the amplitudes themselves. This is often exemplified by a discussion of the *double-slit experiment*. Because of its general importance for our understanding of the fundamentals, we shall examine the interference pattern of an admittedly idealized thought experiment.

Consider the Euclidean plane with points (x, y) in Cartesian coordinates. Two slits centered at (a, L) and $(-a, L)$ are placed opposite a screen, the x-axis. To avoid cumbersome calculations we do not use ordinary slits having well-defined widths but rather Gaussian slits[8]: these are given by Gauss functions (with respect to the variable x with maxima at $x = -a$ and $x = a$ respectivly and with variance $s > 0$) and used to represent the initial values of the single-particle wave function with regard to its dependence on x. In addition, the momentum in the y-direction is assumed to be p:

$$\psi_\pm(x, y, 0) = K(x \pm a, s) \exp(ipy) \tag{1.5.8}$$

($K(x, z)$ is the $(d = 1)$-transition function with complex time z). We might at first suppose that the wave packet in the region $0 < y < L$ is due to the passage of the particle through only one of the slits at time $t = 0$ (simply because the other one is closed or the passage has been recorded by some extra device). Then the wave function at time $t > 0$ is given by

$$\psi_\pm(x, y, it) = K(x \pm a, s + it) \exp(ipy - itp^2/2). \tag{1.5.9}$$

The intensity $|\psi_\pm|^2$ at the screen is Gaussian with variance $(2s)^{-1}(s^2 + t^2)$. However, intensities do not add when both slits are open and if the particle passes these slits with equal probability. Quantum mechanics tells us that the wave function actually is

$$\psi = 2^{-1/2}\left(\psi_+ + \psi_-\right) \tag{1.5.10}$$

---

[7] We must not expect to be able to superpose any two (pure) states in a many-particle theory as was first demonstrated by Wick, Wightman, and Wigner [1.13], the effect being attributed to the presence of superselection rules. Thus, states that may be superposed, in a more general framework, fall into *sectors*, each sector being characterized by some "charge". For a discussion of superspositions and sectors in quantum theory from an algebraic point of view see [1.14].

[8] Indeed, the use of a step function in place of the Gaussian would mean that we are leaving the domain of elementary functions as soon as $t > 0$. See also [1.15].

and thus the intensity at the screen is given by the following expression:

$$|\psi(x,0,it)|^2 \propto \left| \exp\left\{ -\frac{(x-a)^2}{2(s+it)} \right\} + \exp\left\{ -\frac{(x+a)^2}{2(s+it)} \right\} \right|^2 . \qquad (1.5.11)$$

Figure 1.4 shows a snapshot of the predicted interference pattern. It is rather close to what one observes in optical diffraction experiments. We conclude that the situation cannot be understood from the point of view of classical statistics using the particle picture throughout.

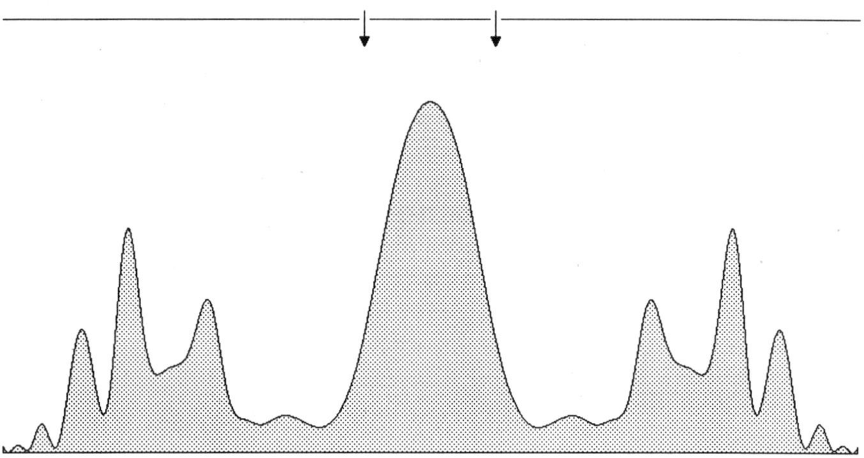

Fig. 1.4. Interference pattern from a thought experiment: a massive particle passes through two slits with equal probability

Now that we have a better grasp of the peculiarities of quantum mechanics as opposed to diffusion, let us envisage a more general situation where one deals with several free particles, all of the same mass (for simplicity). The principal change here is that the dimension d of the configuration space assumes values greater than three. Usually, $d = 3n$ if n is the number of particles involved. It is for this reason that, from now on, we want to keep d arbitrary and shall treat x as an element of $\mathbb{R}^d$. Crudely, for "reasonable" initial values $\psi(x,0)$ we may expect the solution $\psi(x,it)$ of the Schrödinger equation to be analytic with respect to its dependence on time and defined everywhere in the halfplane $\mathrm{Re}\, z > 0$ such that

$$\psi(x,z) = \int dx'\, K(x-x',z)\psi(x',0) , \qquad (1.5.12)$$

where

$$K(x,z) = (2\pi z)^{-d/2} \exp\left(-\frac{x^2}{2z} \right) . \qquad (1.5.13)$$

This is to be expected since the condition $\mathrm{Re}\,z > 0$ makes the integral (1.5.12) converge fast: the integrand is exponentially damped. There is little chance to give meaning to the integral if $\mathrm{Re}\,z < 0$ because the integrand would then grow exponentially. Clearly, $\mathrm{Re}\,z = 0$ represents a borderline case, the situation of ordinary quantum mechanics. Now, to be on the safe side, convergence of the integral (1.5.12) in the usual sense (i.e., absolute integrability) is guaranteed through the condition $\int dx\,|\psi(x,0)| < \infty$ imposed on the initial data. This is more than one normally requires in quantum mechanics[9].

On the halfaxis $z = s > 0$, which lies within the domain of holomorphy, solutions of the Schrödinger equation become solutions of the diffusion equation with diffusion constant $D = \frac{1}{2}$ (or $D = \hbar^2/(2m)$ on reintroducing the mass m and Planck's constant). While in quantum mechanics the time evolution is formulated in terms of a unitary group $e^{it\Delta/2}$, the corresponding operators $e^{s\Delta/2}$ in diffusion theory form what is called a *semigroup* since, for almost all initial data, a propagation backward in time is impossible. Though the loss of the group property along the halfaxis $s > 0$ may seem severe, it is compensated, however, by an important gain:

1. The kernel $K(x,s)$ has a strictly positive Fourier transform (compare (1.4.3)). It is for this reason that

$$(\phi, e^{s\Delta/2}\phi) \equiv \int dx \int dx'\, \bar{\phi}(x) K(x - x', s)\phi(x') \geq 0. \qquad (1.5.14)$$

   The integral vanishes only if $\phi(x) = 0$ almost everywhere.
2. The kernel itself is strictly positive: $K(x, s) > 0$. Consequently, $\phi(x) \geq 0$ implies $[e^{s\Delta/2}\phi](x) = \int dx'\, K(x - x', s)\phi(x') \geq 0$. The semigroup $e^{s\Delta/2}$ of operators is thus said to be *positivity preserving*.
3. The kernel is normed, i.e., $\int dx\, K(x, s) = 1$, implying stationarity of the constant function $\phi(x) = 1$.

Remark. It should be evident by now that, according to the point of view taken in this book, quantum mechanics as well as quantum field theory makes use of an *imaginary* time variable, whereas it appears vital to statistical mechanics that time is *real*. It goes without saying that the opposite statement repeatedly found in the literature can be maintained just as well, i.e., the way time is looked at and treated, whether it is real or imaginary, is largely a matter of taste. Still deeper-lying and perhaps justifying our present position, there is the fact that the structure of the Minkowski space underlying quantum field theory deviates from that of a four-dimensional Euclidean space. As has often been observed, the discrepancy can formally be resolved by introducing an imaginary time coordinate $x^4 := ix^0$ (x^0 is the Minkowskian time) though one may justly argue against a "formal replacement" because it certainly hides more than it reveals. However, new emphasis and justification has been given to the concept of imaginary time by the advent of Euclidean field theory since, here, it is the process of *analytic continuation*

---

[9] For $\psi(\cdot, 0) \in L^2(\mathbb{R}^d)$, however, it is still true that $\psi(\cdot, it)$ exists, in the L^2 sense, as a boundary value of $\psi(\cdot, s + it)$.

which connects the Euclidean and the Minkowski structure. In a way, relativistic invariance sheds more light onto the question: *What is time?* Yet, a precise answer cannot be given; the matter is being debated among scientists and philosophers.

Whenever we have a solution of the Schrödinger equation before us, we may look upon it as an analytic continuation, in some sense or another, of a solution of the diffusion equation. We thus connect the evolution of a quantum system of Schrödinger particles with the classical, but indeterministic, motion of a single Brownian particle living in the same configuration space.

The role played by the interaction potential, though crucial, has not yet been clarified. Before turning to this important topic we need to discuss the Wiener process.

1.6 The Wiener Process

1.6.1 The Analysis of Random Paths

Consider a Brownian particle starting at $x = 0 \in \mathbb{R}^d$ at time $s = 0$. Throughout the remainder of the book we let the diffusion constant be $D = \frac{1}{2}$, a convention which, if necessary, can always be achieved by an appropriate choice of the time unit. The position of the particle at some later time $s > 0$ is a *random variable* X_s taking values in $\mathbb{R}^d$. If $d > 1$, X_s is also called a *random vector* to stress that it has several components.

The concept of a *random variable* is central to probability theory and the approach preferred by most mathematicians uses a measure theoretic setting [1.16,17]. In this, one starts with the notion of a probability space $(\Omega, \mathcal{F}, P)$, where

1. Ω is the sample space with elements ω called *samples* or *simple events*.
2. $\mathcal{F}$ is σ-field of subsets of Ω called events (a σ-field is closed under complementation, countable intersections and unions).
3. P is a probability measure on $\mathcal{F}$. To any $B \in \mathcal{F}$, it assigns a number $P(B)$ between 0 and 1 called the probability of the event B.

A function $X \colon \Omega \to \mathbb{R}^d$ is called a random variable if it is measurable, i.e., if for any Borel set $A \subset \mathbb{R}^d$, the set $X^{-1}(A) \equiv \{\omega \,|\, X(\omega) \in A\}$ belongs to $\mathcal{F}$. If we talk about a collection (finite or infinite) of random variables such as $\{X_s \,|\, s > 0\}$, it is always understood that they have a common underlying probability space $(\Omega, \mathcal{F}, P)$. For an account of the basic notions of measure theory see [1.18].

The probabilistic model of Brownian motion takes the set of sample paths as the underlying space Ω. For a construction of the σ-field $\mathcal{F}$ and the probability measure P consistent with the above mathematical framework see below. In essence, the random variable X_s assigns the vector $x = \omega(s) \in \mathbb{R}^d$ to any path $\omega \in \Omega$. We may thus say that X_s evaluates ω at s and write $X_s(\omega) = \omega(s)$. In what follows we shall often write $X_s \in A$ to mean that the random variable X_s assumes some value in a Borel set $A \in \mathbb{R}^d$. Better: "$X_s \in A$" stands for the set $\{\omega \,|\, X_s(\omega) \in A\}$ element of $\mathcal{F}$ and $P(X_s \in A)$ is the probabilist's way of writing $P\big(X_s^{-1}(A)\big)$.

Though vectors $x \in \mathbb{R}^d$ are possible values of the position X_s, we should not expect single points to occur with nonzero probability. It it thus necessary to consider subsets $A \subset \mathbb{R}^d$ of finite volume, say, a ball or a hypercube.

The event $X_s \in A$ simply means that the Brownian particle has passed the "window" A at time s as sketched in Fig. 1.5.

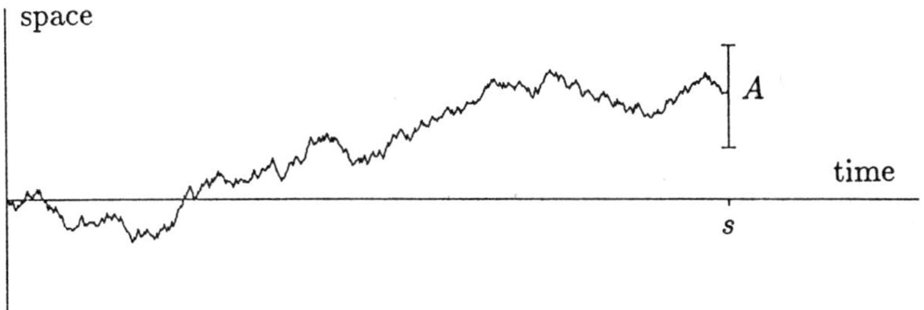

Fig. 1.5. A Brownian path starting at $x = 0$ and passing the "window" A at time s

For the situation at fixed time s, all information about the random variable X_s is coded in the probabilities $P(X_s \in A)$. The probability measure $A \mapsto P(X_s \in A)$ that assigns numbers in $[0, 1]$ to suitably choosen subsets $A \subset \mathbb{R}^d$ is called the *distribution of* X_s. Interpreting our previous results in this light we have

$$P(X_s \in A) \;=\; \int_A dx\, K(x, s) \tag{1.6.1}$$

with density

$$K(x, s) = (2\pi s)^{-d/2} \exp\left(-\frac{x^2}{2s}\right). \tag{1.6.2}$$

Thus, X_s is said to have a normal distribution or, equivalently, it is called a *Gaussian variable*. Notice that, by convention, the Gauss function describing the density is always centered and normalized.

The obvious generalization of a finite collection of random variables is a map $s \mapsto X_s$, where s ranges over some interval. Any such map will be called a *stochastic process* (in continuous time)[1.19]. There are prominent cases where the time interval is $[0, 1]$, $[0, \infty)$, or $(-\infty, \infty)$. For definiteness, we take $[0, \infty)$ for the Wiener process pertaining to a Brownian particle starting at the origin at time $s = 0$.

For a stochastic process to be defined, it is necessary to be able to determine the probabilities of *general* events. This certainly requires more than just knowing the set of distributions $P(X_s \in A)$. The essential step toward the goal is to consider compound events of the form

$$\text{``}X_{s_1} \in A_1 \text{ and } X_{s_2} \in A_2 \text{ and } \ldots X_{s_n} \in A_n\text{''},$$

where $0 < s_1 < s_2 < \cdots < s_n$, $A_i \subset \mathbb{R}^d$, and $n > 0$, and to devise a rule that determines its probability. In plain words, the compound event informs us that the Brownian particle has passed the windows $A_1, \ldots, A_n$ at specified times, a situation depicted in Fig. 1.6.

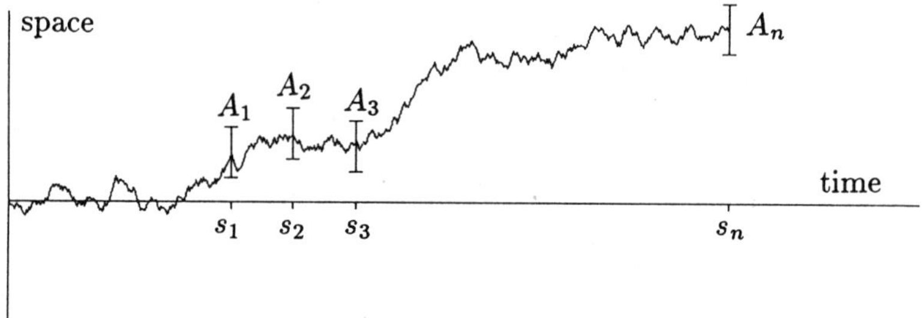

Fig. 1.6. A Brownian particle passes the windows $A_1, A_2, A_3 \ldots, A_n$ at the times $s_1, s_2, s_3, \ldots, s_n$.

The probability of a compound event is denoted by

$$P(X_{s_1} \in A_1, \ldots, X_{s_n} \in A_n). \tag{1.6.3}$$

Varying $A_1, \ldots, A_n$ we get the *joint distribution* of the random variables X_s, $s \in S = (s_1, s_2, \ldots, s_n)$. This is also called a *distribution* of the process with base S and may be abbreviated as $P_S(A)$, where $A = A_1 \times \cdots \times A_n \subset \mathbb{R}^{nd}$. The distribution is said to be *finite-dimensional* (of order n) since the base is finite (has n elements).

Given a set of finite-dimensional distributions $P_S(A)$, defined for all n, of some stochastic $\mathbb{R}^d$-valued process, they are said to be *consistent* if

$$P_S(A_1 \times \cdots \times \mathbb{R}^d \times \cdots \times A_n) = P_{S'}(A_1 \times \cdots \times \hat{A}_k \times \cdots \times A_n), \tag{1.6.4}$$

where $S = (s_1, \ldots, s_n)$, $S' = (s_1, \ldots, \hat{s}_k, \ldots, s_n)$, and $1 \leq k \leq n$ (the hat ˆ over some variable means elimination). Phrased differently, consistency states that adding the information $X_s \in \mathbb{R}^d$ (for some $s \in S$) does not change the event at all. The distributions of the Wiener process given below satisfy this condition as can be easily verified. Given a set of consistent finite-dimensional distributions $P_S(A)$, one may always associate some stochastic process X_s with it. This essentially states that there exists some measure space $(\Omega, \mathcal{F}, P)$ of sample paths such that $X_s(\omega) = \omega(s)$ for all $\omega \in \Omega$ and $P_S(A) = P(B_S)$, where $A = A_1 \times \cdots \times A_n$ and $B_S = \bigcap_{s \in S} X_s^{-1}(A_s) \in \mathcal{F}$. A standard construction (due to Kolmogorov) takes $\mathcal{F}$ as the smallest σ-field (subsets of Ω) containing all cylinders $\{\omega \mid \omega(s) \in A_s, s \in S\}$ with finite base S (see [1.17] for details). However, the measure space is never unique owing to the freedom of adding sets of measure zero: a disturbing fact at first sight! Different realizations of a stochastic process X_s are called *versions* of the process. We anticipate that it is of minor importance which version of the Wiener process has been picked. The field of measurable sets, however, should be chosen large enough so as to avoid undefined path integrals lateron. Whatever the choice, we are assured that it does not affect the distributions with finite (or countable) base.

The stochastic process X_s is said to be the *Wiener process*[10] if the finite-dimensional distributions are of the form

$$P(X_{s_1} \in A_1, \ldots, X_{s_n} \in A_n) =$$

$$\int_{A_n} dx_n \cdots \int_{A_2} dx_2 \int_{A_1} dx_1 \, K(x_n - x_{n-1}, s_n - s_{n-1})$$

$$\cdots K(x_2 - x_1, s_2 - s_1) \, K(x_1, s_1) \qquad (1.6.5)$$

and if the initial distribution is

$$P(X_0 \in A_0) = \begin{cases} 1 & \text{if } A_0 \ni 0 \\ 0 & \text{otherwise.} \end{cases} \qquad (1.6.6)$$

Hence, the particle starts at the origin: equivalently, one writes $X_0 = 0$. In a slightly different notation, (1.6.5) reads:

$$P(X_{s_1} \in dx_1, \ldots, X_{s_n} \in dx_n) = \prod_{k=1}^{n} dx_k \, K(x_k - x_{k-1}, s_k - s_{k-1})$$

$$(1.6.7)$$

$(x_0 = 0, s_0 = 0)$ with density $K(x, s)$ given by (1.6.2).

The fact that the right hand side of (1.6.7) is a product tells us that, given the present position X_s of the Brownian particle, the distribution of $X_{s'}$ at some later time s' (the future) is completely determined and does not further depend on the past history of the path taken by the particle. A stochastic process with this property is said to have no memory and is termed a *Markov process*. The Markov property is simply the probabilistic analogue of a property familiar from the theory of deterministic dynamical systems: given the initial data, then, by solving the equation of motion, the future state of the system can be obtained without knowing what happened in the past. The present state already contains all information relevant for the future.

Note that the definition of a Markov process places no restriction either on the initial distribution $P(X_0 \in A_0)$ or on the *transition probabilities*

$$P(X_{s'} \in A \mid X_s = x) = \int_A dx' \, K(x' - x, s' - s) \quad (s' > s \geq 0), \quad (1.6.8)$$

except normalization. But in Brownian motion the initial distribution is (1.6.6) and the density $K(x, s)$, or *transition function*, is given by (1.6.2).

The example illustrates well the notions of conditional probability and stationarity: $P(X_{s'} \in A \mid X_s = x)$ gives the probability of the event $X_{s'} \in A$ at time $s' > s$ under the assumption $X_s = x$ (i.e., the particle started at x at time s). In general, the transition probabilities of a Markov process are allowed to depend on s and s'. But if they merely

---

[10] The theory has its origin in a fundamental paper by N. Wiener [1.20] in 1923. A brief account of its significance for Brownian motion and quantum physics may be found in [1.21] and an extensive discussion in [1.22]. For a detailed mathematical analysis see [1.23] and [1.5].

depend on the difference $s' - s$, they are said to be *stationary*. This is obviously the case in Brownian motion. A process having stationary transition probabilities is called *temporally homogeneous*. This by no means implies that the distribution $P(X_s \in A)$ is independent of s.

It is common practice to restrict oneself to the class of homogeneous Markov processes since they have a property similar to *autonomous* deterministic dynamical systems, in particular to conservative Hamiltonian systems (i.e., where the Hamiltonion does not vary with time).

There are rare cases where all finite-dimensional distributions of a given stochastic process X_s on $-\infty < s < \infty$ turn out to be invariant with respect to time shifts, i.e., $P_{S^t}(A) = P_S(A)$ for $S^t = \{ s+t \mid s \in S \}$ and all $t \in \mathbb{R}$. Then the distribution $P(X_s \in A)$ does not depend on s. Such processes are said to be *invariant*. A Markov process with stationary transition probabilities is invariant precisely if its initial distribution does not change as the system evolves. This is the probabilistic analogue of a situation in dynamical systems, where one deals with a stationary state. Brownian motion is not invariant in this sense since the initial distribution changes. Moreover, it cannot be made invariant by choosing a different initial state because there is no invariant state in the system.

Finally, we call a Markov process *spatially homogeneous* whenever its transition probabilities are invariant with respect to translations in space: $P(X_{s'} \in A+x \mid X_s = a+x) = P(X_{s'} \in A \mid X_s = a)$. In fact, the Wiener process is spatially homogeneous.

Scale invariance. The Wiener process X_s has a nice property that follows from investigating the initial distribution and the transition probabilities with regard to their behavior under a change of scales. Notice that the transition function $K(x, s)$ depends on the quotient x^2/s only. We thus consider $Y_s = \ell X_{s/\ell^2}$ for $\ell > 0$. Then Y_s is again a Wiener process (a different version so to speak) since there is no way to physically distinguish the two processes, X_s and Y_s: all their sample paths start at the origin with the same transition probabilities. This is a remarkable fact. For it tells us that Brownian motion shows the same behavior on any length scale (provided time is rescaled simultaneously). Borrowing a notion from the theory of fractal sets we may say that Brownian motion is *self-similar*. Self-similarity becomes even more pronounced when the trail (a curve in $\mathbb{R}^d$) of a typical Brownian path is considered: whether we look at it using a microscope or from a distance, we always observe the same fractal structure.

1.6.2 Multidimensional Gaussian Measures

We saw above that the finite-dimensional distributions of order n of the Wiener process are built from n Gaussian transition functions. In fact, these distributions are Gaussian measures, a property shared by the so-called *Gaussian processes*. Equation (1.6.5) may be written in a more compact form that is characteristic of what is meant by a multidimensional Gaussian measure. To do so, fix n, put $N = nd$, and let $x = \{x_1, x_2, \ldots, x_n\}^T$ be the multivector corresponding to an element of the sample space $\mathbb{R}^N$ for $X_S = \{X_{s_1}, \ldots, X_{s_n}\}$, where $S = (s_1, \ldots, s_n)$. We shall also write $dx = dx_1 dx_2 \cdots dx_n$ and consider the positive quadratic form

$$x^T Q x = \frac{x_1^2}{s_1} + \frac{(x_2 - x_1)^2}{s_2 - s_1} + \cdots + \frac{(x_n - x_{n-1})^2}{s_n - s_{n-1}} \tag{1.6.9}$$

$(0 < s_1 < s_2 < \cdots < s_n)$. This defines Q as a real $N \times N$ matrix which is positive symmetric. By a change of coordinates, Q can be made diagonal, i.e., $Q = M^T D M$, where

$$
M = \begin{pmatrix}
\mathbb{1} & & & 0 \\
-\mathbb{1} & \mathbb{1} & & \\
& \ddots & \ddots & \\
0 & & -\mathbb{1} & \mathbb{1}
\end{pmatrix}
\tag{1.6.10}
$$

$$
D = \mathrm{diag}(s_1^{-1}\mathbb{1}, (s_2 - s_1)^{-1}\mathbb{1}, \ldots, (s_n - s_{n-1})^{-1}\mathbb{1}).
\tag{1.6.11}
$$

To simplify the represention, matrices were written using $d \times d$ blocks; $\mathbb{1}$ denotes the d-dimensional unit matrix. From $\det M = 1$ we infer that

$$
\det Q = \det D = [s_1(s_2 - s_1) \cdots (s_n - s_{n-1})]^{-d}.
\tag{1.6.12}
$$

For any Cartesian product $A = A_1 \times A_2 \times \cdots \times A_n$ of sets $A_i \in \mathbb{R}^d$ we may now write:

$$
\boldsymbol{P}(X_S \in A) = \left[\det\left(\frac{Q}{2\pi}\right) \right]^{\frac{1}{2}} \int_A dx \, \exp\{-\tfrac{1}{2}x^T Q x\}.
\tag{1.6.13}
$$

There are certain reasons for extending of this formula (if not circumventing it entirely):

1. The (Borel) subset $A \subset \mathbb{R}^N$ may be arbitrary, not just a Cartesian product of sets in $\mathbb{R}^d$. In this case, the event $X_S \in A$ is no longer of the form '$X_{s_1} \in A_1$ and $X_{s_2} \in A_2$ and $\ldots X_{s_n} \in A_n$'. The justification for this extension comes from the fact that A may always be approximated by (finite) unions of Cartesian products.

2. By considering limits of (1.6.13) as $n \to \infty$ one can attribute a distribution to X_S, where the base S is countable. However, the explicit construction requires some theory of Gaussian measures on linear topological spaces (see [1.24]) and an application of the Bochner–Minlos theorem [1.25,28], which we shall not delve into. One of the problems here is that $\det(Q/(2\pi))$ in (1.6.13) may not approach a limit, and one way to get around this difficulty is to consider the Fourier transform instead:

$$
\int \boldsymbol{P}(X_S \in dx)e^{ipx} = \exp\left\{-\tfrac{1}{2}pQ^{-1}p^T\right\}
$$

$$
\begin{aligned}
pQ^{-1}p^T &= s_1(p_1 + \cdots + p_n)^2 \\
&\quad + (s_2 - s_1)(p_2 + \cdots + p_n)^2 \\
&\quad + \cdots + (s_n - s_{n-1})p_n^2 \\
p &= \{p_1, \ldots, p_n\}, \qquad (p_i \in \mathbb{R}^d).
\end{aligned}
\tag{1.6.14}
$$

The formula makes sense in the general situation

$$0 \leq s_1 \leq s_2 \leq \cdots \leq s_n, \tag{1.6.15}$$

and hence even in the degenerate case, where equality occurs at some places in (1.6.15). This helps whenever $s_i - s_{i-1}$ approaches zero. The Fourier transform, also called the *characteristic function* of the measure, is defined in the limit $n \to \infty$, provided $\lim s_n = s < \infty$ and $\|p\|^2 = \sup_{k \geq 1} \left(\sum_{i=k}^{\infty} p_i \right)^2 < \infty$ since $pQ^{-1}p^T \leq s\|p\|^2$.

3. To restrict oneself to sets of Brownian paths that depend on a countable number of coordinates does not seem sufficient in some applications. As a challenge we quote the following formula from Sect. 2.4:

$$\boldsymbol{P}\left(\sup_{0 \leq s \leq t} |X_s| \leq a \right) = \sum_{n=1}^{\infty} 2(-1)^{n+1} \exp\left\{ -t\frac{n^2\pi^2}{2a^2} \right\}. \tag{1.6.16}$$

This expresses, as a Jacobian theta function, the probability that the process with $d = 3$ remains bounded in some time interval. The set that has been assigned a measure by the above formula may also be written

$$\left\{ \omega \,\Big|\, \sup_{0 \leq s \leq t} |\omega(s)| \leq a \right\} = \bigcap_{0 \leq s \leq t} \left\{ \omega \,\big|\, |\omega(s)| \leq a \right\} \tag{1.6.17}$$

and arises as a noncountable intersection of 'simple' measurable sets. It is hard to imagine how theta functions as in (1.6.16) come up as limits of Gaussian integrals (1.6.13) by way of the "time slicing" method (see below).

From the standpoint of path integrals, there are certain advantages to always working with the smoothest possible version of the Wiener process. One version that is both convenient and smooth (though not the smallest version) has all sample paths continuous. Not only is it economic to deal with a rather small sample space Ω, but this also eliminates many events from $\mathcal{F}$ with zero probability. For instance, sets of discontinuous paths have zero measure.

If a process admits a continuous version, intuition tells us that it suffices to work with finite-dimensional distributions and get the others as limits. It is instructive to watch how this in done in practice. A prominent method, known as "time slicing", repeatedly bisects a given time interval $[0, t]$ and after the nth step arrives at subintervals separated by equidistant points $s_k = kt2^{-n}$, $k = 1, \ldots, 2^n - 1$. One then approximates the path integral based on $[0, t]$ by integrating only with respect to the coordinates $x_k = \omega(s_k)$ of a path ω and finally passes to the limit $n \to \infty$. Let us consider an example.

The probabilities $\boldsymbol{P}(B_n)$ converge for any decreasing sequence $B_0 \supset B_1 \supset B_2 \supset \cdots$ towards $\boldsymbol{P}(B)$, where $B = \bigcap_n B_n$. In particular, if

$$B_n = \left\{ \omega \,\big|\, |\omega(kt2^{-n})| \leq a, \ k = 0, \ldots, 2^n \right\} \tag{1.6.18}$$

for some $a > 0$, then

$$B = \bigcap_n B_n = \bigcap_{s/t \in D} \left\{ \omega \,\big|\, |\omega(s)| \leq a \right\}, \tag{1.6.19}$$

where D is the set of dyadic rationals in the interval $[0,1]$ (i.e., those $r \in [0,1]$ that have a finite binary representation): D is strictly less than the set of all reals in the same interval. Hence, B may be strictly larger than the set defined in (1.6.17) and it is not a priori clear whether the two events get the same probability. If they don't, it would be very disturbing; it would almost entirely invalidate previous and future attempts to calculate path integrals by the time slicing method. Here continuity comes to our rescue. With continuous sample paths, the two sets coincide. The reason: there is no continuous function $\omega(s)$ satisfying $|\omega(s)| \le a$ at every rational s/t that possibly violates the same condition at some real s/t. It is thus fair to say that it is continuity that makes time slicing a safe and successful strategy.

1.6.3 Increments

There is a simple but important application of (1.6.13) for $n = 2$. Let $G \subset \mathbb{R}^d$ be arbitrary and consider the set

$$A = \{\, x_1, x_2 \mid x_2 - x_1 \in G \,\} \subset \mathbb{R}^{2d}. \tag{1.6.20}$$

Geometrically speaking, A may be viewed as some cylinder in $2d$ dimensions whose basis is G. Formula (1.6.13) may then be used to calculate $P(\{X_{s_1}, X_{s_2}\} \in A)$, which obviously answers the question: *what is the probability that the increment $x_2 - x_1$ of a Brownian path between the times s_1 and s_2 assumes some value within the region G?* The event $\{X_{s_1}, X_{s_2}\} \in A$ may thus be written as $X_{s_2} - X_{s_1} \in G$ and the simple substitution $y = x_2 - x_1$ (replacing x_2) leads to the formula

$$P(X_{s_2} - X_{s_1} \in G) = \int_G dy\, K(y, s_2 - s_1), \tag{1.6.21}$$

where we have used the fact that $\int dx_1\, K(x_1, s_1) = 1$. What we have shown here can also be expressed as

$$P(X_{s_2} - X_{s_1} \in G) = P(X_{s_2 - s_1} \in G) \tag{1.6.22}$$

provided $0 \le s_1 \le s_2$, that is, *the random variables $X_{s_2} - X_{s_1}$ and $X_{s_2 - s_1}$ have the same distribution*[11]. Any attempt fails that seeks to introduce the velocity of a Brownian particle as a random variable. To see this consider the approximate velocity $V_s(\tau) = \tau^{-1}(X_{s+\tau} - X_s)$, where $s > 0, \tau > 0$. The two events $V_s(\tau) \in G$ and $X_{s+\tau} - X_s \in \tau G$ coincide and occur with the probability

$$P(V_s(\tau) \in G) = (2\pi\tau)^{-d/2} \int_{\tau G} dx\, e^{-(2\tau)^{-1}x^2}$$

$$= \left(\frac{\tau}{2\pi}\right)^{d/2} \int_G dv\, e^{-\tau v^2/2}. \tag{1.6.23}$$

---

[11]Warning: this does not imply the equality $X_{s_2} - X_{s_1} = X_{s_2 - s_1}$ since the two random variables give rise to different joint distributions involving X_s with arbitrary $s > 0$.

When $\tau \to 0$, what happens is that the density $[\tau/(2\pi)]^{d/2}\exp(-\tau v^2/2)$ becomes broader. At the same time it flattens, i.e., there is no limit distribution. As for the Brownian experiment, the smaller the precision of the time measurement, τ, the more the tendency of the approximate velocity $V_s(\tau)$ to assume large values (in norm) with high probability. This certainly sounds strange from the point of view of special relativity. To stress the strangeness of the situation let us introduce the velocity of light, c. Then, given $\epsilon > 0$, there is always some τ sufficiently small such that $P(|V_s(\tau)| > c) > 1 - \epsilon$, that is, the probability for the approximate velocity (with an appropriate precision τ) of a Brownian particle to exceed the velocity of light can be made arbitrary close to one. It is therefore necessary to emphasize that the Wiener process as a model for particle motion merely represents an idealized physical situation and completely disregards the laws of special relativity. Here we do not dare tackle the problem of formulating a relativistic analogue of the Wiener process[12].

We slightly extend the above discussion by considering the set

$$A = \{\, x_1, \ldots, x_n \mid x_i - x_{i-1} \in G_i,\ i = 2, \ldots, n \,\} \subset \mathbb{R}^{nd}, \tag{1.6.24}$$

where $G_i \subset \mathbb{R}^d$. It follows from (1.6.13) that

$$P(\{X_{s_1}, \ldots, X_{s_n}\} \in A) = \prod_{i=2}^{n} \int_{G_i} dy\, K(y, s_i - s_{i-1}) \tag{1.6.25}$$

$(0 < s_1 < \cdots < s_n)$ or to put it differently:

$$P(X_{s_i} - X_{s_{i-1}} \in G_i\,;\ i = 2, \ldots, n) = \prod_{i=2}^{n} P(X_{s_i} - X_{s_{i-1}} \in G_i). \tag{1.6.26}$$

How may the content of this formula be phrased in plain words? The reader who is familiar with the basic concepts of probability theory will recall that a sequence of random variables, $Y_1, Y_2, \ldots$, be it finite or infinite, is termed *independent* if all its finite-dimensional distributions factorize:

$$P(Y_1 \in A_1, \ldots, Y_n \in A_n) = P(Y_1 \in A_1) \cdots P(Y_n \in A_n). \tag{1.6.27}$$

In accordance with this concept we say that a stochastic process has *independent increments* if (1.6.26) holds for all n. The analysis has shown that the Wiener process is a homogeneous Gaussian process with independent increments.

---

[12]For attempts in this direction see [1.26] and [1.27].

1.7 Expectation Values

It is not difficult to check that the expectation value (or mean value) of the position of a Brownian particle stays constant and hence is zero:

$$E(X_s) = \int dx\, x K(x, s) = 0. \tag{1.7.1}$$

This essentially follows from a symmetry argument, i.e., from the relation $K(x, s) = K(-x, s)$, and may be interpreted by saying that the Brownian paths have no preferred directions in space. Though the Wiener process does not remember previous directions, in a sense it recalls the starting position $x = 0$ since future positions are constrained by (1.7.1); $E(X_s)$ ought to be interpreted as a *conditional expectation*, i.e., as the mean value of X_s *given the information* $X_0 = 0$. The origin of space plays a distiguished role only by convention. Nevertheless, equivalence of all points in space is restored by the formula

$$\int dx'\, x' K(x' - x, s' - s) = x \qquad (0 \le s \le s'), \tag{1.7.2}$$

which determines the mean value of $X_{s'}$ given the information $X_s = x$.

The mean square deviation has been considered previously (see (1.4.4)). Since the random vector X_s has several components which we denote by X_{sk} $(k = 1, \ldots, d)$, we may wish to generalize the former result in the following way:

$$E(X_{sk} X_{s'k'}) = G(s, s')\delta_{kk'}. \tag{1.7.3}$$

The expression to the left is called the *covariance matrix* of the process. The special formula (1.7.3), which from now on will be conveniently abbreviated as $E(X_s X_{s'}) = G(s, s')\mathbb{1}$, follows from a symmetry argument. Taking traces on both sides and comparing with (1.4.4) yields $G(s, s) = s$ (since $D = \frac{1}{2}$). This settles the case $s = s'$.

Next, let $0 \le s \le s'$. The two increments $X_s - X_0$ and $X_{s'} - X_s$ are independent. Consequently, their joint expectation value factorizes with both factors being zero. This tells us that

$$E(X_s X_{s'}) = E\big((X_s - X_0)(X_{s'} - X_s)\big) + E(X_s^2) = s\mathbb{1}, \tag{1.7.4}$$

where we have used $E(X_0) = 0$. To complete the argument we remind ourselves that $E(X_s X_{s'})$ is symmetric with respect to an interchange of s and s' and write the result as

$$E(X_s X_{s'}) = \min(s, s')\mathbb{1} \tag{1.7.5}$$

so that $G(s, s') = \min(s, s')$ in (1.7.3).

The Covariance Operator of the Wiener Process. We interrupt the present line of thought by reminding ourselves that the expression $G(s, s') = \min(s, s')$ is in fact the Green's function for the classical problem of the swinging rope. A few remarks about the surprising connection seem to be in order. Points along the semi-infinite rope are specified by their coordinate $s \geq 0$. Think of the rope as being suspended at one end and capable of performing free oscillations in some transverse direction. One then associates the differential operator $Df(s) = -f''(s)$ with the problem and imposes the boundary condition $f(0) = 0$. The operator is assumed to act on $L^2(0, \infty)$. Its spectrum is continuous. To any spectral value λ^2 of D there exists some mode of oscillation given by the generalized eigenfunction

$$e_\lambda(s) = \sqrt{\frac{2}{\pi}} \sin \lambda s \qquad (\lambda \geq 0). \tag{1.7.6}$$

It is easy to check that

$$\langle s | D^{-1} | s' \rangle := \int_0^\infty d\lambda \, \lambda^{-2} e_\lambda(s) e_\lambda(s') = \min(s, s'). \tag{1.7.7}$$

In writing down formulas like (1.7.7) we follow Dirac's convention, i.e., $\langle s | D^{-1} | s' \rangle$ denotes the kernel of the operator D^{-1}:

$$D^{-1} f(s) = \int_0^\infty ds' \, \langle s | D^{-1} | s' \rangle f(s') \qquad (f \in L^2). \tag{1.7.8}$$

The kernel of D^{-1} is also called the *Green's function* of the differential operator D (any differential expression supplemented by boundary conditions).

Suppose X_s is some general Gaussian process (one-dimensional for simplicity) whose covariance is of the form $G(s, s') = \langle s | D^{-1} | s' \rangle$. Then D^{-1} is called the *covariance operator* of the process. The process is completely determined by its covariance operator or else by the covariance function $\boldsymbol{E}(X_s X_{s'}) = G(s, s')$. This may be shown as follows. For real $p_1, \ldots, p_n$ the random variable

$$Y = p_1 X_{s_1} + p_2 X_{s_2} + \cdots p_n X_{s_n} \tag{1.7.9}$$

is normally distributed with $\boldsymbol{E}(Y) = 0$ and $\boldsymbol{E}(Y^2) = \sum_{j,k} p_j p_k G(s_j, s_k)$. Thus

$$\boldsymbol{E}\left(\exp i \sum p_j X_{s_j} \right) = \exp \left\{ -\tfrac{1}{2} \sum p_j p_k G(s_j, s_k) \right\}. \tag{1.7.10}$$

All finite-dimensional distributions of the process X_s may be obtained by applying a Fourier transformation to (1.7.10) with respect to $p = (p_1, \ldots, p_n)$.
Remark. Assume $s_j \neq s_k$ if $j \neq k$. The condition $\boldsymbol{E}(Y^2) \geq 0$ ($= 0$ only if $p_i = 0$, $i = 1, \ldots, n$) implies that the numbers $g_{jk} = G(s_j, s_k)$ may be regarded as elements of some positive $n \times n$ matrix. Since this is the case for all time sequences $s_1, \ldots, s_n$ (all n), it is clear that the operator D has to be positive. If the positivity condition is violated, there is no Gaussian process attached to D.

When restricted to $d = 1$ the result (1.7.5) may be reformulated in two ways:

$$\boldsymbol{E}\big((X_{s'} - X_s)(X_{s'} - X_s) \big) = |s' - s| \tag{1.7.11}$$

$$\boldsymbol{E}\big((X_{s'} - X_s)(X_{t'} - X_t) \big) = \big| [s, s'] \cap [t, t'] \big| \tag{1.7.12}$$

$(s' > s,\ t' > t)$. These formulas demonstrate once more the lack of differentiability for the Wiener process. Nevertheless, in a vague manner using Dirac's δ-function one has

$$\frac{\partial}{\partial s}\frac{\partial}{\partial s'}\min(s,s') = \delta(s-s').\tag{1.7.13}$$

It goes without saying that the "derivative" $W_s = \dot{X}_s$ is merely a formal object (like any Schwartz distribution if considered a "function") with covariance

$$\boldsymbol{E}(W_s W_{s'}) = \delta(s-s').\tag{1.7.14}$$

It is termed *white noise* by physicists, thereby emphasizing that the Fourier transformed covariance is constant. To give precise meaning to W_s one introduces the concept of a *generalized stochastic process* (see [1.28, Chap.3] for details), i.e., in our case some smearing with real functions $f \in L^2(0,\infty)$ is necessary to obtain bona fide random variables

$$W(f) := \int_0^\infty f(s)W_s ds \equiv \int_0^\infty f(s)dX_s.\tag{1.7.15}$$

The expression to the left is also referred to as a *stochastic integral*[13]. The properties of the Wiener process are echoed by stating that the variables $W(f)$ are linear in f, Gaussian, and satisfy

$$\boldsymbol{E}\big(W(f)\big) = 0, \qquad \boldsymbol{E}\big(W(f)^2\big) = \|f\|^2 := \int_0^\infty ds\, f(s)^2.\tag{1.7.16}$$

"Polarizing" the second formula we get the covariance:

$$\boldsymbol{E}\big(W(f)W(g)\big) = (f,g) := \int_0^\infty ds\, f(s)g(s)\tag{1.7.17}$$

$(f,g \in L^2,\ \text{real})$. Even more startling is the fact (easy to prove) that complete information about the stochastic integral $W(f)$ is coded in the generating functional

$$\boldsymbol{E}\big(\exp\{iW(f)\}\big) = \exp\{-\tfrac{1}{2}\|f\|^2\}.\tag{1.7.18}$$

So far, time has been nonnegative. However, as may be seen from the above, this restriction is in fact unnecessary. To extend white noise to negative times we simply regard $W(f)$ as a generalized stochastic process on $L^2(-\infty,\infty)$, setting

---

[13] Normally, in analysis, a function $X(s)$ which is locally of bounded variation induces a measure $dX(s)$ and $\int f(s)dX(s)$ makes sense as a Stieltjes integral. Notice, however, that almost any path of the Brownian motion is of unbounded variation on any time interval [1.23. p.49]. Inspite of this, the stochastic integral (1.7.15) is valid even when the process X_s does not have a version, where all paths are (locally) of bounded variation. For more details about stochastic integrals see [1.29] and [1.30]

$$\|f\|^2 = \int_{-\infty}^{\infty} ds\, f(s)^2 \tag{1.7.19}$$

in (1.7.18). The stochastic integral $W(f)$, now avoiding unnatural restrictions, can be understood as a universal frame for building many Gaussian processes of interest, i.e., a particular process often arises as $Y_t = W(f_t)$ with f_t a family of real square-integrable functions. The simplest example is provided by the Wiener process X_t where

$$X_t = W(f_t) \qquad f_t(s) = \begin{cases} 1 & \text{if } 0 \le s \le t \\ 0 & \text{otherwise.} \end{cases} \tag{1.7.20}$$

For more examples see Appendix B.

Lack of Ergodicity. With X_s the standard one-dimensional Brownian motion we define the mean with respect to time as the random variable

$$Y_t = \frac{1}{t} \int_0^t ds\, X_s \tag{1.7.21}$$

and ask whether Y_t converges (in distribution) as t tends to infinity. The answer is no. To see this we write $Y_t = W(\bar{f}_t)$ setting $\bar{f}_t = t^{-1} \int_0^t dt'\, f_{t'}$, where f_t is given by (1.7.20) so that

$$\bar{f}_t(s) = \begin{cases} 1 - t/s & \text{if } 0 \le s \le t \\ 0 & \text{otherwise} \end{cases} \tag{1.7.22}$$

and

$$\|\bar{f}_t\|^2 = \int_0^t ds\, (1 - s/t)^2 = t/3. \tag{1.7.23}$$

From (1.7.18) we infer that, for real p and $t \to \infty$,

$$\boldsymbol{E}\big(\exp\{ipY_t\} \big) = \exp\{-p^2 t/6\} \quad \to \quad \begin{cases} 1 & \text{if } p = 0 \\ 0 & \text{otherwise.} \end{cases} \tag{1.7.24}$$

Since the limit characteristic function is discontinuous, there can be no probability distribution associated with it and hence no random variable Y_∞. What actually happens is this: the (Gaussian) distribution of Y_t flattens and wants to approach a uniform distribution which does not exist because the space accessible to the Brownian particle has infinite volume. It should therefore be obvious that recipes from ergodic theory cannot be applied in the present context: the mean values with respect to an ensemble of paths cannot be calculated as and replaced by the corresponding time means using one single path. Means with respect to an infinite period of time simply do not exist.

Fractional Brownian Motion. Mandelbrot and Van Ness [1.31] have introduced the concept of fractional Brownian motion B_t as a kind of fractional integral of white noise letting $B_t - B_0 = W(f_t - f_0)$, where $t \ge 0$ and[14]

---

[14]One defines $(t - s)_+ = t - s$ if $t \ge s$ and $= 0$ otherwise.

$$f_t(s) = \Gamma(\alpha)^{-1}(t-s)_+^{\alpha-1} \tag{1.7.25}$$

In order that $f_t(s) - f_0(s)$ be square-integrable, it is necessary and sufficient that $\frac{1}{2} < \alpha < \frac{3}{2}$. The model system reduces to ordinary Brownian motion if $\alpha = 1$ and $B_0 = 0$. Increments $B_{t'} - B_t$ are of the order $|t' - t|^{\alpha-1/2}$. So the variance of increments follows a power law with exponent $2\alpha - 1$, and there is no intrinsic time scale. It is easy to convince oneself that the fractional Brownian motion has *dependent* increments unless $\alpha = 1$. This unusual behavior leads to large-time correlations of increments which, normally, is in conflict with the assumptions of statistical physics (except at a point of second-order phase transition).

1.8 The Ornstein–Uhlenbeck Process

The setback of Einstein's theory of Brownian motion, the lack of any sort of velocity of the particle, has prompted physicists to formulate theoretical alternatives. One such alternative, now called the dynamical theory of Brownian motion [1.32], was put forward by Ornstein and Uhlenbeck [1.33]. It is the simplest theory of its kind that includes Einstein's model (which then becomes valid on a large-time scale).

For the sake of simplicity, we shall study the one-dimensional case only. The extension to arbitrary dimensions d is straightforward. Again, let the random variable $x(t)$ denote the position of a Brownian particle at time t. But now, we assume that there exists the velocity $v(t) = \dot{x}(t)$ satisfying a Langevin equation of the following type:

$$dv(t) = -\gamma v(t)dt + \sigma dX_t \qquad (X_t = \text{Wiener process}, \ \gamma > 0). \tag{1.8.1}$$

This ansatz may be interpreted by saying that the force acting on the particle consists of two parts: the (nonprobabilistic) friction term $-\gamma v(t)$ and the stochastic term $\sigma W(t)$, where the derivative of the Wiener process, $W(t) = \dot{X}_t$, is nothing but white noise.

Though, at first sight, the mathematical interpretation of (1.8.1) seems vague, a rigorous theory of stochastic differential equations is available [1.34,35] and could, in principle, be applied here. Instead, we continue discussing some physical aspects[15] of the Ornstein–Uhlenbeck theory:

1. The Brownian particle is viewed as moving in a homogeneous medium (some caricature of the molecular chaos) that acts on it via a deterministic and a stochastic force. The background is at rest and defines a preferred system of reference, thus explaining the lack of invariance with respect to the Galilei group.

---

[15] For an account of the various ways in which stochastic differential equations are used in physics and chemistry see [1.36].

2. The friction term in (1.8.1) leads to a time constant $\tau = \gamma^{-1}$, typically of the order of 10^{-8} seconds. The time scale for following the particle's motion is supposed to be of the same order or larger.

Given $v(0) = v_0$, we write down the solution of the initial-value problem as if we had started from an ordinary differential equation:

$$v(t) = e^{-\gamma t} v_0 + \sigma \int_0^t e^{-\gamma(t-s)} dX_s \qquad (t \geq 0). \tag{1.8.2}$$

This makes sense if we interpret the second term on the right hand side as a *stochastic integral*:

$$\int_0^t e^{-\gamma(t-s)} dX_s = W(f_t), \qquad f_t(s) = \begin{cases} e^{-\gamma(t-s)} & \text{if } 0 \leq s \leq t \\ 0 & \text{otherwise.} \end{cases} \tag{1.8.3}$$

To make life as simple as possible we let the initial velocity v_0 be some deterministic quantity. In this case, the velocity at some later time is normally distributed with mean value

$$\boldsymbol{E}(v(t)) = e^{-\gamma t} v_0 \tag{1.8.4}$$

and covariance

$$\boldsymbol{E}\big(v(t)v(t')\big) = \sigma^2 \boldsymbol{E}\Big(W(f_t)W(f_{t'})\Big) = \sigma^2 G(t, t'), \tag{1.8.5}$$

where $v(t) = e^{-\gamma t} v_0 + u(t)$ and

$$G(t, t') = (f_t, f_{t'}) = \frac{1}{2\gamma}\left(e^{-\gamma|t-t'|} - e^{-\gamma(t+t')}\right). \tag{1.8.6}$$

If $\gamma \to 0$, $G(t, t')$ approaches $\min(t, t')$, showing that the Wiener process comes out as a limit of the Ornstein–Uhlenbeck velocity process. If $\gamma > 0$ and $t \gg \tau$ ($\gamma t \gg 1$), the Brownian particle does no longer remember its initial velocity, i.e., $v(t)$ behaves asymptotically like a Gaussian random variable with mean zero and variance $\sigma^2/(2\gamma)$. From standard thermodynamics, we borrow the relation

$$\tfrac{1}{2} m \boldsymbol{E}(v^2) = \tfrac{1}{2} k_\mathrm{B} T, \qquad \text{i.e., } \sigma^2 = 2\gamma k_\mathrm{B} T / m , \tag{1.8.7}$$

that connects the temperature T with the constant σ.

Next, we adopt the starting value $x(0) = 0$ for the position of the Brownian particle and get

$$x(t) = \int_0^t dt\, v(t) , \tag{1.8.8}$$

with $v(t)$ given by (1.8.2). As a result, $x(t)$ is normally distributed with mean

$$\boldsymbol{E}(x(t)) = \gamma^{-1}(1 - e^{-\gamma t})v_0 \qquad (1.8.9)$$

and covariance $\boldsymbol{E}\big(y(t)y(t')\big) = \sigma^2 \hat{G}(t,t')$, where $x(t) = \gamma^{-1}(1 - e^{-\gamma t})v_0 + y(t)$,

$$\hat{G}(t,t') = \int_0^t ds \int_0^{t'} ds'\, G(s,s') = \gamma^{-2}\min(t,t') + g(t,t'), \qquad (1.8.10)$$

and

$$g(t,t') = -\tfrac{1}{2}\gamma^{-3}\left(e^{-\gamma|t-t'|} + e^{-\gamma(t+t')} + 2 - 2e^{-\gamma t} - 2e^{-\gamma t'}\right). \qquad (1.8.11)$$

The Ornstein–Uhlenbeck theory differs from Einstein's theory of Brownian motion by the extra term $g(t,t')$ in (1.8.10). However, for large times (either t or t'), this term becomes negligible compared to $\gamma^{-2}\min(t,t')$. To complete the contact with Einstein's theory it is necessary to relate the diffusion constant D to the parameters of the Ornstein–Uhlenbeck theory. The desired relation follows from the asymptotic behavior of the variance:

$$D = \lim_{t\to\infty} \frac{\sigma^2}{2t}\hat{G}(t,t) = \frac{\sigma^2}{2\gamma^2}. \qquad (1.8.12)$$

In conclusion, then, we maintain that, whenever t is large, i.e., if $t \gg \gamma^{-1}$, both theories give identical results. In particular, the time evolution of the position $x(t)$ as predicted by the Ornstein–Uhlenbeck velocity process, for large t, becomes indistiguishable from that predicted by the Wiener (position) process provided $x(0) = \gamma^{-1}v_0$. Moreover, the Einstein relation $D = k_{\mathrm{B}}T/f$ holds, where the friction constant is given by $f = m\gamma$.

The Covariance Operator of the Ornstein–Uhlenbeck Process. It is instructive to learn that the covariance $G(t,t')$ may also be interpreted as the Green's function for some differential operator. Let D be the operator on $L^2(0,\infty)$ corresponding to the expression $-d^2/dt^2 + \gamma^2$ and to the boundary condition $f(0) = 0$. The operator D is positive and its spectrum is continuous. To any spectral value $\lambda^2 + \gamma^2$ of D there corresponds the generalized eigenfunction $e_\lambda(t) = \sqrt{2/\pi}\sin\lambda t$ $(\lambda > 0)$, and with the help of standard Fourier analysis we find the kernel of the operator D^{-1}:

$$\langle t|D^{-1}|t'\rangle := \int_0^\infty d\lambda\, \frac{e_\lambda(t)e_\lambda(t')}{\lambda^2 + \gamma^2} = \frac{1}{2\gamma}\left(e^{-\gamma|t-t'|} - e^{-\gamma(t+t')}\right). \qquad (1.8.13)$$

Comparison with (1.8.5) and (1.8.6) shows that $\sigma^2 D^{-1}$ is in fact the covariance operator of the Ornstein–Uhlenbeck velocity process.

1.8.1 The Oscillator Process

There is another important process which comes up in the context of quantum physics and which is very closely related to the Ornstein–Uhlenbeck process. It is called the oscillator process. The easiest approach is to write it as a stochastic integral:

$$Q_t = \int_t^\infty e^{-k(s-t)} dX_s = W(g_t)$$

$$g_t(s) = \begin{cases} e^{-k(s-t)} & \text{if } s \geq t \\ 0 & \text{otherwise.} \end{cases}$$

$$(1.8.14)$$

Again, we treat the one-dimensional case only. Since the formula makes sense for all $t \in \mathrm{I\!R}$, the oscillator process Q_t belongs to the special class of two-sided processes. Its role will soon become apparent, once we know more about the path integral for the harmonic oscillator in quantum mechanics. There the Hamiltonian is of the form

$$H = \tfrac{1}{2}(-d^2/dx^2 + k^2x^2 - k) \qquad (k > 0) \tag{1.8.15}$$

with k the frequency of the oscillator.

Obviously, $\boldsymbol{E}(Q_s) = 0$ and it is a relatively easy matter to evaluate the covariance:

$$\boldsymbol{E}(Q_s Q_{s'}) = (g_s, g_{s'})$$
$$= e^{k(s+s')} \int_{\max(s,s')}^\infty dt\, e^{-2kt} = (2k)^{-1} e^{-k|s'-s|}. \tag{1.8.16}$$

The Covariance Operator of the Oscillator Process. We shall be brief and simply quote the relevant formula:

$$\langle s|D^{-1}|s'\rangle = (2k)^{-1}e^{-k|s-s'|} = (2\pi)^{-1}\int_{-\infty}^\infty d\lambda\, \frac{e^{i\lambda(s-s')}}{\lambda^2 + k^2}. \tag{1.8.17}$$

It is thus seen that D coincides with the differential operator $-d^2/ds^2 + k^2$ acting on $L^2(-\infty, \infty)$. As a consequence, D^{-1} is the covariance operator of the oscillator process.

As $k \to \infty$, $(k/2)e^{-k|s-s'|}$ tends to $\delta(s - s')$ implying that, formally, white noise arises as a limit of kQ_s. In fact, the oscillator process (including white noise) may be shown to be the only *invariant* (with respect to time translations) Gaussian Markov process up to a change of scales [1.22. Corollary 4.11].

The Mehler Formula. Let us evaluate some distributions of the oscillator process that will quickly prove to be relevant in the context of the quantum mechanical oscillator. To start with, consider $\boldsymbol{P}(Q_s \in dx)$. From the generating function

$$\boldsymbol{E}\Big(\exp(ipQ_s)\Big) = \exp(-\tfrac{1}{2}p^2\|g_s\|^2) \tag{1.8.18}$$

and the identity $\|g_s\|^2 = (2k)^{-1}$ it follows, by a Fourier transformation, that

$$\boldsymbol{P}(Q_s \in dx) = dx\,(k/\pi)^{1/2}\exp(-kx^2) \equiv dx\,\Omega(x)^2 \tag{1.8.19}$$

irrespective of s. Notice the connection with the ground state wave function $\Omega(x)$ of the harmonic oscillator. Let us turn next to the joint distribution of Q_s and $Q_{s'}$, where $s < s'$. Define $\nu = k(s' - s)$ and

$$A = \begin{pmatrix} (g_s, g_s) & (g_s, g_{s'}) \\ (g_{s'}, g_s) & (g_{s'}, g_{s'}) \end{pmatrix} = \frac{1}{2k} \begin{pmatrix} 1 & e^{-\nu} \\ e^{-\nu} & 1 \end{pmatrix}, \quad \mathbf{x} = \begin{pmatrix} x \\ x' \end{pmatrix}, \mathbf{p} = \begin{pmatrix} p \\ p' \end{pmatrix}. \quad (1.8.20)$$

By Fourier transforming the generating function

$$\boldsymbol{E}\left(\exp(ipQ_s + ip'Q_{s'})\right) = \exp\left(-\tfrac{1}{2}\mathbf{p}^T A \mathbf{p}\right) \tag{1.8.21}$$

with respect to p and p' we are led to the Gaussian distribution

$$\boldsymbol{P}(Q_s \in dx, Q_{s'} \in dx') = \frac{dx\, dx'}{2\pi\sqrt{\det A}} \exp\left(-\tfrac{1}{2}\mathbf{x}^T A^{-1}\mathbf{x}\right), \tag{1.8.22}$$

where

$$A^{-1} = \frac{2k}{1 - e^{-2\nu}} \begin{pmatrix} 1 & -e^{-\nu} \\ -e^{-\nu} & 1 \end{pmatrix}, \qquad \det A = \frac{1 - e^{-2\nu}}{4k^2}. \tag{1.8.23}$$

To obtain the transition function of the oscillator process we write

$$\boldsymbol{P}(Q_s \in dx, Q_{s'} \in dx') \;=\; \boldsymbol{P}(Q_{s'} \in dx' \mid Q_s = x)\boldsymbol{P}(Q_s \in dx) \tag{1.8.24}$$

and cast the result into a special form:

$$\boldsymbol{P}(Q_{s'} \in dx' \mid Q_s = x) \;=\; dx'\, \Omega(x')\langle x', s' | x, s\rangle \Omega(x)^{-1}, \qquad \nu = k(s' - s). \tag{1.8.25}$$

The result is known as the *Mehler formula*:

$$\langle x', s' | x, s\rangle = \left(\frac{k}{\pi(1 - e^{-2\nu})}\right)^{1/2} \exp\left\{-\frac{k(x^2 + x'^2)}{2\tanh\nu} + \frac{kxx'}{\sinh\nu}\right\}. \tag{1.8.26}$$

Here,

$$\langle x', s' | x, s\rangle = \langle x' | e^{-(s'-s)H} | x\rangle, \qquad H = \frac{1}{2}\left(-\frac{d^2}{dx^2} + k^2 x^2\right), \tag{1.8.27}$$

taking into account that quantum mechanics has its own notation; $\langle x', s' | x, s\rangle$ is called the *transition amplitude* for the harmonic oscillator and it is precisely this quantity that will later be written as a path integral when the ground has been prepared for a more systematic treatment. We may also mention that, in some applications, we need only know the values of the transition amplitude along the "diagonal" $x = x'$ which simplifies matters:

$$\langle x, s' | x, s\rangle \;=\; \left(\frac{k}{\pi(1 - e^{-2\nu})}\right)^{1/2} \exp\left\{-kx^2\tanh\frac{\nu}{2}\right\}. \tag{1.8.28}$$

In particular, the so-called *partition function* is expressed as

$$e^{-\beta F} = \int dx\, \langle x | e^{-\beta H} | x\rangle$$

$$= \int \boldsymbol{P}(Q_\beta \in dx \mid Q_0 = x) \;=\; (1 - e^{-k\beta})^{-1}, \tag{1.8.29}$$

where $F = F(\beta)$ denotes the *free energy* of the oscillator.

Exercise 1. If X is a real Gaussian random variable with mean zero and variance $E(X^2) = \sigma^2 > 0$, one introduces the nth 'Wick power' $:X^n:$ via the generating function

$$\sum_{n=0}^{\infty} \frac{t^n}{n!} :X^n: = \exp\{tX - \tfrac{1}{2}t^2\sigma^2\}.$$

Prove $E(:X^n:) = 0$ $(n > 0)$ and the orthogonality relation

$$E(:X^n: :X^m:) = n!\sigma^{2n}\delta_{n,m} \qquad (n, m \geq 0).$$

Exercise 2. Under the assumptions of the preceding exercise, let $\mathcal{H}$ be the Hilbert space of functions $f: \mathbb{R} \to \mathbb{C}$ with respect to the inner product $(f, g) = E(\bar{f}(X)g(X))$ and consider normalized Wick powers:

$$e_n(X) = (n!)^{-1/2}\sigma^{-n} :X^n:.$$

Show that these functions form a (orthonormal) basis in $\mathcal{H}$.

Exercise 3. For Q_s a real Gaussian process with mean zero, and V, some real function, assume that there be an expansion

$$V(Q_s) = \sum_{n=0}^{\infty} c_n(s) :Q_s^n:$$

with coefficients $c_n(s)$. Write the coefficients as expectations to prove that each c_n is constant, i.e., independent of the time s, if the process is invariant (with respect to time translations).

Exercise 4. If, in Exercise 3, Q_s is taken to be the oscillator process with covariance $(2k)^{-1}e^{-k|s-s'|}$, then from the relation to the quantum harmonic oscillator conclude that

$$c_n = (n!)^{-1/2}(2k)^{n/2}(\Omega, V\Omega_n),$$

where $\Omega_n(x)$ is the nth normalized eigenstate of the oscillator Hamiltonian (so that $\Omega_0 = \Omega$), and V means multiplication by $V(x)$.

2 The Feynman–Kac Formula

> The physicist cannot understand the mathematician's care in
> solving an idealized physical problem. The physicist knows the
> real problem is much more complicated. It has already been sim-
> plified by intuition which discards the unimportant and often
> approximates the remainder.
>
> *Richard Feynman*

The main questions raised so far concerned the Brownian motion as a real
phenomenon. The problem of relating it to quantum mechanics was only
touched upon. Now that we have the material to penetrate deeper into
the subject, we shall, in the present chapter, give an account of the functional integral description of interacting quantum mechanical systems using
Brownian motion as a *tool*. In doing so we no longer mention the Brownian
particle: it is treated merely as a formal object and is not considered any
part of the observable reality. Also, quantum mechanical probability should
not be confused with the statistics of the underlying Brownian motion.

The basic instrument, well adapted to our needs, is an intriguing integral
called the *path integral*. In ordinary integration theory one talks about some
domain of integration consisting of infinitely many points. By analogy, path
integration is carried out with respect to domains that are sets of (infinitely
many) Brownian paths with a certain characteristic. It is thus the entire
path that now plays the role of a single point in some abstract space. In
fact the ideas for introducing a measure will be the same as in Sect. 1.6.1,
but the notation and emphasis will be different: it is rather the language
of integration theory (not that of probability theory) that we prefer in the
present context. For its scope is much wider, and it also allows us to study
integrals that do not arise from probability measures.

The main result, the formula of Feynman and Kac, relates the Schrödinger semigroup e^{-tH} with generator $H = -\frac{1}{2}\Delta + V$ to some integral
over Brownian paths. There are various proofs of this formula. One proof
uses the Lie–Trotter product equality [2.1], another the integral equation
satisfied by e^{-tH} [1.22], and still another makes extensive use of the theory
of stochastic processes [2.2, 1.5]. We shall present the first two proofs.

2.1 The Conditional Wiener Measure

Given the dimension d of the configuration space, we can, for $x, x' \in \mathbb{R}^d$ and times s, s' such that $0 < s < s'$, choose, as the underlying space, the set Ω of all paths $\omega : [s, s'] \to \mathbb{R}^d$, where $\omega(s) = x$ and $\omega(s') = x'$, i.e., we assume that the initial and the final points of all paths are the same and momentarily fixed. We also say that those paths are pinned down at both ends (see Fig. 2.1).

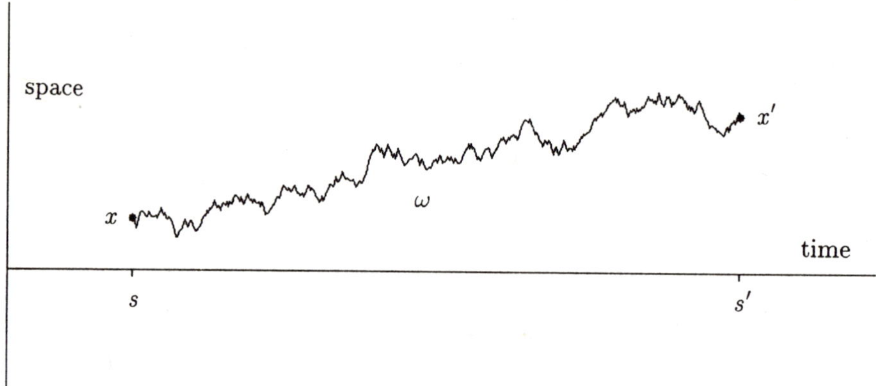

Fig. 2.1. A Brownian path with fixed initial and final points

We are going to define a measure μ on Ω such that the total mass becomes

$$\mu(\Omega) = \int_\Omega d\mu(\omega) = K(x' - x, s' - s) > 0, \qquad (2.1.1)$$

where $K(x, s) = (2\pi s)^{-d/2} \exp(-(2s)^{-1} x^2)$. For this reason, μ cannot be interpreted as a probability measure.

Next, we assign a measure to certain subsets of Ω, the so-called cylinder sets. The basic idea is to choose, for arbitrary n, a time sequence $s_1, s_2, \ldots, s_n$ such that $s < s_1 < \cdots < s_n < s'$. To each s_i we attach a 'window' $A_i \subset \mathbb{R}^d$ (i.e., some Borel set) and consider the set of all paths passing each window at the prescribed time. For brevity, the set so obtained will be denoted Ω_A, where the index A stands for the Cartesian product $A_1 \times \cdots \times A_n$ while the dependence on the base $S = (s_1 \ldots, s_n)$ is suppressed. Formally,

$$\Omega_A = \{\, \omega \in \Omega \mid \omega(s_i) \in A_i, \ i = 1, \ldots, n \,\}. \qquad (2.1.2)$$

Quite analogous to the construction of finite-dimensional distributions of the Wiener process (compare (1.6.5)) we put

$$\mu(\Omega_A) = \int_{\Omega_A} d\mu(\omega) = \int_{A_n} dx_n \cdots \int_{A_1} dx_1 \, K(x_1, \ldots, x_n), \qquad (2.1.3)$$

where

$$K(x_1, \ldots, x_n) = \prod_{i=0}^{n} K(x_{i+1} - x_i, s_{i+1} - s_i) \qquad (2.1.4)$$

$(x_0 = x, \; x_{n+1} = x', \; s_0 = s, \; s_{n+1} = s')$. Thus $\mu(\Omega_A)$ may be viewed as an integral over points in the subset A of $\mathbb{R}^{nd}$. Varying n, the base S, and $(A_i)_{i=1,\ldots,n}$ one obtains, in a sense, sufficiently many cylinder sets to give meaning to μ as a measure on Ω called the *conditional Wiener measure*[16]. Both the underlying space Ω and the measure μ depend on x, x', s, s' as parameters which are kept fixed during the construction. Later on they are allowed to vary; we shall then be forced to indicate the dependence and write:

$$\Omega = \Omega_{x,s}^{x',s'}, \qquad \mu = \mu_{x,s}^{x',s'}. \qquad (2.1.5)$$

2.1.1 The Path Integral

The next step is to give meaning to the *path integral*

$$I(f) = \int d\mu(\omega) \, f(\omega) \qquad (2.1.6)$$

for suitable real functions f on Ω. In a most direct approach, the idea is to approximate $I(f)$ by ordinary integrals over $\mathbb{R}^{nd}$ and to take the limit of these integrals as n tends to infinity[17]. More explicitly, we shall, in place of (2.1.6), often write

$$\int_{(x,s)}^{(x',s')} d\mu(\omega) \, f(\omega) = \int d\mu_{x,s}^{x',s'}(\omega) \, f(\omega) \qquad (2.1.7)$$

though the notation used on the left-hand side is slightly misleading: (x, s) and (x', s') are not really lower and upper limits of some "definite" path integral but parameters of the measure μ; in fact, each path ω has a hidden dependence on (x, s) and (x', s') which should never be forgotten.

Integration theory starts out by considering *simple functions* f for which $I(f)$ may be evaluated immediately; measurable functions will be defined as

---

[16]Notice first that the "distributions" for n and $n+1$ are *consistent* in the sense of (1.6.4). This allows us to extend μ to a σ-additive measure on $\mathcal{F}$, which is the smallest σ-field containing all cylinders $\{\, \omega \mid \omega(s) \in A_s, s \in S \,\}$.

[17]In the mathematics literature, this is known as the "sequential" definition of the Wiener integral. Besides, there is an indirect definition, avoiding any kind of limit, which proves existence but is nonconstructive: it uses Minlos's theorem (see Exercise 4 at the end of this section).

limits of simple functions [1.18]. The simplest example of a simple function is the characteristic function of a measurable set. For instance we may take

$$f_A(\omega) = \begin{cases} 1 & \text{if } \omega \in \Omega_A \\ 0 & \text{otherwise,} \end{cases} \tag{2.1.8}$$

where the set Ω_A has been defined in (2.1.2.). As the result of the integration we obtain

$$\int d\mu(\omega)\, f_A(\omega) = \mu(\Omega_A) \tag{2.1.9}$$

which is trivial since it follows from the definitions. Indeed, the function f_A is very special: it may be written as a product of elementary characteristic functions, i.e.,

$$f_A(\omega) = \prod_{i=1}^{n} \chi_{A_i}(\omega(s_i)), \tag{2.1.10}$$

where

$$\chi_B(x) = \begin{cases} 1 & \text{if } x \in B \\ 0 & \text{otherwise} \end{cases} \qquad (B \subset \mathbb{R}^d). \tag{2.1.11}$$

The essential property of the function f_A is that it depends only on a *finite set of path coordinates* which are the positions $\omega(s_1), \ldots, \omega(s_n)$. Other functions having this property appear as a superposition of characteristic functions, and any function so obtained is said to be *piecewise constant*. In more detail the construction is as follows.

We first fix a natural number n, then fix a basis $S = (s_1, \ldots, s_n)$, and finally vary the sets A_i in $A = A_1 \times \cdots \times A_n$. By forming finite superpositions $g = \sum_A c_A f_A$ with real coefficients c_A we generate, for our purpose, sufficiently many piecewise constant functions g that depend on finitely many path coordinates. By the linearity of the path integral we may thus write

$$\begin{aligned}
\int d\mu\, g &= \sum_A c_A\, \mu(\Omega_A) \\
&= \sum_A c_A \int_{A_n} dx_n \cdots \int_{A_1} dx_1\, K(x_1, \ldots, x_n) \\
&= \int dx_n \cdots \int dx_1\, K(x_1, \ldots, x_n) \sum_A c_A \prod_{i=1}^{n} \chi_{A_i}(x_i) \\
&= \int dx_n \cdots \int dx_1\, K(x_1, \ldots, x_n) g(x_1, \ldots, x_n),
\end{aligned} \tag{2.1.12}$$

where – by abuse of notation – we also regard g as a function on $\mathbb{R}^{nd}$:

$$g(x_1, \ldots, x_n) = \sum_A c_A \prod_{i=1}^{n} \chi_{A_i}(x_i). \tag{2.1.13}$$

Simple functions form a real-linear vector space L of bounded functions on Ω. It is closed under the lattice operations

$$(f \vee g)(\omega) = \max \left(f(\omega), g(\omega) \right) ,$$
$$(f \wedge g)(\omega) = \min \left(f(\omega), g(\omega) \right) ,$$

(2.1.14)

and we may take absolute values since $|f| = f \vee 0 - f \wedge 0$. The path integral $I(f)$ on $f \in L$ is a linear functional satisfying

$$f \geq 0 \Rightarrow I(f) \geq 0 , \tag{2.1.15}$$

$$f_n \downarrow 0 \Rightarrow I(f_n) \downarrow 0 , \tag{2.1.16}$$

where "$f_n \downarrow 0$" means that the sequence decreases pointwise to zero. Also

$$|I(f)| \leq I(|f|), \qquad I(|f + g|) \leq I(|f|) + I(|g|), \tag{2.1.17}$$

and moreover $I(|f|) = 0$ holds if and only if $f = 0$ except on a set of measure zero: one writes $f = 0$ μ-a.e. (almost everywhere) in this case. Regarding f and f' *essentially the same function* if $f - f' = 0$ μ-a.e., one defines the distance between f and g by the equation

$$\|f - g\| = I(|f - g|). \tag{2.1.18}$$

A sequence f_n is called a *Cauchy sequence* if the distance $\|f_n - f_m\|$ becomes arbitrarily small.

The general task is to extend $I(f)$ to a larger space of functions f having all the properties of L (except piecewise constancy), and being, in addition, closed on taking limits. The extension of L is then the class of *integrable functions*. The concept is this: a real-valued function f defined μ-a.e. on Ω (and hence not necessarily in L) is said to be *integrable* if there exists a Cauchy sequence $f_n \in L$ such that $f_n \to f$ pointwise except on a set of measure zero.

Among the set of all integrable functions, there is the subset $\hat{L}$ of those functions which depend on finitely many positions of the Brownian path. Obviously, any simple function is of this type; hence $L \subset \hat{L}$ and any $f \in \hat{L}$ may be approximated by simple functions f_n. For any such function f, the path integral reduces to an ordinary finite-dimensional integral. As an immediate consequence of (2.1.12) we may state:

Suppose the function $f : \Omega \to {\rm I\!R}$ is such that it depends only on finitely many coordinates of the path ω. If these are the positions $x_i = \omega(s_i) \in {\rm I\!R}^d$ at the times s_i, $i = 1, \ldots, n$ and if $K(x_1, \ldots, x_n)$ is given by (2.1.4), then

$$\int d\mu \, f = \int dx_n \cdots \int dx_1 \, K(x_1, \ldots, x_n) f(x_1, \ldots, x_n) \tag{2.1.19}$$

provided the integral on the right-hand side exists.

Since the natural number n assumes arbitrarily large (finite but unbounded) values, one may justly speak of the path integral as an extension, to infinite dimensions, of the traditional notion of a finite-dimensional integral. For a general function f on Ω it may become a painful (often impossible) task to determine the path integral, though, in many instances, its existence can be demonstrated by referring to Lebesgue's theorem of bounded convergence [1.18, Sect. 26]. Almost invariably one resorts to the method of time slicing, i.e., one constructs a sequence $f_n \in \hat{L}$ converging to f such that, for fixed n, f_n depends on the positions $\omega(s_k)$, where $s_k = k(s' - s)2^{-n}$ and $k = 1, \ldots, 2^n - 1$. If f_n is bounded (i.e., $|f_n| < c$ for all n), Lebesgue's theorem applies (since the constant function is integrable) with two consequences: (1) f is integrable, (2) $I(f_n) \to I(f)$.

Exercise 1. Consider d-dimensional Brownian motion. Let f be the characteristic function of the set of paths

$$\left\{ \omega \mid \sup_{s \le t \le s'} |\omega(t)| < a \right\}$$

for some $a > 0$. For $n \ge 1$, let the functions $f_n \in L$ be defined by

$$f_n(\omega) = \prod_{t \in S_n} \chi_A\big(\omega(t)\big), \qquad S_n = \{\, k2^{-n}(s' - s) \mid k = 1, \ldots, 2^n - 1 \,\},$$

where A is the ball in $\mathbb{R}^d$ of radius a. Prove monotonicity, i.e., $f_1 > f_2 > f_3 > \cdots > f$, and show that f_n is a Cauchy sequence converging to f.

Exercise 2. Consider the function

$$f(\omega) = g\left(\int_s^{s'} dt\, h\big(\omega(t)\big) \right),$$

where $g \colon \mathbb{R} \to \mathbb{R}$ is positive increasing and $h \colon \mathbb{R}^d \to \mathbb{R}$ is bounded continuous. Approximate the integral with respect to the time variable t by a Riemann sum to get $f_n \in \hat{L}$ as the function

$$f_n(\omega) = g\left(\sum_{t \in S_n} 2^{-n} h\big(\omega(t)\big) \right)$$

(S_n is the same in Exercise 1.). Prove that Lebesgue's theorem applies in this case, i.e., f_n is a Cauchy sequence converging to f and dominated by a constant function (though the constant will generally depend on the difference $s' - s$).

Exercise 3. Let the function $f(\omega)$ be given as a product, i.e., $f(\omega) = \prod_{i=1}^d f_i(\omega_i)$, where ω_i stands for the ith component of the path $\omega \in \Omega_{x,s}^{x',s'}$. Prove that the integral over all d-dimensional paths, $I(f)$, factorizes as $\prod_i I_i(f_i)$, where each factor represents an integral over one-dimensional paths: this explains why the conditional Wiener measure μ in d dimensions is in fact a product of Wiener measures in one dimension.

Exercise 4. It will be shown in Sect. 4.4 that the conditional Wiener measure has the characteristic functional

$$F(f) := \int d\mu_{x,s}^{x',s'}(\omega) \, \exp\left\{ i \int_s^{s'} dt \, f(t)\omega(t) \right\} =$$

$$\left(2\pi(s'-s)\right)^{-d/2} \exp\left\{ -\frac{(x'-x)^2}{2(s'-s)} + i \int_s^{s'} dt \, f(t) \frac{(s'-t)x + (t-s)x'}{s'-s} \right.$$

$$\left. - \int_s^{s'} dt \int_s^t dt' \, f(t)f(t') \frac{(s'-t)(t'-s)}{s'-s} \right\}$$

($f\colon [s,s'] \to \mathbb{R}^d$ of type L^∞). Argue why this formula serves as an alternative definition of the measure $d\mu_{x,s}^{x',s'}$. Hint: Show that $F(f)$ is continuous and positive-definite, i.e., $\sum_{i,j} \bar{z}_i z_j F(f_i - f_j) \geq 0$; apply Minlos's theorem [1.22, Theorem 2.2].

2.1.2 The Stochastic Formulation

From the viewpoint of probability theory, it seems preferable to always work with *expectations*, mainly for reasons of economy of concepts and notation. There is a remarkable passage from expectations $\boldsymbol{E}(f(X))$ with respect to the Wiener process X_t to the path integral $\int d\mu \, f$ which is not surprising but still well worth study. Let us therefore assume that the function $f(X)$ depends on the section $\{\, X_t \mid s \leq t \leq s' \,\}$ with $s > 0$ and s' fixed. Recalling that the Brownian particle starts out at the origin at time zero, first moves to (x,s) via the transition function $K(x,s)$, and then continues to (x',s'), where x and x' may take on all possible values in $\mathbb{R}^d$, we obtain the following expression for the expectation of $f(X)$ which connects it to the path integral between the spacetime points (x,s) and (x',s'):

$$\boldsymbol{E}(f(X)) = \int dx' \int dx \int_{(x,s)}^{(x',s')} d\mu(\omega)f(\omega)K(x,s). \tag{2.1.20}$$

The formula suggests writing $\boldsymbol{E}(f(X)) = \int dx \, \boldsymbol{E}_{x,s}(f(X))K(x,s)$ letting $\boldsymbol{E}_{x,s}$ denote the expectation with respect to the random motion of a Brownian particle that starts at (x,s) instead of $(0,0)$. With this definition, there is no need to further restrict s to positive values: the initial time s may be taken arbitrary in the identity

$$\boldsymbol{E}_{x,s}(f(X)) = \int dx' \int_{(x,s)}^{(x',s')} d\mu(\omega) \, f(\omega) \,, \tag{2.1.21}$$

which holds true provided $f(X)$ depends on $\{X_t | s \le t \le s'\}$. In particular, choosing the constant function 1 we would get

$$\boldsymbol{E}_{x,s}(1) = \int dx' \int_{(x,s)}^{(x',s')} d\mu(\omega) = 1. \tag{2.1.22}$$

Next, we start thinking about whether relation (2.1.21) serves to uniquely define the path integral. The answer is: it does. One way to demonstrate this fact is to replace $f(X)$ in (2.1.21) by $\exp\left\{-ip'X_{s'}\right\}f(X)$ with $p' \in \mathbb{R}^d$ so that

$$\boldsymbol{E}_{x,s}\left(\exp\left\{-ip'X_{s'}\right\}f(X)\right) = \int dx'\, e^{-ip'x'} \int_{(x,s)}^{(x',s')} d\mu(\omega)\, f(\omega).$$

$$\tag{2.1.23}$$

By an inverse Fourier transformation,

$$\int_{(x,s)}^{(x',s')} d\mu(\omega)\, f(\omega) = (2\pi)^{-d} \int dp'\, e^{ip'x'}\, \boldsymbol{E}_{x,s}\left(\exp\left\{-ip'X_{s'}\right\}f(X)\right)$$

$$= \boldsymbol{E}_{x,s}\left(\delta(X_{s'} - x')f(X)\right). \tag{2.1.24}$$

Path integrals with respect to the conditional Wiener measure enter the formalism of quantum mechanics as kernels of operators. It is a good idea to complete and consolidate our present knowledge about the relationship between path integrals and (unconditioned) expectations by pretending that there is an operator $U_{s',s}$ acting on the Hilbert space $L^2(\mathbb{R}^d)$ whose kernel $\langle x'|U_{s',s}|x\rangle$ coincides with the path integral of some complex function $u(\omega)$:

$$\langle x'|U_{s',s}|x\rangle = \int_{(x,s)}^{(x',s')} d\mu(\omega)\, u(\omega). \tag{2.1.25}$$

If we choose arbitrary wave functions $\psi, \psi' \in L^2$ (i.e., 'states' of a fictitious quantum mechanical system), we would get

$$(\psi', U_{s',s}\psi) = \int dx \int dx'\, \bar{\psi}'(x')\langle x'|U_{s',s}|x\rangle\psi(x). \tag{2.1.26}$$

Letting $f(X) = \bar{\psi}'(X_{s'})u(X)\psi(X_s)$ in (2.1.21) and integrating both sides with respect to x, taking into account (2.1.25) and (2.1.26), we get

$$\int dx\, \psi(x)\boldsymbol{E}_{x,s}\left(u(X)\bar{\psi}'(X_{s'})\right) = (\psi', U_{s',s}\psi). \tag{2.1.27}$$

Unfortunately, we are facing the problem of now having two different notions of "expectation". On the one hand there is the probabilistic interpretation of the term and on the other there is the quantum mechanical interpretation. The conflict is exemplified in (2.1.27); suppose that $\psi' = \psi$ and $\|\psi\| = 1$. Then $(\psi, U_{s',s}\psi)$ is said to be the *expectation* of $U_{s',s}$ with respect to the state represented by ψ, whereas $\boldsymbol{E}_{x,s}$ is said to be the expectation with respect to Brownian motion. Equation (2.1.27), however, reconciles the divergent definitions by connecting both concepts.

The simplest case of an operator whose kernel is known to us and which may also be represented by a path integral arises if $u(\omega) = 1$:

$$\langle x'|U_{s',s}|x\rangle = \int_{(x,s)}^{(x',s')} d\mu(\omega) \;=\; K(x' - x, s' - s) \tag{2.1.28}$$

$$U_{s',s} = e^{-(s'-s)H_0}, \qquad H_0 = -\tfrac{1}{2}\Delta.$$

This formula deals with the free motion of one (or several) Schrdinger particles: the operators $U_{t,0}$ ($t \geq 0$) form the Schrdinger semigroup of a noninteracting system. Equation (2.1.27) now reads:

$$\int dx\, \psi(x)\boldsymbol{E}_{x,0}\big(\bar{\psi}'(X_t)\big) = (\psi', e^{-tH_0}\psi). \tag{2.1.29}$$

The next idea we shall try is to replace $\psi(x)$ and $\psi'(x')$ in (2.1.27) by plane waves e^{ipx} and $e^{ip'x'}$ writing

$$U_{s',s}(p',p) = (2\pi)^{-d} \int dx' \int dx\, e^{i(px - p'x')} \langle x'|U_{s',s}|x\rangle\,, \tag{2.1.30}$$

which produces the result

$$U_{s',s}(p',p) = (2\pi)^{-d} \int dx\, e^{ipx}\boldsymbol{E}_{x,s}\Big(u(X)\exp(-ip'X_{s'})\Big)$$

$$= (2\pi)^{-d} \int dx\, e^{i(p-p')x}\boldsymbol{E}_{x,s}\Big(u(X)\exp\Big\{-ip'\int_s^{s'}dX_t\Big\}\Big). \tag{2.1.31}$$

Example. For $d = 3$, let $A(t) \in \mathbb{R}^3$ be some vector potential (dependent on the time variable t but not on the space variable x) and consider $H(t) = \frac{1}{2}(i\nabla + A(t))^2$, the quantum mechanical Hamiltonian of a particle of unit charge moving under the influence of the potential A. Notice that the magnetic field $B = \nabla \times A$ vanishes and that $[H(t), H(t')] = 0$. Also, conservation of energy does not generally hold for this model. Suppose we want to determine the operator $U_{s',s} = \exp\{-\int_s^{s'} dt\, H(t)\}$ so that $\psi_t = U_{t,0}\phi$ solves the problem

$$H(t)\psi_t = -\frac{\partial}{\partial t}\psi_t, \qquad \psi_0 = \phi. \tag{2.1.32}$$

We show that the solution involves the function

$$u(X) = \exp\Big\{i\int_s^{s'} A(t)\cdot dX_t\Big\}, \tag{2.1.33}$$

where X_t is three-dimensional Brownian motion. This may be seen as follows: we first check that $u(X)$ depends on $\{X_t | s \leq t \leq s'\}$ as required; we then insert $u(X)$ into (2.1.31) and evaluate the expectation $E_{x,s}(\exp\{iW(f)\})$, where $W(f)$ is the stochastic integral (see Sect. 1.7) and $f(t) = A(t) - p'$. It follows from (1.1.18) that

$$E_{x,s}\left(u(X)\exp\left\{-i\int_s^{s'} p' \cdot dX_t\right\}\right) = \exp\left\{-\tfrac{1}{2}\int_s^{s'} dt\,(p' - A(t))^2\right\}.$$

$$(2.1.34)$$

The resulting expression is independent of x and therefore

$$U_{s',s}(p',p) = \exp\left\{-\tfrac{1}{2}\int_s^{s'} dt\,(p - A(t))^2\right\}\delta(p' - p). \qquad (2.1.35)$$

which is the standard answer to the problem using the conventional Fourier transformation approach.

Exercise 1.   Under the assumptions above, specialize (2.1.35) to the situation where $A(t) = -\mathcal{E}t$ with $\mathcal{E} \in \mathrm{I\!R}^3$, and prove that

$$\frac{\langle x'|U_{s',s}|x\rangle_\mathcal{E}}{\langle x'|U_{s',s}|x\rangle_0} = \exp\left\{\frac{i}{2}(s' + s)(x - x') \cdot \mathcal{E} - \frac{1}{24}(s' - s)^3\mathcal{E}^2\right\}.$$

The analytic continuation $\mathcal{E} \to -iE$, $s' \to it'$, $s \to it$ will change the above formula which then directly addresses the time evolution problem of a Schrdinger particle moving in a constant electric field E. Demonstrate the existence of some one-parameter unitary group $e^{-it\hat{H}}$ ($t \in \mathrm{I\!R}$) such that

$$\langle x'|U_{it',it}|x\rangle_{-iE} = e^{-it'x'E}\langle x'|e^{-i(t'-t)\hat{H}}|x\rangle e^{itxE}.$$

Find an explicit expression for the generator $\hat{H}$, reintroduce the Planck constant $\hbar$, the mass m, and the charge q, and prove that $H(t) = V(t)\hat{H}V(t)^{-1}$ for some unitary $V(t)$, a gauge transformation.

2.2 The Integral Equation Method

Suppose we are dealing with a quantum mechanical system given in terms of the Hamiltonian $H = H_0 + V$, where $H_0 = -\tfrac{1}{2}\Delta$. The dimension d of the configuration space may be arbitrary, and the scalar potential $V(x)$ is assumed to be independent of time. We are going to represent the kernel of $\exp(-sH)$ ($s > 0$) as a path integral: the resulting identity is usually referred to as the *Feynman–Kac formula*. One proof, due to B. Simon, has the merit of avoiding time slicing. It starts from the observation that the difference

$$A(t) := e^{-tH_0} - e^{-tH} \tag{2.2.1}$$

solves the problem

$$\frac{d}{dt} A(t) = V e^{-tH_0} - H A(t), \qquad A(0) = 0, \tag{2.2.2}$$

which is readily turned into an integral equation for e^{-tH}:

$$e^{-tH_0} - e^{-tH} = \int_0^t ds\, e^{-(t-s)H} V e^{-sH_0}. \tag{2.2.3}$$

Taking kernels on both sides yields

$$K(x' - x, t) - \langle x'|e^{-tH}|x\rangle = \int_0^t ds \int dy\, \langle x'|e^{-(t-s)H}|y\rangle V(y) K(y - x, s). \tag{2.2.4}$$

This has to be confronted with the observation that, if ω is any Brownian path leading from $(x, 0)$ to (x', t), we have

$$1 - \exp\left\{-\int_0^t ds\, V(\omega(s))\right\} = \int_0^t ds\, \frac{d}{ds} \exp\left\{-\int_s^t ds'\, V(\omega(s'))\right\}$$

$$= \int_0^t ds\, V(\omega(s)) \exp\left\{-\int_s^t ds'\, V(\omega(s'))\right\}. \tag{2.2.5}$$

Taking the path integral with respect to the conditional Wiener measure $\mu_{x,0}^{x',t}$ on both sides leads to the following identity:

$$K(x' - x, t) - \int d\mu_{x,0}^{x',t}(\omega) \exp\left\{-\int_0^t ds\, V(\omega(s))\right\}$$

$$= \int_0^t ds \int d\mu_{x,0}^{x',t}(\omega) V(\omega(s)) \exp\left\{-\int_s^t ds'\, V(\omega(s'))\right\}$$

$$= \int_0^t ds \int dy \int d\mu_{y,s}^{x',t}(\omega) \exp\left\{-\int_s^t ds'\, V(\omega(s'))\right\} V(y) K(y - x, s), \tag{2.2.6}$$

where we have deliberately changed the order of integration. For arbitrary $t \geq 0$, let Q_t be the unique operator whose kernel is given by

$$\langle x'|Q_{s'-s}|x\rangle = \int d\mu_{x,s}^{x',s'}(\omega) \exp\left\{-\int_s^{s'} dt\, V(\omega(t))\right\}. \tag{2.2.7}$$

Invariance of the path integral with respect to time translation guarantees that the right hand side is merely a function of the difference $s' - s \geq 0$. Comparing (2.2.4) and (2.2.6) we learn that both e^{-tH} and Q_t satisfy the same integral equation. Since the solution is unique[18] we may now state our result:

Let $H = -\frac{1}{2}\Delta + V$ be some Schrdinger operator with $V(x)$ a bounded continuous function on $\mathbb{R}^d$. If $s' > s$, then the operator $e^{-(s'-s)H}$ has the kernel

$$\langle x'|e^{-(s'-s)H}|x\rangle = \int_{(x,s)}^{(x',s')} d\mu(\omega) \, \exp\left\{ -\int_s^{s'} dt\, V\big(\omega(t)\big) \right\} \qquad (2.2.8)$$

(Feynman–Kac formula). *The path integral extends over all Brownian paths* $\omega : (x,s) \rightsquigarrow (x',s')$.

The conditions on the potential formulated above are far from optimal. They have been chosen to make the integral equation method work straightforwardly. Neither the attractive Coulomb potential nor the harmonic oscillator seem to be included. The general strategy is as follows: suppose we are given some potential V which is not bounded continuous. Using smoothing techniques (e.g., forming convolutions of V with sufficiently smooth functions) one attempts to construct a sequence V_n of bounded continuous potentials such that

$$f_n(\omega) = \exp\left\{-\int_s^{s'} dt\, V_n\big(\omega(t)\big)\right\} \xrightarrow[\mu\text{-a.e.}]{} f(\omega) = \exp\left\{-\int_s^{s'} dt\, V\big(\omega(t)\big)\right\},$$

$$(2.2.9)$$

One then tries to prove that the sequence f_n is Cauchy (with respect to the distance given by the path integral) and tends to f μ-a.e., or else one resorts to the Lebesgue bounded convergence theorem. Frequently, the method works even though pointwise convergence $V_n \to V$ may fail[19], but, generally speaking, the proofs are long-winded, intricate, and very seldomly illuminating.

It is relatively easy to see that one condition can immediately be relaxed: the potential does not have to be bounded, it need only be *bounded below*. This permits us to apply the Feynman–Kac formula to potentials like $|x|^n$. We may even allow for the value $+\infty$ on (Borel) sets in $\mathbb{R}^d$. The reason: suppose that $V(x) = +\infty$ if $x \in A$; this simply means that the function $f(\omega)$ defined above vanishes as soon as the paths ω enters the set A, i.e., if

---

[18]Existence and uniqueness may be demonstrated provided the potential $V(x)$ is a bounded continuous function. The method is to replace V by λV with $\lambda \in \mathbb{R}$ in the integral equation (2.2.3), to apply the technique of successive approximation, and to express the solution as a power series with respect to λ.

[19]Some results in this direction, though not optimal, have been obtained by B. Simon [1.22, Chap. II.6].

$\omega(t) \in A$ for some $t \in [s, s']$. Phrased differently, the support of f consists of paths ω having all their coordinates $\omega(t)$ in A^c, the complement of A.

An important property of e^{-tH} implied by the Feynman–Kac formula (2.2.8) is that the operator is *positivity preserving* (see the analogous statement for e^{-tH_0} in Sect. 1.5), and another is that the kernel is symmetric:

$$\langle x'|e^{-tH}|x\rangle = \langle x|e^{-tH}|x'\rangle \geq 0. \tag{2.2.10}$$

Symmetry follows from the fact that, if $\omega\colon (x,0) \rightsquigarrow (x',t)$ is a Brownian path, so is $\omega'\colon (x',0) \rightsquigarrow (x,t)$ where $\omega'(s) := \omega(t-s)$, and

$$I(f) = \int d\mu_{x,0}^{x',t}(\omega)\, f(\omega) = \int d\mu_{x',0}^{x,t}(\omega')\, f(\omega). \tag{2.2.11}$$

Thus, if the function f obeys $f(\omega) = f(\omega')$ (the technical term here is *invariance with respect to time reversal*), then the integral $I(f)$ is symmetric with respect to interchanging the parameters x and x'. This condition is obviously satisfied for integrals of the type $\int_0^t ds\, V(\omega(s))$, but symmetry is violated whenever some time-dependent $V(x,t)$ replaces $V(x)$.

2.2.1 Stochastic Representation of Operator Norms

The analysis of operators requires that we be able to assess the "size" of an operator A, in particular in situations where A is an integral operator with kernel $\langle x'|A|x\rangle$. For most purposes, it is convenient to work with p-norms of A (as an operator on L^p), but for our purpose[20] it proves convenient to restrict to $p = \infty$. We thus define

$$\|A\| = \sup_{x'} \int dx\, |\langle x'|A|x\rangle| \tag{2.2.12}$$

and note that $\|A + B\| \leq \|A\| + \|B\|$ (triangle inequality) and also $\|AB\| \leq \|A\|\,\|B\|$. If the kernel is positive symmetric, we may also write

$$\|A\| = \sup_{x} \int dx'\, \langle x'|A|x\rangle. \tag{2.2.13}$$

As an easy result, taking $H_0 = -\frac{1}{2}\Delta$ we have that

$$\|e^{-H_0 t}\| = \sup_{x} \int dx'\, K(x' - x, t) = 1. \tag{2.2.14}$$

More generally, taking $H = H_0 + V$ one infers from (2.2.8) and (2.2.13) that

$$\|e^{-tH}\| = \sup_{x} \boldsymbol{E}_{x,0}\left(\exp\left\{-\int_0^t ds\, V(X_s)\right\}\right). \tag{2.2.15}$$

---

[20]For a more detailed operator analysis based on the stochastic representation see [2.3–5].

This representation of the norm as an expectation with respect to the Wiener process X_t has nice features. To motivate these, a few remarks are in order.

A potential is said to be *attractive* if it is negative and vanishes at infinity: this is the favorite situation for studying bound states and is of direct physical interest since it addresses the existence of atoms, ions, molecules, etc.; the associated energy levels are expected to be negative. So, if we prove that, for some attractive potential V, the spectrum of $H = H_0 + V$ has no negative part, there can be no bound state whatever with nonzero binding energy, and therefore $H \geq 0$. This is certainly the case if the norm $\|e^{-tH}\|$ is a bounded function of t (the converse may not be true). On the other hand, if there are bound states, then, by the spectral decomposition theorem [2.6], we expect the norm $\|e^{-tH}\|$ to be dominated, for sufficiently large t, by the lowest–lying energy level $E_0 < 0$ of the system in the sense that the norm grows *exponentially* like $e^{-E_0 t}$. We now state the consequences of (2.2.15) under the assumption $V \leq 0$:

1. $\|e^{-tH}\|$ is an *increasing function of t*. This is clearly so since, for any path ω starting at $(x,0)$, the function $f(t) = -\int_0^t ds\, V\big(\omega(s)\big)$ is increasing.

2. If $\|e^{-tH}\|$ is bounded, the following limit exists:

$$L = \lim_{t\to\infty} \|e^{-tH}\| = \sup_x \boldsymbol{E}_{x,0}\Big(\exp\big\{-\textstyle\int_0^\infty ds\, V(X_s)\big\}\Big). \qquad (2.2.16)$$

   $L < \infty$ implies $H \geq 0$ and the absence of bound states.

3. Suppose we compute

$$\gamma = \sup_x \boldsymbol{E}_{x,0}\Big(-\textstyle\int_0^\infty ds\, V(X_s)\Big)$$

$$= \sup_x \int dx' \int_0^\infty ds\, K(x - x', s)\{-V(x')\}$$

$$= \sup_x \int dx' \int_0^\infty ds\, \langle x|e^{-sH_0}|x'\rangle\{-V(x')\} = \|H_0^{-1}V\|. \qquad (2.2.17)$$

Then there are ways to get two a priori estimates:

$$e^\gamma \leq L \leq 1 + \gamma L, \qquad (2.2.18)$$

and $\gamma < 1$ implies that $L < (1-\gamma)^{-1}$ (Portenko's lemma [2.7,8]) and the absence of bound states.

The lower bound in (2.2.18) follows from Jensen's inequality[21]:

$$\boldsymbol{E}_{x,0}\Big(\exp\big\{-\textstyle\int_0^t ds\, V(X_s)\big\}\Big) \geq \exp \boldsymbol{E}_{x,0}\Big(-\textstyle\int_0^t ds\, V(X_s)\Big). \qquad (2.2.19)$$

The upper bound in (2.2.18) is a special case of a resolvent estimate to be discussed below.

---

[21]See Appendix C for details.

2.2.2 Stochastic Representation of Green's Functions

For E such that $H + E > 0$, the resolvent $(H + E)^{-1}$ exists; its kernel, the Green's function of $H + E$, is positive symmetric and can be represented as a path integral:

$$\langle x'|(H + E)^{-1}|x\rangle = \int_0^\infty dt\, e^{-Et} \int_{(x,0)}^{(x',t)} d\mu(\omega)\, \exp\left\{-\int_0^t ds\, V(\omega(s))\right\}.$$

$$(2.2.20)$$

Note that (2.2.20) follows directly from the Feynman–Kac formula (2.2.8).

Our goal is a stochastic representation of the norm $\|(H + E)^{-1}\|$. We begin by integrating both sides of (2.2.20) with respect to x' to obtain

$$\int dx'\, \langle x'|(H + E)^{-1}|x\rangle = \int_0^\infty dt\, e^{-Et} \boldsymbol{E}_{x,0}\left(\exp\left\{-\int_0^t ds\, V(X_s)\right\}\right).$$

$$(2.2.21)$$

We then take the supremum with respect to x; so the norm is

$$\|(H + E)^{-1}\| = \sup_x \int_0^\infty dt\, e^{-Et} \boldsymbol{E}_{x,0}\left(\exp\left\{-\int_0^t ds\, V(X_s)\right\}\right). \quad (2.2.22)$$

In particular, if $V = 0$, we would get $\|(H_0+E)^{-1}\| = E^{-1}$ because $\boldsymbol{E}_{x,0}(1) = 1$.

On the other hand, a Laplace transformation turns the integral equation (2.2.3) into the resolvent equation

$$\begin{aligned}
(H_0 + E)^{-1} - (H + E)^{-1} &= (H + E)^{-1}V(H_0 + E)^{-1} \\
&= (H_0 + E)^{-1}V(H + E)^{-1}
\end{aligned} \quad (2.2.23)$$

which is normally used to expand the resolvent into a series. The basic bound for control over the solution is

$$\begin{aligned}
\|(H + E)^{-1}\| &\leq \|(H_0 + E)^{-1} - (H_0 + E)^{-1}V(H + E)^{-1}\| \\
&\leq \|(H_0 + E)^{-1}\| + \|(H_0 + E)^{-1}V(H + E)^{-1}\| \quad (2.2.24) \\
&\leq E^{-1} + \|(H_0 + E)^{-1}V\|\,\|(H + E)^{-1}\|
\end{aligned}$$

with the obvious consequence that

$$E\|(H + E)^{-1}\| \leq \left\{1 - \|(H_0 + E)^{-1}V\|\right\}^{-1} \quad (2.2.25)$$

provided

$$\|(H_0 + E)^{-1}V\| < 1. \quad (2.2.26)$$

This is of interest when $V \leq 0$. As follows from their stochastic representations, both $\|(H_0 + E)^{-1}V\|$ and $\|(H + E)^{-1}\|$ are decreasing functions of the variable E, tending to zero as E becomes large. So, if we determine

$$\underline{E} = -\inf\left\{ E > 0 \mid \|(H_0 + E)^{-1}V\| < 1 \right\}, \tag{2.2.27}$$

then there can be no bound state of H with energy less than $\underline{E}$. We conclude that (2.2.27) provides a lower bound for the ground state energy and, if $\underline{E} = 0$, there are no bound states at all in the system. Moreover, if $\gamma = \|H_0^{-1}V\| < 1$, then $L = \lim_{E\downarrow 0} E\|(H + E)^{-1}\| \le (1 - \gamma)^{-1}$, which is Portenko's result mentioned above.

Exercise 1. Prove that the three-dimensional attractive Yukawa potential $V(x) = -gr^{-1}e^{-\mu r}$ $(r = |x|, g > 0, \mu > 0)$ admits no bound state when the coupling g is smaller than $\mu/4$ (by the way, the optimal value is 0.84μ). For this result, use the representation

$$V(x) = -4\pi g \int_0^\infty ds\, e^{-s\mu^2/2} K(x, s)$$

to calculate $\gamma = \|H_0^{-1}V\| = 4g/\mu$ and apply Portenko's lemma.

Exercise 2. A Schrdinger particle of unit negative charge is attracted by a positive charge distributed in 3-space with total charge $q = \int dx\, \rho(x)$. For the solution $V(x)$ of the Poisson equation $\Delta V = 4\pi\rho \ge 0$ vanishing at infinity, prove that

$$\|(H_0 + E)^{-1}V\| \le 4q(2E)^{-1/2} \qquad (E > 0).$$

Thus, (2.2.26) holds provided $E > 8q^2$, and no energy levels occur below $-8q^2$ (which is off the optimal $-q^2/2$ by a factor 16).

2.3 The Lie–Trotter Product Method

As indicated before, time slicing is the preferred way to approximate path integrals. It seems natural to resort to this method since it reminds us of the approximation of ordinary integrals by Riemann sums. This section addresses the relationship of time slicing and operator products.

We begin by assuming that our underlying time interval $[s, s']$ has been divided into $n + 1$ subintervals $[s_k, s_{k+1}]$ of equal lengths, i.e.,

$$s_k = s + k\tau \quad , \quad k = 0, \ldots, n \quad , \quad \tau = \frac{s' - s}{n + 1}, \tag{2.3.1}$$

noting that the time step τ goes to zero as n becomes large. We are particularly interested in situations where the function to be integrated is an exponential of an integral with respect to time:

$$f(\omega) = \exp\left\{ -\int_s^{s'} dt\, V(\omega(t)) \right\}. \tag{2.3.2}$$

Time slicing then corresponds to the replacement of the t-integral by a Riemann sum, thus producing an approximate factorizing function f_n dependent on finitely many path coordinates:

$$f_n(\omega) = \exp\left\{ -\sum_{k=0}^n \tau V(\omega(s_k)) \right\}. \tag{2.3.3}$$

The challenge is to calculate $I_n = \int d\mu\, f_n$, where one integrates over all paths $\omega\colon (x,s) \rightsquigarrow (x',s')$, and to pass to the limit $I = \lim_{n\to\infty} I_n$. This may be attacked in the following manner:

$$\begin{aligned}
I_n &= \int dx_n \cdots \int dx_1\, K(x' - x_n, \tau) e^{-\tau V(x_n)} K(x_n - x_{n-1}, \tau) \\
&\quad \cdots e^{-\tau V(x_2)} K(x_2 - x_1, \tau) e^{-\tau V(x_1)} K(x_1 - x, \tau) e^{-\tau V(x)} \\
&= \int dx_n \cdots \int dx_1\, \langle x'|e^{\tau\Delta/2}|x_n\rangle e^{-\tau V(x_n)} \langle x_n|e^{\tau\Delta/2}|x_{n-1}\rangle \\
&\quad \cdots e^{-\tau V(x_2)} \langle x_2|e^{\tau\Delta/2}|x_1\rangle e^{-\tau V(x_1)} \langle x_1|e^{\tau\Delta/2}|x\rangle e^{-\tau V(x)} \\
&= \int dx_n \cdots \int dx_1\, \langle x'|e^{\tau\Delta/2}e^{-\tau V}|x_n\rangle \cdots \langle x_1|e^{\tau\Delta/2}e^{-\tau V}|x\rangle \\
&= \langle x'|T_{n+1}|x\rangle, \qquad T_{n+1} := \left(e^{\tau\Delta/2}e^{-\tau V} \right)^{n+1}.
\end{aligned} \tag{2.3.4}$$

By returning to the operator calculus we have somehow managed to find a short and elegant expression for the nth approximand. This entails that we interpret $e^{-\tau V}$ as a multiplication operator:

$$[e^{-\tau V}\phi](x) = e^{-\tau V(x)}\phi(x) \qquad (\phi \in L^2(\mathbb{R}^d)). \tag{2.3.5}$$

The result so obtained suggests that we next prove the existence of $T = \lim_{n\to\infty} T_n$ and, moreover, that

$$T = e^{-(s'-s)H}, \qquad H = -\tfrac{1}{2}\Delta + V. \tag{2.3.6}$$

This then would provide an alternative proof of the Feynman–Kac formula (2.2.8).

2.3.1 The Lie–Trotter Product Formula

What sense does it make to say that the sequence

$$\left(e^{A/n} e^{B/n} \right)^n \tag{2.3.7}$$

approaches a limit if A and B are operators? No doubt the sequence is constant with the obvious limit e^{A+B}, provided A and B commute. Granted certain technical assumptions one may in fact prove the *Lie–Trotter product formula*

$$e^{A+B} = \lim_{n\to\infty} \left(e^{A/n}e^{B/n}\right)^n \tag{2.3.8}$$

in the general case, where $[A, B] \neq 0$. The meaning of convergence is notoriously vague in our statements.

The simplest way to prove (2.3.8) is to assume that both A and B are bounded operators (with respect to some operator norm that seems convenient). Then the correct technical term is *norm convergence*. We shall present the details of the proof for pedagogical reasons and use the abbreviations

$$C = e^{A/n+B/n}, \qquad D = e^{A/n}e^{B/n}. \tag{2.3.9}$$

Then

$$\begin{aligned}
\|C\| &\leq \exp\{\tfrac{1}{n}\|A + B\|\} \leq \exp\{\tfrac{1}{n}(\|A\| + \|B\|)\} \\
\|D\| &\leq \|e^{A/n}\|\,\|e^{B/n}\| \\
&\leq \exp\{\tfrac{1}{n}\|A\|\}\exp\{\tfrac{1}{n}\|B\|\} = \exp\{\tfrac{1}{n}(\|A\| + \|B\|)\}.
\end{aligned} \tag{2.3.10}$$

We are done once we have shown that $\|C^n - D^n\|$ tends to zero. To prove this we write

$$C^n - D^n = \sum_{k=1}^{n} C^{k-1}(C - D)D^{n-k} \tag{2.3.11}$$

and estimate each term of the sum:

$$\begin{aligned}
\|C^n - D^n\| &\leq \sum_{k=1}^{n} \|C\|^{k-1}\|C - D\|\,\|D\|^{n-k} \\
&\leq n\|C - D\|\exp\left\{\tfrac{n-1}{n}(\|A\| + \|B\|)\right\} \\
&\leq n\|C - D\|\exp\left\{(\|A\| + \|B\|)\right\}.
\end{aligned} \tag{2.3.12}$$

On the other hand, both C and D admit convergent series expansions. Hence,

$$C - D = \tfrac{1}{n^2}R, \qquad R = \tfrac{1}{2}[B, A] + O(\tfrac{1}{n}) \tag{2.3.13}$$

as n gets large. As R is a bounded operator, $\|C - D\| = O(n^{-2})$ and $\|C^n - D^n\| = O(n^{-1})$ which completes the proof.

To be more specific let us take $A = (s' - s)\Delta/2$ and $B = -(s' - s)V$; so $A + B = -(s' - s)H$, where H is the Hamiltonian and $s' > s$. In view of (2.3.4),

$$T_n = \left(e^{A/n}e^{B/n}\right)^n. \tag{2.3.14}$$

Although neither A nor B are bounded operators, the Lie–Trotter product formula

$$T = \lim_{n\to\infty} T_n = e^{-(s'-s)H} \tag{2.3.15}$$

can be shown to hold provided, crudely speaking, the potential V is bounded below. The problem posed here, namely to get optimal results with regard to weakening the conditions on the potential, provides an irresistible challenge to mathematicians [2.1,9,10].

2.3.2 Miscellaneous Remarks and Results

In the sequel we shall refer to the kernel of $e^{-(s'-s)H}$ as the *transition amplitude* associated with the underlying Schrdinger equation and shall frequently write

$$\langle x', s'|x, s\rangle := \langle x'|e^{-(s'-s)H}|x\rangle, \tag{2.3.16}$$

the reason being that the concept, as will be explained later, makes sense even for systems that are not homogeneous with respect to time: generally, the transition amplitude is allowed to have a separate dependence on s and s'.

Are there explicitly known transition amplitudes for nontrivial potentials? One such case mentioned before (in Sect. 1.8) is Mehler's formula for the isotropic d-dimensional harmonic oscillator, i.e., for

$$H = -\tfrac{1}{2}\Delta + V \quad , \qquad V(x) = \tfrac{1}{2}k^2 x^2, \tag{2.3.17}$$

where k is the frequency of the oscillator, one has

$$
\begin{aligned}
\langle x', s'|x, s\rangle &= \langle x'|e^{-(s'-s)H}|x\rangle \\[2mm]
&= \int d\mu_{x,s}^{x',s'}(\omega)\, \exp\left\{-\tfrac{1}{2}k^2 \int_s^{s'} dt\,\omega(t)^2\right\} \\[2mm]
&= \left[\frac{k}{2\pi\sinh\nu}\right]^{d/2} \exp\left\{-\frac{k(x^2+x'^2)}{2\tanh\nu} + \frac{kx\cdot x'}{\sinh\nu}\right\}
\end{aligned}
\tag{2.3.18}
$$

with $\nu = k(s'-s)$ the intrinsic time variable. An easy straightforward proof makes use of operator techniques [2.11]. Another proof utilizes the Lie–Trotter product formula [2.12]. The last one can be reformulated: the harmonic oscillator then emerges from the theory of spin chains. We shall present this version of the proof in Sect. 2.8.

When k tends to zero in (2.3.18), the expression on the left-hand side approaches the transition amplitude $K(x'-x, s'-s)$ of a free particle as it should. Also: the transition amplitude of the harmonic oscillator, when considered a function of ν, gives rise to an analytic continuation into the halfplane $\operatorname{Re}\nu > 0$ with *essential* singularities occurring on the imaginary axis at the points $\nu = in\pi$, $n \in \mathbb{Z}$. Therefore, as far as transition amplitudes are concerned, it is good practice to stay away from the imaginary-time axis!

Suppose we are mainly interested in knowing the bottom of the spectrum of H and the associated ground state. What sense does it make to continue e^{-tH} to imaginary times and then try to extract the necessary information from the unitary group e^{-itH}? Not only would such a procedure be tedious, but it would also be unsuccessful. Why? The main portion of the argument is left to the judgement of the reader, but, heuristically, the essence of it may be seen from the following remark. Let $\phi(x)$ be the (unique) ground state of H with eigenvalue E. For sufficiently large values of t, one expects an exponential fall-off or growth like[22]

$$\langle x'|e^{-tH}|x\rangle \approx \phi(x')\overline{\phi(x)}e^{-tE} \geq 0 \, , \tag{2.3.19}$$

implying that ϕ may be chosen to be a positive function: $\phi(x) \geq 0$. There simply *does not exist* a corresponding asymptotic formula for the kernel of e^{-itH}.

The above assertion about the asymptotic behavior of e^{-tH} can be made precise[23]:

$$\begin{aligned}
E &= -\lim_{t\to\infty} t^{-1} \log \int dx \, \langle x|e^{-tH}|0\rangle \\
&= -\lim_{t\to\infty} t^{-1} \log \boldsymbol{E}\left(\exp\left\{-\int_0^t ds\, V(X_s)\right\}\right) .
\end{aligned} \tag{2.3.20}$$

Formula (2.3.20) gives the ground energy E in terms of an expectation $\boldsymbol{E}(\cdot)$ with respect to the d-dimensional Wiener process X_s. Ordinarily, E is given by the Rayleigh–Ritz principle, i.e., as the solution of the variational problem $(\phi, H\phi) = \min!$ (assuming $\|\phi\| = 1$). The claim is that the limit in (2.3.20) and the variational problem provide the same answer and is a typical "large deviation" result[24]. From (2.3.19) assuming $\phi(x) \geq 0$ and t large, $\int dx' \, \langle x'|e^{-tH}|x\rangle \approx c\phi(x)e^{-tE}$, where $c = \int dx\, \phi(x)$ and thus

$$\frac{\phi(x)}{\phi(0)} = \lim_{t\to\infty} \frac{\boldsymbol{E}\left(\exp\left\{-\int_0^t ds\, V(x + X_s)\right\}\right)}{\boldsymbol{E}\left(\exp\left\{-\int_0^t ds\, V(X_s)\right\}\right)} \tag{2.3.21}$$

provided the limit makes sense. This clearly provides a companion to (2.3.20) and may be looked at as *a stochastic representation of the ground state.*

---

[22] If P is the projector onto the ground state, it has the kernel $\langle x'|P|x\rangle = \phi(x')\bar{\phi}(x)$ and $\int dx\, |\phi(x)|^2 = 1$.

[23] See Simon [1.22. Theorem 6.3] for sufficient conditions on the potential.

[24] The stochastic theory of Donsker and Varadhan [2.13] on large deviations is an extremely flexible and versatile tool with important applications in various parts of physics. Actually, we have encountered in (2.3.20) just the tip of the wealth of impressive results concerning the large-time behavior of stochastic expectations.

It is quite common in quantum mechanics to have an excluded domain of the configuration space, say $G^c \subset \mathrm{I\!R}^d$, and, formally, this is taken care of by declaring that

$$V(x) = +\infty \quad \text{if } x \in G^c, \tag{2.3.22}$$

where G^c denotes the complement of the "allowed" open set G. When the Schrdinger equation is being converted into the Feynman–Kac formula, the resulting path integral implements the condition (2.3.22) in a very simple manner, namely by eliminating all paths from the integral that enter the domain G^c within the given period of time, say $[0, t]$. Mathematically, this can also be achieved by restricting the path integral to the subset

$$\mathcal{G} = \bigcap_{0 \leq s \leq t} \{ \omega \mid \omega(s) \in G \} \tag{2.3.23}$$

and essentially amounts to multiplying the integrand by the characteristic function $\chi_{\mathcal{G}}(\omega)$ associated with the set $\mathcal{G}$. In such situations, $\langle x' | e^{-tH} | x \rangle = 0$ unless $x \in G$ and $x' \in G$ since x and x' are the endpoints of every acceptable path. Needless to say, by the spectral decomposition theorem, all eigenfunctions of e^{-tH} vanish in G^c.

As indicated by the above argument, enlarging a domain G affects the transition amplitude and the ground state energy in a predictable way. To see the details of it, take any Hamiltonian, say $H = -\frac{1}{2}\Delta + V$, and restrict it to the subset G of the configuration space, thus defining the Hamiltonian $H(G)$ which varies with G. Obviously,

$$\langle x' | e^{-tH(G)} | x \rangle = \int_{(x,0)}^{(x',t)} d\mu(\omega)\, \chi_{\mathcal{G}}(\omega) \exp\left\{ -\int_0^t ds\, V\big(\omega(s)\big) \right\}, \tag{2.3.24}$$

where $\mathcal{G}$ is related to G by (2.3.23). If $G \subset G'$, then $\mathcal{G} \subset \mathcal{G}'$ and thus $\chi_{\mathcal{G}} \leq \chi_{\mathcal{G}'}$. In effect,

$$\langle x' | e^{-tH(G)} | x \rangle \leq \langle x' | e^{-tH(G')} | x \rangle \chi_G(x) \chi_G(x'), \tag{2.3.25}$$

where χ_G designates the characteristic function of the set G. Let $E(G)$ denote the bottom of the spectrum of the Hamiltonian $H(G)$. Then $E(G) \geq E(G')$ as may be inferred from (2.3.20): *enlarging the accessible domain of the configuration space generally lowers the ground state energy.*

In particular, if $V = 0$ in (2.3.24), $H(G)$ is synonymous with $-\frac{1}{2}\Delta_G$, where Δ_G is said to be the *Laplacian on $G \in \mathrm{I\!R}^d$ with Dirichlet boundary conditions.* To get the flavor of the type of results that are to come out of the inequality (2.3.25), let us compare the interesting situation where the domain G has finite volume $|G|$ to the easy case where $G' = \mathrm{I\!R}^d$. Taking $x' = x$ in (2.3.25), we get the inequality

$$\langle x | e^{t\Delta_G/2} | x \rangle \leq (2\pi t)^{-d/2} \chi_G(x) \tag{2.3.26}$$

and so immediately conclude that

$$(2\pi t)^{d/2} \, \mathrm{tr} \, e^{t\Delta_G/2} \leq |G|. \tag{2.3.27}$$

At first glance this does not tell us very much. However, if $E(G)$ is the ground state energy of the operator $-\frac{1}{2}\Delta_G$, a weakening of (2.3.27) leads to the promising result

$$(2\pi t)^{d/2} e^{-tE(G)} \leq |G| \tag{2.3.28}$$

valid for all $t > 0$. Since the function on the left-hand side assumes its supremum at $tE(G) = d/2$,

$$E(G)|G|^{2/d} \geq \frac{\pi}{e} d. \tag{2.3.29}$$

To get this nice geometric estimate, we hardly used more than the representation (2.3.24); the rest of the proof was painless.

If H has a ground state, it corresponds to the maximal eigenvalue of e^{-tH}. As the kernel $\langle x'|e^{-tH}|x\rangle$ is strictly positive (on $G \times G$ if necessary), an extension of the classical Perron–Frobenius result (originally proven for matrices with strictly positive entries) to integral operators (see [1.10. Theorem 3.3.2]) establishes two facts:

1. The ground state is unique.
2. The ground state wave function ϕ satisfies $\phi(x')\overline{\phi(x)} > 0$ (on $G \times G$) and can therefore be chosen strictly positive (on G).

For illustration we take Mehler's formula for the harmonic oscillator and study its asymptotic behavior for large $\nu = k(s' - s)$:

$$\langle x', s'|x, s\rangle \approx \left[\frac{k}{\pi} e^{-\nu}\right]^{d/2} \exp\{-\tfrac{1}{2}kx'^2 - \tfrac{1}{2}kx^2\}. \tag{2.3.30}$$

As expected, the wave function ϕ and the energy E of the ground level are:

$$\phi(x) = (k/\pi)^{d/4} e^{-kx^2/2}, \qquad E = \tfrac{d}{2}k. \tag{2.3.31}$$

The stochastic expectation

$$\mathbf{E}\left(\exp\left\{-\tfrac{1}{2}k^2 \int_0^t ds\, X_s^2\right\}\right) = (\cosh kt)^{-d/2} \tag{2.3.32}$$

yields an even simpler expression with the asymptotic behavior $\sim e^{-tE}$.

It is true to say that path integrals cannot be evaluated in most cases. So, why do we recommend doing quantum mechanics the hard way? It seems though that the path integral method has certain merits, the main advantage being that integrals can be transformed, approximated, split into parts, and estimated. We can apply inequalities and convex analysis and may even try to numerically determine the value of path integrals by a digital computer. On the mathematical side, the stochastic representation of semigroups has led to more insight into the properties of Schrödinger operators [2.14–17].

2.3.3 Several Particles with Different Masses

This chapter has so far dealt with the Feynman–Kac formula either for a single particle of mass 1 or for a system of identical particles with common unit mass: the n-particle system in a 3-dimensional space is formally treated like a single-particle system in $3n$ dimensions. This unifying description can always be achieved – even though each particle may possess a mass of its own – by an appropriate scaling of the space variables associated with each particle. Frequently, it is not desirable to hide the masses in this manner. How would we proceed if we are dealing with n particles, of masses $m_1, \ldots, m_n$, influenced by mutual and external forces?

Clearly, there must be rules to construct some *anisotropic* $3n$-dimensional Wiener process X_s from a mass matrix $M = \mathrm{diag}(m_1, \ldots, m_n)$. The process X_s will be called *adapted* since it is specific to the problem under study. The suggested and obvious procedure to define it is: go back to the beginning of Chap. 1 and start afresh, this time using the modified diffusion equation

$$\frac{\partial}{\partial s} f(x, s) = \sum_{k=1}^{n} \frac{1}{2m_i} \Delta_i f(x, s), \tag{2.3.33}$$

where we have introduced the multivector $x = \{x_1, \ldots, x_n\}^T \in \mathbb{R}^{3n}$ so that

$$x^T M x = \sum_{k=1}^{n} m_k x_k^2. \tag{2.3.34}$$

Fortunately, previous formulas will only change in the obvious way.

To the n-particle system there corresponds a single Brownian particle in $3n$ dimensions with transition probabilities

$$P(X_s' \in dx' | X_s = x) \;=\; dx'\, K_M(x' - x, s' - s) \tag{2.3.35}$$

$(dx = dx_1 dx_2 \cdots dx_n)$ and the transition function

$$K_M(x, s) = \left[\det\left(\frac{M}{2\pi s} \right) \right]^{\frac{1}{2}} \exp\left\{ -\frac{x^T M x}{2s} \right\}. \tag{2.3.36}$$

Not much has changed: the inverse mass matrix replaces the diffusion constant D (normally taken to be $\frac{1}{2}$). The conditional adapted Wiener measure $\mu_{x,s}^{x',s'}$ is obtained by simply replacing $K(x, s)$ by $K_M(x, s)$.

Once the system is given the potential

$$V(x) = \sum_{k} V_k(x_k) + \sum_{j<k} V_{jk}(x_j - x_k), \tag{2.3.37}$$

the outer appearance of the Feynman–Kac formula is the same as before.

At this point it is perfectly legitimate to ask: what are the goals and what is to be learned from path integrals? Most likely, the main goal is insight, not numbers. Certain projects that do provide insight will be persued in the remaining sections of this chapter.

Exercise.   For Δ_G the Laplacian on $G \in \mathbb{R}^d$ with Dirichlet boundary conditions, find a way to prove

$$(2\pi t)^{d/2} \langle x | e^{t\Delta_G/2} | x \rangle \xrightarrow[t \to 0]{} \chi_G(x)$$

($x \notin \partial G$, χ_G = characteristic function). Hint: use Brownian motion to represent the left-hand side as a probability. For finite t the right-hand side is an upper bound, and the convergence is monotonic. If $|G| < \infty$, the operator $-\frac{1}{2}\Delta_G$ has an infinity of eigenvalues which we designate

$$E_0 \leq E_1 \leq E_2 \leq \cdots$$

and list according to their multiplicities. By the above limit,

$$(2\pi t)^{d/2} \sum_{n=0}^{\infty} e^{-tE_n} \xrightarrow[t \to 0]{} |G|.$$

Use the Tauberian theorem ([1.22, Theorem 10.3]) from the theory of Laplace transforms to prove that

$$(2\pi/E)^{d/2} \sum_{n=0}^{\infty} \Theta(E - E_n) \xrightarrow[E \to \infty]{} |G|/\Gamma(1 + d/2)$$

(Θ = step function, Γ = gamma function), which is a famous classical result due to Weyl [2.18] about the asymptotic distribution of eigenvalues.

2.4 The Brownian Tube

Let us take up a case study that concerns Brownian motion as a *real* phenomenon in three space dimensions. As before, the diffusion constant is assumed to be $\frac{1}{2}$. Suppose the particle starts at a certain point belonging to the interior of some sphere, $|x| < a$. How would we determine the probability for not leaving the spacetime tube $|x| \leq a$, $s < t < s'$? The question raised addresses the diffusion of a classical particle in a vessel with absorbing walls.

If Ω denotes the set of all paths with fixed endpoints $w(s) = x$ and $w(s') = x'$ (so that $|x| < a$ and $|x'| < a$) and if μ is the associated conditional Wiener measure, our first task would be to determine $\mu(\Omega_a)$ (i.e., the path integral $\int d\mu\, \chi$ with χ the characteristic function of Ω_a) for the set

$$\Omega_a = \{ w \in \Omega \mid |w(t)| \leq a, \ s \leq t \leq s' \} \tag{2.4.1}$$

and our second to integrate the result with respect to x'. Notice that (2.4.1) places conditions on infinitely many coordinates of the path: no finite-dimensional integral can be found to represent our path integral.

We may however perform our task comfortably by passing to the corresponding quantum mechanical problem which deals with a different particle, i.e., with a Schrdinger particle of unit mass moving in a potential

$$V(x) = \begin{cases} 0 & \text{if } |x| \leq a \\ +\infty & \text{otherwise} \end{cases} \tag{2.4.2}$$

so that

$$\chi(\omega) = \exp\left\{ -\int_s^{s'} dt\, V(\omega(t)) \right\} = \begin{cases} 1 & \text{if } \omega \in \Omega_a \\ 0 & \text{otherwise.} \end{cases} \tag{2.4.3}$$

If $H = -\frac{1}{2}\Delta + V$, we may now write

$$\mu(\Omega_a) = \int d\mu\, \chi = \int_{(x,s)}^{(x',s')} d\mu(\omega) \exp\left\{ -\int_s^{s'} dt\, V(\omega(t)) \right\} \tag{2.4.4}$$
$$= \langle x'|e^{-(s'-s)H}|x\rangle,$$

where the last line follows from the Feynman–Kac formula.

The basic idea then is to use the spectral decomposition of the operator H. Rotational invariance permits us to reduce the amount of work necessary to obtain the spectrum, which is discrete in this case, and the eigenfunctions: in spherical coordinates r, θ and ϕ, a separation of variables leads to the spherical harmonics $Y_{\ell m}(\theta, \phi)$ on the one hand and to the radial equation

$$\left(\frac{d^2}{dr^2} + \frac{2}{r}\frac{d}{dr} - \frac{\ell(\ell+1)}{r^2} + 2E \right) u(r) = 0 \tag{2.4.5}$$

on the other. Solutions of (2.4.5) that are regular at $r = 0$ and vanish at $r = a$ are the so-called spherical Bessel functions[25] $j_\ell(z)$ with $z^2 = 2Er^2$ and E a member of the set $\{\, \frac{1}{2}\lambda_{n\ell}^2 a^{-2} \mid n, \ell+1 \in \mathbb{N}, -\ell \leq m \leq \ell \,\}$ (the eigenvalues of $-\frac{1}{2}\Delta$ for the domain $|x| \leq a$ with Dirichlet boundary conditions), where $\lambda_{n\ell}$ stands for the nth zero of the function $j_\ell(z)$:

$$j_\ell(\lambda_{n\ell}) = 0, \qquad 0 < \lambda_{1\ell} < \lambda_{2\ell} < \cdots. \tag{2.4.6}$$

Thus the eigenfunctions are

$$\phi_{n\ell m}(x) = \begin{cases} c_{n\ell}^{-1/2} j_\ell(\lambda_{n\ell} r/a) Y_{\ell m}(\theta, \phi) & \text{if } r < a \\ 0 & \text{if } r \geq a, \end{cases} \tag{2.4.7}$$

$$c_{n\ell} = \int_0^a r^2 dr\, j_\ell(\lambda_{n\ell} r/a)^2. \tag{2.4.8}$$

---

[25]Spherical Bessel functions are related to ordinary Bessel functions of half-integer order by $j_\ell(z) = (\pi/2z)^{1/2} J_{\ell+1/2}(z)$. They are oscillating elementary functions involving only powers and trigonometric functions.

Setting $t = s' - s$ we have

$$\langle x'|e^{-tH}|x\rangle = \sum_{n\ell m} \phi_{n\ell m}(x')\bar{\phi}_{n\ell m}(x)\exp\{-t(2a^2)^{-1}\lambda_{n\ell}^2\}. \qquad (2.4.9)$$

Integration with respect to x' projects onto the $\ell = 0$ part:

$$\int dx'\,\langle x'|e^{-tH}|x\rangle = \sum_{n=1}^{\infty} c_{n0}^{-1} j_0(\lambda_{n0}r/a)I_n \exp\{-t(2a^2)^{-1}\lambda_{n0}^2\} \quad (r < a)$$

$$I_n = \int_0^a r^2 dr\, j_0(\lambda_{n0}r/a). \qquad (2.4.10)$$

Since $j_0(z) = z^{-1}\sin z$, $\lambda_{n0} = n\pi$, and consequently

$$c_{n0} = \frac{a^2}{n^2\pi^2}\int_0^a dr\,[\sin(n\pi r/a)]^2 = \frac{a^3}{2n^2\pi^2} \qquad (2.4.11)$$

$$I_n = \frac{a}{n\pi}\int_0^a rdr\,\sin(n\pi r/a) = (-1)^{n+1}\frac{a^3}{n^2\pi^2}, \qquad (2.4.12)$$

the result is

$$\int dx'\,\langle x'|e^{-tH}|x\rangle = \sum_{n=1}^{\infty} 2(-1)^{n+1}\frac{\sin(n\pi r/a)}{n\pi r/a}\exp\left\{-t\frac{n^2\pi^2}{2a^2}\right\} \quad (r < a)$$

$$=: P_a(r,t). \qquad (2.4.13)$$

The function $P_a(r,t)$, here represented by an infinite sum, belongs to the class of Jacobian theta functions [2.19]. Compare (2.4.13) to the situation where there are no walls, formally, when $a = \infty$:

$$P_\infty(r,t) = \int dx'\,\langle x'|e^{-tH_0}|x\rangle = \int dx'\,K(x'-x,t) = 1. \qquad (2.4.14)$$

At first sight, it does not seem obvious at all that P_a tends to P_∞ when a gets large. Fortunately, Poisson's summation formula applied to the above sum will convert it into a different sum from which one gets an affirmative answer. Details of the calculation are left to the reader.

To summarize: $P_a(r,t)$, as given by (2.4.13), represents the probability for the event that the Brownian particle, which starts at a distance r away from the center of the vessel, will pass the tube, without being absorbed in time t while performing a random walk. The probability $P_a(r,t)$ as a function of time is depicted in Fig. 2.2 for various values of the parameter r/a. The scale of time is set by the lifetime $\tau = 2(a/\pi)^2 = 1/E_0$, and it so happens that τ coincides with the inverse of the ground state energy E_0.

It follows from (2.4.13) that there is zero chance of survival[26] if the particle starts at the boundary, i.e., if $P_a(a,t) = 0$ for $t > 0$. The probability

---

[26] This somewhat surprising effect is due to considering the continuum limit. Things would come out different on a lattice.

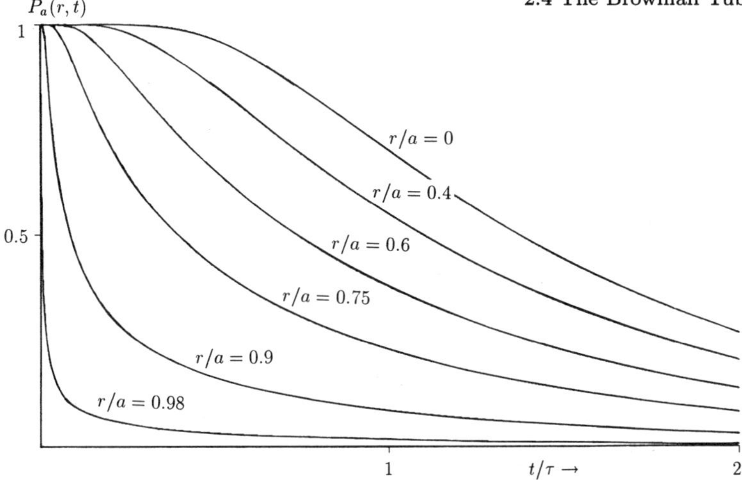

Fig. 2.2. The dependence of the probability $P_a(r,t)$ on t/τ for selected values of r/a.

of survival, for fixed t, reaches its maximum when $r = 0$. In any case, the probability $P_a(r,t)$ decays exponentially at the rate $e^{-t/\tau}$ with τ the lifetime given above. For a general diffusion constant D (mass $m = (2D)^{-1}$ of the associated Schrdinger particle) we would, however, find that

$$\tau = \frac{a^2}{\pi^2 D}. \tag{2.4.15}$$

The important general message, implicit in these results, is:

> *The lifetime of a Brownian particle moving in a medium with absorbing boundaries is given by the inverse ground state energy of a related quantum mechanical system.*

To put it all in one word: the solution of a dynamical problem (lifetime) follows from the solution of another static problem (energy).

Exercise 1. For a Brownian particle moving in some bounded region G with absorbing boundary, consider the probability

$$P(x,t) = \int_G dx' \, \langle x'|e^{-tH}|x\rangle \qquad (x \in G, \, t > 0)$$

of survival assuming that the particle starts at x. As before, H is taken to be the Hamiltonian of a fictitious Schrdinger particle of unit mass moving in a potential V which is zero in G and $+\infty$ elsewhere. Show that $P(x,t)$ is a decreasing function of the time variable t approaching zero as $t \to \infty$. Fix x and introduce a probability density $\rho(t)$ setting

$$P(x,t) = 1 - \int_0^t dt' \, \rho(t').$$

As for Brownian motion, $\int_a^b dt \, \rho(t)$ represents the probability of leaving G at some time $t \in [a,b]$ (first exit time). Prove that the *mean exit time* $\langle t \rangle = \int_0^\infty dt \, t\rho(t)$ is given by the following formula:

$$\langle t \rangle = \int_0^\infty dt \int dx' \, \langle x' | e^{-tH} | x \rangle = \int dx' \, \langle x' | H^{-1} | x \rangle.$$

Exercise 2. In the foregoing exercise, let G be the ball in $\mathbb{R}^3$ with radius a. Use (2.4.13) and

$$\sum_{n=1}^\infty n^{-3}(-1)^{n+1}\sin(n\pi u) = \pi^3 u(1-u^2)/12 \qquad (-1 \le u \le 1)$$

(where $u = r/a$ and $r = |x|$) to show that

$$\langle t \rangle = (a^2 - r^2)/3$$

provided $0 \le r \le a$. Averaging over the ball yields $\overline{r^2} = 3a^2/5$, and hence $\overline{\langle t \rangle} = 2a^2/15$ which is slightly smaller than the lifetime $\tau = 2a^2/\pi^2$ obtained from the exponential decay $P(x,t) \propto e^{-t/\tau}$. Generalize the formula for $\langle t \rangle$ so that it holds for an arbitrary diffusion constant D.

2.5 The Golden–Thompson–Symanzik Bound

We now turn to the problem of estimating traces relating to partition functions of quantum statistical mechanics. Preferably, we think of a "large" system containing many particles enclosed in a box. So let there be some Hamiltonian $H = -\frac{1}{2}\Delta + V$ whose spectrum consists solely of eigenvalues E_n, $n = 0, 1, 2, \ldots$, counted according to their multiplicity and growing sufficiently fast with n such that, for each $\beta > 0$, the following trace exists:

$$\operatorname{tr} e^{-\beta H} = \sum_{n=0}^\infty e^{-\beta E_n}. \tag{2.5.1}$$

From the theory of integral operators (or else from the spectral decomposition theorem) it follows that

$$\operatorname{tr} e^{-\beta H} = \int dx \, \langle x | e^{-\beta H} | x \rangle. \tag{2.5.2}$$

To simplify matters we may, in the Feynman–Kac formula for $\langle x | e^{-\beta H} | x \rangle$, write every path $(x,0) \rightsquigarrow (x,\beta)$ as $x + \omega(t)$, where $\omega \colon (0,0) \rightsquigarrow (0,\beta)$ is another Brownian path: the translation by a vector x establishes a one-to-one correspondence between two sets of paths, $\Omega_{x,0}^{x,\beta}$ and $\Omega_{0,0}^{0,\beta}$, since there is no preferred origin of space in Brownian motion. Also, the conditional Wiener measure $d\mu_{x,0}^{x,\beta}$ changes into the measure $d\mu_{0,0}^{0,\beta}$, so that

$$\langle x|e^{-\beta H}|x\rangle = \int_{(0,0)}^{(0,\beta)} d\mu(\omega)\, \exp\left\{ -\int_0^\beta dt\, V\big(x+\omega(t)\big) \right\}. \qquad (2.5.3)$$

With respect to the inner time integral extending over the interval $[0,\beta]$, we apply Jensen's inequality (see Appendix C):

$$\exp\left\{ -\int_0^\beta dt\, V\big(x+\omega(t)\big) \right\} \le \frac{1}{\beta}\int_0^\beta dt\, e^{-\beta V(x+\omega(t))}. \qquad (2.5.4)$$

The trick now is to change the order of integration when (2.5.4) is inserted into (2.5.2):

$$
\begin{aligned}
\mathrm{tr}\, e^{-\beta H} &\le \int dx \int_{(0,0)}^{(0,\beta)} d\mu(\omega)\, \frac{1}{\beta}\int_0^\beta dt\, e^{-\beta V(x+\omega(t))} \\
&= \int_{(0,0)}^{(0,\beta)} d\mu(\omega)\, \frac{1}{\beta}\int_0^\beta dt \int dx\, e^{-\beta V(x+\omega(t))} \\
&= \int_{(0,0)}^{(0,\beta)} d\mu(\omega)\, \frac{1}{\beta}\int_0^\beta dt \int dx\, e^{-\beta V(x)} \\
&= K(0,\beta) \int dx\, e^{-\beta V(x)}. \qquad (2.5.5)
\end{aligned}
$$

The result is known as the *Golden–Thompson–Symanzik bound*[27]:

$$\mathrm{tr}\, e^{-\beta H} \le (2\pi\beta)^{-d/2} \int dx\, e^{-\beta V(x)}, \qquad (\beta > 0). \qquad (2.5.6)$$

To find out whether this bound is reasonable, choose the harmonic oscillator in d dimensions, $H = -\frac{1}{2}\Delta + \frac{1}{2}k^2 x^2$, with eigenvalues $E(n_1,\ldots,n_d) = (n_1 + \cdots + n_d + d/2)k$ $(n_i \ge 0)$ such that

$$\mathrm{tr}\, e^{-\beta H} = \left\{ \sum_{n=0}^{\infty} e^{-(n+1/2)\beta k} \right\}^d = [2\sinh(\beta k/2)]^{-d}. \qquad (2.5.7)$$

By contrast, the upper bound (2.5.6) yields

$$(2\pi\beta)^{-d/2} \int dx\, e^{-\beta k^2 x^2/2} = (\beta k)^{-d}, \qquad (2.5.8)$$

which is obviously correct (set $u = \beta k/2$) since $u \le \sinh u$ provided $u \ge 0$. The bound is reasonable when βk is small, it becomes worse when βk grows.

The foregoing discussion bears no relation to classical statistical mechanics. In fact this is not quite so: surprisingly enough, by simply writing

---

[27]For the history if the subject, the reader is referred to the original work [2.20–22]. In its first version, the proof was based on the trace inequality $\mathrm{tr}\,(e^{A+B}) \le \mathrm{tr}\,(e^A)\,\mathrm{tr}\,(e^B)$ valid for self–adjoint operators A and B.

$$(2\pi\beta)^{-d/2} = (2\pi)^{-d} \int dp \, \exp\{-\tfrac{1}{2}\beta p^2\} \tag{2.5.9}$$

and thus introducing the total momentum $p \in \mathbb{R}^d$ we encounter the classical Hamiltonian $\hat{H}(p,q) = \tfrac{1}{2}p^2 + V(q)$ and are able to cast the inequality (2.5.6) into a new form that reveals the relation to classical physics:

$$\mathrm{tr}\, e^{-\beta H} \leq (2\pi)^{-d} \int dpdq \, e^{-\beta\hat{H}(p,q)}. \tag{2.5.10}$$

The integral is now over the entire phase space. The precise way in which the expression on the right-hand side depends on Planck's constant ($\hbar$ or $h = 2\pi\hbar$) is easily exhibited:

$$\mathrm{tr}\, e^{-\beta H} \leq h^{-d} \int dpdq \, e^{-\beta\hat{H}(p,q)}. \tag{2.5.11}$$

The reason: a dimensional argument says that a factor $\hbar^{-d}$ is missing in (2.5.10). The Liouville measure $dpdq$ on phase space has the physical dimension $(action)^d$. After an appropriate normalization, one obtains the semi-classical measure $h^{-d}dpdq$ whith no physical dimension. In fact, it measures phase space volumes in units of h^d.

In principle, it always seems possible to turn the inequality (2.5.6) into an equality by introducing the concept of the *effective potential*. The price to be paid is that the auxiliary potential $V_{\mathrm{eff}}(x)$ generally depends on β. For the moment, we treat β as a fixed parameter. So we hide the β dependence and define:

$$\begin{aligned}
\exp\{-\beta V_{\mathrm{eff}}(x)\} &= (2\pi\beta)^{d/2}\langle x|e^{-\beta H}|x\rangle \\
H_{\mathrm{eff}}(p,q) &= \tfrac{1}{2}p^2 + V_{\mathrm{eff}}(q).
\end{aligned} \tag{2.5.12}$$

It is then straightforward to establish the equality

$$\mathrm{tr}\, e^{-\beta H} = h^{-d} \int dpdq \, \exp\{-\beta H_{\mathrm{eff}}(p,q)\} \tag{2.5.13}$$

demonstrating the formal equivalence of a classical and a quantum partition function. We shall come back to this matter in Sect. 4.2.

There are other situations where the effective potential plays a profound role. Consider for instance a gas of molecules interacting via a two-body potential $V(x)$. The equation of state (i.e., the pressure p as a function of the specific volume v and the temperature T) is assumed to admit a *virial expansion*:

$$p = k_B T\left[v^{-1} + B(T)v^{-2} + C(T)v^{-3} + \cdots\right]. \tag{2.5.14}$$

This expansion enjoyed considerable success, particularly because, in the classical situation, the second virial coefficient comes out to be a simple integral (see [2.23]):

$$B_{\mathrm{cl}}(T) = \tfrac{1}{2} \int dx \left(1 - \exp\left(-\beta V(x)\right)\right). \tag{2.5.15}$$

In particular, for a hard-core potential, $1 - e^{-\beta V(x)} = \chi_G(x)$, where χ_G designates the characteristic function of the ball $G = \{x \mid |x| \le a\}$, so that $B_{\text{cl}}(T) = \frac{1}{2}|G|$, which means half the volume of the excluded region, a constant with respect to T.

On the other hand, we have that, in the quantum situation, the second virial coefficient consists of a direct and an exchange part,

$$B_{\text{qu}}(T) = B_{\text{di}}(T) \pm B_{\text{ex}}(T) \qquad (+\text{Bose}, -\text{Fermi}), \tag{2.5.16}$$

where the direct part obeys a formula which, compared to (2.5.15), looks quite similar:

$$B_{\text{di}}(T) = \tfrac{1}{2} \int dx \left(1 - \exp\left(- \beta V_{\text{eff}}(x) \right) \right). \tag{2.5.17}$$

Thus, the passage from classical to quantum gases requires, among other things, the calculation of the effective potential. By a reasoning analogous to the one yielding the estimate (2.3.26) one may show that, for the hard-core potential considered above, $1 - \exp\left(- \beta V_{\text{eff}}(x) \right) \ge \chi_G(x)$ and so $B_{\text{di}}(T) \ge \frac{1}{2}|G| = B_{\text{cl}}(T)$.

The exchange part in (2.5.16) is much more challenging. It involves the integral $\int dx \, \langle x | e^{-\beta H} | -x \rangle$ (with H the Hamiltonian of relative motion) which can also be written as a path integral and has been the subject of an impressive investigation by Lieb [2.24] (using path integral estimates) who showed that the exchange part drops very fast, i.e., at the rate e^{-cT} with increasing temperature, where, in the hard-core case, the coefficient in the exponent is given by $c = (2\hbar)^{-2}(\pi a)^2 m k_B$ with m the molecule's mass. The reason for this rapid fall-off may be seen in the fact that $-x$ is always in the optical shadow of the ball G centered at the origin when light is emitted from a source at x.

In principle, the above formulas and inequalities apply to N-particle systems with no change. Yet there are questions. What difference does it make to attribute individual masses, say $m_1, \ldots, m_N$, to the particles? How do we accommodate the concept of particle statistics and what role does the interaction play in this game?

The classical Hamiltonian of N particles usually has many components; each component is either a one-body kinetic energy, a one-body external potential, or a two-body relative potential:

$$\hat{H}(p, q) = \sum_{i=1}^{N} \left(\frac{p_i^2}{2m_i} + V_i(q_i) \right) + \sum_{i<j} V_{ij}(q_i - q_j). \tag{2.5.18}$$

A compact way of writing such a complicated expression is

$$\hat{H}(p, q) = \tfrac{1}{2} p^T M^{-1} p + V(q), \tag{2.5.19}$$

where p und q are treated as vectors with $3n$ components and M is a (diagonal) mass matrix. Associated with M there is the Gauss function[28] $K_M(x, \beta)$ pertinent to the motion of a fictitious Brownian particle in $3N$ dimensions. In addition, by invoking the correspondence principle, there is a Schrdinger operator H describing the very same N-particle system from the viewpoint of quantum mechanics, and, as before, we have

---

[28]See the parallel discussion at the end of Sect. 2.3 and consult (2.3.34).

$$\operatorname{tr} e^{-\beta H} \leq K_M(0,\beta) \int dq\, e^{-\beta V(q)} = \left[\det\left(\frac{M}{2\pi\beta} \right) \right]^{\frac{1}{2}} \int dq\, e^{-\beta V(q)}.$$

$$(2.5.20)$$

Not much has changed even though a nontrivial mass matrix occurs in all our expressions, and conventional wisdom requires that the square root in (2.5.20) be replaced by a Gaussian integral:

$$\sqrt{\det(M/2\pi\beta)} = \int dp\, \exp(-\tfrac{1}{2}\beta p^T M p). \qquad (2.5.21)$$

Not unexpectedly, the effect of this replacement is that the integration is now over the phase space:

$$\operatorname{tr} e^{-\beta H} \leq h^{-3N} \int dp\,dq\, e^{-\beta \hat{H}(p,q)}. \qquad (2.5.22)$$

A system manifests itself as being *thermodynamical* if there are many particles of the same kind (equal masses) in the system and if their macroscopic (large N) behavior is governed by intensive variables like the temperature T. In order to be able to apply the above formalism with $\beta = (k_B T)^{-1}$ and k_B the Boltzmann constant, some caution has to be exercised.

On our present understanding, particles of the same kind cannot be distinguished one from another, they obey either Bose or Fermi statistics. Taking whatever the statistics into account will reduce the trace of the operator $e^{-\beta H}$ to a certain subspace of states (either the symmetric or antisymmetric wave functions) whereas, in the formulas above, the trace has been taken with respect to *all* states with no attention to the statistics, be it Bose or Fermi, of the particles.

Nevertheless, one could argue that *Maxwell–Boltzmann statistics* rather than Bose–Fermi statistics provides a more manageable description of many-particle systems: traces are then taken with no restrictions and are easier to manipulate. This is no doubt true. From the quantum statistics point of view, however, this seems a frivolous attempt since the Maxwell–Boltzmann statistics does not occur in nature. So, at first sight, it is rather disturbing to have a nonrealistic model of a thermodynamic system. We shall not go into the fundamentals of statistics here except to say the Maxwell–Boltzmann statistics provides a justifiable approximation at sufficiently high temperature (see Exercise 1 below).

Let us review some facts from thermodynamics. The basic postulate, due to Gibbs, is that the equilibrium state, for the N-particle system in contact with a heat bath at the temperature T, is described by the probability distribution[29]

---

[29]We deliberately include a nonclassical factor h^{-3N}. Note however that it cancels a similar factor in Z_N and so, after all, drops out completely.

$$d\sigma(p,q) = \hat{Z}_N(\beta)^{-1} \frac{dpdq}{N!h^{3N}} e^{-\beta\hat{H}(p,q)} \tag{2.5.23}$$

called the *canonical ensemble*. The constant

$$\hat{Z}_N(\beta) = \int \frac{dpdq}{N!h^{3N}} e^{-\beta\hat{H}(p,q)} \tag{2.5.24}$$

has been introduced to normalize the density. It is frequently called the *classical partition function* of N identical particles. Oddly enough, the origin of the factor $1/N!$ is the postulate of symmetrization or antisymmetrization of wave functions. In total, the inclusion of the factor $1/(N!h^{3N})$ accounts for certain aspects of quantum physics to facilitate comparison: one would like the classical thermodynamical quantities to coincide with their quantum mechanical counterparts for small enough β.

To write the classical partition function, $\hat{Z}_N(\beta)$, in a more explicit manner consider a gas of N particles of mass m enclosed in a volume Ω: in (2.5.14) we let each external potential $V_i(q_i)$ be zero for $q_i \in \Omega$ and $+\infty$ otherwise. Requiring that there be a common pair potential V, we obtain

$$\hat{Z}_N(\beta) = \left(N!\lambda^{3N}\right)^{-1} \int_\Omega dq_1 \cdots \int_\Omega dq_N \, \exp\left\{ - \beta \sum_{i,j} V(q_i - q_j)\right\},$$

$$\tag{2.5.25}$$

where $\lambda = h(2\pi m k_B T)^{-1/2}$. To evaluate integrals of this type, when both the particle number N and the volume Ω get large, is a highly nontrivial and formidable task.

To the classical Gibbs distribution and its partition function there correspond quantum mechanical quantities which are

$$\rho = (Z_N(\beta)N!)^{-1}e^{-\beta H}, \qquad Z_N(\beta) = (N!)^{-1}\mathrm{tr}\, e^{-\beta H}, \tag{2.5.26}$$

and ρ is called the *statistical operator*. In essence, the Golden–Thompson–Symanzik inequality compares two partition functions: $Z_N(\beta) \leq \hat{Z}_N(\beta)$. They play a fundamental role on our way to constructing thermodynamical potentials. For instance, the (Helmholtz) *free energy* F has a classical and a quantum mechanical definition. In fact the defining relations are

$$e^{-\beta\hat{F}(\beta)} = \hat{Z}_N(\beta), \qquad e^{-\beta F(\beta)} = Z_N(\beta) \tag{2.5.27}$$

with the consequence that $F(\beta) \geq \hat{F}(\beta)$. For this inequality to be valid, almost nothing needs to be known about the pair potential. However, the correspondence principle and the Maxwell–Boltzmann statistics have been invoked.

In this context, another evasive attempt to deal with a large system is to neglect two-body forces altogether, keeping only the one-body potential (e.g., the walls of the container). The resulting simplified system is an *ideal*

gas^{30}. The dependence on the number of particles, N, can then be made explicit: $F = Nf$ and $\hat{F} = N\hat{f}$, where f and $\hat{f}$ denote the the free energies per particle to be calculated solely from the one-particle model. In view of the occasionally justified ideal-gas approximation, it makes sense to talk about the free energy connected with some Schrdinger operator $H = -\frac{1}{2}\Delta + V$ pertaining to a *single* particle, as is done, for instance, later in this book. Under general circumstances, i.e., when two-body forces are no longer neglegible, the proportionality relation $F \sim N$ holds only asymptotically. It is then reasonable to define[31] the free energy per particle as $f = \lim_{N\to\infty} F/N$. Again, it is easily confirmed that $f \geq \hat{f}$ holds in general.

As we lower the temperature, the approximation $f \approx \hat{f}$ becomes worse. The reason: the boson or fermion statistics is no longer capable of being evaded. Rather, the true behavior of the system, even on a macroscopic scale, is dictated by genuine quantum effects in the vicinity of zero temperature.

Exercise 1.   The grand-canonical distribution is in many ways more useful than the canonical distribution. It introduces a new variable z called the *fugacity*. Depending on the statistics being used, there are three (quantum) grand-canonical partition functions:

$$Z^{(+)}(\beta, z) = 1 + \sum_{N=1}^{\infty} z^N Z_N^{(+)}(\beta) \qquad \text{Bose,}$$

$$Z^{(-)}(\beta, z) = 1 + \sum_{N=1}^{\infty} z^N Z_N^{(-)}(\beta) \qquad \text{Fermi,}$$

$$Z(\beta, z) = 1 + \sum_{N=1}^{\infty} z^N Z_N(\beta) \qquad \text{Maxwell–Boltzmann.}$$

The partition functions with subscript N refer to the canonical ensemble of N identical particles. In the ideal-gas approximation with single-particle energies ϵ_k, $k \in \mathbb{N}$, the grand-canonical partition functions are taken to be

$$Z^{(+)}(\beta, z) = \prod_{k=1}^{\infty} \left(1 - ze^{-\beta\epsilon_k}\right)^{-1}$$

$$Z^{(-)}(\beta, z) = \prod_{k=1}^{\infty} \left(1 + ze^{-\beta\epsilon_k}\right)$$

$$Z(\beta, z) = \prod_{k=1}^{\infty} \exp\left(ze^{-\beta\epsilon_k}\right).$$

When $z > 0$, prove that

$$Z^{(-)}(\beta, z) < Z(\beta, z) < Z^{(+)}(\beta, z).$$

When $z = e^{\beta\mu}$ (μ is the chemical potential), demonstrate the equivalence of the three descriptions at high temperature by showing that $Z^{(+)}(\beta, e^{\beta\mu}) \approx Z^{(-)}(\beta, e^{\beta\mu}) \approx Z(\beta, e^{\beta\mu})$ as $\beta \to 0$. Prove also that

$$Z_N(\beta) = \frac{1}{N!} \left(\sum_{k=1}^{\infty} e^{-\beta\epsilon_k}\right)^N.$$

---

[30]A better way of saying this is: the ideal gas represents the real gas at zero density.
[31]The so-called thermodynamic limit requires that, as N grows, the volume Ω accessible to the particles grows simultaneously while the density $N/|\Omega|$ stays fixed.

This ultimately justifies the inclusion of the factor $1/N!$ in the definition of Z_N as well of $\hat{Z}_N$. Why does the thermodynamic limit fail (even for the ideal gas) when the factor $1/N!$ is not there?

Exercise 2. Let $\hat{f}(\beta)$ be the classical free energy of a single particle (of unit mass) moving in a potential $V(x)$ such that $\int dx \, \exp\{-\beta V(x)\}$ exists for all $\beta > 0$. Let E denote the ground state energy of the corresponding quantum problem. Use (2.5.6) to prove that

$$E \geq \sup_{\beta > 0} \hat{f}(\beta).$$

In particular: calculate explicitly the right-hand side for $V(x) = \lambda|x|^p$, where $\lambda > 0$, $p > 0$, and $x \in \mathbb{R}^d$. Prove also that the lower bound for E as given by the above formula reduces to the former bound (2.3.29) if $V(x) = 0$ for $x \in G$ and $V(x) = \infty$ otherwise.

Exercise 3. Suppose that U and V are two potentials satisfying $U(x) \leq V(x)$ for all $x \in \mathbb{R}^d$. Prove that the same is true for the corresponding effective potentials: $U_{\text{eff}}(x) \leq V_{\text{eff}}(x)$. In particular,

$$a \leq V(x) \leq b \qquad \text{implies} \qquad a \leq V_{\text{eff}}(x) \leq b.$$

Let g be some element of the Euclidean group in d dimensions so that $gx = Rx + a$, where R denotes an element of the orthogonal group. The Euclidean group acts on potentials via the definition $(gV)(x) = V(g^{-1}x)$. Prove that the following is true: if V_{eff} is the effective potential associated with V, then gV_{eff} is associated with gV. To put it in more formal terms, $(gV)_{\text{eff}} = gV_{\text{eff}}$. In particular, if V is invariant, so is V_{eff}.

Exercise 4. Let $V(x; G)$ stand for the potential that is zero in $G \subset \mathbb{R}^d$ and $+\infty$ otherwise. Write $V_{\text{eff}}(x; G)$ for the associated effective potential. It seems natural to compare this situation with the one where G has been replaced by its complement G^c in $\mathbb{R}^d$. A trivial argument tells us that the equality

$$e^{-\beta V(x;G)} + e^{-\beta V(x;G^c)} = 1$$

holds for all $\beta > 0$. But the same is true for the effective potentials:

$$e^{-\beta V_{\text{eff}}(x;G)} + e^{-\beta V_{\text{eff}}(x;G^c)} = 1.$$

Prove it!

Exercise 5. Suppose that the atoms of a gas interact via a two-body potential V of the hard-core type: it takes the value zero for $|x| > a$ and $+\infty$ otherwise. Since x is the relative coordinate of two atoms, each atom has radius $a/2$. Since we want the reduced mass to be 1, each atom is assigned the mass 2. Starting from the definition (2.5.17), demonstrate that the direct second virial coefficient is given by the following sum:

$$B_{\mathrm{di}} = \tfrac{1}{2}(2\pi\beta)^{3/2} \sum_{n,\ell}(2\ell+1)\exp\left\{-\beta\frac{\lambda_{n\ell}^2}{2a^2}\right\},$$

where $\lambda_{n\ell}$ denotes the nth zero of the spherical Bessel function $j_\ell(z)$. Hint: use (2.4.9) and the preceding exercise. Why is the above representation of B_{di} is less useful at high temperature? From a previous exercise (at the end of Sect. 2.3) it may be inferred that

$$B_{\mathrm{di}} \to (2\pi/3)a^3 \qquad \text{as} \qquad \beta \to 0 \; ;$$

in other words, B_{di} approaches its classical value at high temperature.

2.6 Hamiltonians and Their Associated Processes

The guiding idea in the theory of path integrals so far has been that the Wiener process is central to dealing with questions and actually solving problems in operator theory, where the prominent operator is the Laplacian. By analogy, we propose to introduce a host of new processes where $-\tfrac{1}{2}\Delta$ is replaced by $-\tfrac{1}{2}\Delta + V$. Though this will provide another level of sophistication, we have seen the construction before: the standard example is the oscillator process, which relates probability theory to the quantum harmonic oscillator. In fact the oscillator process is intimately related to the Ornstein–Uhlenbeck velocity process as a dynamical theory of Brownian motion and was therefore the subject of some consideration in Sect. 1.8.

To simplify the exposition, we require that there be some ground state $\Omega \in L^2$ of the Hamiltonian H such that $\Omega(x) > 0$ if x is in a subset A of $\mathbb{R}^d$. Hence, $H\Omega = E\Omega$ and $\|\Omega\| = 1$. We take $E = 0$ (which is no restriction) and define the Markov process X_s for $s \in \mathbb{R}$ (taking values in A) by setting

$$P(X_s \in dx) \;=\; dx\, \Omega(x)^2 \tag{2.6.1}$$

$$P(X_{s'} \in dx' | X_s = x) \;=\; dx'\, \Omega(x')\langle x'|e^{-(s'-s)H}|x\rangle\Omega(x)^{-1} \tag{2.6.2}$$

$(s < s', x \in A)$. Let us examine the ensuing finite-dimensional distributions:

$$P(X_{s_1} \in dx_1, \ldots, X_{s_n} \in dx_n) \;=$$

$$dx_1 \cdots dx_n\, \Omega(x_n)\Omega(x_1) \prod_{k=2}^{n} \langle x_k|e^{-(s_k-s_{k-1})H}|x_{k-1}\rangle \tag{2.6.3}$$

$(s_1 < \cdots < s_n)$. Looking closely, we observe that the process X_s is invariant since each finite-dimensional distribution remains unchanged if s_k is replaced by $s_k + t$.

The above formulas look deceivingly simple. However, besides the fact that the process X_s seems well defined, it is not really all that simple: as is often the case, no explicit expression is available, be it for the ground state $\Omega(x)$ or the transition amplitude $\langle x'|e^{-(s'-s)H}|x\rangle$.

There is no simple relation of the process we have been constructing to ordinary Brownian motion. Whereas, in the Wiener case, to incorporate a potential V into the diffusion equation means admitting absorption of the Brownian particle, by contrast, in the present situation, the particle though moving randomly is never absorbed. Notice, however, that space regions, where the the potential takes large values, are not easily penetrated. Excluded regions are those for which the potential is infinite: if the particle hits the boundary, it will be reflected. Consequently, the role played by the potential is rather close to what we expect from either classical or quantum mechanics.

2.6.1 Correlation Functions

To avoid confusion, it is often helpful in quantum mechanics to distinguish the position operator q (a vector) from the variable $x \in \mathbb{R}^d$ though the connection is trivial: $[q\psi](x) = x\psi(x)$. To a real function $F(x)$ we may associate

1. an operator $A = F(q)$ called an *observable* in the sense of quantum mechanics;
2. a random variable $\mathcal{A} = F(X_0)$ called an *observable* in the sense of probability theory, where the variable X_0 is distributed according to (2.6.1).

Now, examine the time evolution in the two situations. As concerns the quantum situation, we would choose the Heisenberg picture rather than the more frequently used Schrdinger picture. Here, the dynamics enters via the time dependence of operators (states remain unchanged), and, owing to Heisenberg's rule, one has

$$A_t = e^{itH} A e^{-itH} = F(e^{itH} q\, e^{-itH}) = F(q_t). \tag{2.6.4}$$

On the other hand we know that $\mathcal{A}_s = F(X_s)$, where X_s is the Markov process canonically associated with the Hamiltonian H.

What is common to both theories is the concept of *correlation*. The nth order correlation function in the sense of statistics is the following expectation

$$C(s_1, \ldots, s_n) = \boldsymbol{E}(\mathcal{A}_{s_1} \cdots \mathcal{A}_{s_n}). \tag{2.6.5}$$

Because random variables commute, that is to say $[\mathcal{A}_s, \mathcal{A}_{s'}] = 0$, the function $C(s_1, \ldots, s_n)$ is symmetric under permutations of its arguments. Moreover, if $s_1 < s_2 < \cdots < s_n$, it follows from (2.6.3) that

$$C(s_1, \ldots, s_n) = (\Omega, A e^{-(s_n - s_{n-1})H} A \cdots e^{-(s_2 - s_1)H} A\Omega). \qquad (2.6.6)$$

In our approach, H was taken to be positive. So we may analytically continue the probabilistic correlation function to complex variables $z_k = s_k - s_{k-1}$ provided $\operatorname{Re} z_k > 0$. This is to demonstrate that the boundary values, i.e., the correlation function at the points $s_k = it_k$ $(t_k \in \mathbb{R})$, are directly related to the Heisenberg picture of quantum mechanics, for we know that

$$C(it_1, \ldots, it_n) = (\Omega, A_{t_1} \cdots A_{t_n} \Omega). \qquad (2.6.7)$$

In fact the expression on the right-hand side is the nth order quantum correlation function of the observable A. In general, $[A_t, A_{t'}] \neq 0$, and so the complex-valued function (2.6.7) is no longer symmetric under permutations. It way well be said that the study of quantum correlation function gives sufficient knowledge about the system. This does not seem to be so well known to practitioners in quantum mechanics, but it is fairly well known in the context of field theory: here it comes under the heading of *Wightman's reconstruction theorem* [2.25].

Summarizing: quantum correlation functions are analytically continued statistical correlations obtained from an associated stochastic process. From this connection follows the striking fact that, via the process of analytic continuation, *symmetric real functions get mapped onto nonsymmetric complex functions*. This is how classical statistics influences quantum time evolution (even if you don't care about path integrals).

The foregoing discussion would get us quickly into the subject of field theory; the temptation to proceed in this manner has to be resisted; we postpone the exposition of this important subject to later sections. While thumbing through these sections you will realize how many statements in quantum field theory are paralleled by similar statements in ordinary quantum mechanics. For instance, in the theory of a single spinless particle, if we replace *field* by *position*, we may treat expectation values like $(\Omega, q_{t_1} \ldots q_{t_n} \Omega)$ as we would treat Wightman functions: everything that can be said about the system's behavior is coded in this set of functions (think why) despite the fact that we made a special (though natural) choice of the observable. We may also treat the associated set of classical correlations as we would treat Schwinger functions in field theory. Finally, field theory provides a particular model that is easy from the mathematical and the physical point of view alike: it is called the *free field*. The corresponding model for quantum mechanics is the harmonic oscillator.

2.6.2 The Oscillator Process Revisited

Under favorable conditions, we were able to attach some process X_s to any potential. It is then natural to ask: when is X_s a Gaussian process? A moment's reflection will tell us that the potential has to be a second order polynomial. Then, by a proper choice of coordinates, the polynomial can be brought into its standard form, i.e., the one used in Mehler's formula. The resulting process is then called the oscillator process Q_s (in d dimensions).

Mehler's formula (2.3.18) allows us to write down explicit expressions for the finite-dimensional distributions of the oscillator process. To do so it suffices to consider the one-dimensional case only, and so the prototype potential is $V(x) = \frac{1}{2}k^2x^2 - \frac{1}{2}k$ with $x \in \mathbb{R}$. Here is a repeat of the defining relations of the one-dimensional oscillator process:

$$\boldsymbol{P}(Q_s \in dx) = dx \, (k/\pi)^{1/2} \exp(-kx^2)$$

$$\boldsymbol{P}(Q_{s'} \in dx' | Q_s = x) = dx' \left(\frac{k/\pi}{1 - e^{-2\nu}}\right)^{1/2} \exp\left\{-k\frac{(x' - e^{-\nu}x)^2}{1 - e^{-2\nu}}\right\},$$

$$(2.6.8)$$

where $\nu = k(s' - s) > 0$, and, as it turns out, the covariance is given by an extremely simple expression (see (1.8.16)):

$$\boldsymbol{E}(Q_sQ_{s'}) = (2k)^{-1}e^{-k|s'-s|}. \tag{2.6.9}$$

Adopting the above terminology, the covariance is a special correlation function of the oscillator process, in fact the only one that needs to be known, as the process is Gaussian. What could be the possible relation of this function to the quantum harmonic oscillator? The answer is: for H the Hamiltonian and Ω the ground state, we have that

$$(\Omega, qe^{-(s'-s)H}q\Omega) = (2k)^{-1}e^{-k(s'-s)} \qquad (s' > s). \tag{2.6.10}$$

It is a most trivial task to analytically continue on both sides to imaginary times to get the result:

$$(\Omega, q_tq_{t'}\Omega) = (2k)^{-1}e^{ik(t'-t)}. \tag{2.6.11}$$

As predicted by the general theory and to illustratate our former assertions, the correlation function so obtained, $(\Omega, q_tq_{t'}\Omega)$, is neither real nor symmetric with respect interchanging t and t'. Higher order correlation functions are given in terms of second order correlations functions; there is no immediate need to consider them now.

Exercise 1. The 'smearing' (call it *smoothing* if you like) of Euclidean fields with real test functions f has its oscillator analogue:

$$Q(f) = \int_{-\infty}^{\infty} ds \, f(s)Q_s.$$

Observe first that this is but another instance of the stochastic integral, i.e., use (1.8.14) to show that $Q(f) = W(\hat{f})$, where $\hat{f}(s) = \int_0^{\infty} dt \, f(s - t)e^{-kt}$, and prove that the characteristic functional takes a Gaussian form:

$$\boldsymbol{E}\big(\exp\{iQ(f)\}\big) = \exp\big\{-\tfrac{1}{2}(f, D^{-1}f)\big\}.$$

The operator $D = -d^2/ds^2 + k^2$ acts on $L^2(-\infty, \infty)$. Recall (from Sect. 1.8) that D^{-1} is the covariance operator of the oscillator process with kernel $\langle s|D^{-1}|s'\rangle = (2k)^{-1}e^{-k|s-s'|}$.

2.6.3 Nonlinear Transformations of Time

The oscillator process Q_s and the Wiener process X_t are more closely related than might have been expected from their different definitions. For it turns out that there exists a nonlinear transformation of time and some accompanying scale transformation of space that connect the two:

$$t = (2k)^{-1}e^{-2ks} \quad , \quad X_t = e^{-ks}Q_s \qquad (s \in \mathbb{R}). \tag{2.6.12}$$

To prove this we argue as follows: X_t $(t \in \mathbb{R}_+)$ as defined by (2.6.12) is a Gaussian process such that

$$(1) \quad X_0 = \lim_{s \to \infty} e^{-ks}Q_s = 0$$

$$(2) \quad E(X_t X_{t'}) = (2k)^{-1}e^{-k(s+s'+|s'-s|)} = \min(t, t'). \tag{2.6.13}$$

These relations indeed characterize the Wiener process. If you feel the need to check the result using Mehler's formula, do the following arithmetic:

$$P(Q_{s'} \in dx'|Q_s = x) = dy'(2\pi(t'-t))^{-1/2}\exp\left\{-\frac{(y'-y)^2}{2(t'-t)}\right\}, \tag{2.7.14}$$

where $y = e^{-ks}x$, $y' = e^{-ks'}x'$, and $dy' = e^{-ks'}dx'$. Your findings then easily extend to the d-dimensional case. Relation (2.6.12) exists even though the two Hamiltonians that we have assigned to the Wiener and the oscillator process have in fact quite different spectra: $\mathrm{spec}(-\frac{1}{2}d^2/dx^2) = \mathbb{R}_+$ and $\mathrm{spec}(-\frac{1}{2}d^2/dx^2 + \frac{1}{2}k^2x^2 - \frac{1}{2}k) = \{nk \mid n = 0, 1, 2, \ldots\}$. This unusual feature encountered here, unprecedented in ordinary quantum mechanics, is due to the nonlinear character of the transformation of time. Normally, nonlinear spacetime transformations are not considered in standard text books on quantum mechanics though they arise in the context of classical physics[32]. It is quite conceivable that nonlinear transformations of path integrals will play a role in our future manipulations and calculations. Some elementary examples of nonlinear transformations of time may be found in Appendix B.

---

[32]Prominent examples are the Kustaanheimo–Stiefel transformation in the treatment of the $1/r$ potential and coordinate transformations in general relativity. Several attempts have been made to introduce similar concepts into quantum physics. They are often motivated by group theory. We quote but a few examples: conformal invariance [2.26], dynamical symmetry [2.27,28], and the Kustaanheimo–Stiefel transformation with regard to the $1/r$ problem [2.29–31]. The solution to the latter problem is marked by a path-dependent time transformation [2.61–66] based on the notion of *stopping times* from stochastic calculus and on the theorem of Dambis and Dubins–Schwarz [2.67, Sect. 3.4] which states under what conditions a transformed process is in fact ordinary Brownian motion.

2.6.4 The Perturbed Harmonic Oscillator

To adhere to the harmonic oscillator model in atomic and solid state physics has several benefits. However, as is often the case, oscillators are perturbed, and a more realistic model (in one dimension for simplicity) might use the following Hamiltonian:

$$H = H_0 + \lambda V, \qquad H_0 = \tfrac{1}{2}\left(-\frac{d^2}{dx^2} + k^2 x^2 - k\right). \qquad (2.6.15)$$

Here, the pertubation is described by the potential $\lambda V(x)$. Most likely, we won't be able to solve the problem exactly. So let's see whether the oscillator process can be of some help: if we use the finite-dimensional distributions of the process Q_s in conjunction with the Lie–Trotter equality for operator products (along the lines of Section 2.3), we would obtain another Feynman–Kac formula:

$$(\Omega, e^{-tH}\Omega) = E\left(\exp\left\{-\lambda \int_0^t ds\, V(Q_s)\right\}\right). \qquad (2.6.16)$$

By construction, Ω is the ground state of the *unperturbed* oscillator. Equation (2.6.16) provides the key to calculating the ground state energy of the perturbed system:

$$E(\lambda) = -\lim_{t\to\infty} t^{-1}\log E\left(\exp\left\{-\lambda \int_0^t ds\, V(Q_s)\right\}\right). \qquad (2.6.17)$$

Though we appreciate having a handy explicit formula for E, we have somehow lost contact with the conventional operator approach to the problem. Just for the sake of better insight let us use perturbation theory, thereby pretending that $E(\lambda)$ admits a convergent power series expansion with respect to the coupling λ. We aim at showing that the expansion

$$\log E(\exp(\lambda X)) = \sum_{n=1}^{\infty} \frac{\lambda^n}{n!} C_n(X) \qquad (2.6.18)$$

of an expectation into cumulants $C_n(X)$ (X may be any random variable) in fact reproduces the familiar perturbation expansion of the energy $E(\lambda)$ provided X is taken to be

$$X = -\int_0^t ds\, V(Q_s). \qquad (2.6.19)$$

The most convenient way to obtain the cumulants of a random variable is to define them recursively:

$$C_1(X) = E(X)$$

$$C_n(X) = E(X^n) - \sum_{k=1}^{n-1}\binom{n-1}{k-1} E(X^{n-k}) C_k(X) \qquad (n \geq 2) \qquad (2.6.20)$$

We use the abbreviation $X_* = X - E(X)$, so that

$$
\begin{aligned}
C_2(X) &= E(X_*^2) \quad =: \mathrm{Var}(X) \\
C_3(X) &= E(X_*^3) \\
C_4(X) &= E(X_*^4) - 3E(X_*^2)^2, \qquad \text{etc.}
\end{aligned}
\tag{2.6.21}
$$

Recall that we proposed (2.6.19) as our choice for X. So it should be noted that $E(X) = (\Omega, V\Omega)t$ since

$$
E\left(\int_0^t ds\, V(Q_s) \right) = \int_0^t ds\, E\big(V(Q_s)\big) = (\Omega, V\Omega) \int_0^t ds \; .
\tag{2.6.22}
$$

Continuing, we would get

$$
E(X_*^n) = n!(-1)^n \times
$$

$$
\int_0^t ds_n \cdots \int_0^{s_3} ds_2 \int_0^{s_2} ds_1\, (\Omega, V_* e^{-(s_n - s_{n-1})H_0} V_* \cdots e^{-(s_2 - s_1)H_0} V_* \Omega),
$$

$$
\tag{2.6.23}
$$

where $V_* = V - (\Omega, V\Omega)$. Asymptotically,

$$
\begin{aligned}
\lim_{t\to\infty} t^{-1} C_1(X) &= (\Omega, V\Omega) \\
\lim_{t\to\infty} t^{-1} C_2(X) &= (\Omega, V_* H_0^{-1} V_* \Omega) \\
\lim_{t\to\infty} t^{-1} C_3(X) &= (\Omega, V_* H_0^{-1} V_* H_0^{-1} V_* \Omega), \qquad \text{etc.}
\end{aligned}
\tag{2.6.24}
$$

Collecting terms, we thus obtain the desired expansion of the ground state energy:

$$
\begin{aligned}
E(\lambda) = {}& \lambda(\Omega, V\Omega) - \lambda^2(\Omega, V_* H_0^{-1} V_* \Omega) \\
& + \lambda^3(\Omega, V_* H_0^{-1} V_* H_0^{-1} V_* \Omega) + O(\lambda^4).
\end{aligned}
\tag{2.6.25}
$$

More is true: the first order term in this expansion provides a rigorous upper bound of $E(\lambda)$. To understand why, we need to prove

$$
(\Omega, e^{-tH}\Omega) \ge e^{-t\lambda(\Omega, V\Omega)} ,
\tag{2.6.26}
$$

which, in view of (2.6.16), is nothing but a particular instance of Jensen's inequality (see Appendix C):

$$
E\left(\exp\left\{ -\lambda \int_0^t ds\, V(Q_s) \right\} \right) \ge \exp E\left(-\lambda \int_0^t ds\, V(Q_s) \right).
\tag{2.6.27}
$$

Exercise 1. Use (2.2.17) to demonstrate that the ground state energy $E(\lambda)$ is a concave function of the coupling λ, i.e., $d^2 E(\lambda)/d\lambda^2 \le 0$.

Exercise 2. Use Exercise 1 to show that

$$\boldsymbol{E}\big(\exp\big\{-\lambda\textstyle\int_0^t ds\, Q_s\big\}\big) = \exp\big\{\lambda^2 k^{-3}(e^{-kt}-1+kt)\big\}$$
$$\approx \exp\{\lambda^2 k^{-2}t\} \qquad (t\to\infty).$$

This, by the way, proves that a linear perturbation $V(x) = \lambda x$ (e.g., a constant electric field applied to an ion in a crystal) of the harmonic oscillator lowers the bottom energy by the amount $E(\lambda) = -\lambda^2 k^{-2}$ (indeed a concave function as predicted).

Exercise 3.   Argue why perturbation theory fails if the perturbing potential is of the form λx^{2n} ($\lambda > 0$, $n \geq 2$). For $n = 2$, this is the famous anharmonic oscillator model (articles referring to it are legion; see for instance [2.32,33]). Illustrate the failure by discussing the behavior of the analytic function $f(z) = \int_{-\infty}^{\infty} dx\, \exp(-x^2 - zx^4)$ (Re $z > 0$) in the neighborhood of $z = 0$.

2.7 The Thermodynamical Formalism

The considerations of this chapter have centered around the question: can the Feynman–Kac formula be embedded into probability theory? A related question is: can this formula also be understood from the point of view of classical statistical mechanics provided we reinterpret *time* as *space*? The new topic eluded to is called *thermodynamical formalism.*

Time slicing provides a natural way to get us quickly into the subject of one-dimensional spin lattices. The idea is this: a discrete set of points s_i on the time axis is considered a "lattice" (one-dimensional, of course), while the path coordinates $\omega(s_i)$ are treated as "spin variables" attached to the lattice points. As it turns out, an interaction arises solely between neighboring spins. Since it tends to align all spins on the lattice, the interaction is said to be *ferromagnetic.*

We now turn to the details of the description. Starting from some time interval $[s, s']$ for the set-up of the Feynman–Kac formula, we shall, as in Sect. 2.3, assume that the interval has been chopped into $n + 1$ pieces of equal length τ; so $s' - s = (n + 1)\tau$. We envisage a model which, in the language of quantum mechanics, refers to N particles with the mass matrix $M = \mathrm{diag}(m_1, \ldots, m_N)$ and the potential $V(\xi)$, where

$$\xi = \{x_1, \ldots, x_N\}^T \in \mathbb{R}^{3N}.$$

In the language of probability theory, this model relates to a Brownian particle moving in $3N$ dimensions. It corresponds to the Gauss function

$$K_M(\xi, s) = \left[\det\left(\frac{M}{2\pi s}\right)\right]^{\frac{1}{2}} \exp\left\{-\frac{\xi^T M \xi}{2s}\right\}, \tag{2.7.1}$$

from which the transition amplitude results (the subscript indicates the dependence on the potential V):

$$\langle \xi', s' | \xi, s \rangle_V = \lim_{n \to \infty} \int d\xi_n \cdots \int d\xi_1 \prod_{k=0}^{n} K_M(\xi_{k+1} - \xi_k, \tau) e^{-\tau V(\xi_k)}.$$

$$(2.7.2)$$

The construction requires that the following boundary conditions be satisfied:

$$\xi_0 = \xi, \qquad \xi_{n+1} = \xi'. \tag{2.7.3}$$

Needless to say, $\langle \xi', s' | \xi, s \rangle_0 = K_M(\xi' - \xi, s' - s)$ if $V = 0$.

The basic idea is to regard (2.7.2) and the implied procedure $n \to \infty$ as the *thermodynamic limit* (i.e., a transition to an infinitely extended lattice) of some finite spin chain with points labeled by $k = 0, 1, 2, \ldots, n+1$ and with 'spin variables' $\xi_k \in \mathrm{I\!R}^{3N}$ such that, at both ends of the chain, the variables ξ_0 and ξ_{n+1} are kept fixed, i.e., assume values prescribed by the boundary conditions (2.7.3). Correspondingly, the $3nN$-dimensional integral in (2.7.2) will be interpreted as the partition function of the finite spin lattice. For clarity, we rewrite (2.7.2) in the following way:

$$\langle \xi', s' | \xi, s \rangle_V = \lim_{n \to \infty} \left[\det \left(\tfrac{M}{2\pi\tau} \right) \right]^{(n+1)/2} \int e^{-I_n(\xi_1, \ldots, \xi_n)} \prod_{k=1}^{n} d\xi_k, \qquad (2.7.4)$$

where the argument of the exponential, the "action", is defined by

$$I_n(\xi_1, \ldots, \xi_n) = \sum_{k=0}^{n} \tau \left\{ \tfrac{1}{2} v_k^T M v_k + V(\xi_k) \right\}, \qquad v_k = \frac{\xi_{k+1} - \xi_k}{\tau}. \tag{2.7.5}$$

This expression has been written suggestively: we view v_k as a discrete version of the derivative with respect to time. In fact we may look at it as representing a lattice version of the velocity of the associated Brownian particle whose kinetic energy is $\tfrac{1}{2} v_k^T M v_k$.

As $n \to \infty$, the time step approaches zero and, moreover, the "action" I_n formally tends to the time integral of the total energy of the underlying N-particle system:

$$I(\omega) = \lim_{n \to \infty} I_n = \int_{s}^{s'} dt \left\{ \tfrac{1}{2} \dot{\omega}(t)^T M \dot{\omega}(t) + V(\omega(t)) \right\}. \tag{2.7.6}$$

If this were to make sense, $I(\omega)$ would express the classical action along the path ω, and, for any differentiable path, there would be no problem. But not so with Brownian paths: they cannot be differentiated.

The above consideration keeps us from performing the limit $n \to \infty$ directly on I_n as it stands. Nonexistence of the derivative $\dot{\omega}$ is accompanied and compensated by the fact that the measure

$$\prod_{k=1}^{n} \left[\det\left(\tfrac{M}{2\pi\tau}\right)\right]^{1/2} d\xi_k$$

has no limit[33] as n tend to infinity. Under these circumstances, a safe procedure would be to always keep the lattice spacing finite and to consider solely *action sums* E_n instead of *action integrals* $I(\omega)$.

Even in the thermodynamic limit, when $n \to \infty$, we want the lattice to be there. This then requires that the lattice spacing (no longer to be identified with the time step) is kept at a fixed value, say 1, and, to make up for that, we are forced to introduce a coupling constant λ which we allow to vary with the size of the lattice. We thus define

$$E_n(\lambda) = \sum_{i=0}^{n+1} \lambda V(\xi_i) + \sum_{i=0}^{n} \tfrac{1}{2}(\xi_{i+1} - \xi_i)^T M(\xi_{i+1} - \xi_i). \qquad (2.7.7)$$

Indeed, setting

$$\beta_n^{-1} = \lambda_n^{1/2} = \tau = \frac{s' - s}{n + 1} \qquad (2.7.8)$$

we almost get what we want, i.e., the equality $\beta_n E_n(\lambda_n) = I_n$. Actually, the correct relation comes out slightly different, namely $\beta_n E_n(\lambda_n) = I_n + \tau V(\xi')$. Fortunately, as n gets large, the extra $\tau V(\xi')$ may be neglected.

Motivated by the standard terminology in spin-lattice theory we wish to reinterpret (2.7.7): $E_n(\lambda)$ is now taken to be the *energy* of a fictitious spin system with dynamical variables ξ_k ($k = 0, \ldots, n + 1$). Consider the first term in (2.7.7). As it is written, it is a sum, of similar contributions, over the lattice points and thus affects the distribution of spin values at each point individually, regardless of the rest. By contrast, the second term in (2.7.7) is a sum over neighboring spins. It therefore represents the "interaction energy" and is at the focus of our attention.

Disregarding the endpoint contributions, we now see that the energy E_n does not distiguish lattice points: the infinitely extended system is, by definition, translationally invariant, a fact reflecting the homogeneity of the underlying N-particle system with respect to time.

Knowing the energy E_n does not determine everything: we ought to have some idea about the a priori probability for the occurrence of spin values in the system (i.e., if the energy were not there). If there is no reason to prefer one choice over the other, an adequate choice would always be the Lebesgue measure $d\xi$ at each lattice point. Then the partition function, for finite size n, is given by

$$Z_n(\beta, \lambda; \xi, \xi') = \int e^{-\beta E_n(\lambda)} \prod_{i=1}^{n} d\xi_i. \qquad (2.7.9)$$

---

[33] As is known from measure theory, there does not exist a Lebesgue measure, i.e., a translational invariant measure, on any real-linear space of infinite dimension.

The extra dependence on ξ and ξ' reflects the influence of the boundary conditions (2.7.3). However, the integral in (2.7.9) is not quite the expression that we encounter on the right-hand side of (2.7.4) since the normalization is different. By forming quotients we ultimately get rid of the irrelevant constant of normalization. This prompts us to compare two situations: one for the actual potential V that we might be interested in, and one for $V = 0$:

$$\frac{\langle \xi', s' | \xi, s \rangle_V}{\langle \xi', s' | \xi, s \rangle_0} = \lim_{n \to \infty} \frac{Z_n(\beta_n, \lambda_n; \xi, \xi')}{Z_n(\beta_n, 0; \xi, \xi')}. \tag{2.7.10}$$

The procedure to construct transition amplitudes in quantum mechanics by way of a thermodynamic limit as in (2.7.10) has unusual features: as implied by (2.7.8) and (2.7.10), the parameters β_n^{-1} and λ_n are necessarily *dependent on the volume* (i.e., on n). They both tend to zero as the size of the systems grows, and they do so in a specified manner: it causes the difference of two free energies to converge. In fact, this is the only way by which quantum mechanics may be recovered from statistical mechanics and which avoids path integrals altogether.

Fig. 2.3. A typical spin configuration of a finite ferromagnetic chain

To emphasize once more the relation to statistical mechanics, we recall the basic properties of the spin systems encountered in the present context:

1. On the lattice, there exist solely short-range interactions (i.e., between nearest neighbors). Their origin is the kinetic energy (not the potential) of the underlying quantum mechanical system. In spite of their short-range nature, these interactions influence the system's behavior globally. There are also contributions of a *local* nature to the energy: they affect the spins locally and originate from the potential (not the kinetic energy).

2. The model systems are one-dimensional and translationally invariant. They belong to the class of ferromagnetic models, for the kinetic energy is always positive. Neighboring spins tend to parallelize though they are forced to interpolate between the prescribed spin values ξ and ξ' at both ends of the chain. Figure 2.3 illustrates this fact.

3. The bulk free energy (i.e., the leading term in $-\beta^{-1} \log Z_n$ as $n \to \infty$) is proportional to n and does not depend on the boundary conditions.

It drops out in (2.7.10). It is the next-to-leading term that we are concerned with.

From the condensed matter physics point of view, one-dimensional systems are dull. For their critical point is always at zero temperature – formally, at $\beta = \infty$ – and, hence, no phase transition (in the true sense of the word) can occur. In (2.7.10), $\lim_n \beta_n = \infty$, meaning that the spin chain tends to its critical point. It is only at the critical point that we encounter the equivalence with quantum mechanics. But notice the delicacy of the limit since three things happen at the same time: (1) the lattice grows indefinitely; (2) β tends to infinity; (3) the coupling λ approaches zero. Everything has to be organized so as to yield the desired quantum mechanical amplitude.

Exercise 1.   What sense does it make to adopt *periodic boundary* conditions when working with spin chains? Does this perhaps provide a method to construct quantum-mechanical traces like $\operatorname{tr} e^{-tH}$?

2.8 A Case Study: the Harmonic Spin Chain

We now turn to a solvable model for a spin chain to see the thermodynamical formalism at work. The example is more than just a testing ground for ideas, it also provides an alternative derivation of Mehler's formula for the harmonic oscillator. The setting is as follows. Consider spin variables $x_k \in \mathbb{R}^d$ at lattice points $k \in \{0, 1, \ldots, n+1\}$ and attribute the energy

$$E_n(\lambda) = \tfrac{1}{2} \sum_{i=0}^{n} (x_{i+1} - x_i)^2 + \tfrac{1}{2}\lambda k^2 \sum_{i=0}^{n+1} x_i^2 \tag{2.8.1}$$

to any spin configuration that satisfies the boundary conditions

$$x_0 = x, \ x_{n+1} = x' \qquad (x, x' \in \mathbb{R}^d, \text{ fixed}). \tag{2.8.2}$$

The main reason for success is the fact that the energy is a quadratic form: it yields manageable thermodynamical functions. It also explains why we refer to the model as the *harmonic spin chain*. The theory of the preceding section relates this model to the quantum mechanical harmonic oscillator with the Hamiltonian $H = -\tfrac{1}{2}\Delta + \tfrac{1}{2}k^2 x^2$ in d dimensions.

Thermodynamical considerations are based on the partition function,

$$Z_n(\beta, \lambda; x, x') = \int dx_1 \cdots \int dx_n \, e^{-\beta E_n(\lambda)}, \tag{2.8.3}$$

for which it is most natural to rewrite the energy as

$$E_n(\lambda) = \tfrac{1}{2}\zeta^T Q\zeta - \zeta^T b + \tfrac{1}{2}(1 + \lambda k^2)b^T b, \tag{2.8.4}$$

where we have introduced the vectors

$$\zeta = \{x_1, x_2 \ldots, x_n\}^T \in \mathbb{R}^{nd}$$
$$b = \{x, 0, \ldots, 0, x'\}^T \in \mathbb{R}^{nd}. \tag{2.8.5}$$

Therefore, ζ contains the free spin variables and b is a fixed vector implementing the boundary conditions. There appears a matrix Q of dimension nd in (2.8.4) which we define as $Q = R \otimes \mathbb{1}$, i.e., it is written in terms of some $n \times n$ matrix R and the d-dimensional unit matrix $\mathbb{1}$:

$$Q = \begin{pmatrix} 2c\mathbb{1} & -\mathbb{1} & & \\ -\mathbb{1} & 2c\mathbb{1} & \ddots & \\ & \ddots & \ddots & -\mathbb{1} \\ & & -\mathbb{1} & 2c\mathbb{1} \end{pmatrix} \qquad R = \begin{pmatrix} 2c & -1 & & \\ -1 & 2c & \ddots & \\ & \ddots & \ddots & -1 \\ & & -1 & 2c \end{pmatrix}. \tag{2.8.6}$$

For convenience, we ave introduced the parameter $c = 1 + \tfrac{1}{2}\lambda k^2$. In what follows, we shall also use the parameter u as given by $c = \cosh u$.

It is now a fairly elementary task to evaluate the Gaussian integral with respect to ζ and to show that the result is the following rather explicit representation of the partition function:

$$Z_n(\beta, \lambda; x, x') = \left[\det \tfrac{\beta Q}{2\pi}\right]^{-1/2} \exp\{-\tfrac{1}{2}\beta\, b^T (2c - 1 - Q^{-1})b\}. \tag{2.8.7}$$

The next step is to simplify the determinant:

$$\det \frac{\beta Q}{2\pi} = \left(\frac{\beta}{2\pi}\right)^{nd} (\det R)^d. \tag{2.8.8}$$

We are left with the problem of calculating $d_n := \det R$. The strategy is to set up a recursion scheme,

$$d_0 = 1, \qquad d_1 = 2\cosh u, \qquad d_{n+1} + d_{n-1} = 2d_n \cosh u, \tag{2.8.9}$$

and to solve the difference equation by an ansatz: $d_n = ae^{nu} + a'e^{-nu}$. The real coefficients a and a' are to be computed from d_0 and d_1:

$$a + a' = 1, \qquad ae^u + a'e^{-u} = 2\cosh u. \tag{2.8.10}$$

So the final answer is

$$d_n \equiv \det R = \frac{\sinh(n+1)u}{\sinh u}. \tag{2.8.11}$$

Here is our routine for evaluating $Q^{-1}b$: the set of linear equations $Qy = b$ is equivalent to

$$2y_k \cosh u = \begin{cases} x + y_2 & \text{if } k = 1 \\ y_{k-1} + y_{k+1} & \text{if } k = 2, \ldots, n-1 \\ y_{n-1} + x' & \text{if } k = n. \end{cases} \qquad (2.8.12)$$

Again, the ansatz $y_k = re^{ku} + se^{-ku}$ will solve the problem, whereby the constant vectors r and s follow from the first and the last line of (2.8.12):

$$r + s = x, \qquad re^{(n+1)u} + se^{-(n+1)u} = x'. \qquad (2.8.13)$$

So the result is

$$y_k = \frac{x \sinh(n - k + 1)u + x' \sinh ku}{\sinh(n+1)u}, \qquad k = 1, \ldots, n. \qquad (2.8.14)$$

We may then use this to compute

$$b^T Q^{-1} b = xy_1 + x' y_n = \frac{(x^2 + x'^2) \sinh nu + 2xx' \sinh u}{\sinh(n+1)u}, \qquad (2.8.15)$$

and with help of the equality

$$\sinh nu = \sinh(n+1)u \cosh u - \cosh(n+1)u \sinh u \qquad (2.8.16)$$

we write the final result as

$$Z_n(\beta, \lambda; x, x') = [(2\pi/\beta)^n w]^{d/2} \exp\{-\tfrac{1}{2}\beta[(x^2 + x'^2)v - 2xx'w]\}$$
$$v = \frac{\sinh u}{\tanh(n+1)u} + \tfrac{1}{2}\lambda k^2 \qquad (2.8.17)$$
$$w = \frac{\sinh u}{\sinh(n+1)u}, \qquad \cosh u = 1 + \tfrac{1}{2}\lambda k^2.$$

If $\lambda \to 0$ (i.e., $u \to 0$), both v and w tend to $(n+1)^{-1}$. Therefore,

$$Z_n(\beta, 0; x, x') = \left[\frac{(2\pi/\beta)^n}{n+1}\right]^{d/2} \exp\left\{-\frac{\beta(x'-x)^2}{2(n+1)}\right\}. \qquad (2.8.18)$$

How is this to be connected with the quantum harmonic oscillator? By the rules for the approach to quantum mechanics laid down in Sect. 2.7, the thermodynamic limit $n \to \infty$ has to be performed in such a way that both β^{-1} and λ tend to zero. Specifically, this requires that we take

$$\beta = \beta_n := (n+1)(s'-s)^{-1}, \qquad \lambda = \lambda_n := (n+1)^{-2}(s'-s)^2. \quad (2.8.19)$$

We write $\nu = k(s'-s)$ for short. The large-n behavior of u, v, and w can be inferred from (2.8.17) and (2.8.19):

$$(n+1)u \to \nu, \quad (n+1)v \to \frac{\nu}{\tanh \nu} \quad (n+1)w \to \frac{\nu}{\sinh \nu}. \qquad (2.8.20)$$

Consequently, the following limit exists:

$$\lim_{n \to \infty} \frac{Z_n(\beta_n, \lambda_n; x, x')}{Z_n(\beta_n, 0; x, x')} = \left[\frac{\nu}{\sinh \nu} \right]^{d/2} \frac{\exp\left\{ -\frac{k(x^2 + x'^2)}{2 \tanh \nu} + \frac{kxx'}{\sinh \nu} \right\}}{\exp\left\{ -\frac{k(x - x')^2}{2\nu} \right\}}. \quad (2.8.21)$$

The expression so obtained has to be multiplied by $K(x' - x, s' - s)$ (the transition amplitude without a potential; see (2.7.10)) to produce the transition amplitude for the harmonic oscillator, i.e., *Mehler's formula*:

$$\langle x', s' | x, s \rangle = \left[\frac{k}{2\pi \sinh \nu} \right]^{d/2} \exp\left\{ -\frac{k(x^2 + x'^2)}{2 \tanh \nu} + \frac{kxx'}{\sinh \nu} \right\}$$

$$= \langle x' | e^{-(s' - s)H} | x \rangle, \qquad H = -\tfrac{1}{2}\Delta + \tfrac{1}{2}k^2 x^2 \qquad (2.8.22)$$

$$k(s' - s) = \nu > 0$$

A final question may be of interest in its own right: what is the free energy per lattice site of the harmonic chain? So we calculate

$$f(\beta, \lambda) = - \lim_{n \to \infty} (\beta n)^{-1} \log Z_n(\beta, \lambda; x, x') = \frac{d}{2}\beta^{-1} \left(u + \log \frac{\beta}{2\pi} \right)$$

$$(2.8.23)$$

(recall that $2 \sinh(u/2) = \lambda^{1/2} k > 0$). As was to be expected, the free energy f does not depend on the boundary conditions, i.e., on x, x'. If the coupling vanishes, we would get

$$f(\beta, 0) = \frac{d}{2}\beta^{-1} \log \frac{\beta}{2\pi} . \qquad (2.8.24)$$

Is there any relation of the free energy to the ground state energy $E = \frac{d}{2}k$ of the harmonic oscillator? It certainly does not come here as a surprise that E arises as a limit:

$$E = \lim_{\substack{\beta \to \infty \\ \lambda \to 0 \\ \beta^2 \lambda \to 1}} \frac{f(\beta, \lambda) - f(\beta, 0)}{\lambda} . \qquad (2.8.25)$$

2.8.1 The Inverted Harmonic Oscillator

Consider the classical Lagrangian

$$L(\dot{q}, q) = T - V = \tfrac{1}{2}\dot{q}^2 + \tfrac{1}{2}k^2 q^2 \qquad (q \in \mathbb{R}^d) \qquad (2.8.26)$$

for the harmonic oscillator potential $V = -\tfrac{1}{2}k^2 q^2$ with opposite sign. Comparing trajectories $q(t)$ with prescribed endpoints $q(s) = x$ and $q(s') = x'$, we know from classical mechanics that the action[34]

---

[34]The real-time action S correspondes to the Euclidean action in field theory. As is typical for the real-time or Euclidean formulation, the classical solution is always obtained by appeal to the variational principle $S = minimum$.

$$S[q] = \int_s^{s'} dt\, L(\dot{q}, q) \tag{2.8.27}$$

is minimal (not just stationary) if $q(t)$ solves the Euler–Lagrange equation $\ddot{q} = k^2 q$ for $s \le t \le s'$. Since the oscillator potential has the "wrong" sign, the solution is given in terms of hyperbolic instead of trigonometric functions:

$$q(t) = \frac{x \sinh\{k(s' - t)\} + x' \sinh\{k(t - s)\}}{\sinh\{k(s' - s)\}}. \tag{2.8.28}$$

For $k = 0$ it reduces to the straight path from x to x', which is traversed with constant speed.

By a direct computation one shows that the minimal action thus determined is

$$S_{\min} = \frac{k(x^2 + x'^2)}{2 \tanh \nu} - \frac{kxx'}{\sinh \nu}, \qquad \nu = k(s' - s). \tag{2.8.29}$$

So Mehler's formula (2.8.22) for the quantum oscillator takes the form

$$\langle x', s' | x, s \rangle = D^{1/2} \exp(-\hbar^{-1} S_{\min}), \tag{2.8.30}$$

where

$$D = \det\left(\frac{-1}{2\pi\hbar} \frac{\partial^2 S_{\min}}{\partial x_i' \partial x_k} \right)_{i,k=1,\dots,d} \tag{2.8.31}$$

is the so-called Van Vleck determinant of the action $S_{\min}$. Also, the dependence on $\hbar$ has been made explicit. The lesson is that, for quadratic actions, the classical solution (2.8.28) not only dominates the path integral but also provides an exact representation of the transition amplitude.

As regards our ansatz (2.8.26–27), suppose that V is replaced by an arbitrary (inverted) potential. Then the formulas (2.8.30–31) form the basis of the celebrated WKB approximation. For more details and references see [2.12].

Exercise 1. Let $Z_n(\beta, \lambda)_{\mathrm{per}}$ denote the partition function of the harmonic chain of length n with periodic boundary conditions, i.e.,

$$Z_n(\beta, \lambda)_{\mathrm{per}} = \int dx\, Z_{n-1}(\beta, \lambda; x, x) \exp\{\beta \tfrac{1}{2} k^2 x^2\}.$$

Demonstrate that this function is simpler in the sense that it factorizes as $A_n(\beta) B_n(\lambda)$, where $A_n(\beta) = (2\pi/\beta)^{nd/2}$ and

$$B_n(\lambda) = \left(e^{nu/2} - e^{-nu/2} \right)^{-d} \xrightarrow[n \to \infty]{} \left(e^{\nu/2} - e^{-\nu/2} \right)^{-d},$$

provided $nu \to \nu$ as in (2.8.20). From the definitions (2.8.22) infer that $\operatorname{tr} e^{-(s'-s)H} = (e^{\nu/2} - e^{-\nu/2})^{-d}$.

Exercise 2.   Under the assumptions of the previous exercise, conclude that there exists a representation of the form $Z_n(\beta, \lambda)_{\mathrm{per}} = \operatorname{tr} L^n$, where L is the so-called *transfer operator* of the harmonic chain with kernel

$$\langle x'|L|x\rangle = \exp\left\{ - \tfrac{1}{2}\beta\big[(x'-x)^2 + \tfrac{1}{2}k^2(x^2 + x'^2)\big]\right\} \qquad (x, x' \in \mathbb{R}^d).$$

Prove that the spectrum of L (a trace-class operator) consists of eigenvalues

$$E(n_1, \ldots, n_d) = \prod_{i=1}^{d} (2\pi/\beta)^{1/2} \exp\left\{ - u(n_i + \tfrac{1}{2})\right\} \qquad (n_i = 0, 1, 2, \ldots),$$

where $\cosh u = 1 + \tfrac{1}{2}\lambda k^2$.

Exercise 3.   A more general harmonic oscillator Hamiltonian reads

$$H = -\tfrac{1}{2}\nabla \cdot M^{-1}\nabla + \tfrac{1}{2}x \cdot Ax, \qquad (x \in \mathbb{R}^d),$$

where both M and A are strictly positive (symmetric) matrices. Changing coordinates, reduce this to the former situation. Hint: introduce the matrix

$$K = \left(M^{-1/2}AM^{-1/2}\right)^{1/2} > 0$$

and diagonalize. Prove the validity of the trace formula

$$\operatorname{tr} e^{-\beta H} = \det\left(e^{\beta K/2} - e^{-\beta K/2}\right)^{-1}.$$

Formulate conditions on K so that the result extends to infinite dimensions $(d \to \infty)$.

2.9 The Reflection Principle

In quantum mechanics, a boundary surface is viewed as some perfectly rigid, inpenetrable wall beyond which the wave function of a Schrdinger particle vanishes identically. More generally, the motion of N particles may be impaired by the presence of an excluded region A_- of the configuration space $\mathbb{R}^{3N}$. We shall suppose that forces are absent besides those exerted on the particles by the presence of the *walls*.

2.9.1 Reflection Groups of Order Two

We begin by assuming that the configuration space has dimension d and that the excluded region is in fact a halfspace $A_- \in \mathrm{I\!R}^d$. We aim at showing that the transition amplitude follows solely from geometric considerations. Think of some hyperplane A_0 separating the allowed region A_+ and the forbidden region A_-. Connected with the hyperplane there is a reflection I acting on $\mathrm{I\!R}^d$ that maps A_+ onto A_- and vice versa.

Clearly, what is needed to cause wave functions to vanish in A_- is a $(0, \infty)$-potential:

$$V(x) = \begin{cases} 0 & \text{if } x \in A_+ \\ \infty & \text{otherwise.} \end{cases} \tag{2.9.1}$$

The formalism of Brownian motion also provides an interpretation of the infinite potential: by the Feynman–Kac formula there is zero probability for the corresponding Brownian particle to enter the forbidden region. For it will be absorbed when hitting the boundary (i.e., the hyperplane A_0). In performing the necessary calculation, we are asked to determine the conditional Wiener measure of the set of all paths $\omega : (x, s) \rightsquigarrow (x', s')$ that have prescribed endpoints in A_+ and never leave the allowed region.

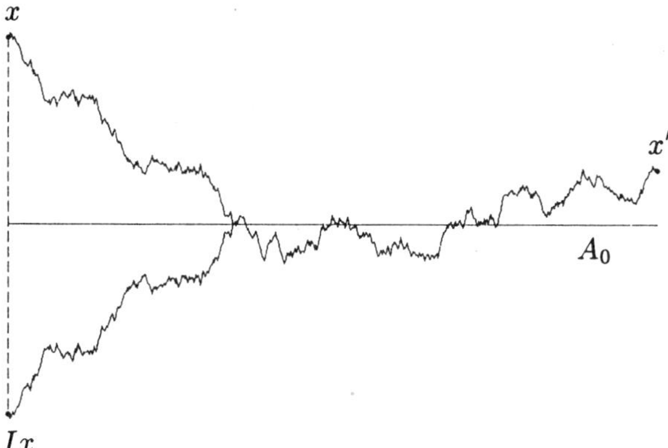

Fig. 2.4. The reflection of a Brownian path at some hyperplane

As regards the set of *unrestricted paths*, Ω, we know that its measure is

$$\mu(\Omega) = \int d\mu(\omega) = K(x' - x, s' - s). \tag{2.9.2}$$

From Ω we have to subtract the subset Ω' of those paths that enter the boundary A_0 during the given interval of time, $s \leq t \leq s'$. Therefore, the transition amplitude is simply

$$\langle x', s' | x, s \rangle = \mu(\Omega \backslash \Omega') = \mu(\Omega) - \mu(\Omega'). \tag{2.9.3}$$

It remains to show that the set Ω' can be characterized explicitly. Indeed, for each $\omega \in \Omega'$, there exists some moment of time,

$$t_\omega = \inf\{\, t \mid s < t < s',\ \omega(t) \in A_0 \,\}, \qquad (2.9.4)$$

when the path first hits the the boundary A_0. This point, also called the *first entrance time into* A_-, divides the path ω into two halves. The approach to the problem consists in reflecting the first half at the hyperplane A_0 – this is illustrated in Fig. 2.4 – to obtain the transformed path $\omega_I : (Ix, s) \rightsquigarrow (x', s')$. Formally,

$$\omega_I(t) = \begin{cases} I\omega(t) & \text{if } s \le t \le t_\omega \\ \omega(t) & \text{if } t_\omega \le t \le s' \end{cases} \qquad (\omega \in \Omega'). \qquad (2.9.5)$$

The essential point here is that the reflection I at the hyperplane establishes a one-to-one correspondence between the set Ω' and the set of all paths from Ix to x'. To convince the sceptic we argue as follows.

1. *Ix and x' belong to disjoint halfspaces; so any path starting in A_- and arriving in A_+ unavoidably crosses the separating hyperplane.* This uses the fact that Brownian paths are continuous.
2. The two sections arising from bisecting a given path are, statistically speaking, uncorrelated (in fact increments are). As a consequence, applying a reflection to one section of it does not change the statistical weight the whole path gets from the Wiener measure. This uses the fact that Brownian motion is a Markov process.

From (2.9.3) and (2.9.5) we infer that

$$\langle x', s' | x, s \rangle = \begin{cases} K(x' - x, s' - s) - K(x' - Ix, s' - s) & \text{if } x, x' \in A_+ \\ 0 & \text{otherwise.} \end{cases}$$

$$(2.9.6)$$

It would be desirable to have a direct check as to why the expression on the right-hand side is in fact positive. A moment's reflection will tell us that positivity is implied by the inequality $(x' - Ix)^2 > (x' - x)^2$, which is valid provided x and x' belong two different halfspaces.

Example 1.   Consider a Schrdinger particle of unit mass on the halfaxis. In more detail, we let the reflection I of the real axis be such that $Ix = -x \in \mathbb{R}$. Hence, $A_0 = \{0\}$. We take $A_+ = \{x > 0\}$ as the allowed and $A_- = \{x < 0\}$ as the forbidden region. By the reflection principle (2.9.6),

$$\langle x' | e^{-tH} | x \rangle = K(x' - x, t) - K(x' + x, t) \qquad (x > 0, x' > 0), \qquad (2.9.7)$$

where $H = -\frac{1}{2}d^2/dx^2 + V$ and $V(x) = 0$ $(x > 0)$ resp. $V(x) = \infty$ $(x \le 0)$. It should be clear that the result (2.9.7) may also be obtained via the spectral decomposition of the operator H. But the present derivation has a more intuitive appeal.

The preceding example also serves to draw attention to certain related path integrals when the underlying space is one-dimensional. To start with, consider the functional

$$\inf \omega = \inf\{\, \omega(s) \mid 0 \leq s \leq t \,\}, \qquad (t \text{ fixed}). \qquad (2.9.8)$$

Let the reflection of the real line be given by $Ix = 2a - x$ so that the fixed point is $x = a$. In view of the above example, the set of admissible paths may now be characterized by the condition $\inf \omega > a$. So, provided $\min(x', x) > a$,

$$\int d\mu(\omega)\, \Theta(\inf \omega - a) = K(x' - x, t) - K(x' + x - 2a, t) \qquad (2.9.9)$$

with $\Theta(x)$ the step function and $d\mu = d\mu_{x,0}^{x',t}$ the conditional Wiener measure. Differentiation with respect to a then leads to the equation

$$\int d\mu(\omega)\, \delta(a - \inf \omega) = \begin{cases} (d/da)K(x' + x - 2a, t) & \text{if } a < \min(x', x) \\ 0 & \text{otherwise.} \end{cases}$$

$$(2.9.10)$$

As is immediately verifiable, for fairly general functions $f(a)$,

$$\int d\mu(\omega) f(\inf \omega) = \int_{-\infty}^{\min(x',x)} da\, f(a) \frac{d}{da} K(x' + x - 2a, t). \qquad (2.9.11)$$

Let all space variables change sign: the companion formula so obtained is

$$\int d\mu(\omega) f(\sup \omega) = - \int_{\max(x',x)}^{\infty} da\, f(a) \frac{d}{da} K(x' + x - 2a, t). \qquad (2.9.12)$$

Example 2. Consider two particles on the line; let x_1 and x_2 be the co-ordinates of the first and second particle respectively. The configuration space is two-dimensional: it consists of pairs $(x_1, x_2) \in \mathbb{R}^2$. No one particle is allowed to overtake the other, i.e., we require the motion to respect the general rule $x_1 < x_2$ which defines the allowed halfspace $A_+ \subset \mathbb{R}^2$. What is relevant here is the existence of the reflection $I\{x_1, x_2\} = \{x_2, x_1\}$, implying that

$$\begin{aligned} \langle x_1', x_2' | e^{-tH} | x_1, x_2 \rangle =\; & K(x_1' - x_1, t) K(x_2' - x_2, t) \\ & - K(x_1' - x_2, t) K(x_2' - x_1, t) \end{aligned} \qquad (2.9.13)$$

This formula is valid provided $x_1 < x_2$, $x_1' < x_2'$. The result looks as if the Fermi statistics were effective here.

2.9.2 Reflection Groups of Infinite Order

An interesting situation arises when a particle (or a general quantum mechanical system) with configuration space $\mathbb{R}^d$ is not permitted to leave the region S_0 between two parallel hyperplanes A_0 and A_1. Again, we may approach the problem geometrically, looking for a suitable group of reflections first.

Let σ_0 and σ_1 denote the reflections of the space $\mathbb{R}^d$ leaving the hyperplanes A_0 and A_1, respectively, fixed, and let G denote the group generated by σ_0 and σ_1 (so G is of infinite order and contains translations as well as reflections). The images A_n ($n \in \mathbb{Z}$) of A_0 under the action of G are parallel hyperplanes deviding the entire space into slices S_n of equal width such that each slice S_n is bounded by two hyperplanes, A_n and A_{n+1}. Assuming $\{\, x_n \in S_n \mid n \in \mathbb{Z}\,\}$ to be the images of $x = x_0 \in S_0$ under G, we conclude that x_n and x_{n-1} are connected by the reflection at the hyperplane A_n. This defines the setup and the notation.

We turn to the Brownian paths $(x, s) \rightsquigarrow (x', s')$, where it is assumed that $x, x' \in S_0$. Let Ω stand for the set of unrestricted paths and Ω_n^i ($i = 0$ or 1) for the set of all Brownian paths $(x_n, s) \rightsquigarrow (x', s')$ leaving the slice S_n via the hyperplane A_{n+i} (what happens to the path from then on is of no concern). From the reflection principle, we derive the following two recursion formulas:

$$\mu(\Omega_{n+1}^0) = K(x' - x_n, s' - s) - \mu(\Omega_n^0), \qquad n < 0$$
$$\mu(\Omega_{n-1}^1) = K(x' - x_n, s' - s) - \mu(\Omega_n^1), \qquad n > 0. \tag{2.9.14}$$

To achieve our goal, which is to obtain an expression for the transition amplitude $\langle x, s | x', s' \rangle$, we need to eliminate from Ω the (disjoint) sets Ω_0^0 and Ω_0^1 and to determine the Wiener measure of the remainder using the recursion (2.9.14):

$$\langle x, s | x', s' \rangle = \mu(\Omega) - \mu(\Omega_0^0) - \mu(\Omega_0^1)$$
$$= \sum_{n=-\infty}^{\infty} (-1)^n K(x' - x_n, s' - s). \tag{2.9.15}$$

One pertinent fact about this sum is the property of growth, $|x' - x_n| = O(n)$, which settles the convergence problem.

Example 3.   Consider a particle enclosed in a box Λ, where

$$\Lambda = \{\, x \in \mathbb{R}^d \mid -L_j/2 \le x_j \le L_j/2, \ j = 1, \ldots, d\,\}.$$

Imagine the exterior space to be filled with some medium that acts as a heat bath at the temperature T, while absorption is being neglected. Suppose we wish to calculate the partition function

$$Z_\Lambda(\beta) = \int_\Lambda dx \langle x|e^{-\beta H}|x\rangle \qquad (\beta^{-1} = k_B T). \qquad (2.9.16)$$

The standard technique, *separation of coordinates*, tells us to decompose the Hamiltonian as $H = H_1 + \cdots + H_d$ and to study each one-dimensional problem separately:

$$H_j = -\frac{1}{2}\frac{d^2}{dx_j^2} + V_j(x_j) , \qquad V_j(x_j) = \begin{cases} 0 & \text{if } |x_j| < L_j/2 \\ \infty & \text{otherwise.} \end{cases} \qquad (2.9.17)$$

Once this is done, the result can be written as a product:

$$\langle x|e^{-\beta H}|x\rangle = \prod_{j=1}^{d} \langle x_j|e^{-\beta H_j}|x_j\rangle . \qquad (2.9.18)$$

In order to solve the one-dimensional problem, we appeal to the previous result (2.9.15). If the point x_j is in $[-L/2, L/2]$ and the group of reflections is applied to it, we get the sequence $x_{jn} = (-1)^n x_j + nL_j$ $(n \in \mathbb{Z})$, and the following formula holds:

$$\langle x_j|e^{-\beta H_j}|x_j\rangle = \sum_n (-1)^n K(x_j - x_{jn}, \beta)$$

$$= (2\pi\beta)^{-1/2} \sum_{k=-\infty}^{\infty} \left[\exp\left\{ -\frac{(2kL_j)^2}{2\beta} \right\} - \exp\left\{ -\frac{[(2k+1)L_j - 2x_j]^2}{2\beta} \right\} \right].$$

$$(2.9.19)$$

Integration over the interval $-L_j/2 \le x_j \le L_j/2$ poses no problem. So, as regards the partition function, the final answer is[35]

$$Z_\Lambda(\beta) = \prod_{j=1}^{d} \tfrac{1}{2}\big(\theta(a_j) - 1\big), \qquad a_j = \frac{1}{L_j}\left(\frac{\pi\beta}{2}\right)^{1/2}, \qquad (2.9.20)$$

where use has been made of the Jacobian theta function

$$\theta(a) = \sum_{n=-\infty}^{\infty} \exp(-\pi n^2 a^2) \qquad (2.9.21)$$

and of its functional equation $\theta(a) = a^{-1}\theta(a^{-1})$. The energy distribution of the canonical ensemble has the following characteristic function

---

[35]For more physical insight, we reintroduce Planck's constant $\hbar$ and the mass m of the particle. We then interpret the dimensionless quantity a_j as the quotient $\lambda/(2L_j)$, where $\lambda = \hbar\sqrt{2\pi\beta/m}$. Thus, λ is the de Broglie wavelength at the temperature T. Just to get an idea: for electrons at room temperature, λ would be of the order of 10^{-7}cm. Hence, a box of size $L_1 \cdots L_d$ is considered to be large when $L_j \gg \lambda$ for all j.

$$\langle e^{itH}\rangle_\Lambda = \frac{Z_\Lambda(\beta - it)}{Z_\Lambda(\beta)}. \tag{2.9.22}$$

For a large box, i.e., if $L_j \to \infty$ $(j = 1,\ldots,d)$, this expression has an interesting limiting behavior: from $\theta(a) = a^{-1} + O(1)$ for $a \to 0$ it follows that

$$\lim_{\Lambda\uparrow}\langle e^{itH}\rangle_\Lambda = (1 - it/\beta)^{-d/2}. \tag{2.9.23}$$

We immediately recognize the characteristic function of the Γ-distribution:

$$(1 - it/\beta)^{-u} = \int_0^\infty d\mu(E)\, e^{itE} \qquad (u = d/2)$$
$$d\mu(E) = \frac{\beta^u}{\Gamma(u)} E^{u-1} e^{-\beta E}\, dE. \tag{2.9.24}$$

The limit distribution $d\mu(E)$ of energies E, known as the *Maxwell distribution* in kinetic theory, emerges from (2.9.22) also by passing the classical limit[36]. It holds for any diluted gas, be it classical or quantum mechanical. Notice that the familiar Maxwellian result recaptured here involves neither operator techniques nor spectral decomposition theory. Rather, the argument is of a geometrical nature and borrows from probability theory.

With growing dimension and increasingly complex geometry of the fundamental domain $F \in \mathbb{R}^d$ accessible to the quantum mechanical system, it becomes more and more difficult to write the transition amplitude as a sum over some discrete group G and establish the validity of such a representation by combinatorical methods. Sometimes, however, the problem can be solved by an ansatz

$$\langle x', s'|x, s\rangle = \sum_{g\in G} \chi(g) K(x' - gx, s' - s) \qquad (x, x' \in F) \tag{2.9.25}$$

and by asking whether the following set of conditions is satisfied:

(A) G is a discrete subgroup of the full Euclidean group[37] in d dimensions and $F \in \mathbb{R}^d$ an open set with the following property: from $gx \in F$ with $g \in G$ and $x \in F$ it follows that $g = 1$, i.e., any orbit (under G) starting in F will never return to F.

(B) With $\bar{F}$ the closure of F in $\mathbb{R}^d$ we have $G\bar{F} = \mathbb{R}^d$, i.e., the images of $\bar{F}$ with respect to the action of G cover the entire space.

(C) The function $\chi : G \to \mathbb{R}$ obeys the relations $\chi(gg') = \chi(g)\chi(g')$, valid for all $g, g' \in G$, and $\chi(1) = 1$. What this amounts to may be phrased as: χ is a real character of the group G.

---

[36]This, of course, is the limit $\hbar \to 0$. However, in view of the preceding footnote, it is also the limit of vanishing de Broglie wavelength: $\lambda \to 0$. All that counts here is that $a_j = \lambda/(2L_j) \to 0$.

[37]This is the group of transformations of the d-dimensional space leaving distances unchanged.

(D) With ∂F the boundary[38] of F in $\mathbb{R}^d$ and $G(x) = \{\, g \in G \mid gx = x \,\}$ the following is true: if $x \in \partial F$ then $\sum_{g \in G(x)} \chi(g) = 0$.

Condition (A) and $\chi(1) = 1$ guarantee the correct behavior of the amplitude (2.9.25) if $s' = s$:

$$\langle x', s | x, s \rangle = \sum_{g} \chi(g) \delta(x' - gx) = \delta(x' - x) \quad (x, x' \in F). \tag{2.9.26}$$

Conditions (A–C) are designed so as to produce the Chapman–Kolmogorov equations:

$$\int_F dy \, \langle x', s' | y, t \rangle \langle y, t | x, s \rangle = \langle x', s' | x, s \rangle \tag{2.9.27}$$

$(x, x' \in F, \ s < t < s')$. To convince ourselves, we proceed as follows. For fixed x, x', s, s', t we define the function $f \colon \mathbb{R}^d \to \mathbb{R}$ by

$$f(y) = \sum_{h \in G} \chi(h) K(h^{-1} x' - y, s' - t) K(y - x, t - s). \tag{2.9.28}$$

Ignoring the subtle behavior on boundaries, the entire space is decomposed into disjoint regions kF $(k \in G)$. Therefore,

$$\sum_{k \in G} \int_{kF} dy \, f(y) = \int dy \, f(y). \tag{2.9.29}$$

Recalling that the function $K(x, s)$ indeed satisfies the Chapman–Kolmogorov equation, we can perform the integral on the right-hand side to obtain

$$\sum_{h \in G} \chi(h) K(h^{-1} x' - x, s' - s) = \sum_{h \in G} \chi(h) K(x' - hx, s' - s) \tag{2.9.30}$$
$$= \langle x', s' | x, s \rangle,$$

where we have used $|h^{-1} x' - x| = |x' - hx|$. The left-hand side of (2.9.29) can be reshaped by introducing new summation variables, $g = k^{-1}$ and $g' = hk$, and by taking $\chi(h) = \chi(g')\chi(g)$ (as implied by (C)) into account:

$$\int_F dy \sum_{g,g' \in G} \chi(g')\chi(g) K(x' - g'y, s' - t) K(y - gx, t - s) \tag{2.9.31}$$
$$= \int_F dy \, \langle x', s' | y, t \rangle \langle y, t | x, s \rangle.$$

For the final answer, use has been made of $|h^{-1} x' - ky| = |x' - g'y|$ and $|ky - x| = |y - gx|$. Having proved the two statements (2.9.26) and (2.9.27) we know now that the amplitude (2.9.25) is indeed the kernel of a semigroup of operators:

---

[38]No element of ∂F belongs to F.

$$\langle x', s' | x, s \rangle = \langle x' | e^{-(s'-s)H} | x \rangle \tag{2.9.32}$$

By a direct inspection of (2.9.25) we are assured that $H = -\frac{1}{2}\Delta$ holds within the interior of the fundamental domain F.

What can be said about the behavior at the boundary? We turn to condition (D) for help. This condition guarantees that the amplitude $\langle x', s' | x, s \rangle$ vanishes whenever $x \in \partial F$ or $x' \in \partial F$ (Dirichlet boundary condition). This is precisely what is needed to reconcile the mathematical model (2.9.25) with our intuitive idea of a confined physical system.

Example 4.   Consider a particle enclosed in an equilateral triangle $\triangle$ of side a. Think of the triangle as part of the plane $\mathbb{R}^2$ and take $x = 0$ as its center. The reflections σ_i $(i = 1, 2, 3)$ at the three bounding straight lines generate a reflection group G of planar transformations such that $(\sigma_i \sigma_k)^3 = 1$. As before, G is a discrete group of infinite order. By letting G act on the basic triangle $\triangle$, we obtain a collection of congruent triangles paving the plane. Their center points are the images of the origin and can be explicitly characterized as

$$n_1 e_1 + n_2 e_2, \qquad n_1, n_2 \in \mathbb{Z}, \qquad n_1 + n_2 \neq 2 \,(\mathrm{mod}\, 3). \tag{2.9.33}$$

Here, e_1 and e_2 are basis vectors of the plane satisfying

$$e_1^2 = e_2^2 = a^2/3, \quad e_1 \cdot e_2 = -a^2/6. \tag{2.9.34}$$

Suppose that the Hamiltonian is $H = -\frac{1}{2}\Delta + V$, where the potential

$$V(x) = \begin{cases} 0 & \text{if } x \in \triangle \\ \infty & \text{otherwise} \end{cases} \tag{2.9.35}$$

confines the particle as requested. Conditions (A–D) are taken care of by setting

$$\langle x' | e^{-\beta H} | x \rangle = \sum_{g \in G} \mathrm{sign}(g) K(x' - gx, \beta) \qquad (x, x' \in \triangle). \tag{2.9.36}$$

We agree on setting $\mathrm{sign}(g) = +1(-1)$ for $g \in G$ an even (odd) product of reflections σ_i. A simpler expresssion arises if $x = x' = 0$. At first,

$$\langle 0 | e^{-\beta H} | 0 \rangle = (2\pi\beta)^{-1} \sum_{n_1, n_2 = -\infty}^{\infty} \sigma(n_1 + n_2) \exp\left\{ -\frac{1}{2\beta}(n_1 e_1 + n_2 e_2)^2 \right\},$$

$$\tag{2.9.37}$$

where $\beta > 0$ and

$$\sigma(n) = \begin{cases} 1 & \text{if } n = 0 \,(\mathrm{mod}\, 3) \\ -1 & \text{if } n = 1 \,(\mathrm{mod}\, 3) \\ 0 & \text{if } n = 2 \,(\mathrm{mod}\, 3). \end{cases} \tag{2.9.38}$$

We also note that $(n_1 e_1 + n_2 e_2)^2 = a^2(n_1^2 + n_2^2 - n_1 n_2)/3$. Introduction of new summation variables, $m = n_1 - n_2$ and $n = n_1 + n_2$, furnishes us with the more convenient representation $\langle 0|e^{-\beta H}|0\rangle = (2\pi\beta)^{-1} B$, where

$$B = \sum_{\substack{n,m=-\infty \\ n+m=\text{even}}}^{\infty} \sigma(n) \exp\left\{-\frac{a^2}{24\beta}(n^2 + 3m^2)\right\}. \tag{2.9.39}$$

But the main trick is to write

$$\sigma(n) = \frac{\mathrm{Re}\,\exp(i2\pi n/3 + i\pi/6)}{\mathrm{Re}\,\exp(i\pi/6)} \tag{2.9.40}$$

so that $B = \mathrm{Re}(Ce^{i\pi/6})/\mathrm{Re}\,e^{i\pi/6}$ for

$$C = \sum_{\substack{n,m=-\infty \\ n+m=\text{even}}}^{\infty} \exp\left\{-\frac{a^2}{24\beta}(n^2 + 3m^2) + i\frac{2}{3}n\pi\right\}. \tag{2.9.41}$$

As C is real, $B = C$. To enforce the condition "$n + m =$even" we introduce the factor $\frac{1}{2}(1 + e^{i\pi(n+m)})$, thus arriving at

$$B = \sum_{n,m=-\infty}^{\infty} f(2\pi n, 2\pi m), \tag{2.9.42}$$

where f is considered a function of two real variables:

$$f(t,s) = \tfrac{1}{2}(1 + e^{i(t+s)/2}) \exp\left\{-\frac{a^2(t^2 + 3s^2)}{24\beta(2\pi)^2} + i\frac{t}{3}\right\}. \tag{2.9.43}$$

The idea here is to apply the Poisson summation formula

$$\sum_{n,m=-\infty}^{\infty} f(2\pi n, 2\pi m) = \sum_{n,m=-\infty}^{\infty} \tilde{f}(n,m), \tag{2.9.44}$$

where $\tilde{f}$ denotes the Fourier transform of f:

$$\tilde{f}(u,v) = \frac{1}{(2\pi)^2} \int_{-\infty}^{\infty} dt \int_{-\infty}^{\infty} ds\, e^{-i(ut+vs)} f(t,s). \tag{2.9.45}$$

To actually perform the Fourier transformation is an elementary task. The result of the calculation shows that the zero-to-zero amplitude can be brought into the following form:

$$\langle 0|e^{-\beta H}|0\rangle = \frac{2}{a^2}\sqrt{3} \sum_{\substack{m,n=-\infty \\ n+m=\text{even}}}^{\infty} e^{-\beta E_{mn}} \tag{2.9.46}$$

$$E_{mn} = 2\big(m^2 + 3(n - \tfrac{2}{3})^2\big)\left(\frac{\pi}{a}\right)^2$$

which is of particular interest since it realizes the spectral decomposition of the amplitude. It would be nice to obtain all eigenvalues E of the Hamiltonian H this way. From the decomposition (2.9.46), however, we can read off only eigenvalues E_{mn} corresponding to eigenfunctions that do not vanish at the origin. As regards the ground state of H, the condition is certainly fulfilled, and so its energy, as implied by (2.9.46), is

$$E_{00} = \frac{8}{3} \left(\frac{\pi}{a} \right)^2 . \tag{2.9.47}$$

The foregoing examples have shown how the geometry of the problem makes it possible to predict the structure of the solution. The purpose of the discussion was to emphasize the role the reflection group G played in connection with each fundamental domain[39]. The final answer to the problem involves some character χ of the group[40]. The remarkable observation to be made is that, in each individual case so far considered, the character function $\chi(g)$ on $g \in G$ coincides with the sign function[41] on the group, $\mathrm{sign}(g) = \pm 1$.

Exercise 1.   To calculate the ground state energy for the equilateral triangle of side a, student A used the following cut-and-paste strategy: he first cut the triangle into two equal halves and then reassembled the pieces so as to form a rectangle of sides $b = a/2$ and $c = \sqrt{3}\,a/2$. For the ground state of a Schrödinger particle enclosed in the box $0 \le x \le b$, $0 \le y \le c$, he immediately wrote down the product wave function $\sin(\pi x/b)\sin(\pi y/c)$ and got its energy as

$$E = \frac{1}{2} \left(\left(\frac{\pi}{b} \right)^2 + \left(\frac{\pi}{c} \right)^2 \right) = \frac{8}{3} \left(\frac{\pi}{a} \right)^2 ,$$

claiming that this be the answer for the triangle, too. Student B, not at all impressed by this calculation, claims that the agreement with (2.9.47) ought to be regarded as an accident. Who is right?

Exercise 2.   Let G be the group generated by the n reflections at n orthogonal hyperplanes such that $n \le d$. Show that this group is Abelian and of the order 2^n. Moreover, it is isomorphic to the product group $\mathbb{Z}_2 \times \mathbb{Z}_2 \times \cdots \times \mathbb{Z}_2$ with n factors. Prove that

---

[39]A group of transformations of a Euclidean space is called a *reflection group* if it is generated by a set of reflections. In the simplest of all cases, i.e., for the halfspace, the reflection group has two elements and, hence, is isomorphic to $\mathbb{Z}_2$.

[40]A character χ is simply a one-dimensional representation of the group G. So it satisfies $\chi(gg') = \chi(g)\chi(g')$ and $\chi(1) = 1$. In the present context, characters have to be real.

[41]The function $\mathrm{sign}(g)$ takes the value $+1$ if g is a product of an even number of reflections and -1 otherwise.

$$\langle x'|e^{-tH}|x\rangle = \sum_{g\in G} \text{sign}(g)\, K(x' - gx, t) \qquad (x, x' \in A_+),$$

where A_+ denotes the intersection of n halfspaces connected with the hyperplanes (in suitably chosen coordinates: $x_1 > 0, \ldots, x_n > 0$). The Hamiltonian is $H = -\frac{1}{2}\Delta + V$, where the potential V is zero in A_+ and $+\infty$ otherwise.

Exercise 3. Consider a wedge $A_+ \subset \mathbb{R}^2$ that is bounded by two straight lines meeting at an angle $2\pi/n$. Determine the group of reflections and the transition amplitude of a particle confined to the wedge.

Exercise 4. Let $d\mu$ be the conditional Wiener measure of one-dimensional Brownian paths $\omega: (0,0) \rightsquigarrow (0, t)$. Then $\sup \omega$ is a random variable estimating excursions from $x = 0$ into the region $x > 0$ during a period of length t. Prove by an explicit calculation that

$$\int d\mu(\omega)\, (\sup \omega)^2 \Big/ \int d\mu(\omega) = t/2.$$

2.10 Feynman Versus Wiener Integrals

Path integration or *functional integration* was introduced into physics, for the first time and in a loose sense, by Dirac [2.34] and then by Feynman [2.35]. The idea has become only slowly absorbed because it was so hard to grasp, and it is only in our own time that we have come to understand at all appreciatively *Feynman integrals* as they are now called. Also the influence of their poetry has been immeasurable.

Yet, our pleasure is marred owing to the radical defect of mathematical justification: it is hopeless to proceed by picking out a solved integral here and another there, arriving at some preconceived idea as to what mathematics it might refer to and then applying it as a kind of axiomatic setup to which the rest must fit.

The mathematical and physical questions Feynman's idea raised have been immense and provided many puzzles for posterity. The origin of these has always been the emphasis on the Schrdinger equation rather than on the diffusion equation. So the resulting integrals were of the "oscillatory type" with no probability measure lurking behind the scene. Naturally, the concepts that are easiest to deal with are those of measure theory and probability theory: they are of no avail here.

Despite the severe lack of mathematical rigor, the Feynman integral not only proved to be a useful tool with regard to applications, but it also provided a challenging alternative formulation of quantum mechanics itself, as opposed to the traditional operator formalism.

2.10.1 Summing over Histories in Configuration Space

A heuristic discussion of the Feynman integral starts from the classical action along a trajectory $q(t)$ in d dimensions[42],

$$S\{q\} = \int_0^t ds\, L\big(\dot{q}(s), q(s)\big) \qquad (t \text{ fixed}) \tag{2.10.1}$$

considered as a functional, where $L(\dot{q}, q)$ is the classical Lagrangian. A sufficiently general class of models to choose from would be

$$L(\dot{q}, q) = \tfrac{1}{2}\dot{q}^2 + \dot{q} \cdot A(q) - V(q). \tag{2.10.2}$$

Then quantum time evolution is described by a unitary group $e^{-(i/\hbar)tH}$ whose kernel – also called the *propagator* of the system – is written formally as a *sum over histories*:

$$\langle x'|e^{-(i/\hbar)tH}|x\rangle = \int_{(x,0)}^{(x',t)} dq\, e^{(i/\hbar)S\{q\}}. \tag{2.10.3}$$

In fact, one aims to "integrate" over all trajectories $q\colon (x,0) \rightsquigarrow (x',t)$ with respect to the complex 'measure' dq. To give meaning to dq and to the Feynman integral (2.10.3) as a whole, one resorts to the method of time slicing (also called the *sequential definition* or Feynman's *polygonal approximation*). On writing the time integral (2.10.1) as a sum over discrete times $t_k = kt/n$, i.e.,

$$S_n\{q\} = \sum_{k=1}^{n} (t/n)L(\dot{q}_k, q_k), \quad \dot{q}_k = \frac{q_k - q_{k-1}}{t/n}, \quad q_k = q(t_k), \tag{2.10.4}$$

one "defines"

$$\langle x'|e^{-(i/\hbar)tH}|x\rangle = \lim_{n\to\infty} \int_{\mathrm{I\!R}^{nd}} d^n q\, e^{(i/\hbar)S_n\{q\}}. \tag{2.10.5}$$

Here, the approximate complex measure is

$$d^n q = \prod_{k=1}^{n-1} (iht/n)^{-d/2} dq_k, \tag{2.10.6}$$

---

[42]The term *trajectory* is reserved here for Feynman paths as opposed to Brownian paths. The difference arises because the concept of time is different in both theories. As for Feynman integrals, *time* always means *physical time*: it is also the sense in which Newton, Heisenberg, and Minkowski used the term. In Sect. 1.5, it was called *imaginary time*.

where dq_k is the ordinary d-dimensional Lebesgue measure and $h = 2\pi\hbar$. The propagator so obtained appears as a limit of finite-dimensional integrals[43]. How can this rule of calculation be justified? Answer: it arises from the Lie–Trotter formula[44] and the fact that, with $H_0 = -\frac{1}{2}\hbar^2\Delta$, the kernel of $\exp\{-(i/\hbar)(t/n)H_0\}$ is

$$\langle x'|e^{i\hbar(t/n)\Delta/2}|x\rangle = (iht/n)^{-d/2}\exp\left\{\frac{it}{2\hbar n}\left(\frac{x'-x}{t/n}\right)^2\right\}. \qquad (2.10.7)$$

There are certain obstacles and shortcomings of the Feynman integral[45] as opposed to the Wiener integral:

1. Even for finite n, the right-hand side of (2.10.5) represents an *improper* integral in the sense that the integrand lacks absolute integrability though the definition makes sense granted certain conditions on the classical Lagrangian. We refer the reader to [2.46] for a formulation that avoids this misbehavior by giving each particle a *complex* mass.

2. The real embarrassment comes from the fact that the sequence of approximating complex measures, $d^n q$, fails to approach a limit measure dq (the main difficulty here is: measure on what?).

3. There is a latent assumption in the approach that there be a class of admissible trajectories in the limit when $n \to \infty$. But what is this class? Whatever it is, trajectories are not *random functions*.

4. There is nothing in dq_n nor in the phase $\exp\{(i/\hbar)S_n\}$ that may prevent a trajectory $q(s)$ (or rather its polygonal approximation) to go astray, thus causing the integrand in (2.10.5) to oscillate wildly: Feynman integrals simply have no built-in damping mechanism whatsoever.

5. If we let the potential $V(q)$ tend to $+\infty$ in some region of space, the integrand in (2.10.5) will vary more and more rapidly: it is solely the effect of interference that will wipe out the contribution of such a region. Individual trajectories enter large-V regions freely, and, what is worse, it does not make any difference (as regards the behavior of trajectories) whether the potential assumes large positive or large negative values. By contrast, ordinary reasoning says that changing the sign of a potential may turn a physically stable situation into an unstable one. It is not at all clear how this is reflected by the Feynman integral.

6. Even for smooth potentials (e.g., the harmonic oscillator), the kernel $\langle x'|\exp\{-(i/\hbar)tH\}|x\rangle$ does not exist for all x, x', and t. This problem however can be overcome by integrating with wave packets. The Feynman integral is then written, not for the kernel, but for $\psi_t = \exp\{-(i/\hbar)tH\}\psi_0$, and convergence is claimed in the L^2-sense only [2.36].

---

[43]Compare the sequential definition (2.1.19) of the Wiener integral.
[44]Details may be found, for instance, in the opening chapter of Schulman's book [2.12].
[45]For rigorous work addressing some of these questions see [2.36–45].

At this point it is worth pointing out that Feynman integrals and Wiener integrals also share a difficulty (that may be overcome). The origin of it is a well known topic from standard quantum mechanics: when dealing with vector potentials $A(q)$, the term $\dot{q}A(q)$ is ambiguous (not in classical mechanics, of course, but in quantum mechanics). The ambiguity follows from the fact that the momentum and the position are noncommuting observables by necessity and it forces us to make a choice. The ambiguity arises in the context of path integration as well (as it should) and is reflected by a rather strange phenomenon. The vector potential contribution to the classical action is $\int dq \cdot A(q)$, and time slicing converts the integral into the sum

$$\sum_{k=1}^{n} (q_k - q_{k-1}) \cdot A(q_k). \tag{2.10.8}$$

At least, this is what we have suggested in (2.10.4). Why evaluate $A(x)$ at the endpoint q_k? Do we get the the same result if we evaluate the vector potential at q_{k-1} or at some point in between and later on pass to the limit $n \to \infty$? The answer is no. The reason: the error terms don't go to zero as they would for a Riemann sum. From the physics point of view, the correct way is to adopt the midpoint rule, that is to take $x = \frac{1}{2}(q_k + q_{k-1})$ and to write

$$\sum_{k=1}^{n} (q_k - q_{k-1}) \cdot A\big(\tfrac{1}{2}(q_k + q_{k-1})\big) \tag{2.10.9}$$

instead of (2.10.8). We shall come back to this question in Chap. 6. Both the warning and the prescription apply to Feynman integrals and Wiener integrals alike.

2.10.2 The Method of Stationary Phase

In analysis, the *method of stationary phase* looks for the asymptotic behavior of oscillatory integrals. In the simplest (one-dimensional) case, it is exemplified by the assertion that

$$\int_{\mathbb{R}} dq \, e^{it\alpha(q)} \xrightarrow[t\to\infty]{} \left(\frac{2\pi i}{t\alpha''(q_0)}\right)^{1/2} e^{it\alpha(q_0)}, \qquad \alpha'(q_0) = 0. \tag{2.10.10}$$

The tacit assumption here is that the real function $\alpha(q)$ has continuous second order derivatives. Moreover, it is required that there be a unique point q_0 where the first, but not the second, derivative vanishes.

The above assertion says that, as t gets large, the dominant contribution comes from regions where the phase $\alpha(q)$ is stationary; for this to be true it is irrelevant whether q_0 corresponds to a maximum or a minimum. The asymptotic expression in (2.10.10) is derived from expanding the function

$\alpha(q)$ into a Taylor series to second order around q_0, dropping cubic and higher order terms. There are easy extensions of the formula (2.10.10) to situations where there are several stationary points in the domain of integration and when the integral is n-dimensional.

In the context of Feynman integrals, the role of parameter t is played by $\hbar^{-1}$. Hence, the method of stationary phases seeks the dominant contribution to the 'sum over histories' in the classical limit, i.e., when Planck's constant goes to zero, the only difference being that the integral is of infinite dimension. Specifically, if the function $\alpha(q)$ is replaced by the functional $S\{q\}$, stationarity means that one looks for a trajectory $q(t)$ obeying the classical Euler–Lagrange equations,

$$\frac{d}{dt}\left(\frac{\partial L}{\partial \dot{q}}\right) - \left(\frac{\partial L}{\partial q}\right) = 0. \tag{2.10.11}$$

So the message is this: close to the classical limit, the dominant contribution to the Feynman integral comes from the classical path. To actually calculate this contribution is the subject of the so-called WKB approximation which we won't go into now, except to say that it calls for an extension of the method of stationary phase to infinite dimensions[46].

2.10.3 Summing over Histories in Phase Space

To someone committed to the Hamiltonian formalism of classical mechanics it would seem natural to consider histories in *phase space* rather than in configuration space. Feynman was first in persuing this idea, which he published in 1951 as an appendix to work on quantum electrodynamics [2.47]. The subject was again taken up by Tobocman [2.48] in 1956, and since then the idea developed slowly until more serious attempts were made to put the theory on a firm ground [2.49–55].

The ultimate goal of functional integration based on trajectories in phase space is *geometric quantization*, i.e., a formulation of how one should quantize a classical system without having to specify the coordinates of the underlying manifold. However, in spite of all the effort in this direction, the role of canonical transformations in functional integration is not yet clear, especially in view of the fact that the group of classical symmetries differs considerably from the group of quantum symmetries[47]. Ambiguities due to operator ordering are not so easily detected in Feynman's phase space integrals. But they do arise and are analogous to (and often worse than) those occuring in Feynman's configuration space integrals when dealing with

---

[46]See [2.40] and the literature quoted therein for a rigorous theory.

[47]Weyl's construction associates to any real function on phase space (a classical observable) some self-adjoint operator on the Hilbert space of states, i.e., a quantum version of the same observable. Yet, the Lie algebra of classical observables (with respect to the Poisson bracket) is by no means isomorphic to its quantum mechanical counterpart, i.e., the Lie algebra of self-adjoint operators (with respect to the commutator)[2.56].

vector potentials. Therefore, our confidence in the validity of formal manipulations, e.g., canonical transformations applied to the integrand, is greatly impaired[48] by existent innate ambiguities.

Assume that the classical system has d degrees of freedom. So its phase space is $2d$-dimensional. Trajectories in phase space are pairs of functions[49]: $p(s), q(s)$, where $0 \leq s \leq t$, such that $q(0) = x$ and $q(t) = x'$. The propagator is formally written as a *sum over histories in phase space*:

$$\langle x' | e^{-(i/\hbar)tH} | x \rangle = \int dpdq \, \exp\left\{ (i/\hbar)\int_0^t ds \left(p\dot{q} - \hat{H}(p,q) \right) \right\}, \qquad (2.10.12)$$

where $\hat{H}(p,q)$ is the classical Hamiltonian, e.g.,

$$\hat{H}(p,q) = \tfrac{1}{2}\left(p - A(q) \right)^2 + V(q). \qquad (2.10.13)$$

There is a bewildering wealth of possible interpretations of the integral in (2.10.12). We mention just one of them: upon choosing discrete times $t_k = kt/n$ and phase space variables $p_k = p(t_k)$, $q_k = q(t_k)$, the "measure" $dpdq$ becomes (essentially) the product of Liouville measures:

$$h^{-d}dp_n \prod_{k=1}^{n-1} h^{-d}dp_k dq_k. \qquad (2.10.14)$$

Notice that this measure is no longer complex, but real-positive. Also, phase space volumes are measured in units of $h = 2\pi\hbar$. The action integral $\int_0^t ds \, (p\dot{q} - \hat{H})$ is interpreted as

$$\sum_{k=1}^{n} \frac{t}{n} \left\{ p_k \frac{q_k - q_{k-1}}{t/n} - \hat{H}\left(p_k, \tfrac{1}{2}(q_k + q_{k-1}) \right) \right\}. \qquad (2.10.15)$$

To justify the discretized version of (2.10.12), one performs the pseudo-Gaussian integration with respect to $p_1, \ldots, p_n$ to arrive at (2.10.5).

Notice that the midpoint rule is still effective in (2.10.15). Moreover, there are several other conventions which seem to be hidden in the action integral. Altering these conventions often changes the answer in an undesired direction with little hope that the decrepancy will disappear in the limit $n \to \infty$. Therefore, it should not come as a surprise that the result may come out plain wrong when a canonical transformation is applied to the action integral without paying attention to how it affects the implicit conventions. It is mainly for this reason that the theory of *summing over histories in phase space* has not not yet matured to the same degree as the theory of *summing over histories in configuration space*.

---

[48]See, however, the careful discussion by Daubechies and Klauder [2.54] using the coherent-state representation.

[49]In Feynman's formulation, the momentum $p(t)$ is not fixed at the endpoints of the time interval. In the coherent-state representation, however, one fixes $p(0) = p$ and $p(t) = p'$.

It it worthwhile mentioning that Feynman integrals have also been considered in polar coordinate systems (leading to the notion of the "radial Feynman integral"), in curved spaces, and in homogeneous spaces G/H, where G and H are groups [2.57–60].

2.10.4 The Feynman Integrand as a Hida Distribution

One rigorous method of defining oscillatory Feynman integrals uses white-noise calculus [2.74-76]. The basic idea is this: since the main obstacle in interpreting (2.10.3) is the absence of any "flat" measure in infinite dimensions, one takes the (normalized) Gaussian measure of the Wiener process, $d\mu(\omega)$, as the reference measure and compensates for the change by writing

$$\langle x'|e^{-itH}|x\rangle = \int d\mu(\omega)\, I(\omega) \exp\left\{-i \int_0^t ds\, V(x + \omega(s))\right\}, \qquad (2.10.15)$$

where $H = -\frac{1}{2}\Delta + V$, $\hbar = 1$, and

$$I(\omega) = \delta(\omega(t) - x' + x) \exp\left\{\frac{1+i}{2} \int_0^t ds\, \dot\omega(s)^2\right\}. \qquad (2.10.16)$$

Though $I(\omega)$ is not an ordinary white noise functional, it makes sense as a Hida distribution (see [2.77] for an outline of the general theory). To understand this better and to make the construction more precise let the quantum system have n degrees of freedom and consider the Gel'fand triple

$$(\mathcal{S}_n) \subset (L_n^2) \subset (\mathcal{S}_n)' \qquad (2.10.17)$$

consisting of three spaces:

1. $(L_n^2) = L^2(\mathcal{S}_n', d\mu)$ is the Hilbert space of square-integrable functionals $\phi: \mathcal{S}_n' \to \mathbb{C}$, where $\mathcal{S}_n'$ (the dual of $\mathcal{S}_n$) designates the space of Schwartz distributions $\omega: \mathbb{R} \to \mathbb{R}^n$. Here, $\mathbb{R}$ and $\mathbb{R}^n$ are interpreted as *time* and *configuration space* respectively. Any $\phi \in (L_n^2)$ may be decomposed as

$$\phi(\omega) = \sum_{k=0}^{\infty} \phi_k(\omega) = F_0 + \sum_{k=1}^{\infty} :\omega^{\otimes k}:(F_k) \qquad (F_0 \in \mathbb{C}) \qquad (2.10.18)$$

with L^2–kernels $F_k: \mathbb{R}^k \to \mathbb{C}^{nk}$ such that $\sum_k k! \|F_k\|^2 < \infty$. The kth Wick power is given by the Hermite polynomial $:\omega(f)^k: =: \omega^{\otimes k}:(f^{\otimes k})$, where $f: \mathbb{R} \to \mathbb{C}^n$ is square-integrable, and

$$\sum_{k=0}^{\infty} \frac{i^n}{n!} :\omega(f)^n: \; = \; :\exp i\omega(f): \; = \; \frac{\exp i\omega(f)}{\int d\mu(\omega)\, \exp i\omega(f)} \qquad (2.10.19)$$

(see Exercise 1 of Sect. 1.8). Equation (2.10.18) establishes an isomorphism between (L_n^2) and the (symmetric) Fock space over $L^2(\mathbb{R}, \mathbb{C}^n)$.

2. $(\mathcal{S}_n)$ is a dense subspace of (L^2) consisting of functionals ϕ whose components F_k are test functions in the sense of Schwartz and nonzero only for a finite number of k's. Elements of $(\mathcal{S}_n)$ are called *Hida test functions*. Notice that the constant functional $\phi(\omega) = 1$ belongs to $(\mathcal{S}_n)$. Notice also that $(\mathcal{S}_n)$ is in fact an algebra with respect to pointwise multiplication of functionals.

3. $(\mathcal{S}_n)'$ is the dual of $(\mathcal{S}_n)$. Elements of $(\mathcal{S}_n)'$ are called *Hida distributions*. Any $\phi \in (\mathcal{S}_n)'$ admits a decomposition (2.10.18) such that each kernel F_k is a (complex) Schwartz distribution.

Functionals ϕ may be characterized either by their *S–transform*

$$(S\phi)(f) = \int d\mu(\omega)\, \phi(\omega) \colon \exp \omega(f) \colon \, , \qquad (2.10.20)$$

or by their *T–transform*

$$(T\phi)(f) = \int d\mu(\omega)\, \phi(\omega) \exp i\omega(f) \, , \qquad (2.10.21)$$

Donker's delta function $\phi(\omega) = \delta(\omega(t) - a)$ where $a \in \mathbb{R}^n$ and $t > 0$, which occurs in (2.10.16), provides a particular interesting example of a Hida distribution, whose S–transform may be calculated:

$$(S\phi)(f) = (2\pi t)^{-n/2} \exp\left\{ -\frac{1}{2t}\left(a - \int_0^t ds\, f(s) \right)^2 \right\} . \qquad (2.10.22)$$

For a discussion of white noise calculus as applied to Feynman integrals and of T–transforms see [2.78].

3 The Brownian Bridge

Associated with a potential V and any Brownian path $\omega\colon (x,s) \rightsquigarrow (x',s')$ there is the integral

$$I = \int_s^{s'} dt\, V\big(\omega(t)\big),$$

which enters the expression for the transition amplitude of a Schrdinger particle moving in that potential. It is often desirable to study the integral I as a function of the time variable s', and sometimes the problem arises of determining the derivative dI/ds'. Then a natural guess would be that the derivative of the above integral is the integrand taken at $t = s'$, and hence $V\big(\omega(s')\big)$. A moment's reflection, however, reveals that the answer is totally false, since $\omega(s') = x'$, suggesting the result $dI/ds' = V(x')$, a constant with respect to s', which is embarrassing. So what went wrong? Why is manipulating the integral so prone to error?

The puzzle is resolved once we realize that each path ω has a hidden dependence on the time variable s', a fact that is in no way reflected in the above expression for the integral I and, consequently, was overlooked in our premature computation. Question: is there a better way to do the calculation? Yes, there is, and the main prerequisite for understanding how and why the variable s' enters the integrand is a knowledge of the *scaling method*. This method is designed so as to exhibit the complete dependence of I on x, x', s, and s'. The next section introduces the necessary notions.

3.1 The Canonical Scaling of Brownian Paths

If we define the most probable path by

$$\bar{x}(t) = \int d\mu_{x,s}^{x',s'}(\omega)\,\omega(t) \Big/ \int d\mu_{x,s}^{x',s'}(\omega) \tag{3.1.1}$$

$(s < t < s')$, a straightforward calculation[50] shows that $\bar{x}(t)$ is in fact the straight-line path from x to x':

$$\bar{x}(t) = \frac{(s' - t)x + (t - s)x'}{s' - s}. \tag{3.1.2}$$

So the most probable path looks as you would expect it to. Moreover, the straight line is traversed by the Brownian particle with constant velocity and may be viewed as the path of classical geometric optics.

The result (3.1.2) then suggests decomposing any Brownian path ω : $(x, s) \rightsquigarrow (x', s')$ in the following manner:

$$\omega(t) = x + (x' - x)\tau + \ell\bar{\omega}(\tau) \tag{3.1.3}$$
$$t = s + (s' - s)\tau \qquad (0 \le \tau \le 1) \tag{3.1.4}$$
$$\ell^2 = s' - s > 0, \tag{3.1.5}$$

where $\bar{\omega}(\tau)$ is a path satisfying $\bar{\omega}(0) = \bar{\omega}(1) = 0$ and describing the deviation from the straight-line path as depicted in Fig. 3.1. The factor ℓ, a parameter with the dimension of a length, serves to normalize the random path $\bar{\omega}$. The set of $\bar{\omega}$'s coincides with the set of sample paths of a some Gaussian random process called the *Brownian bridge*, which we shall describe in a minute. Since we claim that the Brownian bridge is independent of the parameters x, x', s, and s', so is each sample path $\bar{\omega}$. By writing (3.1.3–5) we have achieved our goal of exhibiting the dependence of any unscaled path $\omega(t)$ on these parameters.

As for the decomposition (3.1.3), certain aspects of it ought to be kept in mind:

1. Though the parameter τ adopts the role of time, it does not carry any physical dimension: it is merely a number between 0 and 1 independent of the inherent time scale of the problem.
2. The formulas are usually written such that they hide their dependence on the characteristic mass m of the system and of Planck's constant $\hbar$. Readers who find our definition (3.1.5) of the length parameter ℓ obscure may find its general form

$$\ell^2 = \hbar(s' - s)/m \tag{3.1.6}$$

   comprehensible. The parameter ℓ sets the scale of random deviations from the most probable path (i.e., the straight line). Naturally, there are large deviations if the difference $s' - s$ becomes large. The Brownian bridge is constructed such that all its coordinates carry no physical dimension whatsoever. They are thus unaffected by the choice of scales.

---

[50]For fixed t, write each ω as a concatenated path $(x, s) \rightsquigarrow (y, t) \rightsquigarrow (x', s')$ with $y \in \mathbb{R}^d$. Then represent $\int d\mu(\omega)\,\omega(t)$ as an integral with respect to y and perform the y-integration.

3. For m large, or if $\hbar \to 0$, that is to say, in the classical limit, we may regard $\ell = 0$ as a reasonable approximation. This then has the following obvious effect: no matter which path of the Brownian bridge is chosen, the original path $\omega(t)$ reduces to its linear component: $\omega(t) = x + (x' - x)\tau$. As Planck's constant is turned on, the path $\omega(t)$ starts to fluctuate around the straight line connecting x and x'. Therefore, one of the goals of the scaling method is to isolate quantum fluctuations from the rest, i.e., to represent such fluctuations by summing over paths of the Brownian bridge[51].

4. With regard to applications in statistical mechanics, it is perfectly legitimate to interpret the length parameter ℓ in a different way: one simply replaces $s'-s$ by $\hbar(k_B T)^{-1}$ in the above formulas, where k_B and T designate the Boltzmann constant and the temperature respectively, so that

$$\ell^2 = \hbar^2/(mk_B T). \tag{3.1.7}$$

Here, the classical limit ($\hbar \to 0$) coincides with the high-temperature limit ($T \to \infty$). Frequently, one writes $\lambda = (2\pi)^{1/2}\ell$ and calls λ the *thermal wavelength* of the system.

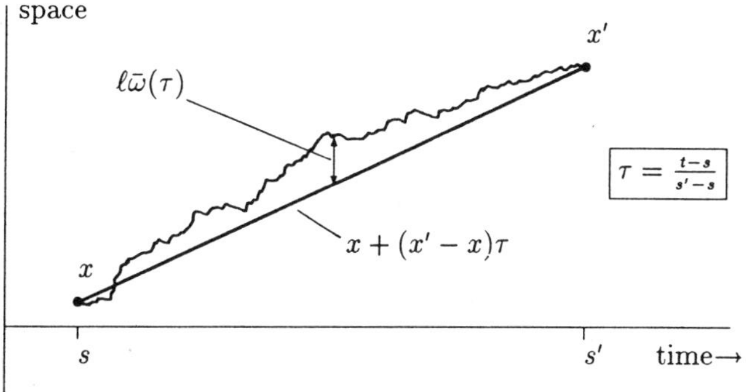

Fig. 3.1. The decomposition of a Brownian path into a linear part and a sample path $\bar{\omega}$ of the Brownian bridge

---

[51]This has to be contrasted with the method of *stationary phase* applied to Feynman integrals (see Sect. 2.10 for details), which predicts that, as $\hbar \to 0$, the integral centers around the classical path, i.e., the solution of the classical equation of motion.

3.1.1 The Process $\bar{X}_\tau$

The probabilistic idea behind the decomposition (3.1.3) consists of two basic steps: (1) if X_t is normalized d-dimensional Brownian motion starting at $X_0 = 0$, a standard scaling argument says that

$$X_\tau^* = \ell^{-1}\big(X_{s+\tau\ell^2} - X_s\big) \tag{3.1.8}$$

is again normalized Brownian motion starting at $X_\tau^* = 0$; (2) if one considers the Gaussian process

$$\bar{X}_\tau = X_\tau^* - \tau X_1^* \qquad (0 \le \tau \le 1), \tag{3.1.9}$$

it is easily verified that this process has mean zero and covariance

$$\boldsymbol{E}(\bar{X}_\tau \bar{X}_{\tau'}) = \big(\min(\tau,\tau') - \tau\tau'\big)\mathbb{1} \tag{3.1.10}$$

which characterizes $\bar{X}_\tau$ completely. In a sense, the new process describes *circular* Brownian motion on $\mathbb{R}^d$ since $\bar{X}_0 = \bar{X}_1 = 0$ and is called the *Brownian bridge*.

Now, if $\bar{\omega}(\tau)$ is a sample path of $\bar{X}_\tau$, the definitions (3.1.8) and (3.1.9) show that, whenever $\ell^2 = s' - s$ and $t = s + \tau(s' - s)$, the function $\omega(t) = x + (x' - x)\tau + \ell\bar{\omega}(\tau)$ is a sample path of X_t satisfying $\omega(s) = x$ and $\omega(s') = x'$, and vice versa.

The Green's function $G(\tau,\tau') = \min(\tau,\tau') - \tau\tau'$ arises in the theory of the classical vibrating string: τ and τ' represent coordinates of a string of length 1 which is held fixed at the positions $\tau = 0$ and $\tau = 1$. The differential equation

$$f''(\tau) + \lambda f(\tau) = 0, \qquad\qquad f(0) = f(1) = 0 \tag{3.1.11}$$

has eigenvalues $\lambda_n = \pi^2 n^2$ ($n \in \mathbb{N}$) and eigenfunctions $\phi_n(\tau) = \sqrt{2}\sin(\pi n\tau)$ such that[52]

$$\langle\tau|D^{-1}|\tau'\rangle = \sum_{n=1}^{\infty} \lambda_n^{-1}\phi_n(\tau)\phi_n\tau') = G(\tau,\tau'), \tag{3.1.12}$$

where the operator D on $L^2(0,1)$ is the expression $-d^2/d\tau^2$ supplemented by the Dirichlet boundary conditions $f(0) = f(1) = 0$. Recall from classical mechanics that eigenvalue problem (3.1.11) is equivalent to

$$f(\tau) = \lambda \int_0^1 d\tau'\, G(\tau,\tau')f(\tau'). \tag{3.1.13}$$

Yet, the relationship between the vibrating string and the Brownian bridge is entirely formal.

---

[52]See (4.1.5) for a slightly more explicit calculation.

3.1.2 Rescaling of Path Integrals

The chief objective of the scaling method is to reduce the problem of summing over *unscaled* paths $\omega\colon (x,s) \rightsquigarrow (x',s')$ to the problem of summing over *scaled* paths $\bar{\omega}\colon (0,0) \rightsquigarrow (0,1)$ (i.e., sample paths of the Brownian bridge $\bar{X}_\tau$). The conditional Wiener measure for unscaled paths is $d\mu_{x,s}^{x',s'}(\omega)$. As for scaled paths, it is $d\mu_{0,0}^{0,1}(\bar{\omega})$. What sense does it make to say that the two measures coincide *up to scaling*? The answer may be expressed as

$$d\mu_{x,s}^{x',s'}(\omega)\big/ \textstyle\int d\mu_{x,s}^{x',s'}(\omega) = d\mu_{0,0}^{0,1}(\bar{\omega})\big/ \int d\mu_{0,0}^{0,1}(\bar{\omega}), \tag{3.1.14}$$

or to put it differently:

$$d\mu_{x,s}^{x',s'}(\omega) = K(x' - x, s' - s)d\bar{\omega} \tag{3.1.15}$$

$$d\bar{\omega} = (2\pi)^{d/2} d\mu_{0,0}^{0,1}(\bar{\omega}) \tag{3.1.16}$$

(since $K(0,1) = (2\pi)^{-d/2}$). Here we have found it convenient to introduce the measure $d\bar{\omega}$, normalized to 1, of the Brownian bridge. This has the advantage that, whenever we encounter a path integral $\int d\bar{\omega}\, f(\bar{\omega})$, it may also quite naturally be written as an expectation:

$$\int d\bar{\omega}\, f(\bar{\omega}) \equiv \boldsymbol{E}(f(\bar{X})). \tag{3.1.17}$$

If $\bar{\omega}$ is some path of the Brownian bridge, so is $-\bar{\omega}$ and $-\bar{\omega}$ is assigned the same weight. Therefore,

$$\int d\bar{\omega}\, f(\bar{\omega}) = \int d\bar{\omega}\, f(-\bar{\omega}). \tag{3.1.18}$$

Suppose we split the functional f into an even and an odd part. Then only the *even part* will contribute to the path integral $\int d\bar{\omega}\, f(\bar{\omega})$, which sometimes may have beneficial effects.

Occasionally, we shall be dealing with a time-dependent potential $V(x,t)$. Then the Feynman–Kac formula may be written in the following generalized form, where scaling has already been applied to the integrand:

$$\langle x',s'|x,s\rangle = K(x' - x, s' - s) \int d\bar{\omega}\, \exp\left\{-\int_0^1 d\tau\, \bar{V}(\bar{\omega}(\tau),\tau)\right\}$$

$$= K(x' - x, s' - s)\, \boldsymbol{E}\left(\exp\left\{-\int_0^1 d\tau\, \bar{V}(\bar{X}_\tau,\tau)\right\}\right). \tag{3.1.19}$$

This construction brings with it the scaled potential $\bar{V}$, defined by

$$\bar{V}(\xi,\tau) = (s' - s)V(y,t) \qquad\qquad (\xi, y \in \mathbb{R}^d) \tag{3.1.20}$$

$$y = x + (x' - x)\tau + \ell\xi \tag{3.1.21}$$

$$t = s + (s' - s)\tau. \tag{3.1.22}$$

As is obvious form these formulas, we treat x, x', s, and s' as fixed quantities and y, ξ, t, and τ as variables.

The introduction of potentials dependent explicitly on time lacks motivation at this point. At a later stage, however, it will become necessary to allow for such an extension (see Sect. 4.3 on coupled systems for motivation). In view of the prescription (3.1.20) little is gained if we were to restrict ourselves to potentials V constant in t since the scaled potential $\bar{V}$ would be τ-dependent anyway.

3.1.3 The Stochastic Integral with Respect to the Brownian Bridge

The derivative $\bar{X}_\tau/d\tau$ of the Brownian Bridge can only be defined in a generalized sense, and its study parallels our former discussion of *white noise* as a derivative of the Wiener process (Sect. 1.7).

For simplicity, the subsequent formulas apply to the simple case $d = 1$ only: this allows us to neglect the vector character of $\bar{X}_\tau$. The generalization to $d > 1$ is left as an exercise.

The main thing to settle is that the derivative of the covariance exists in a generalized sense. Indeed,

$$\frac{\partial}{\partial\tau}\frac{\partial}{\partial\tau'}\big(\min(\tau,\tau') - \tau\tau'\big) = \delta(\tau - \tau') - 1. \tag{3.1.23}$$

Formally, the derivative $\bar{W}_\tau = d\bar{X}_\tau/d\tau$ has the covariance

$$\boldsymbol{E}(\bar{W}_\tau\bar{W}_{\tau'}) = \delta(\tau - \tau') - 1 \qquad (\tau,\tau' \in [0,1]). \tag{3.1.24}$$

This essentially says that after integration with real functions $f \in L^2(0,1)$ the Gaussian random variable

$$\bar{W}(f) := \int_0^1 f(\tau)\bar{W}_\tau d\tau \equiv \int_0^1 f(\tau)d\bar{X}_\tau \tag{3.1.25}$$

exists. It will be called the *stochastic integral* with respect to the Brownian bridge and is characterized by the following defining properties:

$$\boldsymbol{E}(\bar{W}(f)) = 0, \qquad \boldsymbol{E}(\bar{W}(f)^2) = \mathrm{var}(f), \tag{3.1.26}$$

where

$$\mathrm{var}(f) := \int_0^1 d\tau\, f(\tau)^2 - \left(\int_0^1 d\tau\, f(\tau)\right)^2. \tag{3.1.27}$$

In short, the heart of the matter is in the statement that the generalized process $f \to \bar{W}(f)$ has the generating functional

$$\boldsymbol{E}\big(\exp\{i\bar{W}(f)\}\big) = \exp\{-\tfrac{1}{2}\mathrm{var}(f)\}. \tag{3.1.28}$$

There are occasions on which we shall need this formula.

Exercise 1.   For $\bar{X}_\tau$ the Brownian bridge, prove that

$$X_t = (1-\tau)^{-1}\bar{X}_\tau, \qquad t = (1-\tau)^{-1}\tau \qquad (0 \le \tau < 1)$$

defines a version of the Wiener process (and vice versa). Hint: check the covariance and use the identity

$$\min\left(\frac{\tau}{1-\tau}, \frac{\tau'}{1-\tau'}\right) = \frac{\min(\tau,\tau') - \tau\tau'}{(1-\tau)(1-\tau')}\ .$$

3.2 Bounds on the Transition Amplitude

Only in very rare cases may a path integral actually be evaluated. In all other cases the simplest (and highly recommended) solution to the problem of estimating the transition amplitude $\langle x', s'|x, s\rangle$ is to place lower und upper bounds on it. Wishing that the derivation of such bounds be as general as possible we shall include Hamiltonians $H(t) = -\frac{1}{2}\Delta + V(\cdot, t)$ explicitly depending on the time t into the discussion. Still, in most instances the potential V will be t-independent and the transition amplitude will then preferably be written in a way that suggests possible applications in statistical mechanics:

$$\langle x', s'|x, s\rangle = \langle x'|e^{-\beta H}|x\rangle, \qquad \beta = s' - s > 0. \qquad (3.2.1)$$

Frequently, very specific properties of the potential enter the derivation of bounds. By contrast, there are 'general bounds' that hold for almost any potential.

3.2.1 Defining a Subset of Paths

In describing general bounds one invariably encounters a restricted set of smooth sample paths of the Brownian bridge and averages with respect to that subset. We start out by discussing certain aspects of these peculiar paths and then proceed to demonstrate their usage.

Let A be a random vector taking values $a \in \mathbb{R}^d$. Suppose further that the components of A are independent and normally distributed. Hence, for suitable functions $f: \mathbb{R}^d \to \mathbb{R}$,

$$E(f(A)) = \int da\, K(a, 1)f(a) = (2\pi)^{-d/2}\int da\, e^{-a^2/2}f(a). \qquad (3.2.2)$$

To each event a of A, we associate some paths of the d-dimensional Brownian bridge $\bar{X}_\tau$,

$$\bar{\omega}_a(\tau) = a\sqrt{\tau(1-\tau)} \qquad (0 \le \tau \le 1,\ a \in \mathbb{R}^d). \qquad (3.2.3)$$

By definition, the path automatically satisfies $\bar{\omega}_a(0) = \bar{\omega}_a(1) = 0$ and obviously represents a very smooth function of the time variable τ (see Fig. 3.2).

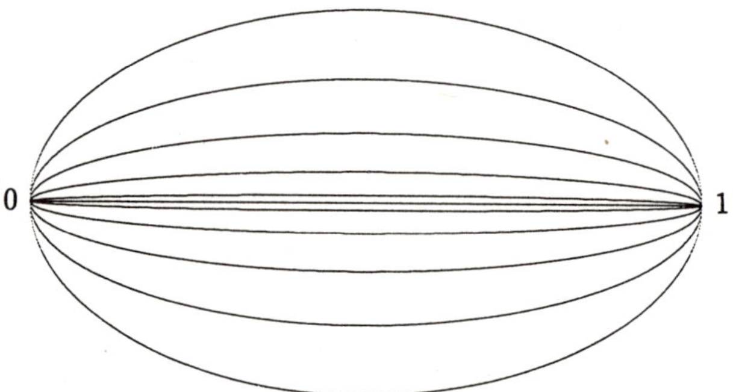

Fig. 3.2. Special paths $\bar{\omega}_a(\tau)$ of the Brownian bridge for various values of the amplitude a

The amplitude a in front of it accounts for the deviation from the trivial path $\bar{\omega} = 0$. What could be the significance of the construction of a d-dimensional subspace of sample paths?

Suppose $f[\bar{\omega}]$ happens to be some functional of the following nature:

$$f[\bar{\omega}] = \int_0^1 d\tau\, f(\bar{\omega}(\tau), \tau), \tag{3.2.4}$$

where f is an ordinary function of $d + 1$ variables. It would then seem desirable to have a simple way of calculating the mean value $\langle f \rangle$. Now, to actually do the calculation we would first interchange the time and the path integration and then proceed as usual:

$$\begin{aligned}
\langle f \rangle &\equiv \int d\bar{\omega}\, f[\bar{\omega}] = \int_0^1 d\tau \int d\bar{\omega}\, f(\bar{\omega}(\tau), \tau) \\
&= \int_0^1 d\tau\, (2\pi)^{d/2} \int dx\, K(x, \tau) f(x, \tau) K(x, 1 - \tau).
\end{aligned} \tag{3.2.5}$$

Upon verifying the relation

$$(2\pi)^{d/2} K(x, \tau) K(x, 1 - \tau) = K\big(x, \tau(1 - \tau)\big) \tag{3.2.6}$$

and substituting $x = a\sqrt{\tau(1 - \tau)}$ one gets

$$\begin{aligned}
\langle f \rangle &= \int_0^1 d\tau \int da\, K(a, 1) f\big(\bar{\omega}_a(\tau), \tau\big) \\
&= \int da\, K(a, 1) f[\bar{\omega}_a] \equiv \boldsymbol{E}\big(f[\bar{\omega}_A]\big),
\end{aligned} \tag{3.2.7}$$

where $E(\cdot)$ means expectation in the sense of (3.2.2). The upshot of these considerations is that the mean value $\langle f \rangle$ with respect to *all* paths of the Brownian bridge reduces to the expectation with respect to a relatively small, i.e., d-dimensional subspace of paths given by (3.2.3).

3.2.2 The Semiclassical Approximation

The scaling procedure when applied to paths $\omega\colon (x,s) \rightsquigarrow (x',s')$ in the Feynman–Kac formula suggests that we factorize the transition amplitude as follows:

$$\langle x', s' | x, s \rangle = K(x' - x, s' - s) \exp \Phi_{x,s}^{x',s'}(V) \qquad (3.2.8)$$

$$\Phi(V) = \log \int d\bar\omega \, \exp\left\{ -\int_0^1 d\tau \, \bar V\big(\bar\omega(\tau), \tau\big) \right\}. \qquad (3.2.9)$$

The functional $\Phi(V)$ involves the scaled potential $\bar V$ introduced by relations (3.1.18–20), any dependence on x, x', s, s' being suppressed in (3.2.9).

The same relations also demonstrate the role of the length parameter ℓ, which sets the scale of quantum fluctuations[53]. Clearly, as ℓ tends to zero, the integrand in (3.2.9) becomes independent of $\bar\omega$, with the effect that the integration can be carried out trivially. The resulting functional is then termed the *semiclassical approximation*[54] of $\Phi(V)$:

$$\begin{aligned}
\Phi_{\mathrm{sc}}(V) &= -\int_0^1 d\tau \, \bar V(0, \tau) \\
&= -\int_s^{s'} dt \, V\big(x + (x' - x)\tfrac{t-s}{s'-s}, \, t\big).
\end{aligned} \qquad (3.2.10)$$

Therefore, one can define

$$V_{x,x'} = \int_0^1 d\tau \, V\big(x + (x' - x)\tau\big) \qquad (3.2.11)$$

if the potential does not explicitly depend on t and write for the kernel of the semiclassical version of $e^{-\beta H}$

$$\langle x' | (e^{-\beta H})_{\mathrm{sc}} | x \rangle = (2\pi\beta)^{-d/2} \exp\{-(2\beta)^{-1}(x' - x)^2 - \beta V_{x,x'}\}. \qquad (3.2.12)$$

Note that there is still a nontrivial, however, hidden, dependence on Planck's constant in this expression. By applying a Laplace transformation (see (2.2.21)) to both sides of (3.2.11) we are able to give meaning to the semiclassical approximation of the Green's function provided $E > -V(x)$ for all x:

---

[53]For the dependence of ℓ on Planck's constant, recall definition (3.1.6).
[54]Not to be confused with the WKB approximation of the transition amplitude as considered in (2.8.30–31).

$$\langle x'|(H+E)_{\text{sc}}^{-1}|x\rangle = \frac{1}{\pi}\left(\frac{k}{2\pi r}\right)^{a} K_{a}(kr), \tag{3.2.13}$$

where $a = d/2 - 1$, $r = |x' - x|$, $k = \{2(V_{x,x'} + E)\}^{1/2}$, and $K_{\nu}(z)$ stands for the modified Bessel function of order ν (compare (2.4.80)). Equation (2.2.13) simplifies if $d = 1$ and $r = 0$ since $V_{x,x} = V(x)$ and so $\langle x|(H+E)_{\text{sc}}^{-1}|x\rangle = \{2[V(x) + E]\}^{-1/2}$.

Later, in Sect. 4.2, we shall write $\Phi(V) = -\beta V_{\text{eff}}(x)$ if $x' = x$ and tackle the problem of determining quantum corrections to the semiclassical approximation $V_{\text{eff}}(x) = V(x)$. Results in this direction were first obtained by Wigner [3.1].

3.2.3 Bounds on the Functional $\Phi(V)$

Let us return to definition (3.2.9) without any approximation applied to it. We may ask: are there reasonable bounds on the functional $\Phi(V)$ not involving path integrals so that their evaluation is no longer atrocious? In answering that question we discover the major advantage of the path integral formulation as well as of the representation obtained by the rescaling procedure: it allows us to immediately apply Jensen's inequality in two ways (for details see Appendix C). First, we have

$$\int d\bar{\omega}\, \exp I(\bar{\omega}) \geq \exp \int d\bar{\omega}\, I(\bar{\omega}). \tag{3.2.14}$$

Second, there is the inequality

$$\exp \int_{0}^{1} d\tau\, f(\tau) \leq \int_{0}^{1} d\tau\, \exp f(\tau). \tag{3.2.15}$$

Both $I(\bar{\omega})$ and $f(\tau)$ are quite arbitrary. However, in view of representation (3.2.9) we take $f(\tau) = -\bar{V}\big(\bar{\omega}(\tau), \tau\big)$ and $I(\bar{\omega}) = -\int_{0}^{1} d\tau\, \bar{V}\big(\bar{\omega}(\tau), \tau\big)$, thus obtaining

$$\Phi(V) \geq -\int d\bar{\omega} \int_{0}^{1} d\tau\, \bar{V}\big(\bar{\omega}(\tau), \tau\big) \tag{3.2.16}$$

$$\Phi(V) \leq \log \int d\bar{\omega} \int_{0}^{1} d\tau\, \exp\Big(-\bar{V}\big(\bar{\omega}(\tau), \tau\big)\Big). \tag{3.2.17}$$

The desired effect has not yet occurred. But the result is such that we now may use (3.2.7):

$$\int_{0}^{1} d\tau\, \boldsymbol{E}\Big(-\bar{V}\big(\bar{\omega}_A(\tau), \tau\big)\Big) \leq \Phi(V) \leq \log \int_{0}^{1} d\tau\, \boldsymbol{E}\Big(\exp\big\{-\bar{V}\big(\bar{\omega}_A(\tau), \tau\big)\big\}\Big).$$

$$\tag{3.2.18}$$

The bounds on $\Phi(V)$ arise from the convexity of the exponential function alone, are written here in terms of the rescaled potential $\bar{V}$ as defined in (3.1.18-20), and require evaluation of $(d+1)$-dimensional integrals which can seldomly be done explicitly, but seems to be quite tractable from the point of view of numerical analysis.

For illustration we propose to inspect simple bounds on the transition amplitude in some detail when V is the repulsive Coulomb potential in three dimensions. We thus consider $V(x) = \alpha r^{-1}$, where $r = |x|$, and take $H = -\frac{1}{2}\Delta + V$ as the Hamiltonian. Since α is positive, a trivial upper bound is provided by $\Phi(V) < 0$. From this and the lower bound given in (3.2.18),

$$\exp\left\{-\beta\alpha \int_r^\infty du \, \frac{1 - e^{-2u^2/\beta}}{u^2}\right\} \leq (2\pi\beta)^{3/2}\langle x|e^{-\beta H}|x\rangle \leq 1, \tag{3.2.19}$$

where for reasons of simplicity we have restricted the kernel of $e^{-\beta H}$ to its diagonal values. The derivation takes a few extra lines. But before the proof continues, let us look at the expression on the left-hand side of (3.2.19): for $\beta \to 0$, the integral approaches the semiclassical value r^{-1}, which is singular at $x = 0$. Nevertheless, the singular behavior at the origin is a spurious thing since, for any $\beta > 0$, the integral actually approaches a finite value if $r \to 0$, and

$$\exp\left\{-\alpha\sqrt{2\pi/\beta}\right\} \leq (2\pi\beta)^{3/2}\langle 0|e^{-\beta H}|0\rangle \leq 1. \tag{3.2.20}$$

In order to prove the validity of the lower estimate in (3.2.19) one has to start from the definition

$$\bar{V}(\xi,\tau) = \beta V(x + (x' - x)\tau + \ell\xi) \qquad (\ell^2 = \beta), \tag{3.2.21}$$

taking $x' = x$ so that $-\beta v(r) \leq \Phi(V)$, where

$$v(r) := \int_0^1 d\tau \, \boldsymbol{E}\left\{V\left(x + A\ell\sqrt{\tau(1-\tau)}\right)\right\}. \tag{3.2.22}$$

Indeed, the result must be rotationally invariant. To determine the expectation value $\boldsymbol{E}(\cdot)$ and the subsequent τ-integral therein we recall the differential equation satisfied by the $1/r$-potential which is $\Delta V(x) = -4\pi\alpha\delta(x)$, i.e., we first compute the density

$$\rho(r) := \alpha \int_0^1 d\tau \, \boldsymbol{E}\left\{\delta\left(x + A\ell\sqrt{\tau(1-\tau)}\right)\right\} \tag{3.2.23}$$

and then proceed to compute $v(r)$ from $\Delta v(r) = -4\pi\rho(r)$. From (3.2.23) and (3.2.2) it follows that

$$\rho(r) = \alpha \int_0^1 \frac{d\tau}{(2\pi\beta\tau(1-\tau))^{3/2}} \exp\left\{-\frac{r^2}{2\beta\tau(1-\tau)}\right\}. \tag{3.2.24}$$

A change of variables, $\tau = (1 + e^{-t})^{-1}$, converts this into a familiar integral:

$$\rho(r) = 2\alpha(2\pi\beta)^{-3/2} \int_{-\infty}^\infty dt \, \cosh(\tfrac{1}{2}t) \exp\left(-(1 + \cosh t)r^2/\beta\right)$$

$$= 4\alpha(2\pi\beta)^{-3/2} e^{-r^2/\beta} K_{1/2}(r^2/\beta). \tag{3.2.25}$$

The result is written in terms of the modified Bessel function $K_\nu(z)$. Its order $\nu = \frac{1}{2}$ is particular to the dimension $d = 3$. Quite generally, for half-integer values of ν, the Bessel functions reduce to elementary functions. As for our case, $K_{1/2}(z) = (2z/\pi)^{-1/2}e^{-z}$, yielding

$$\rho(r) = \alpha(\pi\beta r)^{-1}e^{-2r^2/\beta} > 0. \tag{3.2.26}$$

Notice that the total charge has a constant value: $\int dx\, \rho(r) = \alpha$.

We are left with the problem of solving the Poisson differential equation,

$$\frac{1}{r^2}\frac{d}{dr}r^2\frac{d}{dr}v(r) = -4\pi\rho(r), \tag{3.2.27}$$

taking $\lim_{r\to\infty} rv(r) = \alpha$ as the appropriate boundary condition. This is but an elementary task directly leading us to the result (3.2.19).

3.2.4 Convexity of the Functional $\Phi(V)$

The key to a variety of similar estimates is the convexity of $\Phi(V)$ as a functional of the potential $V(x)$:

$$\Phi(\alpha V_1 + (1 - \alpha)V_2) \leq \alpha\Phi(V_1) + (1 - \alpha)\Phi(V_2) \qquad (0 \leq \alpha \leq 1). \tag{3.2.28}$$

In order to enhance the usefulness of the argument and for clarity, we prove quite generally that $\Psi(I) = \log \int d\bar{\omega}\, \exp I(\bar{\omega})$ is convex in I with no further assumptions about the functional $I(\bar{\omega})$. The basis is Hölder's inequality for integrals:

$$\int d\bar{\omega}\, F_1(\bar{\omega})^\alpha F_2(\bar{\omega})^{1-\alpha} \leq \left(\int d\bar{\omega}\, F_1(\bar{\omega})\right)^\alpha \left(\int d\bar{\omega}\, F_2(\bar{\omega})\right)^{1-\alpha} \qquad (F_i \geq 0).$$

$$\tag{3.2.29}$$

Upon inserting $F_i(\bar{\omega}) = \exp I_i(\bar{\omega})$ into (3.2.29) and taking the logarithm on both sides, one obtains $\Psi(\alpha I_1 + (1 - \alpha)I_2) \leq \alpha\Psi(I_1) + (1 - \alpha)\Psi(I_2)$ as required. The special convexity statement (3.2.28) for $\Phi(V)$ appears then as an easy consequence of the general convexity property of $\Psi(I)$ and the choice $I_i(\bar{\omega}) = -\int_0^1 d\tau\, V_i(\bar{\omega}(\tau), \tau)$.

By appeal to convexity we are able to improve on the bounds for Φ. To do so it is wise to start from the inequality

$$\Psi\left(\int_0^1 d\tau\, I_\tau\right) \leq \int_0^1 d\tau\, \Psi(I_\tau) \tag{3.2.30}$$

which is valid for any one-parameter family of functionals $I_\tau(\bar{\omega})$ and then to choose $I_\tau(\bar{\omega}) = -\bar{V}(\bar{\omega}(\tau), \tau)$. So inequality (3.2.30) actually reads

$$\Phi(V) \leq \int_0^1 d\tau\, \log E\left(\exp\left\{-\bar{V}(\bar{\omega}_A(\tau), \tau)\right\}\right). \tag{3.2.31}$$

Compare the bound (3.2.31) with (3.2.18) and note the difference. Alas! The logarithm is a concave function. Therefore, inequality (3.2.31) implies the upper bound in (3.2.18) but not vice versa.

The improved upper bound on $\Phi(V)$ is due to Symanzik [2.13] and should be considered a major achievement in this field. This bound and previous bounds merge when β tends to zero. They may however turn out to be bad estimates when β becomes large, say, at zero temperature.

Exercise 1.   For a one-dimensional periodic array of δ-scatterers, i.e., for a particle moving on the straight line under the influence of the potential

$$V(x) = \lambda \sum_{n=-\infty}^{\infty} \delta(x - na)$$

compute the semiclassical approximation $\Phi_{\mathrm{sc}}(V)$.

Exercise 2.   Show that the three-dimensional Coulomb potential $V(x) = \alpha r^{-1}$ leads, after some necessary integration, to

$$\Phi_{\mathrm{sc}}(V) = \frac{-\beta\alpha}{|x - x'|} \log \frac{r' + x'e}{r + xe}, \qquad e := \frac{x - x'}{|x - x'|}, \qquad r = |x|, \ r' = |x'|,$$

which isvalid for $x \neq x'$. There is, for this case and to our present knowledge, no exact representation of $\Phi(V)$ in terms of elementary functions or by some elementary integral (serious efforts and partial success in this direction are found in [3.2,3]). The Feynman-Integral for the Coulomb propagator, however, has been evaluated by the Duru-Kleinert transformation method [2.29–30].

Exercise 3.   Inequalities like (3.2.18) place bounds on the kernel of $e^{-\beta H}$ when $x' = x$. Extend the discussion that has lead to (3.2.19) for the repulsive Coulomb potential, so that it applies to the situation $x' \neq x$. One step in the computation is a proof of

$$\rho(x, x') := \alpha \int_0^1 d\tau \, \mathbf{E}\Big(\delta\big(x + (x' - x)\tau + A\ell\sqrt{\tau(1 - \tau)}\big)\Big)$$

$$= \alpha(2\pi\beta)^{-1}(r^{-1} + r'^{-1}) \exp\Big(-\beta^{-1}rr'(1 + \cos\theta)\Big),$$

where $xx' = rr'\cos\theta$, and another is to write down the solution of the Poisson equation

$$\Delta_y v(y - \tfrac{1}{2}\xi, y + \tfrac{1}{2}\xi) = -4\pi\rho(y - \tfrac{1}{2}\xi, y + \tfrac{1}{2}\xi)$$

in terms of which the lower bound is expressed:

$$K(x' - x, \beta) \exp(-\beta v(x, x')) \leq \langle x'|e^{-\beta H}|x\rangle \leq K(x' - x, \beta).$$

3.3 Variational Principles

An entirely different method for placing bounds on the functional $\Phi(V)$ is to compare it to $\Phi(V_0)$, where V_0 is a more tractable potential in the sense that one actually knows how to explicitly compute or estimate $\Phi(V_0)$. We wish to pick the varying potential V_0 to be as "close" as possible to the given potential V. For, as the subsequent analysis indicates, it will be the "distance" between V and V_0 that dictates the precision and the usefulness of the method. Indeed, the upper and lower bounds described below tighten as the two potentials coalesce.

As dependencies on x, s, x', s' (the initial and final parameters) play no role in the argument, it is preferable to always work with the rescaled potentials $\bar{V}$ and $\bar{V}_0$ and to consider the difference

$$\Delta_\tau[\bar{\omega}] = \bar{V}_0\big(\bar{\omega}(\tau), \tau\big) - \bar{V}\big(\bar{\omega}(\tau), \tau\big). \tag{3.3.1}$$

The structure of the bounds, the validity of which will soon be demonstrated, is best described by writing them in the following abbreviated form (compare [3.4]):

$$\int_0^1 d\tau\, \langle \Delta_\tau \rangle_0 \ \leq\ \Phi(V) - \Phi(V_0) \ \leq\ \int_0^1 d\tau\, \log\langle \exp \Delta_\tau \rangle_0. \tag{3.3.2}$$

This notation borrows from the theory of Gibbs states in statistical mechanics: if $F[\bar{\omega}]$ is some functional (or *observable*), its mean value with respect to the Boltzmann weights $\exp(-V_0[\bar{\omega}])$ is given by

$$\langle F \rangle_0 = \frac{\int d\bar{\omega}\, F[\bar{\omega}] \exp(-V_0[\bar{\omega}])}{\int d\bar{\omega}\, \exp(-V_0[\bar{\omega}])}, \qquad V_0[\bar{\omega}] := \int_0^1 d\tau\, \bar{V}_0\big(\bar{\omega}(\tau), \tau\big). \tag{3.3.3}$$

By varying V_0 one generates a host of Gibbs states, each state producing an upper and a lower bound of $\Phi(V)$. As V_0 tends to V in an appropriate sense, the bounds merge, and this is why inequalities (3.3.2) are referred to as *variational principles*. Normally, the variation is carried out in a restricted sense, namely, in that merely a finite set of parameters on which V_0 depends is varied so as to maximize (minimize) the lower (upper) bound.

To prove the validity of (3.3.2) we use convexity, of course. Convexity of $\Phi(V)$, among other things, states that the following ordinary function of a single real variable is convex:

$$f(t) = \Phi\big(V_0 + t(V - V_0)\big) \qquad (t \in \mathbb{R}). \tag{3.3.4}$$

Consequently,

$$f(t) - f(0) \geq tf'(0), \tag{3.3.5}$$

which, for $t = 1$, yields the lower bound in (3.3.2). As for the upper bound, we would set

$$I_\tau[\bar\omega] = \Delta_\tau[\bar\omega] - V_0[\bar\omega] \tag{3.3.6}$$

and invoke the convexity of $\Psi(I) = \log \int d\bar\omega \exp I[\bar\omega]$, i.e., call upon inequality (3.2.26). From the definitions (3.3.1) and (3.3.3),

$$\int_0^1 d\tau \, I_\tau[\bar\omega] = -\int_0^1 d\tau \, \bar V\big(\bar\omega(\tau),\tau\big). \tag{3.3.7}$$

The conclusion that (3.2.31) is nothing but the second inequality of (3.3.2) completes the proof.

To control the error when applying the variational principles to some potential V, it might in some instances be helpful to consider the distance

$$||V - V_0|| := \sup_{y,t} |V(y,t) - V_0(y,t)| \le \infty \tag{3.3.8}$$

so that $-\beta||V - V_0|| \le \Delta_\tau(\bar\omega) \le \beta||V - V_0||$ and hence, by virtue of the bounds (3.3.2),

$$|\Phi(V) - \Phi(V_0)| \le \beta||V - V_0|| \qquad (\beta = s' - s) \tag{3.3.9}$$

which is interesting since it specifies in what sense the functional $\Phi(V)$ may be said to be *uniformly continuous*. Another useful observation follows directly from definition (3.2.9): the functional $\Phi(V)$ is decreasing, i.e., $V \le V_0$ implies that $\Phi(V) \ge \Phi(V_0)$.

Notice that bounds (3.3.2) are reasonable in the sense that they reduce path integration to ordinary $(d+1)$-dimensional integration. This point is best illustrated by taking the harmonic oscillator,

$$V_0(\xi) = \tfrac{1}{2}k^2\xi^2 \qquad (\xi \in \mathrm{I\!R}^d), \tag{3.3.10}$$

as the potential of comparison. Given τ in the interval $[0,1]$, each path $\bar\omega\colon (0,0) \rightsquigarrow (0,1)$ of the Brownian bridge factorizes as

$$\bar\omega\colon (0,0) \rightsquigarrow (\xi,\tau) \rightsquigarrow (0,1). \tag{3.3.11}$$

For to actually perform the path integration, we need to invoke Mehler's formula in three ways:

$$\langle 0,1|0,0\rangle_{\mathrm{Me}} = \left[\frac{k}{2\pi \sinh k}\right]^{d/2}$$

$$\langle \xi,\tau|0,0\rangle_{\mathrm{Me}} = \left[\frac{k}{2\pi \sinh(k\tau)}\right]^{d/2} \exp\left\{-\frac{k\xi^2}{2\tanh(k\tau)}\right\} \tag{3.3.12}$$

$$\langle 0,1|\xi,\tau\rangle_{\mathrm{Me}} = \left[\frac{k}{2\pi \sinh\big(k(1-\tau)\big)}\right]^{d/2} \exp\left\{-\frac{k\xi^2}{2\tanh(k(1-\tau))}\right\}.$$

Indeed, what enters the integral representation of both $\langle \Delta_\tau\rangle_0$ and $\langle \log \Delta_\tau\rangle_0$ is the quotient

$$q(\xi,\tau) := \frac{\langle 0,1|\xi,\tau\rangle_{\mathrm{Me}}\langle \xi,\tau|0,0\rangle_{\mathrm{Me}}}{\langle 0,1|0,0\rangle_{\mathrm{Me}}}, \tag{3.3.13}$$

which compares the amplitude of compound paths to that of all paths. The result may thus be written as

$$\langle \Delta_\tau \rangle_0 = \int d\xi \, q(\xi, \tau)\left(\tfrac{1}{2}k^2\xi^2 - \bar{V}(\xi, \tau)\right), \tag{3.3.14}$$

and a similar formula holds for $\langle \log \Delta_\tau \rangle$.

A remarkable fact about Mehler's formula is that forming quotients, as in (3.3.13), establishes a connection with the transition amplitude of a *free particle*, $K(x, t)$. The reason: if we introduce a new time variable by

$$t = h_k(\tau) := \frac{\sinh(k\tau)\sinh(k(1-\tau))}{k \sinh k} \geq 0, \tag{3.3.15}$$

we see that the definition is equivalent to

$$\frac{1}{h_k(\tau)} = \frac{k}{\tanh(k\tau)} + \frac{k}{\tanh(k(1-\tau))}, \tag{3.3.16}$$

and so the previously declared function q satisfies $q(\xi, \tau) = K(\xi, t)$ when τ and t are related by (3.3.15).

The function $h_k(\tau)$ will later (in Sect. 4.4) be seen to play a fundamental role in a different context, namely, in the theory of the driven harmonic oscillator. It should also be noted that, as k tends to zero, the function approaches a limit: $h_0(\tau) = \tau(1-\tau)$. Nonlinear (even path-dependent) time transformations play a decisive role in other contexts too. See [2.29] and [3.5].

The trouble with the comparison theorem (3.3.2) is, of course, that the list of potentials $V_0(x)$ for which the path integral can be solved explicitly is very short. This limits the scope of the method and, at present, destroys most of our hopes. However, there is nothing in the formalism that prevents us from comparing a given *time-independent* potential $V(x)$ with some family of *time-dependent* potentials $V_0(x, t)$ that appear convenient for some reason. The idea might seem curious, but the gain is flexibility. A tractable class is provided by the potential of the driven oscillator

$$V_0(x, t) = \tfrac{1}{2}k^2x^2 - E(t)x, \tag{3.3.17}$$

where $E(t)$ is a vector whose dependence on t may be arbitrary. As will be shown in Sect. 4.4, models of this type are indeed solvable. By simply varying the oscillator frequency k and the "electric field" $E(t)$ one may, in principle, be able to obtain useful bounds of $\Phi(V)$ from (3.3.2). Nevertheless, the resulting formulas are too complex to be discussed here in full detail.

3.3.1 The Mean Position of a Path

Feynman [3.6] proposed a different route towards a satisfactory semiclassical theory. He observed that the second term in a Taylor expansion of the potential,

$$V\big(\omega(t)\big) = V(\bar{x}) + V'(\bar{x})\big(\omega(t) - \bar{x}\big) + \cdots, \tag{3.3.18}$$

does not contribute to the integral $\int dt\, V\big(\bar{\omega}(t)\big)$, provided $\bar{x}$ has been chosen such that $\int dt\, \big(\omega(t) - \bar{x}\big) = 0$. This reasoning has led him to consider the candidate

$$\bar{x} = \frac{1}{s' - s} \int_s^{s'} dt\, \omega(t) \tag{3.3.19}$$

as the distinguished point in space around which the potential ought to be expanded. Using the scaling technique and writing $\omega : (x, s) \rightsquigarrow (x', s')$ in terms of some path of the Brownian bridge, $\bar{\omega}: (0,0) \rightsquigarrow (0,1)$, we arrive at an equivalent representation, $\bar{x} = \frac{1}{2}(x + x') + \ell y$, where $\ell^2 = s' - s > 0$ and

$$y = \int_0^1 d\tau\, \bar{\omega}(\tau). \tag{3.3.20}$$

If we accept this, $\bar{x}$ becomes dependent on ω and is termed the *mean position of a path*. What we have found is that to any $\bar{x}$ there corresponds a mean position y of the path $\bar{\omega}$. The terminology used does not signify that some sort of probabilistic expectation has been taken; the possible values $y \in \mathbb{R}^d$ still fluctuate around $y = 0$ and may be regarded as the events of a random vector $Y = \int_0^1 d\tau\, \bar{X}_\tau$.

Being a linear function of a Gaussian process, the mean position Y is normally distributed and satisfies

$$\boldsymbol{E}(Y) = 0, \qquad \boldsymbol{E}(Y_i Y_k) = \frac{1}{12}\delta_{ik} \quad (i, k = 1, \ldots, d). \tag{3.3.21}$$

In fact, the number $1/12$ is found by the following simple computation:

$$\int_0^1 d\tau \int_0^1 d\tau'\, \big(\min(\tau, \tau') - \tau\tau'\big) = \frac{1}{12}. \tag{3.3.22}$$

Disregarding effects induced by derivatives $V^{(n)}$ beyond the first establishes an approximate version of the Feynman–Kac formula,

$$\langle x', s' | x, s \rangle \approx K(x' - x, \beta)\boldsymbol{E}\Big(\exp\big\{ - \beta V\big(\tfrac{1}{2}(x + x') + \ell Y\big)\big\}\Big), \tag{3.3.23}$$

$$\beta = s' - s, \quad \ell = (s' - s)^{1/2},$$

where $\boldsymbol{E}(\cdot)$ stands for the expectation with respect to the normal distribution of the random vector Y. The approximate formula (3.3.23) becomes exact for potentials whose second derivative vanishes identically. Yet, it offers no solution to the problem of describing quantum fluctuations as $\hbar$ gets small since it does not even reproduce the semiclassical approximation (3.2.10) in the limit $\ell \to 0$ unless $x' = x$ or $V(x) = a + bx$. We shall not tackle the problem of going beyond the first derivative of the potential. Though Feynman was able to study the effect of the second derivative, no systematic analysis or algorithm for treating the nth derivative has so far emerged from such studies, and the preliminary results in this direction do not look very promising[55].

Exercise 1. Show that the approximate formula (3.3.23) can be turned into a rigorous inequality provided the potential $V(x)$ is a convex (concave) function of x. Hint: use Jensen's inequality

$$\int_s^{s'} dt\, V\big(\bar{\omega}(t)\big) \leq \beta V(\bar{x})$$

(in the convex case).

3.4 Bound States

The main purpose of this section is to demonstrate how path integrals may be used to obtain information about discrete spectra of Schrödinger operators. One of our goals is to estimate the number of bound states.

We assume the configuration space to be d-dimensional and the Hamiltonian to be of the form $H = -\frac{1}{2}\Delta - V$, where

1. $V(x) \geq 0$ for all x,
2. $V(x) \to 0$ if $|x| \to \infty$ (sufficiently fast).

More specific conditions will be needed later on.

To any positive value of the energy there corresponds some scattering state and hence some point in the continuous part of the spectrum. If occurring at all, the eigenvalues E_i turn out to be nonpositive. They describe the possible energy levels for which bound states exist. We shall always assume them to be ordered:

$$E_0 \leq E_1 \leq E_2 \leq \cdots \leq E_{N-1} \leq 0. \tag{3.4.1}$$

Also, we shall assume that each eigenvalue is listed according to its multiplicity. As a consequence, N equals the total number of bound states of the

---

[55]Feynman's expansion method gains flexibility if $V(\bar{x})$ and $V''(\bar{x})$ are replaced by arbitrary functions of $\bar{x}$. The strategy then is to take the resulting expression as the basis for some variational principle. Details may be found in [3.7–10].

system. We require that N be finite. This condition already excludes some potentials of interest in three dimensions, those, for instance, which behave like $1/r$ for large distances: the Coulomb potential does not belong to the class considered.

Along with the model under consideration we want to study the family

$$H_\lambda = -\tfrac{1}{2}\Delta - \lambda V(x), \qquad \lambda \in \mathbb{R} \tag{3.4.2}$$

where we have introduced an additional scaling variable λ.

As λ increases, the total number of bound states, N_λ, changes from zero to any desired natural number (in general). Let the corresponding eigenvalues (ordered as in (3.4.1)) be denoted by $E_i(\lambda)$ where $i = 0, 1, \ldots, N_\lambda - 1$. By the min–max principle of Courant [3.11], each eigenvalue $E_i(\lambda)$ is a decreasing function of λ. As λ tends to zero, each eigenvalue $E_i(\lambda)$ necessarily disappears at some characteristic value $\lambda = \lambda_i \geq 0$ for which $\lim_{\lambda \downarrow \lambda_i} E_i(\lambda) = 0$: for $\lambda < \lambda_i$, the eigenvalue E_i plunges into the continuous spectrum, thereby losing its existence (Fig. 3.3). As a result, there exists an infinite sequence

$$0 \leq \lambda_0 \leq \lambda_1 \leq \lambda_2 \leq \cdots \tag{3.4.3}$$

tending to infinity (in general). We have to bear in mind that, in the case of degeneracies, some of the characteristic values λ_i coincide.

One possible way to determine the number of bound states of $H = H_1$ is to simply count the characteristic values λ_i belonging to the interval $[0, 1)$. In the (λ, E) plane there exists, for any i, some curve $E_i(\lambda) = E$ which meets the line $E = 0$ in λ_i. Choosing a different line $E < 0$ we would get a different set of values λ_i. In other words, it seems natural to regard λ_i more generally as the solution of $E_i(\lambda) = E$ which makes it a function of the energy E. We may thus claim that there exists a normalizable wave function ψ_i such that, for each $E < 0$,

$$(-\tfrac{1}{2}\Delta - \lambda_i V)\psi_i = E\psi_i, \tag{3.4.4}$$

where $\lambda_i = \lambda_i(E)$.

For $E < 0$, the operator $-\tfrac{1}{2}\Delta - E$ can be inverted and the Birman–Schwinger operator

$$B_E = V^{1/2}(-\tfrac{1}{2}\Delta - E)^{-1}V^{1/2} \tag{3.4.5}$$

exists. From (3.4.4) it follows that

$$B_E\Phi_i = \lambda_i^{-1}\Phi_i , \tag{3.4.6}$$

where $\Phi_i = V^{1/2}\Psi_i$ (normalizable provided $V(x)$ is bounded):

> *To any eigenfunction of H_λ with eigenvalue E there corresponds an eigenfunction of B_E with eigenvalue λ^{-1}.*

Thus, for $E < 0$, the *Birman–Schwinger principle* holds:

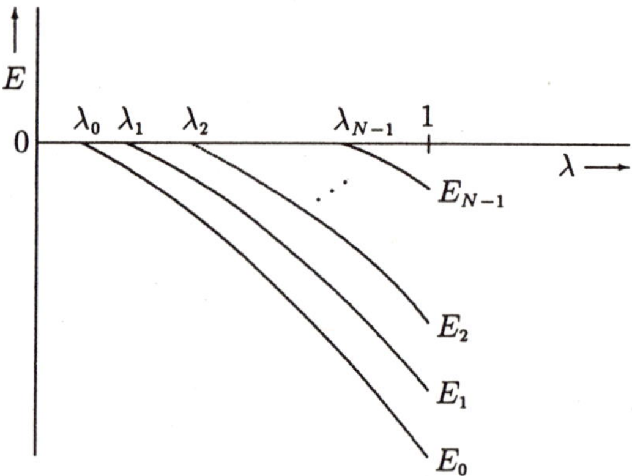

Fig. 3.3. A sketch of the eigenvalues of H_λ considered as functions of λ on the interval $[0,1)$.

$$N(E) = \text{number of eigenvalues } E_i < E \text{ of } H_1$$
$$= \text{number of eigenvalues } \lambda_i^{-1} > 1 \text{ of } B_E.$$

As a consequence, for any positive, increasing function $F : \mathbb{R}_+ \to \mathbb{R}_+$, we have the following chain of inequalities:

$$N(E)F(1) \leq \sum_{i=0}^{N(E)-1} F(\lambda_i^{-1}) \leq \sum_{i=0}^{\infty} F(\lambda_i^{-1}) = \operatorname{tr} F(B_E), \qquad (3.4.7)$$

though the expression on the right may be infinite owing to a bad choice of F. Equality occurs for the step function

$$F(s) = \begin{cases} 1 & \text{if } s > 1 \\ 0 & \text{otherwise.} \end{cases} \qquad (3.4.8)$$

Recall that the main interest is to put bounds on $N \equiv N(E{=}0) \equiv N_{\lambda=1}$. So we need to perform the limit $E \uparrow 0$ in (3.4.7) at the end.

To get the general flavor of the type of bounds that might result from (3.4.7), take $d = 3$, for instance. Then B_E is some integral operator with kernel[56]

$$\langle x'|B_E|x\rangle = V(x')^{1/2}\frac{e^{-\sqrt{-2E}\,|x'-x|}}{2\pi|x'-x|}V(x)^{1/2}. \qquad (3.4.9)$$

Choosing $F(s) = s^2$ we obtain from (3.4.7) the special inequality (*Birman–Schwinger bound*)

$$N \leq \|V\|_R^2 := \lim_{E \uparrow 0} \operatorname{tr} B_E^2 = \frac{1}{(2\pi)^2} \int dx \int dx' \, \frac{V(x')V(x)}{|x'-x|^2}. \qquad (3.4.10)$$

The potential $V \geq 0$ is said to belong to the *Rollnik class* if $\|V\|_R < \infty$ holds. Assuming this to be the case and varying λ we get the general inequality

---

[56]See Remark 6 at the end of this section.

$$N_\lambda \leq \lambda^2 \|V\|_R^2 \tag{3.4.11}$$

i.e., the step function $\lambda \mapsto N_\lambda$ is bounded from above by some parabola. If $\lambda < \|V\|_R^{-1}$, there is no bound state for H_λ, since the inequality implies that $N_\lambda = 0$.

The spectrum of the symmetric operator B_E is discrete and positive with an accumulation point at zero. For negative λ, i.e., for $\lambda = -\alpha < 0$, and $E < 0$

$$V^{1/2}(H_{-\alpha} - E)^{-1}V^{1/2} = (\alpha + B_E^{-1})^{-1}. \tag{3.4.12}$$

The resolvent of $H_{-\alpha}$ may be represented by an integral involving the Schrödinger semigroup:

$$(H_{-\alpha} - E)^{-1} = \int_0^\infty ds\, e^{Es} \exp\{-sH_{-\alpha}\}. \tag{3.4.13}$$

Then the conditional Wiener measure describes the kernel in terms of an integral over all paths $\omega : (x, 0) \rightsquigarrow (x', s)$:

$$\langle x'|(H_{-\alpha} - E)^{-1}|x\rangle$$
$$= \int_0^\infty ds\, e^{Es} \int d\mu_{x,0}^{x',s}(\omega) \exp\left\{-\alpha \int_0^s dt\, V(\omega(t))\right\}. \tag{3.4.14}$$

The integrand assumes nonnegative values and the integration with respect to s converges for large s if $E < 0$ and for small s provided $x' \neq x$. This may be seen from the estimate

$$0 \leq \langle x'|(H_{-\alpha} - E)^{-1}|x\rangle$$
$$\leq \int_0^\infty ds\,(2\pi s)^{-d/2} \exp\left\{Es - \frac{(x' - x)^2}{2s}\right\} < \infty. \tag{3.4.15}$$

Now, multiply both sides of (3.4.14) by $V(x)$ to get:

$$\left[\frac{V(x)}{V(x')}\right]^{1/2} \langle x'|(\alpha + B_E^{-1})^{-1}|x\rangle$$
$$= \int_0^\infty ds\, e^{Es} \int d\mu_{x,0}^{x',s}(\omega) \exp\left\{-\alpha \int_0^s dt\, V(\omega(t))\right\} V(x) \tag{3.4.16}$$

The structure of (3.4.16) is remarkable indeed:

- The left-hand side involves the nonlinear function $F_\alpha(y) = (\alpha + y^{-1})^{-1}$, where the real variable y has been replaced by the positive symmetric operator B_E.
- The right hand side involves the nonlinear function $g_\alpha(u) = \exp\{-\alpha u\}$, where the real variable u has been replaced by the positive integral $\int_0^s dt\, V(\omega(t))$.
- Any further operation in (3.4.16) appears to be linear.

Suppose now that somebody provides us with a *linear* transformation that maps functions $g : \mathbb{R}_+ \to \mathbb{R}$ into functions $F : \mathbb{R}_+ \to \mathbb{R}$, such that, for any $\alpha \geq 0$, g_α is carried[57] into F_α. Then, as $x' \to x$ and for any pair (g, F), we have the following identity:

$$\langle x|F(B_E)|x\rangle = \int_0^\infty ds\, e^{Es} \int d\mu_{x,0}^{x,s}(\omega)\, g\left(\int_0^s dt\, V\big(\omega(t)\big)\right) V(x), \quad (3.4.17)$$

provided the function $g(u)$ vanishes fast enough as u tends to zero so as to guarantee convergence for the final s integration in (3.4.17) for small values of s. The transformation meeting all conditions stated above is given by

$$F(y) = y \int_0^\infty du\, e^{-u} g(yu). \quad (3.4.18)$$

Integrating both sides of (3.4.17) with respect to $x \in \mathbb{R}^d$ now yields an equation for the trace of $F(B_E)$:

$$\operatorname{tr} F(B_E) = \int_0^\infty ds\, e^{Es} \int d\nu_s(\omega)\, V\big(\omega(0)\big) \quad (3.4.19)$$

(*Lieb's formula*), where, for convenience and to elucidate the structure of the integral so obtained, we introduced the product measure

$$d\nu_s(\omega) = dx\, d\mu_{x,0}^{x,s}\, g\left(\int_0^s dt\, V\big(\omega(t)\big)\right) \quad (3.4.20)$$

on the space Ω_s of *all* cyclic paths of period s:

$$\omega(t + s) = \omega(t) \qquad (t \in \mathbb{R}). \quad (3.4.21)$$

The emphasis on cyclic paths is more than just a change of notation and language; it also changes our point of view. For instance, x is no longer treated as a fixed position from where the Brownian path starts but is regarded as a variable of the path itself: x is used in (3.4.20) to abbreviate $\omega(0)$. As for (3.4.19), we avoided usage of the symbol x and wrote $V(\omega(0))$ in place of $V(x)$.

The variable s, although a fixed parameter for the inner integral of (3.4.19), in some sense measures the "length" of a cyclic path. Paths of different lengths get different weights, each weight being determined by $d\nu_s$ on the one hand and by the factor e^{Es} on the other. Choosing a large value for $-E$ will suppress the contribution of very long paths in (3.4.19).

One striking aspect of the measure $d\nu_s$ is its invariance under time translation, i.e., for any measurable function f on Ω_s, we have

---

[57]Implicitly, we view each function $g(u)$ as being approximated by sums of the form $\sum c_\alpha g_\alpha(u)$ in some suitable topology, i.e., we work within a linear space of functions spanned by the basis vectors g_α.

$$\int d\nu_s(\omega)\, f(\omega_u) = \int d\nu_s(\omega)\, f(\omega) \qquad (3.4.22)$$

for all u, where $\omega_u(t) = \omega(t - u)$: shifting the time by a fixed amount u simply means reparametrization of each cyclic path whose geometry and weight remain unchanged. Specifically, we may write

$$\int d\nu_s(\omega)\, V\big(\omega(0)\big) = \int d\nu_s(\omega)\, V\big(\omega(u)\big)$$
$$= s^{-1} \int d\nu_s(\omega) \int_0^s dt\, V\big(\omega(t)\big) \qquad (3.4.23)$$

and thus get another version of Lieb's formula,

$$\operatorname{tr} F(B_E) = \int_0^\infty ds\, s^{-1} e^{Es} \int dx \int d\mu_{x,0}^{x,s}\, f\left(\int_0^s dt\, V\big(\omega(t)\big) \right), \qquad (3.4.24)$$

with $f(u) = u g(u)$, so that

$$F(y) = \int_0^\infty du\, u^{-1} e^{-u} f(yu). \qquad (3.4.25)$$

Version (3.4.24) is preferred over (3.4.19) for various reasons. First, the potential occurs only once on the right-hand side of (3.4.24). Second, the time averaging according to (3.4.23) renders the integrand invariant with respect to time translation. Third, the transformation from f to F has the following property:

> If the n-th derivative $f^{(n)}$ is nonnegative, so is $F^{(n)}$. In particular, if f is nonnegative (increasing, convex), so is F.

The next step is a simple change of representation: we scale each Brownian path in (3.4.24) and effectively replace it by an equivalent path of the Brownian bridge as described in Sect. 3.1. In the present case, the formulas connecting the two processes become much simpler because all paths arising in (3.4.24) are closed. In detail, $\omega(t) = x + s^{1/2}\bar{\omega}(\tau)$ establishes a 1:1 correspondence between paths $\omega : (x,0) \rightsquigarrow (x,s)$ pinned at both ends and paths $\bar{\omega} : (0,0) \rightsquigarrow (0,1)$ of the Brownian bridge that are independent of x and s. Notice that $t = s\tau$ with $0 \le \tau \le 1$ and

$$d\mu_{x,0}^{x,s}(\omega) = (2\pi s)^{-d/2} d\bar{\omega}, \qquad (3.4.26)$$

where $d\bar{\omega}$ (again independent of s) denotes the normalized Wiener measure of the Brownian bridge. As a shorthand we use

$$J_E(s) = (2\pi s)^{-d/2} e^{Es} \qquad (s > 0, E < 0) \qquad (3.4.27)$$

and obtain

$$\operatorname{tr} F(B_E)$$
$$= \int_0^\infty \frac{ds}{s}\, J_E(s) \int d\bar{\omega} \int dx\, f\left(\int_0^1 d\tau\, s V\big(x + s^{1/2}\bar{\omega}(\tau)\big) \right). \qquad (3.4.28)$$

The idea is to concentrate on the subclass of *convex* functions $f(u)$. The reason: we may then apply Jensen's inequality to the integrand. From $J_E(s) > 0$,

$$
\begin{aligned}
\operatorname{tr} F(B_E) &\leq \int_0^\infty \frac{ds}{s}\, J_E(s) \int d\bar\omega \int dx \int_0^1 d\tau\, f\left(sV\left(x + s^{1/2}\bar\omega(\tau)\right)\right) \\
&= \int_0^\infty \frac{ds}{s}\, J_E(s) \int d\bar\omega \int_0^1 d\tau \int dx\, f\left(sV(x)\right) \\
&= \int_0^\infty \frac{ds}{s}\, J_E(s) \int dx\, f\left(sV(x)\right).
\end{aligned}
\tag{3.4.29}
$$

The second line was obtained through an interchange of the x and the τ integration, the third by using $\int d\bar\omega \int d\tau = 1$.

For $E = 0$, the function $J_E(s)$ becomes homogeneous of the order $-d/2$. Therefore, a passage to the limit $E \uparrow 0$ in (3.4.29) simplifies matters considerably: we first change the order of integration and then substitute $u = sV(x)$ to obtain

$$
\begin{aligned}
\lim_{E \uparrow 0} \operatorname{tr} F(B_E) &\leq \int dx \int_0^\infty \frac{du}{u}\, J_0\left(uV(x)^{-1}\right) f(u) \\
&= c_d \int dx\, V(x)^{d/2},
\end{aligned}
\tag{3.4.30}
$$

where the constant

$$
c_d = \int_0^\infty \frac{du}{u}\, (2\pi u)^{-d/2} f(u) = \frac{2}{d} \int_0^\infty du\, (2\pi u)^{-d/2} f'(u)
\tag{3.4.31}
$$

is independent of the details of the potential. Now use (3.4.7) and choose some *positive, increasing, convex* function $f(u)$ satisfying $c_d < \infty$ and $F(1) = 1$. The result so obtained is that the number of bound states, N, for $H = -\frac{1}{2}\Delta - V$ in d dimensions, is bounded in the following manner[58]:

$$
\boxed{N \leq c_d \int dx\, V(x)^{d/2}}
\tag{3.4.32}
$$

---

[58]For applications it seems reasonable to demand that V be in the space $L^{d/2}$. Otherwise the right hand side of (3.4.30) would be infinite or even undefined. Recall hat V belongs to L^p if

$$
\|V\|_p := \left(\int dx\, |V(x)|^p\right)^{1/p} < \infty \qquad (p > 0).
$$

Besides the proof given here which is due to Lieb [3.12,13], there are alternative proofs of bound (3.4.32) by other means (see [3.14–16], for instance), though the description of the constant c_d is not always as satisfactory as in (3.4.31). The spherically symmetric case has been considered separately [3.17] and, of course, yields smaller constants. Much of the material is reviewed in [3.18].

(*Cwickel–Lieb–Rosenbljum bound*). On our way we have neglected two essential questions that ought to be answered:

Question 1. *Can the condition $c_d < \infty$ be satisfied at all?* The problem at hand is to make sure the integrals in (3.4.31) converge at large values of u. Recall that the derivative $f'(u)$ should be increasing. Therefore, unless $f'(u)$ happens to vanish identically (meaning $f(u)$ is constant and hence zero), there exists a $v > 0$ such that $f'(u) \geq f'(v) > 0$ for all $u > v$. So

$$\int_0^\infty du\, u^{-d/2} f'(u) \geq f'(v) \int_v^\infty du\, u^{-d/2} \quad \begin{cases} = \infty & d = 1 \text{ or } 2 \\ < \infty & d \geq 3. \end{cases} \tag{3.4.33}$$

In other words, a reasonable bound of the form (3.4.32) can only be derived for dimensions $d \geq 3$.

Question 2. *Which of the functions f satisfying the properties stated above minimizes the constant c_d?* We suggest starting our search among those functions that are twice differentiable. From convexity ($f''(u) \geq 0$) and

$$f'(u) = \int_0^u dv\, f''(v), \qquad f(u) = \int_0^u dv\, f'(v) \tag{3.4.34}$$

one infers monotonicity ($f'(u) \geq 0$) and positivity ($f(u) \geq 0$). Consider the *exponential integrals*,

$$E_n(u) = \int_1^\infty dt\, t^{-n} e^{-ut} \qquad (n \in \mathbb{Z}, \quad u > 0), \tag{3.4.35}$$

such that

$$E_0(u) = u^{-1} e^{-u}, \qquad E_n'(u) = -E_{n-1}(u). \tag{3.4.36}$$

The special role played by these functions in the present context may be seen from (3.4.25) and

$$F(1) = \int_0^\infty du\, E_0(u) f(u)$$
$$= \int_0^\infty du\, E_1(u) f'(u) = \int_0^\infty du\, E_2(u) f''(u), \tag{3.4.37}$$

where a partial integration has been performed twice in order to represent $F(1)$ in terms of the second derivative f''. Notice that $E_2(u) > 0$. Inspired guessing will then suggest the ansatz

$$f''(u) = E_2(u)^{-1} |\phi(u)|^2, \qquad \phi \in L^2(\mathbb{R}_+) \tag{3.4.38}$$

with no essential loss of generality. It follows that

$$\|\phi\|^2 = \int_0^\infty du\, |\phi(u)|^2 = F(1) \tag{3.4.39}$$

and $F(1) = 1$ means that ϕ is normalized.

Partial integration also serves to rewrite (3.4.31) as

$$c_d = \frac{(2\pi)^{-d/2}}{\frac{d}{2}(\frac{d}{2}-1)} \int_0^\infty du\, u^{1-d/2} f''(u) \,. \tag{3.4.40}$$

The constant c_d, now dependent on ϕ and calculated according to formula (3.4.40), appears as an expectation value of some operator R on $L^2(\mathbb{R}_+)$:

$$c_d = (\phi, R\phi) = \int_0^\infty du\, R(u)|\phi(u)|^2 \tag{3.4.41}$$

$$R(u)^{-1} = (d/2)(2\pi)^{d/2}\, a\, u^a E_2(u), \qquad a = d/2 - 1. \tag{3.4.42}$$

Clearly, R is a positive, though unbounded multiplication operator with a dense domain of definition. The goal of finding the minimal value for c_d is achieved once we have solved a standard problem in Hilbert space theory. The problem is easy and may be phrased as

$$c_d = \inf_{\|\phi\|=1} (\phi, R\phi) = \inf_{u>0} R(u). \tag{3.4.43}$$

Equivalently,

$$c_d^{-1} = (d/2)(2\pi)^{d/2} a\, m_a, \tag{3.4.44}$$

where $a = d/2 - 1$ and

$$m_a = \sup_{u>0} u^a E_2(u). \tag{3.4.45}$$

By inspection of its graph, the function $u^a E_2(u)$ is seen to possess a unique maximum provided $a > 0$. The value of the maximum, however, has to be determined using numerical methods. As for low dimensions, the results are given in Table 3.1.

Table 3.1. The constants c_d for dimensions $d = 3, \ldots, 10$ to six decimal places. Given $a = d/2 - 1$, the function $u^a E_2(u)$ is maximal at $u = u_a$ and the maximum so obtained is m_a.

d	u_a	m_a	c_d
3	0.246\,207	0.258\,871	$3.270\,28 \times 10^{-1}$
4	0.610\,058	0.165\,728	$7.642\,12 \times 10^{-2}$
5	1.019\,154	0.148\,527	$1.814\,31 \times 10^{-2}$
6	1.453\,495	0.164\,612	$4.081\,77 \times 10^{-3}$
7	1.903\,461	0.212\,877	$8.634\,42 \times 10^{-4}$
8	2.363\,884	0.310\,597	$1.721\,48 \times 10^{-4}$
9	2.831\,699	0.500\,220	$3.248\,99 \times 10^{-5}$
10	3.304\,954	0.875\,726	$5.830\,45 \times 10^{-6}$

Let us pause and contemplate one striking aspect of bound (3.4.32): a familiar physical principle of quantization tells us that each cell of the phase space of size $h^d = (2\pi\hbar)^d$ can only be occupied by a single state. This is generally referred to as the semiclassical picture (see [3.19] for a discussion). As for bound states with energies less than zero, the accessible domain in phase space is

$$B = \{p, q \mid \tfrac{1}{2}p^2 - V(q) \le 0\} \subset \mathbb{R}^d \times \mathbb{R}^d. \tag{3.4.46}$$

From the semiclassical treatment, choosing $\hbar = 1$ (as we always do), we obtain, as a guess, the number of bound states,

$$N \approx \frac{1}{(2\pi)^d} \int_B dpdq = \frac{1}{(2\pi)^d} \int dq\, A\left(\sqrt{2V(q)}\right)$$
$$= \frac{A(\sqrt{2})}{(2\pi)^d} \int dq\, V(q)^{d/2}, \tag{3.4.47}$$

where $A(R)$ denotes the volume of the ball with radius R in d dimensions. The calculation, though it does not mention path integrals, gives essentially the same result: a fact that is surprising and beautiful at the same time. However, there are differences worth mentioning. Firstly, the constant in front of the integral – let us denote it by $\hat{c}_d$ – differs from c_d (it is smaller):

$$\hat{c}_d = \frac{A(\sqrt{2})}{(2\pi)^d} = \left[(2\pi)^{d/2}\Gamma(1 + d/2)\right]^{-1}. \tag{3.4.48}$$

Secondly, in the semiclassical picture there is no fundamental distinction between the cases $d \ge 3$ and $d < 3$.

3.4.1 Moment Inequalities for Eigenvalues

Inequality (3.4.32) can be extended in various ways. One way is to look at it as a particular case of similarly structured inequalities for expressions like $\sum_i |E_i|^n$, i.e., for the nth moments, where the sum extends over all eigenvalues $E_i < 0$ of the Schrödinger operator $-\tfrac{1}{2}\Delta - V$ (counting multiplicities). Obviously, the 0th moment is nothing but the number N studied above.

For any $E \le 0$, consider $N(E)$, the number of bound states with energies less than E. On the one hand we have

$$\sum_i |E_i| = -\sum_i E_i = \int_{-\infty}^0 dE\, N(E) , \tag{3.4.49}$$

and, on the other hand, taking (3.4.7) and (3.4.29) into account we get

$$N(E)F(1) \le \operatorname{tr} F(B_E) \le \int_0^\infty \frac{ds}{s}\, J_E(s) \int dx\, f\big(sV(x)\big) \tag{3.4.50}$$

provided f is positive, increasing and convex. Insertion into (3.4.49) invites us to do the E integration (recall that $J_E(s) = e^{Es}(2\pi s)^{-d/2}$),

$$j(s) := \int_{-\infty}^{0} dE\, J_E(s) = \frac{1}{s}(2\pi s)^{-d/2} = 2\pi(2\pi s)^{-(d+2)/2}, \tag{3.4.51}$$

and then to write the result as

$$F(1)\sum_i |E_i| \le \int dx \int_0^{\infty} \frac{du}{u}\, j\bigl(uV(x)^{-1}\bigr) f(u). \tag{3.4.52}$$

Since $j(s)$ is a homogeneous function, we may proceed as before (i.e., as in the proof of (3.4.32) with d replaced by $d+2$), and taking the extra factor 2π in (3.4.51) into account we obtain the bound

$$\boxed{\sum_i |E_i| \le 2\pi c_{d+2} \int dx\, V(x)^{(d+2)/2}.} \tag{3.4.53}$$

The constant in front of the integral does not contain anything new: apart from the factor 2π, its value is given by (3.4.40) with d replaced by $d+2$. Also, the proof has shown that inequality (3.4.53) holds in all dimensions (including $d = 1$ and $d = 2$). As for $d = 3$, the numerical value of the constant is

$$2\pi c_5 = 0.113\,996\ldots. \tag{3.4.54}$$

Along the same lines, one is able to derive upper bounds for higher moments (the reader should find no difficulty to do so).

Remark 1. The above method also applies to estimates of $\sum |E_i|^\gamma$ with arbitrary $\gamma \in \mathbb{R}_+$, i.e., one may write $\sum |E_i|^\gamma = \int d(|E|^\gamma)\, N(E)$ and then use

$$\int d(|E|^\gamma)\, J_E(s) = (2\pi s)^{-d/2} \int_0^{\infty} dt\, \exp(-st^\sigma) \qquad (\sigma = 1/\gamma)$$

$$= (2\pi)^\gamma \Gamma(\gamma + 1)(2\pi s)^{-(d+2\gamma)/2} \tag{3.4.55}$$

to obtain, for $d/2 + \gamma > 1$,

$$\sum |E_i|^\gamma \le c_{\gamma,d} \int dx\, V(x)^{(d+2\gamma)/2} \tag{3.4.56}$$

where

$$c_{\gamma,d} = (2\pi)^\gamma \Gamma(\gamma + 1) c_{d+2\gamma}$$

$$= (2\pi)^{-d/2} \frac{\Gamma(\gamma + 1)}{a(a+1)m_a}, \qquad a = d/2 - 1 + \gamma. \tag{3.4.57}$$

This makes sense even though $d + 2\gamma$ is not necessarily an integer since $m_a = \sup u^a E_2(u)$ is defined for any $a > 0$.

Compare (3.4.56) now to the semiclassical treatment (as suggested in [3.21]), i.e., proceed as in the derivation of (3.4.45) to get

$$\sum |E_i|^\gamma \approx (2\pi)^{-d} \int_B dpdq \, \left(V(q) - \tfrac{1}{2}p^2\right)^\gamma$$

$$= \hat{c}_{\gamma,d} \int dx \, V(x)^{(d+2\gamma)/2} \tag{3.4.58}$$

$$\hat{c}_{\gamma,d} = (2\pi)^{-d/2} \frac{\Gamma(\gamma+1)}{\Gamma(a+2)}.$$

Thus the quotient

$$c_{\gamma,d}/\hat{c}_{\gamma,d} = \Gamma(a)/m_a \tag{3.4.59}$$

is merely a function of $a = d/2 - 1 + \gamma$. Equation (3.4.59) conforms to a result by A.Martin [3.20] who showed that moment estimates with different γ and d but same $\gamma + d/2$ are intimately related.

From the definition (3.4.43) one derives the following bounds:

$$\frac{a^a e^{-a}}{a+2} < m_a < \frac{a^a e^{-a}}{a+1}. \tag{3.4.60}$$

The lower bound follows from the obvious inequality $\sup u^a E_2(u) > a^a E_2(a)$ and from $e^a E_2(a) > (a+2)^{-1}$ (a consequence of the continued fraction representation of E_2 [3.22, p. 229]), the upper bound follows from

$$\sup_{u>0} u^a \int_1^\infty \frac{dt}{t^2} e^{-ut} < \int_1^\infty \frac{dt}{t^2} \sup_{u>0} u^a e^{-ut} = a^a e^{-a} \int_1^\infty \frac{dt}{t^{a+2}}. \tag{3.4.61}$$

Hence, for large a, $m_a = a^{a-1} e^{-a}\left(1 + O(a^{-1})\right)$. Compare this with Stirling's formula for $\Gamma(a)$ to obtain the asymptotic formula

$$c_{\gamma,d}/\hat{c}_{\gamma,d} = (2\pi a)^{1/2}\left(1 + O(a^{-1})\right). \tag{3.4.62}$$

Remark 2. In view of the above one may wonder whether the results can be extended to potentials that do not fall into the category considered. The following argument helps to widen our scope. Take two Schrödinger operators $H_i = -\frac{1}{2}\Delta + V_i$ $(i = 1, 2)$ such that the inequality $V_1(x) \geq V_2(x)$ holds in the sense of ordinary functions. Then the inequality $H_1 \geq H_2$ holds in the sense of operators. By virtue of the min-max principle for eigenvalues, the nth eigenvalue of H_2 lies below its corresponding eigenvalue of H_1. Let N_i be the number of negative eigenvalues of H_i (counting multiplicity). By the preceding argument, $N_1 \leq N_2$. The idea is, given a potential V_1, to construct V_2 as close to V_1 as possible, such that the conditions $V_2(x) \leq 0$ and $V_1(x) \geq V_2(x)$ are met. Obviously, the optimal choice is

$$V_2(x) = V_1^{(-)}(x) := \begin{cases} 0 & \text{if } V_1(x) > 0 \\ V_1(x) & \text{otherwise.} \end{cases} \tag{3.4.63}$$

Call $V_1^{(-)}$ the *negative part* of V_1 and apply the previous results to $V = -V_1^{(-)}$ to obtain upper bounds for either N_1 or for moments of the eigenvalues. This strategy is best demonstrated by giving an example.

Let $N(E)$ denote the number of bound states with energies less than E for the harmonic oscillator $-\frac{1}{2}\Delta + \frac{1}{2}x^2$ in three dimensions. Here we would set $V_1(x) = \frac{1}{2}x^2 - E$, $V = -V_1^{(-)}$, and then perform the necessary integration,

$$\int dx\, V(x)^{3/2} = \frac{\pi^2}{2\sqrt{2}}E^3.   \tag{3.4.64}$$

The result may now be phrased as follows:

$$N(E) \leq \text{number of bound states of } -\tfrac{1}{2}\Delta - V$$

$$\leq c_3 \int dx\, V(x)^{3/2} = c_3 \frac{\pi^2}{2\sqrt{2}}E^3.   \tag{3.4.65}$$

For comparison with the semiclassical picture, we simply replace the constant c_3 by its classical counterpart, $\hat{c}_3 = (3\pi^2)^{-1}\sqrt{2}$, to obtain

$$N(E) \approx \hat{c}_3 \int dx\, V(x)^{3/2} = \tfrac{1}{6}E^3.   \tag{3.4.66}$$

Surprisingly enough, the expression on the right is asymptotically correct, i.e., it represents $N(E)$ for large E. This may seen from the exact result

$$N(E) = \tfrac{1}{6}n(n+1)(n+2), \qquad n = \begin{cases} \lfloor E - \tfrac{1}{2}\rfloor & \text{if } E > \tfrac{3}{2} \\ 0 & \text{otherwise} \end{cases}   \tag{3.4.67}$$

(easily inferred from the well-known eigenvalues $\sum_{i=1}^{3}(n_i + \tfrac{1}{2})$ $(n_i = 0, 1, 2, \ldots)$; in (3.4.67), $\lfloor a \rfloor$ denotes the integer part of a). Figure 3.4 shows the function $N(E)$ together with the semiclassical approximation and the upper bound.

Remark 3.   Replace V by λV in all previous expressions. In particular, estimate (3.4.32) generalizes to

$$N_\lambda \leq c_d \lambda^{d/2} \int dx\, V(x)^{d/2} \qquad (\lambda > 0)   \tag{3.4.68}$$

where, as before, N_λ is the number of bound states of $-\frac{1}{2}\Delta - \lambda V$. Suppose next that λ has been chosen small enough that

$$c_d \lambda^{d/2} \int dx\, V(x)^{d/2} < n   \tag{3.4.69}$$

for some $n \in \mathbb{N}$. It then follows that $N_\lambda < n$, and hence $\lambda_{n-1}^{-1} < \lambda^{-1}$ and

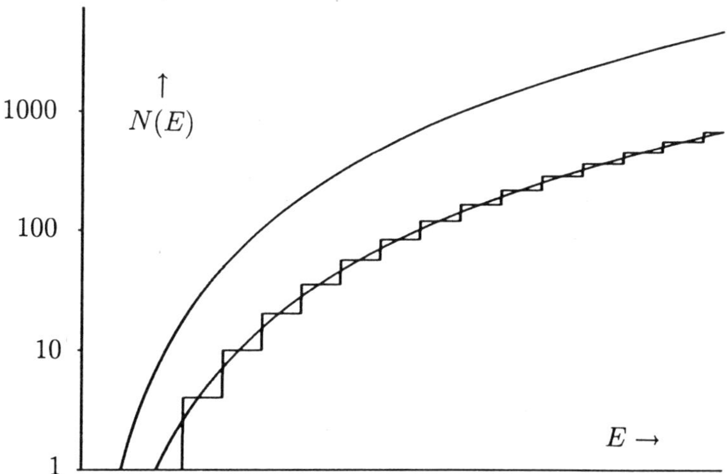

Fig. 3.4. The number of bound states with energies smaller then E for the harmonic oscillator is a step function $N(E)$. It is well approximated by the semiclassical value $E^3/6$ (*lower smooth curve*). The *smooth curve on top* represents the upper bound (3.4.65).

$$\lambda_{n-1}^{-1} < \left(\frac{c_d}{n}\int dx\, V(x)^{d/2}\right)^{2/d} = \left(\frac{c_d}{n}\right)^{2/d}\|V\|_{d/2}. \tag{3.4.70}$$

In particular, setting $n = 1$ we obtain some estimate for the spectral norm of the Birman–Schwinger operator B_E in the limit $E \uparrow 0$ involving the $d/2$-norm of the potential:

$$\|B_0\|_{\mathrm{sp}} = \mathrm{Max}\,\{\lambda_i^{-1}\} = \lambda_0^{-1} < c_d^{2/d}\|V\|_{d/2} \qquad (d \geq 3). \tag{3.4.71}$$

Remark 4.   E. Lieb [3.12,13], whose presentation of the subject was essentially followed, prefers to work with $\lambda = \frac{1}{2}$. Denoting e_i the (negative) eigenvalues of $-\Delta - V$, Lieb writes

$$\sum_i |e_i|^\gamma \leq L_{\gamma,d}\int dx\, V(x)^{\gamma+d/2}, \tag{3.4.72}$$

where $L_{\gamma,d}$ is the smallest constant possible. It follows from (3.4.56) that

$$L_{\gamma,d} \leq (\tfrac{1}{2})^{d/2}c_{\gamma,d} = (\tfrac{1}{2})^{d/2}(2\pi)^\gamma \Gamma(\gamma+1)c_{d+2\gamma} \tag{3.4.73}$$

provided $\gamma - 1 + d/2 > 0$. Therefore, in terms of number,

$$\begin{aligned} L_{0,3} &\leq\quad 2^{-3/2}c_3 &= 0.115\,622 \\ L_{1,3} &\leq\ \ 2^{-3/2}2\pi c_5 &= 0.040\,304 \end{aligned} \tag{3.4.74}$$

Remark 5.   Transformation (3.4.25) maps $f(u) = 1 - e^{-u}$ into $F(y) = \log(1 + y)$. This particular example of a pair (F, f) serves to define the determinant of the operator $1 + B_E$ in terms of a path integral. However, as it stands, it will make sense only for $d = 1$ and $E < 0$:

$$\log \det(1 + B_E) = \operatorname{tr} \log(1 + B_E)$$

$$= \int_0^\infty \frac{ds}{s} \, e^{Es} \int_{-\infty}^\infty dx \int d\mu_{x,0}^{x,s}(\omega) \left\{ 1 - \exp\left(-\int_0^s dt \, V\big(\omega(t)\big) \right) \right\}. \quad (3.4.75)$$

Under the assumptions $V(x) \geq 0$ and $\int dx \, V(x) < \infty$ the integral to the right is convergent. By a formal manipulation using the definition (3.4.5) we obtain the identity

$$\frac{\det(-\tfrac{1}{2}\Delta + V - E)}{\det(-\tfrac{1}{2}\Delta - E)} = \det(1 + B_E) \qquad (3.4.76)$$

which teaches us a lesson, namely, the quotient of two formal determinants may be given a meaning although the operators involved possess continuous spectra. The representation of the quotient as a path integral is best suited to deriving lower and upper bounds:

$$\int dx \left\{ \sqrt{2\big(V(x) - E\big)} - \sqrt{-2E} \right\} \leq \log \det(1 + B_E)$$

$$\leq \frac{\int dx \, V(x)}{\sqrt{-2E}}. \qquad (3.4.77)$$

To get the upper bound we have applied the inequality $1 - e^{-u} \leq u$ to (3.4.75) and evaluated the elementary integral

$$\int_0^\infty ds \, e^{Es} (2\pi s)^{-1/2} = (-2E)^{-1/2}. \qquad (3.4.78)$$

As for the lower bound, the concavity of $1 - e^{-u}$, Jensen's inequality, and the integral

$$\int_0^\infty \frac{ds}{s} \, e^{Es} (2\pi s)^{-1/2} \left(1 - e^{-sV(x)} \right) = \sqrt{2\big(V(x) - E\big)} - \sqrt{-2E} \quad (3.4.79)$$

did the trick. Quotients as in (3.4.76) occur frequently in the literature concerned with the WKB approximation, i.e., they arise in connection with the Van Vleck determinant. Frequently, the variable x is then interpreted as *time*. See [3.23, Chap.7] and [3.24, Appendix].

Remark 6.   The resolvent of the Laplacian in d dimensions can be expressed by the modified Bessel function $K_a(z)$ with $a = d/2 - 1$:

$$\langle x'|(-\Delta + k^2)^{-1}|x\rangle = \tfrac{1}{2}\int_0^\infty ds\, e^{-k^2 s/2}\langle x'|e^{s\Delta/2}|x\rangle \qquad (k>0)$$

$$= \tfrac{1}{2}\int_0^\infty ds\,(2\pi s)^{-d/2}\exp\left\{-\tfrac{1}{2}(k^2 s + r^2 s^{-1})\right\}$$

$$= \frac{1}{2\pi}\left(\frac{k}{2\pi r}\right)^a K_a(kr), \qquad r = |x'-x|. \qquad (3.4.80)$$

In particular, since $K_{\pm 1/2}(z) = \left(\frac{\pi}{2z}\right)^{1/2}e^{-z}$,

$$\langle x'|(-\Delta + k^2)^{-1}|x\rangle = \begin{cases} (2k)^{-1}\exp\{-kr\} & \text{if } d=1 \\ (4\pi r)^{-1}\exp\{-kr\} & \text{if } d=3, \end{cases} \qquad (3.4.81)$$

and, for arbitrary dimension d, the Birman–Schwinger operator B_E has the kernel

$$\langle x'|B_E|x\rangle = V(x')^{1/2}\langle x'|(-\tfrac{1}{2}\Delta - E)^{-1}|x\rangle V(x)^{1/2} \qquad k^2 = -2E > 0$$

$$= \frac{1}{\pi}\left(\frac{k}{2\pi r}\right)^a K_a(kr)\Big(V(x')V(x)\Big)^{1/2}. \qquad (3.4.82)$$

This settles the question as to how one would write down formulas for traces like $\operatorname{tr} B_E^n$ ($d/2 < n \in \mathbb{N}$) without ever mentioning path integrals. The simplest of these formulas arises for $d = n = 1$:

$$\operatorname{tr} B_E = (-2E)^{-1/2}\int_{-\infty}^\infty dx\, V(x). \qquad (3.4.83)$$

We stress that the more sophisticated method of path integration is needed only when one deals with unusual traces like $\operatorname{tr} F(B_E)$ where the function $F(y)$ is of a more general kind (not simply y^n).

Remark 7. Bound (3.4.53) plays a decisive role in a number of physical problems (see [3.25] for a list). Here, we shall mention only two of them.

Firstly, there is a well worked out theory called *stability of matter*, initiated by Dyson and Lenard [3.26], who showed that, for a system of n electrons and m static (i.e., heavy) protons under the influence of their mutual Coulomb forces, the total energy may be bounded from below by some expression of the form $-c(n+m)$, where c is independent of the positions attained by the protons. To be sure, the essential ingredient here is the Pauli principle: bosons behave differently. Now, Lieb and Thirring [3.27] simplified the proof considerably, thereby improving the constant c by many orders of magnitude.

Secondly, Ruelle [3.28] derived upper bounds on the magnitude and number of nonnegative characteristic (Liapunov) exponents for the Navier–Stokes flow of an incompressible fluid in a domain $\Omega \subset \mathbb{R}^d$ with an important bearing on information production (Kolmogorov–Sinai entropy) and Hausdorff dimension of attracting sets. Lieb [3.25] improved these bounds, which are then given in terms of the first moment of the (negative) eigenvalues of the formal Schrödinger operator $H = -\nu\Delta - V$ in dimension d ($= 2$ or 3), where ν denotes the kinematic viscosity and $V(x) \geq 0$ with

$$V(x)^2 = \frac{d-1}{4d} \sum_{i,j=1}^{d} \left(\frac{\partial v_i}{\partial x_j} - \frac{\partial v_j}{\partial x_i} \right)^2 \tag{3.4.84}$$

(v is the velocity field). Thus, the square of the formal potential is proportional to the rate of the energy dissipation in the flow.

3.5 Monte Carlo Calculation of Path Integrals

Numerical integration, sometimes also called *quadrature*, permeates most of the literature on applied analysis and has a long history. The techniques involved are based, in one way or another, on the idea of replacing the integral by a sum with the obvious goal of determining the integral as accurately as possible with a prescribed number of function evaluations. What, in former times, seemed a painful task, nowadays, with the advent of computers, has become a routine calculation.

The situation is not quite so favorable when it comes to multidimensional integrals. Path integrals are even worse in that respect since the underlying space is truly infinite-dimensional. Another fundamental disadvantage of numerical path integration is that it makes extensive use of Monte Carlo methods, so its accuracy increases slowly, i.e., the error decreases roughly as $N^{-1/2}$, where N is the number of sample paths.

A pertinent problem we encounter in numerical path integration is to design a computer program that repeatedly generates Brownian paths between fixed spacetime points, say (x, s) and (x', s') where $x, x' \in \mathbb{R}^d$. One may think of routines that create random paths, all starting at (x, s), using the principles of the random walk. However, these paths will go in all directions and only a minute fraction will actually hit the point x' at time s'. As a consequence, most of the random paths so obtained are useless and have to be discarded. It should be clear that this is bad strategy and would make numerical work too expensive. So we have to look for some alternative.

It is always good practice to reduce the general task to a standard one. Here, the standard problem we face is to invent an algorithm that creates sample paths $\bar{\omega}(\tau)$ of the d-dimensional Brownian bridge $\bar{X}_\tau$, where the time interval is $0 \leq \tau \leq 1$ and

$$E(\bar{X}_\tau \bar{X}_{\tau'}) = \big(\min(\tau, \tau') - \tau\tau' \big)\mathbb{1}. \tag{3.5.1}$$

Recall that any path integral where one integrates over Brownian paths pinned down at both ends, can be rescaled so as to become an integral with respect to the Brownian bridge. The resulting expression will then be of the type

$$\int d\bar{\omega}\, f\left(\int_0^1 d\tau\, g\big(\bar{\omega}(\tau),\tau\big)\right),\qquad (3.5.2)$$

where the outer integral is over paths $\bar{\omega}$ between spacetime points $(0,0)$ and $(0,1)$. Clearly, the particular application we might have in mind will dictate the choice of the functions f and g. The stategy then is to evaluate f for sufficiently many random paths and to interpret $\int d\bar{\omega} f$ as the average value.

The computer program no doubt has to call a routine returning paths of the Brownian bridge. The idea behind such a routine is to repeatedly bisect the unit time interval and then to attribute random vectors, normally distributed, to the midpoints[59]. The desired path of the Brownian bridge is thus approached via a sequence of polygons pinned to $x = 0$ at both ends as required. The basis for such a procedure is the decomposition

$$\bar{X}_{(a+b)/2} = \frac{1}{2}\left\{\bar{X}_a + \bar{X}_b + (b-a)^{1/2}Y\right\},\qquad 0 \le a < b \le 1 \qquad (3.5.3)$$

which, for fixed a and b, introduces another random variable Y characterizing how the actual path deviates at the midpoint $(a+b)/2$ from the value given by the straight line segment between a and b. The advantage of the representation (3.5.3) is that Y is normally distributed with mean zero und covariance $E(Y_i Y_k) = \delta_{ik}$ (as may be inferred from (3.5.1) by a straightforward calculation) irrespective of the choice of a and b. This then suggests constructing the path $\bar{\omega}$ by starting from $\bar{\omega}(0) = \bar{\omega}(1) = 0$ and by repeatingly applying

$$\bar{\omega}\Big(\frac{a+b}{2}\Big) = \frac{1}{2}\big\{\bar{\omega}(a) + \bar{\omega}(b) + (b-a)^{1/2}y\big\} \qquad (3.5.4)$$

to obtain new coordinates of the path from random numbers y_i $(i = 1,\ldots,d)$. The numbers y_i ought to be sufficiently random, uncorrelated, and normally distributed with mean zero and unit variance. Here are the details of the algorithm.

Imagine that we start from the trivial path as the zeroth approximation which is the straight line joining the spacetime points $(0,0)$ and $(0,1)$.

Step 1. Bisecting the time interval $[0,1]$ amounts to choosing $a = 0$ and $b = 1$ in (3.5.4). Pick random numbers $y_1,\ldots,y_d$ to form a vector y, put $\bar{\omega}(\frac{1}{2}) = \frac{1}{2}y$, and join the points

$$(0,0) \;\text{------}\; (\bar{\omega}(\tfrac{1}{2}),\tfrac{1}{2}) \;\text{------}\; (0,1)$$

by straight line segments to obtain the first approximation.

---

[59]This procedure is due to Lévy [3.29], and pioneering numerical work appeared in [3.30].

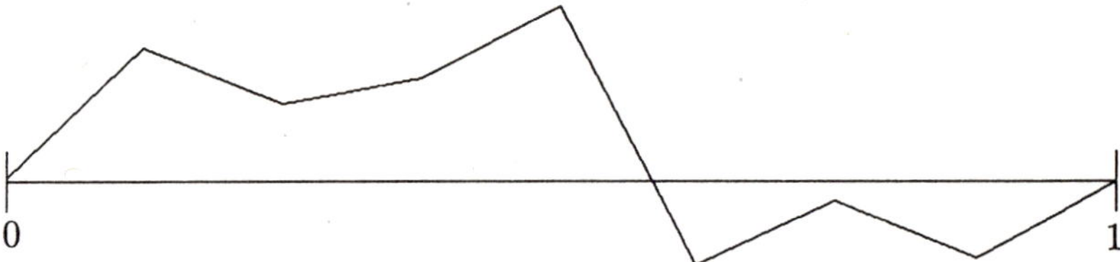

Fig. 3.5. A third-order approximation to a path of the one-dimensional Brownian bridge using 8 time steps

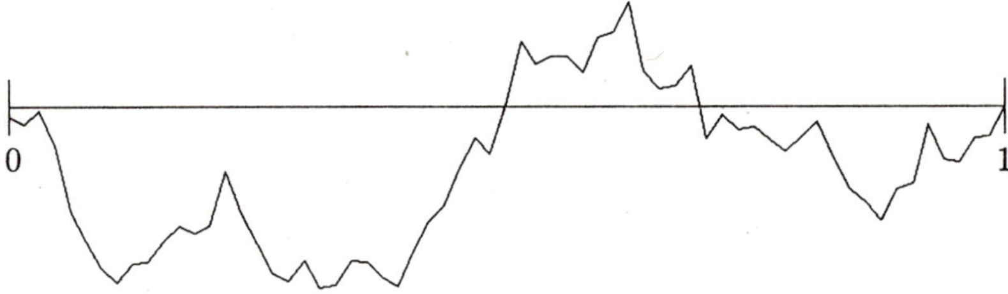

Fig. 3.6. A sixth-order approximation to a path of the one-dimensional Brownian bridge using 64 time steps

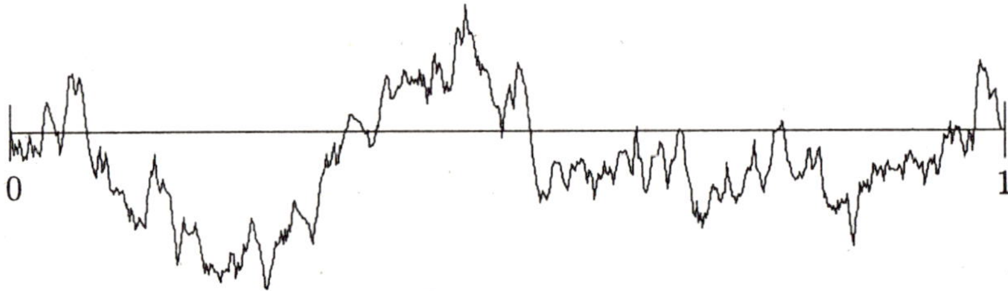

Fig. 3.7. A nineth-order approximation to a path of the one-dimensional Brownian bridge using 512 time steps

Step 2. The two subintervals are divided again and this time equation (3.5.4) will be applied twice: first, one takes $a = 0$, $b = \frac{1}{2}$ and then $a = \frac{1}{2}$, $b = 1$. The number of calls of the random number generator doubles, and one constructs $\bar{\omega}(\frac{1}{4})$ and $\bar{\omega}(\frac{3}{4})$. Join the points

$$(0,0) \quad\text{---}\quad (\bar{\omega}(\tfrac{1}{4}), \tfrac{1}{4}) \quad\text{---}\quad (\bar{\omega}(\tfrac{1}{2}), \tfrac{1}{2}) \quad\text{---}\quad (\bar{\omega}(\tfrac{3}{4}), \tfrac{3}{4}) \quad\text{---}\quad (0,1)$$

by straight-line segments; the polygon so obtained constitutes the second approximation. The process of bijection continues until we reach

Step n. For $k = 1, \ldots, 2^{n-1}$ one applies (3.5.3) with

$$a = \frac{k-1}{2^{n-1}}, \quad b = \frac{k}{2^{n-1}}. \tag{3.5.5}$$

This requires specification of $2^{n-1}d$ random numbers in the nth step. Having gone through all steps and having chosen $(2^n - 1)d$ random numbers altogether, one ends up with $2^n + 1$ spacetime points, which will then be joined by a polygon, the nth-order approximation of the Brownian bridge path.

We will not try to discuss good and bad random number generators but assume that invoking the built-in generator will result in a sequence of random numbers, uniformly distributed within the interval [0,1]. Then there is an elegant way to transform two uniform deviates x_1 and x_2 into normal deviates y_1 and y_2:

$$y_1 = (-2\log x_1)^{1/2}\cos(2\pi x_2)$$
$$y_2 = (-2\log x_1)^{1/2}\sin(2\pi x_2) \tag{3.5.6}$$

This method is known as the *Box–Muller* algorithm. See [3.31] for details and special programming tricks.

The result of a typical computer experiment is shown in Figs. 3.5–7. Having created *one* path of order n, how do we approximate the integral $I = \int_0^1 d\tau\, g(\bar{\omega}(\tau), \tau)$ in (3.5.2)? Let us use the abbreviation $\hat{g}(\tau) = g(\bar{\omega}(\tau), \tau)$. The obvious thing to do is to define the nth-order estimate of I by recursion:

$$I_0 = \tfrac{1}{2}\hat{g}(0) + \tfrac{1}{2}\hat{g}(1)$$
$$I_n = \tfrac{1}{2}I_{n-1} + 2^{-n}\sum_{k=1}^{2^{n-1}} \hat{g}\big(2^{-n}(2k-1)\big) \qquad (n \geq 1), \tag{3.5.7}$$

so that the precision increases as n gets large. Note that we need $2^n + 1$ evaluations of the function g for each path, but only one of the function f. From the integration theory point of view, (3.5.7) corresponds to applying the trapezoidal rule to each subinterval (each of which has length 2^{-n}). Even for smooth functions g, there is no benefit from using higher order Newton–Cotes formulas like Simpson's rule since the typical Brownian path is not differentiable.

It will be instructive to see the numerical algorithm at work, applying it to a path integral where one knows the analytic result beforehand. Such a case is provided by the following integral over the two-dimensional Brownian bridge:

$$I = \int d\bar{\omega}\,\exp\left\{-\frac{1}{2}\int_0^1 d\tau\,\bar{\omega}(\tau)^2\right\} = \frac{1}{\sinh 1} = 0.850\,918\dots \tag{3.5.8}$$

This integral is related to the harmonic oscillator in two dimensions, so the analytic result follows from Mehler's formula.

Table 3.2. The Monte Carlo results for the integral (3.5.8) using two-dimensional paths. The number of time steps and the number of sample paths are varied in this experiment to see their influence on the accuracy of the result. The standard deviation has been put into brackets and refers to the last digit: $0.852(3) = 0.852 \pm 0.003$. Numbers marked by a star disagree with the theoretical prediction by more than one standard deviation.

number of	number of sample paths				
time steps	10	10^2	10^3	10^4	10^5
2	*0.89(3)	*0.889(8)	*0.889(3)	*0.8886(8)	*0.8886(3)
4	0.86(3)	0.860(8)	*0.860(3)	*0.8604(8)	*0.8604(3)
8	0.85(3)	0.854(8)	0.854(3)	*0.8536(8)	*0.8536(3)
16	0.85(3)	0.851(8)	0.851(3)	0.8513(8)	*0.8513(3)
32	0.85(3)	0.851(8)	0.851(3)	0.8509(8)	0.8509(3)

The conclusion to be drawn from Table 3.2 is that the guiding strategy for getting reliable (well-bracketed) results should be to generate and use as many sample paths as can be afforded rather than to enlarge $m = 2^n$, the number of time steps in a drastic manner. With limited resources, say 10^3 samples, it suffices to choose $m = 8$ considering the rather large fluctuation (± 0.003 for the case at hand). It pays off only a little to pass on to $m = 16$ and nothing is gained by a further increase of m. But, of course, the behavior with respect to a change of m depends on the particular path integral under investigation and cannot be predicted in general.

A more formidable task is to determine the asymptotic behavior of $\langle x'|e^{-tH}|x\rangle$ as t tends to infinity and extract from it the ground state energy E. For illustration we will choose the following two-dimensional model: a particle of mass $m = 1$ confined to an equilateral triangle of side $a = 1$. This is Example 4 which has been analyzed in Sect. 2.9 in some detail. There the analytical result was

$$E = 8\pi^2/3 = 26.3189\ldots \tag{3.5.9}$$

Our present objective is to extract that number from a Monte Carlo simulation of Brownian motion in a region with absorbing boundaries. The general scheme is to write

$$\langle 0|e^{-tH}|0\rangle = (2\pi t)^{-1}P(t), \tag{3.5.10}$$

where

$$P(t) = \int d\bar{\omega}\, \chi_\triangle(t^{1/2}\bar{\omega}) \tag{3.5.11}$$

and $\chi_\triangle(\bar{\omega}) = \{\bar{\omega}\,|\,\bar{\omega}(\tau) \in \triangle, 0 \le \tau \le 1\}$, assuming that the center of the triangle $\triangle$ is at $x = 0$, and then to let the computer determine the probability $P(t)$ that the scaled path $t^{1/2}\bar{\omega}$ never leaves the triangle, where $\bar{\omega}: (0,0) \rightsquigarrow (0,1)$ is a sample path of the Brownian bridge. To do so we rely on the aforementioned Monte Carlo algorithm and use it to generate sufficiently many paths $\bar{\omega}$, say N, using 2^n intermediate time steps $\tau_k \in [0, 1]$ and ask whether the scaled path stays well inside the triangle. Let the fraction of these paths be $Z_n(t)$. Then

$$P_n(t) = Z_n(t)/N \tag{3.5.12}$$

is a reasonable approximation to $P(t)$ provided n is large enough. This routine has to be repeated for different increasing values of t until it becomes apparent that $(2\pi t)^{-1}P(t)$ drops exponentially with t. We therefore define

$$E = -\lim_{t \to \infty} \frac{d}{dt} \log\left((2\pi t)^{-1}P(t)\right). \tag{3.5.13}$$

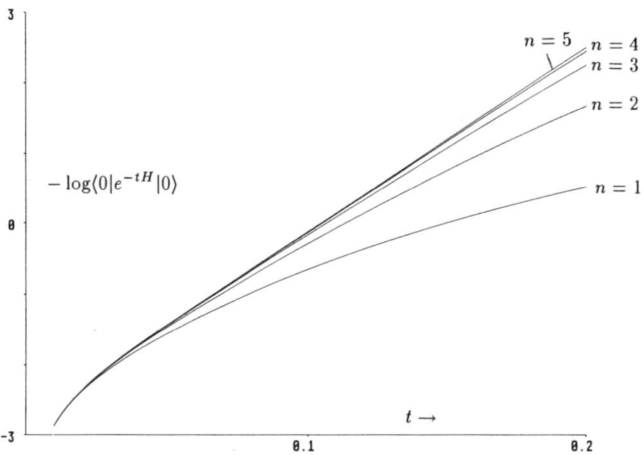

Fig. 3.8. Monte Carlo simulation to determine the ground state energy E of a particle (mass 1) confined to an equilateral triangle (side 1). The plot shows numerical data obtained from 10^5 sample paths using the bisection algorithm. As the number of time steps, 2^n, increases, the approximate values of $-\log\langle 0|e^{-tH}|0\rangle$ fall onto a straight line with slope $E \approx 26.3$, provided $t > 0.1$.

The problem we encounter in any such numerical calculation is that, for fixed n and large t, the function $P_n(t)$ does not decrease exponentially but, rather, algebraically. This forces us to evaluate the limits

$$E = -\lim_{t \to \infty} \lim_{n \to \infty} \frac{d}{dt} \log\left((2\pi t)^{-1}P_n(t)\right) \tag{3.5.14}$$

precisely in that order. Indeed, reversing the order of limits in (3.5.14) we would get

$$0 = \lim_{n \to \infty} \lim_{t \to \infty} \frac{d}{dt} \log\left((2\pi t)^{-1}P_n(t)\right). \tag{3.5.15}$$

It may be instructive to watch the systematic error grow with t in an actual experiment: the numerical results are shown graphically in Fig. 3.8. No matter what n, the error grows considerably as t increases, i.e., the fixed-n

curves spread, indicating a nonuniform convergence for $n \to \infty$, and, unless we do something to stop it, the spreading will actually impede our attempts to extract the correct asymptotic behavior of $\langle 0|e^{-tH}|0\rangle$ from the data. It is therefore absolutely crucial for any successful numerical computation to *extrapolate* all data to $n = \infty$, before any attempt is made to pass on to large values of t.

On the other hand, there is no systematic error to be expected from the fact that the number of samples, N, is held fixed. provided it has been chosen large enough. The reason: the number N merely influences the fluctuations which are ultimately Gaussian and, consequently, negative and positive errors of the same magnitude are equally likely (by the way, they may even grow with t).

The Monte Carlo experiment referred to in Fig. 3.8 used $N = 10^5$ and $n = 1, 2, \ldots, 5$ (and beyond). An extrapolation to $n = \infty$ has then been performed assuming

$$P_n(t) = P(t) + b(t)2^{-n} + c(t)2^{-2n} + \cdots, \tag{3.5.16}$$

where the coefficients $b(t)$ and $c(t)$ depend on t but not on n. The resulting function $(2\pi)^{-1}P(t)$ is well approximated by Ce^{-Et} for $t > 0.1$ and the numerical value of E is found to be 26.30 which is off the analytical result (3.5.9) by less than 0.1%.

Exercise 1. Recall that any path $\bar{\omega}$ of the Brownian bridge belongs to $C_0[0, 1]$, the space of real-linear functions on the interval $[0, 1]$ vanishing at 0 and 1. Following Schauder [3.32] construct a basis in $C_0[0, 1]$ as follows: let $h(t) = \max\big(0, 1-|t|\big)$ and

$$e_r(\tau) = h\big(2^n|\tau-r|\big), \qquad r = \frac{2k + 1}{2^n} ,$$

where r runs through the dyadic rationals satisfying $0 < r < 1$, hence $k = 0, 1, \ldots, 2^{n-1}-1$ and $n = 1, 2, \ldots$. The number $n = n(r)$ is said to be the *order* of the rational r. Prove that the Lévy bisection algorithm as described in this section amounts to an expansion of the form

$$\bar{\omega}(\tau) = \sum_r c_r y_r\, e_r(\tau), \qquad c_r = 2^{-[n(r)+1]/2}$$

such that the (random) coefficients y_r become statistically independent and normally distributed with mean zero and variance one. Prove also that, if the series is truncated after $n(r) = m$, i.e., if one writes

$$\bar{\omega}(\tau) = \sum_{n(r)\leq m} c_r y_r\, e_r(\tau) + \bar{\omega}_m(\tau),$$

the remainder $\bar{\omega}_m(\tau)$ represents an arbitrary path of the Brownian bridge vanishing at all rational points $\tau = r$ with $n(r) \leq m$. Estimate the error when the series expansion with respect to the Schauder basis is truncated after $n(r) = m$ in the path integral

$$\int d\bar{\omega} \,\exp\left\{-\int_0^1 d\tau\, \bar{V}\big(\bar{\omega}(\tau),\tau\big)\right\}.$$

(Assume that $\bar{V}$ has partial derivatives.)

Exercise 2.   The calculation of the (quantum) second virial coefficient of a gas of atoms reduces to the solution of a two-body problem. There is a direct and an exchange term. The latter involves the three-dimensional integral

$$I = \int dx\, \langle x|e^{-\beta H}|-x\rangle,$$

where H is the relative energy and x is the relative coordinate of two atoms of radius $a/2$ (in the hard sphere case) and mass $m = 2$ so that the reduced mass is unity and

$$H = -\tfrac{1}{2}\Delta + V, \qquad V(x) = \begin{cases} 0 & \text{if } |x| > a \\ \infty & \text{otherwise.} \end{cases}$$

Prove that I is merely a function of $\kappa = 2a^2/\beta \,(= ma^2\hbar^{-2}k_\mathrm{B}T)$. So we write $I(\kappa)$. To get a feeling for what is behind the exponential fall-off at large values of κ use a path integral representation in conjunction with a series expansion $\bar{\omega}(\tau) = 2^{-1}y_{1/2}e_{1/2}(\tau) + \cdots$ truncated after the first term (see the previous exercise). Change to more convenient variables: $t = 2\tau - 1$, $u = x/a$, and $v = \tfrac{1}{2}(\ell/a)y_{1/2}$, where $\ell^2 = \beta$. Then the integral becomes

$$I(\kappa) = \left(\frac{\kappa}{2\pi}\right)^3 \int_D du\, dv\, e^{-\kappa(u^2 + v^2)}$$

with the following domain of integration:

$$D = \Big\{ (u,v) \in \mathbb{R}^6 \ \Big| \ |tu + (1 - |t|)v| \geq 1, \quad \forall t \in [-1,1] \Big\}.$$

Show that

$$\inf_{(u,v)\in D} (u^2 + v^2) = 4.$$

Hence $I(\kappa)$ behaves like $e^{-4\kappa}$ (as opposed to the exact $e^{-\pi^2\kappa/4}$ [2.33]) as $\kappa \to \infty$. The asymptotic law may also be written as $e^{-L^2/(2\beta)}$, where L is the length of the shortest path between opposite points (mirror points) not intersecting the ball $|x| \leq a$. In the above estimate, $L = 4a$ since the path is assumed to be a polygon with two segments: $4a$ is half the circumference of a bounding square. Geometry, however, tells us that the shortest path is in fact a halfcircle of radius a, and therefore $L = \pi a$, yielding the exact asymptotic behavior.

4 Fourier Decomposition

> This sum-over-histories way of looking at things is not really
> so mysterious, once you get used to it. Like other profoundly
> original ideas, is has become slowly absorbed into the fabric of
> physics, so that now after thirty years it is difficult to remember
> why we found it at the beginning so hard to grasp.
>
> *Freeman Dyson*

In this chapter we wish to show that the Brownian bridge may be reconstructed from a denumerable set of independent Gaussian random variables
other than the ones used in the bisection algorithm considered in Sect. 3.5.
For it is possible to give meaning to the concept of Fourier decomposition
of Brownian motion and to prove statistical independence of the relevant
(random) Fourier coefficients. The motive for introducing this new structural element is that it allows us to approximate or even evaluate path
integrals in a more systematic fashion.

4.1 Random Fourier Coefficients

Every path of the Brownian bridge tends to the origin both at $\tau = 0$ and
$\tau = 1$. So do all functions $\sin(n\pi\tau)$ with $n \in \mathrm{I\!N}$. This offers a solution to the
problem of decomposing a path (i.e., a fractal curve) into smooth functions,

$$\bar{\omega}(\tau) = \sum_{n=1}^{\infty} \frac{\sqrt{2}}{n\pi} \xi_n \sin(n\pi\tau), \tag{4.1.1}$$

where the Fourier coefficients ξ_n ought to be regarded as random numbers
(or vectors). The extra factor $\sqrt{2}/(n\pi)$ has been introduced for convenience.
Correspondingly, the Brownian bridge $\bar{X}_\tau$ admits a Fourier decomposition
of the form

$$\bar{X}_\tau = \sum_{n=1}^{\infty} \frac{\sqrt{2}}{n\pi} X_n \sin(n\pi\tau) \tag{4.1.2}$$

such that, for each n, the random variable X_n takes the value ξ_n when
$\bar{X}_\tau$ is assigned the path $\bar{\omega}(\tau)$. Notice that X_n and ξ_n are vectors with
d components if our concern is d-dimensional Brownian motion. Next we
remind ourselves that the Fourier coefficients are explicitly given by

$$X_n = n\pi\sqrt{2} \int_0^1 d\tau\, \bar{X}_\tau \sin(n\pi\tau). \tag{4.1.3}$$

Since each random variable X_n is thus linearly dependent on a Gaussian process, it must be Gaussian itself. The next properties to be verified are *normalization* and *independence*:

$$\boldsymbol{E}(X_n X_{n'}) = \delta_{nn'} \mathbb{1} \tag{4.1.4}$$

We may also turn the argument around and equivalently start from (4.1.4) to prove the validity[60] of (3.1.10):

$$
\begin{aligned}
\boldsymbol{E}(\bar{X}_\tau \bar{X}_{\tau'}) &= \sum_{n=1}^{\infty} \frac{2}{n^2\pi^2} \sin(n\pi\tau)\sin(n\pi\tau')\mathbb{1} \\
&= \sum_{n=1}^{\infty} \frac{1}{n^2\pi^2}\big(\cos n\pi(\tau - \tau') - \cos n\pi(\tau' + \tau)\big)\mathbb{1} \\
&= \frac{1}{4}\left\{(\tau - \tau')^2 - 2|\tau - \tau'| - (\tau + \tau')^2 + 2|\tau + \tau'|\right\}\mathbb{1} \\
&= \big(\min(\tau, \tau') - \tau\tau'\big)\mathbb{1}.
\end{aligned}
\tag{4.1.5}
$$

In other words, if the X_n for $n = 1, 2, \ldots$ form a set of independent normalized Gaussian random variables, the process $\bar{X}_\tau$ given by (4.1.2) is a version of the Brownian bridge. The utility of Fourier decomposition in the context of Brownian motion comes from the fact that the Fourier coefficients are statistically independent: the same property would not necessarily hold for a series expansion with respect to some other basis of functions (compare, however, Exercise 1 at the end of Sect. 3.5).

Given representation (4.1.1), it is tempting to extend the Brownian bridge to a periodic function on the entire time axis satisfying

$$\bar{X}_{\tau+2} = \bar{X}_\tau, \qquad \bar{X}_{-\tau} = -\bar{X}_\tau. \tag{4.1.6}$$

Specifically, each path of the extended Brownian bridge will be continuous and periodic.

4.1.1 Fourier Analysis of Time Integrals

Path integrals are susceptible to numerical computation in many ways. One way is to Fourier-transform each path $\bar{\omega}$ of the Brownian Bridge and to treat the Fourier coefficients ξ_n as integration variables. Since integration with respect to the measure $d\bar{\omega}$ means *expectation* in the sense of probability theory, the expectation may also be evaluated using the Gaussian distribution of each variable $\xi_n \in \mathbb{R}^d$. Statistical independence of these variables makes $\bar{\omega}$ a product measure:

---

[60]We refer to Sect. 3.1 for comparison, especially to relation (3.1.12), which one encounters in a different context.

$$d\bar{\omega} = \prod_{n=1}^{\infty} \left[(2\pi)^{-d/2} e^{-\xi_n^2/2} \, d\xi_n \right].$$ (4.1.7)

On a purely conceptual level we may argue that time integrals such as

$$\int_0^1 d\tau \, \bar{V}(\bar{\omega}(\tau), \tau) = F(\xi_1, \xi_2, \ldots)$$ (4.1.8)

can be performed at least in principle, i.e., the above integral transforms into an expression depending smoothly on the variables ξ_n, $n = 1, 2, \ldots$, provided the scaled potential $\bar{V}$ is a sufficiently nice function. For instance, if $\bar{V}(y, \tau)$ happens to be a polynomial in y of order n, so is $F(\xi_1, \xi_2, \ldots)$ in the variables ξ_n. The following short list of integrals (written for $d = 1$) may serve as an illustration:

$$\int_0^1 d\tau \, \bar{\omega}(\tau) = \frac{\sqrt{8}}{\pi^2} \sum_{n=1}^{\infty} \frac{\xi_{2n-1}}{(2n-1)^2}$$ (4.1.9)

$$\int_0^1 d\tau \, \bar{\omega}(\tau)^2 = \frac{1}{\pi^2} \sum_{n=1}^{\infty} \frac{\xi_n^2}{n^2}$$ (4.1.10)

$$\int_0^1 d\tau \, \cos(\pi\tau)\bar{\omega}(\tau)^2 = \frac{1}{\pi^2} \sum_{n=1}^{\infty} \frac{\xi_n \xi_{n+1}}{n(n+1)}.$$ (4.1.11)

Observe that large n sine wave contributions to a typical path become less and less important: their effect on the resulting path integral decreases at the rate n^{-2}. Thus, it seems reasonable to truncate the Fourier expansion after some index $n_{\max}$ hoping that the error is small. Nevertheless, convergence of path integrals as $n_{\max} \to \infty$ may be frustratingly slow for large values of the length parameter ℓ (large $\hbar$ or low temperature T).

The crudest approximation (termed *semiclassical* in Sect. 3.2) neglects the ξ-dependence of integrands altogether and hence amounts to the replacement $F(\xi_1, \xi_2, \ldots) \to F(0, 0, \ldots)$ in (4.1.8). The ξ-integration of e^{-F} can then be carried out trivially. To get a better approximation, one may in a next step keep the dependence on a few variables, say $\xi_1, \ldots, \xi_m$, writing $F(\xi_1, \ldots \xi_m, 0, \ldots, 0)$ in place of (4.1.8). The ξ-integration is no longer trivial but reduces to the task of performing some finite-dimensional integral with respect to the variables kept. However, as m gets large, the calculation of the mth approximate to a path integral becomes undoubtedly unwieldy.

In an mth order approximation, the effect of ξ_n-fluctuations for $n > m$ can partially be taken into account (this method is called *partial averaging* by Coalson, Freeman, and Doll [4.1]) by applying Jensen's inequality to each of the ξ_n-averages, i.e., upon writing

$$\bar{\omega}_p^q(\tau) = \sum_{n=p}^{q} c_n(\tau)\xi_n, \qquad c_n(\tau) = \frac{\sqrt{2}}{n\pi} \sin(n\pi\tau) \qquad (q > p)$$ (4.1.12)

and also

$$d\bar{\omega}_p^q = \prod_{n=p}^{q} (2\pi)^{-d/3} e^{-\xi_n^2/2} d\xi_n \tag{4.1.13}$$

we have that $\bar{\omega} = \bar{\omega}_1^m + \bar{\omega}_{m+1}^\infty$ and $d\bar{\omega} \equiv d\bar{\omega}_1^\infty = d\bar{\omega}_1^m d\bar{\omega}_{m+1}^\infty$. Therefore,

$$\int d\bar{\omega} \, \exp\{-F\} \geq \int d\bar{\omega}_1^m \, \exp\left\{ - \int d\bar{\omega}_{m+1}^\infty F \right\}. \tag{4.1.14}$$

The method derives most of its power from the observation that

$$\int d\bar{\omega}_{m+1}^\infty \bar{V}\big(\bar{\omega}(\tau), \tau\big)$$

$$= (2\pi)^{-d/2} \int_{\mathbb{R}^d} dy \, e^{-y^2/2} \bar{V}\big(\bar{\omega}_1^m(\tau) + \lambda_m(\tau)y, \ \tau\big), \tag{4.1.15}$$

where

$$\lambda_m(\tau)^2 = \sum_{n=m+1}^{\infty} \frac{2}{n^2\pi^2} \sin^2(n\pi\tau)$$

$$= \tau(1-\tau) - \sum_{n=1}^{m} \frac{2}{n^2\pi^2} \sin^2(n\pi\tau). \tag{4.1.16}$$

The proof of (4.1.15) runs as follows. Consider the random variable $Y = \lambda^{-1} \sum_{n>m} c_n X_n$ with $\lambda^2 = \sum_{n>m} c_n^2$ and X_n as in (4.1.2). By construction, Y is Gaussian with mean zero and variance one. Hence, for suitable functions $f \colon \mathbb{R}^d \to \mathbb{R}$,

$$\boldsymbol{E}\big(f(Y)\big) = (2\pi)^{-d/2} \int_{\mathbb{R}^d} dy \, e^{-y^2/2} f(y). \tag{4.1.17}$$

Now take c_n as in (4.1.12) and $f(y) = \bar{V}\big(\bar{\omega}_1^m(\tau) + \lambda_m(\tau)y, \ \tau\big)$ to arrive at relation (4.1.15).

What has been achieved by applying (4.1.15) to the lower bound of (4.1.14) is that an infinite-dimensional Gaussian integral has been reduced to a d-dimensional Gaussian integral. The price to be paid is that, almost invariably, one encounters a more complicated integrand in the τ-integral. For a computer, however, the added complexity does not matter. All that matters is the low dimensionality of the reduced problem. For numerical work using these techniques see [4.1].

4.2 The Wigner–Kirkwood Expansion
of the Effective Potential

Quantum statistical mechanics is predominantly concerned with the density function $\langle x|e^{-\beta H}|x\rangle$, i.e., with the diagonal values of the transition amplitude. Its calculation may for instance serve as the essential step towards determining the free energy F:

$$e^{-\beta F} = \int dx\, \langle x|e^{-\beta H}|x\rangle. \tag{4.2.1}$$

Yet another concept which helps the physical intuition and insight is that of the *effective potential* $V_{\text{eff}}(x)$. Let us recall its definition from Sect. 2.5:

$$\langle x|e^{-\beta H}|x\rangle = (2\pi\ell^2)^{-d/2} e^{-\beta V_{\text{eff}}(x)}.$$

Actually, $\ell^2 = \beta$, since it is assumed that $m = \hbar = 1$. Still, in the present formalism, β^{-1} will be treated as an energy, while ℓ is a length:

$$\beta = (k_B T)^{-1} \quad , \qquad \ell = \hbar(m k_B T)^{-1/2}. \tag{4.2.2}$$

So because they play different roles, it is certainly not wise to eliminate one of these parameters. It should also be kept in mind that $V_{\text{eff}}(x)$ has a hidden nontrivial dependence on β and ℓ.

We will be focusing on the dependence of $V_{\text{eff}}(x)$ on ℓ. Since ℓ dictates the scale of quantum fluctuations, any expansion into powers of ℓ^2 provides higher order corrections to the semiclassical approximation $V_{\text{eff}}(x) = V(x)$. The most natural procedure would then be to appeal to (3.1.19), i.e.,

$$\exp\left\{-\beta V_{\text{eff}}(x)\right\} = \int d\bar{\omega} \, \exp\left\{-\beta \int_0^1 d\tau\, V\big(x + \ell\bar{\omega}(\tau)\big)\right\}, \tag{4.2.3}$$

and then use the Fourier decomposition of the Brownian bridge. The success is limited only by the amount of labor needed to do the necessary integrals. Here we shall be discussing merely the first few terms of that series. The approximation to be constructed should, however, meet the following demands:

1. For potentials at most quadratic in $x \in \mathbb{R}^d$, the approximate version of $V_{\text{eff}}(x)$ becomes exact.
2. For any other potential, the approximation represents $V_{\text{eff}}(x)$ up to an error of the order $\hbar^4$.

We start out from the Taylor expansion

$$\beta V(x + \ell\bar{\omega}(\tau)) = c + v^T \bar{\omega}(\tau) + \tfrac{1}{2}\bar{\omega}(\tau)^T M \bar{\omega}(\tau) + \cdots, \tag{4.2.4}$$

where use is made of the following abbreviations:

$$
\begin{aligned}
c &= \beta V(x) & \text{(a number)} \\
v &= \beta \ell V'(x) & \text{(a vector)} \\
M &= \beta \ell^2 V''(x) & \text{(a matrix)}.
\end{aligned}
$$

From the general representation (4.2.3)

$$
\beta V_{\text{eff}}(x) = c - \log \int d\bar{\omega}\, \exp\left\{ - \int_0^1 d\tau \left(v^T \bar{\omega}(\tau) + \tfrac{1}{2}\bar{\omega}(\tau)^T M \bar{\omega}(\tau) \right) \right\}
$$

$$
+ O(\hbar^4). \qquad (4.2.5)
$$

For reasons of symmetry (reflection $\bar{\omega} \to -\bar{\omega}$) there are only *even* powers of ℓ in the expansion of $V_{\text{eff}}(x)$. The first term neglected in (4.2.5) is of the order $\beta \ell^4$ and involves the fourth derivative $V^{(4)}$. Besides, there is a competing term of the order $\beta^2 \ell^4$ involving the product $V'V'''$. In total, the error is of the order $\hbar^4$ as indicated.

The following calculation, where we plunge into the intricacies of Fourier analysis, are more or less straightforward. Previous formulas such as (4.1.9) and (4.1.10) have obvious d-dimensional analogues which will be needed. Upon introducing vectors a_n and matrices A_n by

$$
a_n = \frac{\sqrt{2}}{\pi^2 n^2}\left[1 - (-1)^n\right] v \quad , \qquad A_n = \mathbb{1} + \frac{1}{\pi^2 n^2} M \qquad (n \in \mathbb{N}) \quad (4.2.6)
$$

the τ-integration in (4.2.5) may be performed yielding

$$
\beta V_{\text{eff}}(x) = c - \sum_{n=1}^{\infty} \log I_n + O(\hbar^4), \qquad (4.2.7)
$$

where the I_n are all Gaussian integrals that pose no problem provided the matrices A_n are nonsingular:

$$
\begin{aligned}
I_n &= (2\pi)^{-d/2} \int d\xi_n\, \exp\{-a_n \xi_n - \tfrac{1}{2}\xi_n^T A_n \xi_n\} \\
&= [\det A_n]^{-1/2} \exp\left\{ \tfrac{1}{2} a_n^T A_n^{-1} a_n \right\}.
\end{aligned}
\qquad (4.2.8)
$$

What we end up with is a sum over n. It naturally leads us to consider two functions $\mathcal{F}$ and $\mathcal{G}$ of a complex variable z:

$$
\mathcal{F}(z) = \tfrac{1}{2} \log \prod_{n=1}^{\infty} \left(1 + \frac{z}{\pi^2 n^2}\right) = \tfrac{1}{2} \log \frac{\sinh \sqrt{z}}{\sqrt{z}}
$$

$$
\qquad (4.2.9)
$$

$$
\mathcal{G}(z) = 4 \sum_{n=1,3,5,\ldots} \left[\pi^2 n^2 (\pi^2 n^2 + z)\right]^{-1} = \frac{1}{2z}\left(1 - \frac{\tanh(\tfrac{1}{2}\sqrt{z})}{\tfrac{1}{2}\sqrt{z}}\right).
$$

To summarize, the result of the calculation can be cast into the form

$$
\beta V_{\text{eff}}(x) = \beta V(x) + \operatorname{tr} \mathcal{F}(M) - v^T \mathcal{G}(M) v + O(\hbar^4), \qquad (4.2.10)
$$

which makes $\mathcal{F}$ and $\mathcal{G}$ functions of the matrix $M = \beta\ell^2 V''(x)$. Both $\operatorname{tr}\mathcal{F}(M)$ and $v^T \mathcal{G}(M)v$ are of order $\hbar^2$, but also include higher orders.

It is sometimes desirable to simplify this formula even further using the leading terms of the Taylor series

$$\mathcal{F}(z) = \frac{1}{12}z + O(z^2), \qquad \mathcal{G}(z) = \frac{1}{24} + O(z). \tag{4.2.11}$$

This way we would get the Wigner–Kirkwood expansion [4.2,3] again correct to order $\hbar^2$:

$$V_{\mathrm{eff}}(x) = V(x) + \frac{\ell^2}{12}\sum_i \frac{\partial^2 V(x)}{\partial x_i^2} - \frac{\beta\ell^2}{24}\sum_i \left(\frac{\partial V(x)}{\partial x_i}\right)^2 + O(\hbar^4). \tag{4.2.12}$$

Caution: both expansion formulas (4.2.10) and (4.2.12) will be useful only in a regime where the nth partial derivatives of the potential, if multiplied by ℓ^n, turn out to be small compared to V itself. Formally,

$$\ell^n |V^{(n)}(x)| \ll |V(x)|. \tag{4.2.13}$$

To illustrate this point we take the Coulomb potential as an example. Then condition (4.2.13) reads $r = |x| \gg \ell$. Quite generally we expect the following to be true: if the potential $V(x)$ happens to develop a singularity at $x = 0$, there is always some vicinity of that point where a series expansion in powers of $\hbar$ loses its meaning. It is ℓ (and, hence, the thermal wavelength $(2\pi)^{1/2}\ell$) that determines the size of that vicinity.

As follows from its derivation, (4.2.10) meets our demands stated at the beginning, whereas (4.2.12) does not: (4.2.10) is exact to the indicated order of $\hbar$, and error terms are absent if the potential is at most quadratic. In the simplest case, where $V(x) = \frac{1}{2}k^2 x^2$ ($x \in \mathbb{R}^d$), the formula (4.2.10) yields

$$V_{\mathrm{eff}}(x) = \frac{d}{2}\beta^{-1}\log\frac{\sinh\nu}{\nu} + k^2 x^2 \frac{\tanh(\nu/2)}{\nu} \qquad (\nu = \beta^{1/2}\ell k), \tag{4.2.14}$$

a result which conforms with what we would get from Mehler's formula. The effective potential of the harmonic oscillator is again a second order polynomial in x. It shows the same rotational symmetry. Yet there are changes: (1) a constant varying with the temperature has been added such that, if $\nu = \beta k$ and $\beta \to \infty$, $\inf V_{\mathrm{eff}}(x) = (d/2)k$ (the ground state energy), and (2) the frequency $k\big((\nu/2)^{-1}\tanh(\nu/2)\big)^{1/2}$ has acquired a temperature dependence.

Notice also that the effective potential (4.2.14) of the oscillator gives the correct free energy: $\beta F/d = \log(2\sinh(\nu/2))$, while (4.2.12) produces an approximation to it:

$$\beta F/d = \log\nu + \tfrac{1}{2}\log(1 - \tfrac{1}{12}\nu^2) + \tfrac{1}{12}\nu^2 + O(\nu^4). \tag{4.2.15}$$

The first term represents the classical expression for $\beta F/d$ provided the phase space volume is measured in units of $(2\pi\hbar)^d$.

For any spherically symmetric potential $V(r)$ in three dimensions the following is true. The matrix M has eigenvalues $\lambda_1 = \beta\ell^2 V''(r)$ and $\lambda_2 = \lambda_3 = -\beta\ell^2 r^{-1} V'(r)$, where $V'(r)$ and $V''(r)$ denote the ordinary derivatives with respect to r. Furthermore, $Mv = \lambda_1 v$, hence $\mathcal{F}(M)v = \mathcal{F}(\lambda_1)v$ and $v^2 = \beta^2\ell^2 V'(r)^2$. As a result, a simplification occurs in (4.2.10):

$$V_{\text{eff}}(x) = V(r) + \beta^{-1}\mathcal{F}\big(\beta\ell^2 V''(r)\big) + 2\beta^{-1}\mathcal{F}\big(-\beta\ell^2 r^{-1}V'(r)\big)$$
$$-\beta\ell^2 V'(r)^2\mathcal{G}\big(\beta\ell^2 V''(r)\big) + O(\ell^4). \tag{4.2.16}$$

Suppose for instance that, asymptotically for large r, the potential $V(r)$ behaves like αr^{-n}. Then, correspondingly, the effective potential shows the following asymptotic behavior:

$$V_{\text{eff}}(x) = \frac{\alpha}{r^n}\left[1 - \frac{1}{12}\left(\frac{\ell}{r}\right)^2\left(n(n-1) + \frac{n^2}{2}\frac{\beta\alpha}{r^n}\right) + O(\ell^4/r^4)\right]. \tag{4.2.17}$$

The case $n = 1$ is peculiar in that the order ℓ^2/r^2 correction drops off faster than usual and ultimately becomes negligible if $\beta\alpha/r \ll 1$, i.e., if $V(x) \ll k_B T$.

4.3 Coupled Systems

In this section we look more closely at path integrals arising from $H = -\frac{1}{2}\Delta + V$ in d dimensions when, for some reason or another, the system is composed of two subsystems indexed 1 and 2. To think of two separate systems, even in the presence of interaction, may be advisable and justified provided one of the systems is sufficiently simple that the path integration can be carried out at least partially. The notation and the techniques we develop here also serve to prepare the ground for the subsequent sections.

Let us thus write $x = (x_1, x_2) \in \mathbb{R}^d$ with $x_1 \in \mathbb{R}^{d_1}$ and $x_2 \in \mathbb{R}^{d_2}$ so that $d_1 + d_2 = d$. The potential is assumed to be given as

$$V(x) = V_1(x_1) + V_2(x_2) + \lambda W(x_1, x_2), \tag{4.3.1}$$

where λW describes the interaction. Frequently, the coupling constant λ is treated as a free parameter. Think of an arbitrary path $\bar{\omega}$ of the Brownian bridge associated with the system as a whole. This path will be uniquely specified by its projections, $\bar{\omega}_1$ and $\bar{\omega}_2$, living in different configuration spaces (of dimension d_1 and d_2), and we may thus write $\bar{\omega} = (\bar{\omega}_1, \bar{\omega}_2)$. Needless to say, both $\bar{\omega}_1$ and $\bar{\omega}_2$ are, in their own territory, unrestricted paths of lower-dimensional versions of the Brownian bridge. Since they are uncorrelated, $d\bar{\omega} = d\bar{\omega}_1 d\bar{\omega}_2$.

The transition amplitude of the compound system acquires the structure of a twofold path integral:

$$\langle x', s'|x, s\rangle = K(x' - x, s' - s)\times$$

$$\int d\bar{\omega}_2 \int d\bar{\omega}_1 \exp\left\{ -\int_0^1 d\tau \left(\bar{V}_1(\bar{\omega}_1(\tau), \tau) + \bar{V}_2(\bar{\omega}_2(\tau), \tau) \right.\right.$$

$$\left.\left. + \lambda \bar{W}(\bar{\omega}_1(\tau), \bar{\omega}_2(\tau), \tau) \right) \right\}, \quad (4.3.2)$$

where for conciseness we have used the scaled potentials $\bar{V}_1$, $\bar{V}_2$, and $\bar{W}$. Obviously, $K(x' - x, s' - s) = K_1(x_1' - x_1, s' - s)K_2(x_2' - x_2, s' - s)$ and a similar product structure of $\langle x', s'|x, s\rangle$ arises if $\lambda = 0$: the two subsystems appear uncorrelated if they do not interact. If $\lambda \neq 0$, the calculation of the transition amplitude may be viewed as a two-step process. The first step consists in performing a partial integration, where one concentrates primarily on properties of the first system. This leaves us with the amplitude

$$\langle x_1', s'|x_1, s\rangle_{\omega_2} = K_1(x_1' - x_1, s' - s)\times$$

$$\int d\bar{\omega}_1 \exp\left\{ -\int_0^1 d\tau \left(\bar{V}_1(\bar{\omega}_1(\tau), \tau) + \lambda \bar{W}(\bar{\omega}_1(\tau), \bar{\omega}_2(\tau), \tau) \right) \right\} \quad (4.3.3)$$

which depends parametrically on $\bar{\omega}_2$, the path the second system might take. Of course, we would always seek to analytically determine or at least control this quantity before we proceed to the second step, which is still another integration yielding the full amplitude:

$$\langle x', s'|x, s\rangle = K_2(x_2' - x_2, s' - s)\times$$

$$\int d\bar{\omega}_2 \langle x_1', s'|x_1, s\rangle_{\omega_2} \exp\left\{ -\int_0^1 d\tau \, \bar{V}_2(\bar{\omega}_2(\tau), \tau) \right\}. \quad (4.3.4)$$

The foregoing procedure is exact and equivalent to any of our previous prescriptions. Yet it reveals that the auxiliary quantity $\langle x_1', s'|x_1, s\rangle_{\omega_2}$ is in fact the transition amplitude of a fictitious system with the time-dependent potential $V_{\omega_2}(x_1, t)$, where – in terms of the unscaled quantities –

$$V_{\omega_2}(x_1, t) = V_1(x_1) + \lambda W(x_1, \omega_2(t))$$
$$\omega_2(t) = x_2 + (x_2' - x_2)\tau + \ell\bar{\omega}_2(\tau) \quad (4.3.5)$$
$$t = s + (s' - s)\tau.$$

It is worth pointing out some vital aspect of (4.3.5). The position $\omega_2(t)$ which $V_{\omega_2}(x_1, t)$ depends upon is a random vector. As a consequence, the potential V_{ω_2} splits into a deterministic and a random part. The fluctuations of the random potential have to be inferred from a study of the complementary subsystem, i.e., the environment. Note, however, that such fluctuations originate from quantum theory, not from classical statistics.

The formalism also makes the potential $V_{\omega_2}(x_1, t)$ dependent on the time t and the position reached by the path $\omega_2\colon (x_2, s) \rightsquigarrow (x_2', s')$ traversed by the second system. This provides sufficient reason to include potentials varying explicitly with time into the general theory of path integration. It gets us into the subject of open systems.

4.3.1 Open Systems

We have argued that it is wise to extend the concept of the potential so as to allow for potentials taking time as a parameter. The extension poses no problem as far as the definition of path integrals is concerned. From the physics point of view, however, there seems to be a problem, since t-dependence of Hamiltonians signals that energy conservation has been sacrificed. As a matter of fact, we do sacrifice conservation of energy for the subsystem, but we do not for the entire system as a whole. In addition, a declaration of a t-dependent potential for some small physical system can hide a declaration of an interaction and hence an exchange of energy with its environment. Hiding details of an interaction with the exterior world often seems unavoidable. If such an influence is present but hidden from the the observer, the small system is said to be *open*.

Before we go on, let us simplify the notation first. We shall write x in place of x_1 and take $H(t) = -\frac{1}{2}\Delta + V(x, t)$ as the t-dependent Hamiltonian of our small system. The transition amplitude as given by the Feynman–Kac formula defines a two-parameter family of evolution operators, $U(s', s)$, such that

$$\langle x', s' | x, s \rangle = \langle x' | U(s', s) | x \rangle. \tag{4.3.6}$$

From an operator point of view, (4.3.6) provides the solution of

$$\left(\frac{\partial}{\partial s'} - H(s') \right) U(s', s) = 0, \quad U(s, s) = 1 \text{ (the unit operator)}. \tag{4.3.7}$$

We must resist the temptation to write the solution of (4.3.7) as $\exp\{-\int_s^{s'} dt\, H(t)\}$ the reason being that, in most cases, $[H(t), H(t')] \neq 0$ if $t \neq t'$. He who wants to associate an explicit expression with the operator $U(s', s)$ may perhaps settle for the following: upon dividing the interval $[s, s']$ into subintervals of equal length,

$$s_k = s + kh \quad, \quad k = 0, 1, \ldots, n \quad, \quad h = \frac{s' - s}{n + 1}, \tag{4.3.8}$$

one can take the following (operator) limit as a definition:

$$U(s', s) = \lim_{n \to \infty} \prod_{k=0}^{n} {}^{>} \exp\left(-hH(s_k) \right). \tag{4.3.9}$$

The symbol $>$ has been attached to the product sign to indicate that a specific order of factors has been chosen. Namely, the product is *time ordered*, meaning that the factor carrying a large time tag stands left of factors with smaller tags. Still another way of representing the same object is a formal expansion into powers of H:

$$U(s',s) = 1 + \sum_{n=1}^{\infty} (-1)^n \int_s^{s'} dt_1 \int_s^{t_1} dt_2 \cdots \int_s^{t_{n-1}} dt_n \, H(t_1)H(t_2)\cdots H(t_n). \quad (4.3.10)$$

However, expansion (4.3.10) is less useful, since each term of the sum defines some unbounded operator making the meaning of convergence rather obscure. Either (4.3.9) or (4.3.10) is conveniently symbolized by writing

$$U(s',s) = \mathrm{T}\exp\left\{ -\int_s^{s'} dt \, H(t) \right\}, \quad (4.3.11)$$

where $\mathrm{T}\exp$ stands for *time ordered exponential*.

There are various contexts in which operators like $U(s',s)$ may occur. The most obvious interpretation is that they describe the time evolution of some diffusion process lacking homogeneity in time (so-called *non-autonomous systems*), i.e., the subject would then be random motion in some absorbing medium with the absorption probability depending on space and time. In quantum mechanics, $U(s',s)$ generalizes the Schrdinger semi-group $e^{-(s'-s)H}$. We would then want to analytically continue to imaginary times, arguing that $U(it',it)$ solves the evolution problem for a quantum mechanical system under the influence of the potential $V(x,it)$. At first sight, the latter construction seems to rely on very peculiar analyticity properties of the potential with regard to its time dependence. To avoid these and a whole lot of confusion we must adopt the philosophy that in reality all nontrivial functions like potentials take *only dimensionless arguments*. In particular, only quotients of time variables are admitted as arguments of V. Such quotients might, for instance, arise as $\tau = t/T$, where T is some intrinsic period of time, or as $\tau = (t-s)/(s'-s)$ when t varies between s and s'. Indeed, the argument τ is then merely a real number and remains so even when T, t, s, s' assume imaginary values. Under these provisions we expect the operators $U(it',it)$ to come out all unitary.

Exercise 1. Consider a time-dependent Hamiltonian of the form $H(t) = -\frac{1}{2}\Delta + V(\cdot,t)$ and its transition amplitude. Extend the Golden–Thompson–Symanzik inequality to this case, i.e., prove that

$$\mathrm{tr}\,\mathrm{T}\exp\left\{ -\int_0^{\beta} dt \, H(t) \right\} \le (2\pi\beta)^{-d/2} \int dx \int_0^1 d\tau \, e^{-\beta V(x,\beta\tau)},$$

where d is the dimension and $\mathrm{T}\exp$ denotes the time-ordered exponential. Hint: represent the trace as an integral with respect to the Brownian bridge and use Jensen's inequality when interchanging the τ-integration and the exponential map.

4.4 The Driven Harmonic Oscillator

There is one problem that admits a complete answer and an analytic solution, and that is the linearly driven harmonic oscillator:

$$H(t) = -\tfrac{1}{2}\Delta + V(\cdot, t) \quad , \qquad V(x, t) = \tfrac{1}{2}k^2 x^2 - E(t)x \tag{4.4.1}$$

($x \in \mathbb{R}^d$) for some vector $E(t)$ changing with time. No matter what d, we will call $E(t)$ *the electric field*. If $d = 3$ we may think of the system as an electron coupled to the radiation field, which to some extent allows for an exchange of energy. The combined system conserves energy, of course.

Given the time interval $[s, s']$, the first step towards a solution of the path integral is to scale down the interval to $[0, 1]$ and the implied passage to the scaled potential $\bar{V}$ (see Sect. 3.1), which allows us to write

$$\bar{V}\big(\bar{\omega}(\tau), \tau\big) = \tfrac{1}{2}k\nu\Big\{ \big(f(\tau) + \ell\bar{\omega}(\tau)\big)^2 + \bar{E}(\tau)^2 \Big\} \tag{4.4.2}$$

$$\ell = (s' - s)^{1/2}, \qquad \nu = k(s' - s)$$

with $\bar{\omega}\colon (0, 0) \rightsquigarrow (0, 1)$ a path of the Brownian bridge and the shorthand

$$\begin{aligned} f(\tau) &= x + (x' - x)\tau - \bar{E}(\tau) \\ \bar{E}(\tau) &= k^{-2}E\big(s + (s' - s)\tau\big). \end{aligned} \tag{4.4.3}$$

The second step is to introduce the Fourier coefficients $f_n \in \mathbb{R}^d$ of $f(\tau)$:

$$f(\tau) = \sum_{n=1}^{\infty} f_n \sqrt{2}\sin(n\pi\tau). \tag{4.4.4}$$

More explicitly,

$$\begin{aligned} f_n &= \frac{\sqrt{2}}{n\pi}\Big(x - (-1)^n x'\Big) - E_n \\ E_n &= \sqrt{2}\int_0^1 d\tau\, \bar{E}(\tau)\sin(n\pi\tau). \end{aligned} \tag{4.4.5}$$

To remove doubts as to why a decomposition such as (4.4.4) is legal, we would like to add one remark. Let $L^2(0, 1)$ denote the Hilbert space of square-integrable functions on $0 \le \tau \le 1$. Then the system of functions $\varphi_n(\tau) = \sqrt{2}\sin(n\pi\tau)$ ($n = 1, 2, \ldots$) forms a basis in that space even though there exist functions $g(\tau) \in L^2(0, 1)$ which do not vanish at the endpoints 0 and 1. The secret of why $g(\tau) = \sum g_n\varphi_n(\tau)$ is still true in that case is that convergence of the sum takes place, not pointwise, but only in the mean, i.e., in the L^2 sense. We thus do not expect the condition $\sum |g_n| < \infty$ to be satisfied but only $\sum |g_n|^2 < \infty$.

4.4.1 From Time Integrals to Sums

We follow the procedure outlined in Sect. 4.1. Let ξ_n be the Fourier coefficients of the path $\bar{\omega}(\tau)$ as given by (4.1.1). It is then desirable to pass from time integrals to sums:

$$\int_0^1 d\tau \left(f(\tau) + \ell\bar{\omega}(\tau) \right)^2 = \sum_{n=1}^{\infty} (f_n + \ell(n\pi)^{-1}\xi_n)^2 \tag{4.4.6}$$

and

$$\tfrac{1}{2}k\nu \int_0^1 d\tau\, \bar{E}(\tau)^2 = \tfrac{1}{2}k\nu \sum_{n=1}^{\infty} E_n^2 =: I \tag{4.4.7}$$

(assuming $I < \infty$) so that

$$\int_0^1 d\tau\, \bar{V}(\bar{\omega}(\tau), \tau) = -I + \tfrac{1}{2}k\nu \sum_{n=1}^{\infty} (f_n + \ell(n\pi)^{-1}\xi_n)^2. \tag{4.4.8}$$

The integral with respect to paths of the Brownian bridge is thus converted into a product of ordinary d-dimensional Gaussian integrals I_n:

$$\int d\bar{\omega}\, \exp\left\{ -\int_0^1 d\tau\, \bar{V}(\bar{\omega}(\tau), \tau) \right\} = e^I \prod_{n-1}^{\infty} I_n. \tag{4.4.9}$$

Each of these integrals can be evaluated immediately:

$$\begin{aligned}
I_n :&= (2\pi)^{-d/2} \int d\xi\, \exp\left\{ -\tfrac{1}{2}\xi^2 - \tfrac{1}{2}k\nu(f_n + \ell(n\pi)^{-1}\xi)^2 \right\} \\
&= \left(1 + \frac{\nu^2}{n^2\pi^2} \right)^{-d/2} \exp\left\{ -\tfrac{1}{2}k\nu f_n^2 \left(1 + \frac{\nu^2}{n^2\pi^2} \right)^{-1} \right\}.
\end{aligned} \tag{4.4.10}$$

As a result, the transition amplitude is given by

$$\langle x', s'| x, s \rangle_E = K(x' - x, s' - s)\, e^I \prod_{n=1}^{\infty} I_n. \tag{4.4.11}$$

This, however, is a rather indirect description. It is also inconvenient.

4.4.2 From Sums Back to Time Integrals

Recall that

$$K(x' - x, s' - s) = (2\pi(s' - s))^{-d/2} \exp\left\{ -\frac{(x' - x)^2}{2(s' - s)} \right\} \tag{4.4.12}$$

and

$$\prod_{n=1}^{\infty} \left(1 + \frac{\nu^2}{n^2\pi^2}\right) = \frac{\sinh \nu}{\nu}. \tag{4.4.13}$$

Insertion of these into (4.4.10) and (4.4.11) yields

$$\langle x', s' | x, s \rangle_E = \left[\frac{k}{2\pi \sinh \nu}\right]^{d/2} \exp\left\{-\frac{k(x'-x)^2}{2\nu}\right\} \times$$

$$\exp\left\{\tfrac{1}{2}k\nu \sum_{n=1}^{\infty}\left[E_n^2 - f_n^2\left(1 + \frac{\nu^2}{n^2\pi^2}\right)^{-1}\right]\right\}. \tag{4.4.14}$$

As a check of the calculation we temporarily switch off the electric field. Then $E_n = 0$ (for all n) in (4.4.5) and therefore $f_n = \frac{\sqrt{2}}{n\pi}\left(x - (-1)^n x'\right)$, which has to be inserted into (4.4.14). To perform the summation we remind ourselves that

$$\frac{1}{\sinh x} = \frac{1}{x} - \sum_{n=1}^{\infty}(-1)^n \frac{2x}{x^2 + n^2\pi^2}$$

$$\frac{1}{\tanh x} = \frac{1}{x} - \sum_{n=1}^{\infty}\frac{2x}{x^2 + n^2\pi^2}. \tag{4.4.15}$$

Not unexpectedly, the result is Mehler's formula!

Next, let us turn on the electric field. By (4.4.14), the transition amplitude considered as a functional of $E(t)$ possesses a representation of the form

$$\langle x', s' | x, s \rangle_E = \langle x', s' | x, s \rangle_0 \exp\left\{\mathcal{B}(E) + \mathcal{L}(E)\right\}, \tag{4.4.16}$$

where the factor in front of the exponential is simply Mehler's expression; $\mathcal{B}(E)$ is bilinear while $\mathcal{L}(E)$ is linear in the electric field (any dependence on x, x', s, s' being suppressed). We treat these two functionals separately.

1. The *bilinear term* is positive definite:

$$\mathcal{B}(E) = \tfrac{1}{2}k\nu^3 \sum_{n=1}^{\infty}\frac{E_n^2}{\nu^2 + n^2\pi^2}. \tag{4.4.17}$$

The key to converting the sum into a time integral is the formula

$$2\sum_{n=1}^{\infty}\frac{\sin n\pi\tau \sin n\pi\tau'}{\nu^2 + n^2\pi^2} = \frac{\cosh[\nu(1 - |\tau - \tau'|)] - \cosh[\nu(1 - |\tau + \tau'|)]}{2\nu \sinh \nu}$$

$$= \frac{\sinh[\nu \min(\tau, \tau')]\sinh[\nu(1 - \max(\tau, \tau'))]}{\nu \sinh \nu} \tag{4.4.18}$$

$(0 \le \tau, \tau' \le 1)$. Hence

$$\mathcal{B}(E) = \int_s^{s'} dt \int_s^t dt'\, E(t)E(t')\, \frac{\sinh[k(s'-t)]\sinh[k(t'-s)]}{k\sinh[k(s'-s)]}. \qquad (4.4.19)$$

Removal of the oscillator potential leads to a simpler expression:

$$\mathcal{B}(E)_{k=0} = \int_s^{s'} dt \int_s^t dt'\, E(t)E(t')\, \frac{(s'-t)(t'-s)}{s'-s}. \qquad (4.4.20)$$

If moreover the electric field is constant in time, the result would be simplest:

$$\mathcal{B}(E=\text{const})_{k=0} = \tfrac{1}{24}(s'-s)^3 E^2. \qquad (4.4.21)$$

2. The *linear term* is also seen to be linear in x and x':

$$\mathcal{L}(E) = k\nu\sqrt{2}\sum_{n=1}^{\infty}\frac{n\pi}{\nu^2+n^2\pi^2}\left(x + (-1)^{n+1}x'\right)E_n. \qquad (4.4.22)$$

Here the following formulas, valid for $0 \le \tau \le 1$, facilitate the conversion (they follow from (4.4.18)):

$$2\sum_{n=1}^{\infty}\frac{n\pi\sin n\pi\tau}{\nu^2+n^2\pi^2} = \frac{\sinh[\nu(1-\tau)]}{\sinh\nu}$$

$$2\sum_{n=1}^{\infty}(-1)^{n+1}\frac{n\pi\sin n\pi\tau}{\nu^2+n^2\pi^2} = \frac{\sinh[\nu\tau]}{\sinh\nu}. \qquad (4.4.23)$$

In fact we may write

$$\mathcal{L}(E) = \int_s^{s'} dt\, E(t)q(t), \qquad (4.4.24)$$

where

$$q(t) = \frac{x\sinh[k(s'-t)] + x'\sinh[k(t-s)]}{\sinh[k(s'-s)]}. \qquad (4.4.25)$$

Notice that $q(t)$ is nothing but the classical path satisfying $\ddot{q} = k^2 q$ with boundary conditions $q(s) = x$ and $q(s') = x'$ (compare (2.8.26–28)). In particular, if k tends to zero,

$$\mathcal{L}(E)_{k=0} = \int_s^{s'} dt\, E(t)\, \frac{(s'-t)x + (t-s)x'}{s'-s}, \qquad (4.4.26)$$

and if in addition the electric field is constant,

$$\mathcal{L}(E=\text{const.})_{k=0} = \tfrac{1}{2}(s'-s)(x'+x)E. \qquad (4.4.27)$$

Exercise 1.   As a simple application of the foregoing discussion consider a Schrdinger particle of mass m and charge e in a constant electric field, i.e., take

$$H = -\frac{\hbar^2}{2m}\Delta - eEx \qquad (d = 3)$$

as its Hamiltonian. Show that the transition amplitude can be analytically continued to imaginary time:

$$\langle x'|\exp(-(i/\hbar)t'H)|x\rangle =$$

$$\left[\frac{m}{2\pi i\hbar t'}\right]^{3/2} \exp\left\{\frac{i}{\hbar}t'\left(\tfrac{1}{2}m\left(\frac{x-x'}{t'}\right)^2 + \tfrac{1}{2}(x+x')eE - \frac{(t'eE)^2}{24m}\right)\right\}.$$

The classical trajectory $q(t)$ solving the problem $m\ddot{q} = eE$ with the boundary conditions $q(0) = x$ and $q(t') = x'$ is

$$q(t) = x + (x' - x)\frac{t}{t'} + \frac{eE}{2m}t(t - t').$$

Prove that for the Lagrangian $L(\dot{q}, q) = \tfrac{1}{2}m\dot{q}^2 + eEq$ and $q(t)$ as above, we have

$$\langle x'|\exp(-(i/\hbar)t'H)|x\rangle = \left[\frac{m}{2\pi i\hbar t'}\right]^{3/2} \exp\left\{\frac{i}{\hbar}S\right\},$$

where $S = \int_0^{t'} dt\, L$ is the classical action. Extend this assertion to the general situation where the electric field changes with time.

Exercise 2.   Consider the Hamiltonian of the linearly driven harmonic oscillator in d dimensions, $H(t) = -\tfrac{1}{2}\Delta + \tfrac{1}{2}k^2x^2 - E(t)x$. By evaluating both sides explicitly, show that the following equality holds:

$$\mathrm{tr}\,\mathrm{T}\exp\left\{-\int_0^\beta dt\, H(t)\right\} = \left[2\sinh\frac{\beta k}{2}\right]^{-d} \boldsymbol{E}\left(\exp\int_0^\beta dt\, E(t)Y_t\right),$$

where $\mathrm{T}\exp$ is the time-ordered exponential, $\boldsymbol{E}$ means *expectation*, and Y_t designates a d-dimensional Gaussian stochastic process (the "oscillator bridge") on the interval $0 \leq t \leq \beta$, with mean zero and covariance

$$\boldsymbol{E}(Y_tY_{t'}) = \frac{e^{-|t-t'|k} + e^{-(\beta-|t-t'|)k}}{2k(1 - e^{-\beta k})}.$$

Hint: use (4.4.16) and represent the trace as an integral over $x \in \mathbb{R}^d$. The expectation in the above trace formula in some sense replaces the integral over all closed paths in $\mathbb{R}^d$.

Exercise 3.   Demonstrate that the "oscillator bridge", as defined by the preceding exercise, is *not a Markov process*, i.e., the future depends not only on the present, but also on the past.

4.5 Oscillating Electric Fields

The formalism developed in the preceding section will now be applied to a charged particle (e.g. an ion in some crystal) confined by a harmonic potential. Assuming it to be initially in the ground state, we shall study the time evolution in an external electric field oscillating with the frequency ω. We are concerned here with two questions: (1) How does the state vary with time and, in particular, what are the probabilities for excitation into higher states? (2) At what rate will energy be transferred from the radiation field to the oscillator?

It certainly suffices to focus attention on one space coordinate only, say x, which is the one that corresponds to the direction of the applied electric field, because the remaining two coordinates decouple from x for a harmonic oscillator potential. We are thus concerned with a one-dimensional reduced problem and assume that its Hamiltonian is of the form $H(t) = -\frac{1}{2}d^2/dx^2 + \frac{1}{2}k^2x^2 - E(t)x$, where $E(t) = \mathcal{E}\sin\omega t$. So the period of field oscillation is $T = 2\pi/\omega$. To facilitate the analysis we permit only time intervals whose lengths are multiples of the given period: $0 \leq t \leq NT$ ($N = 1, 2, \ldots$).

In dealing with a variety of time scales it is convenient to rewrite everything in a dimensionless form. To this end we introduce a dimensionless parameter $\nu = kNT$, where k is the oscillator frequency. Also, the dimensionless time variable $\tau = t/(NT) = it/(NiT)$ is used when dealing with the Brownian bridge. As indicated, τ is unaffected by a transition to imaginary times, which has to be performed at the end of the computation.

It follows from our previous formulas for the driven oscillator (setting $s = 0$, $s' = NT$ everywhere in Sect. 4.4) that there is but one Fourier coefficient of $E(t)$ in the present case:

$$E_n = \begin{cases} 2^{-1/2}k^{-2}\mathcal{E} & \text{if } n = 2N \\ 0 & \text{otherwise.} \end{cases} \tag{4.5.1}$$

We now find the following transition amplitude:

$$\langle x', NT | x, 0 \rangle_\mathcal{E} = \langle x', NT | x, 0 \rangle_0 \exp\left(a + (x - x')b\right), \tag{4.5.2}$$

where

$$a = \frac{NT^3\mathcal{E}^2/4}{4\pi^2 + k^2T^2} \quad , \quad b = \frac{2\pi T\mathcal{E}}{4\pi^2 + k^2T^2} \tag{4.5.3}$$

and

$$\langle x', NT | x, 0 \rangle_0 = \left[\frac{k}{2\pi\sinh\nu}\right]^{1/2} \exp\left\{-\frac{k(x^2 + x'^2)}{2\tanh\nu} + \frac{kxx'}{\sinh\nu}\right\}. \tag{4.5.4}$$

The eigenfunctions of the harmonic oscillator Hamiltonian are normalized Hermite–Bessel functions:

$$\Phi_n(x) = c_n H_n(\sqrt{k}x)e^{-kx^2/2} \qquad (n = 0, 1, 2, \ldots)$$

$$c_n = (k/\pi)^{1/4}(2^n n!)^{-1/2} \tag{4.5.5}$$

$$H_n = \text{the } n\text{th Hermite polynomial .}$$

After a time period of length NT (in the diffusive picture), the ground state has evolved into

$$\Phi(x', NT) = \int_{-\infty}^{\infty} dx \, \langle x', NT | x, 0 \rangle_{\mathcal{E}} \, \Phi_0(x)$$

$$= c_0 \exp\left\{ a - \frac{\nu}{2} + \frac{1 - e^{-2\nu}}{4k}b^2 - \frac{kx'^2}{2} - (1 - e^{-\nu})bx' \right\}. \tag{4.5.6}$$

We can think of such a state as a superposition of eigenstates of the harmonic oscillator:

$$\Phi(x, NT) = \sum_{n=0}^{\infty} A_n(NT)\Phi_n(x), \tag{4.5.7}$$

where the coefficients are given by

$$A_n(NT) = \int_{-\infty}^{\infty} dx \, \Phi_n(x)\Phi(x, NT)$$

$$= c_n c_0 I \exp\left\{ a - \frac{\nu}{2} + \frac{1 - e^{-\nu}}{2k}b^2 \right\}$$

$$I = \int_{-\infty}^{\infty} dx \, H_n(\sqrt{k}x)e^{-k(x-y)^2} \tag{4.5.8}$$

$$= (\pi/k)^{1/2}(2\sqrt{k}y)^n, \qquad y = -\frac{1 - e^{-\nu}}{2k}b.$$

Summarizing:

$$A_n(NT) = \frac{(-1)^n}{n!}e^{a-\nu/2}\left[\frac{1 - e^{-\nu}}{\sqrt{2k}}b\right]^n \exp\left\{\frac{1 - e^{-\nu}}{2k}b^2\right\}. \tag{4.5.9}$$

4.5.1 Poisson Statistics

To perform the necessary analytic continuation is easy. It amounts to the replacements

$$T \to iT \qquad a \to -i\frac{NT\mathcal{E}^2}{\omega^2 - k^2} \qquad b \to i\frac{\mathcal{E}\omega}{\omega^2 - k^2} \qquad \nu \to i2\pi Nk/\omega.$$

As before, $\omega = 2\pi/T$. This way we get the probability $p_n = |A_n(iNT)|^2$ of the nth energy level being occupied:

$$p_n = \frac{1}{n!}\lambda^n e^{-\lambda}, \tag{4.5.10}$$

where

$$\lambda = \mathcal{E}^2 k^{-3} f_N(\omega/k), \qquad f_N(\alpha) = 2\left[\frac{\sin(N\pi\alpha^{-1})}{\alpha - \alpha^{-1}}\right]^2. \tag{4.5.11}$$

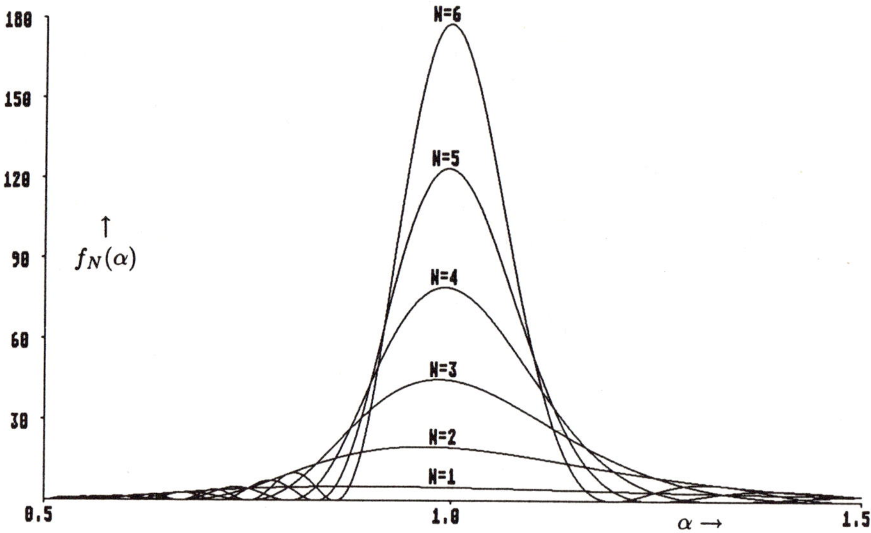

Fig. 4.1. The function $f_N(\alpha)$ of $\alpha = \omega/k$ for various values of N

By convention, the mass m, the charge q, and $\hbar$ have been put to unity. The reader who wants to reintroduce these quantities will find that $\lambda = q^2\mathcal{E}^2/(\hbar m k^3)f_N(\alpha)$. Apparently, the parameters λ, α and f_N keep their dimensionless status.

Result (4.5.10) shows that the energy levels $(n + \frac{1}{2})\hbar k$ of the oscillator are occupied according to the law of Poisson statistics. We quickly convince ourselves that $\sum p_n = 1$ and $\bar{n} = \sum n p_n = \lambda$. Moreover, we find that, in the mean during a time period of length NT, the oscillator has received the energy $\bar{n}\hbar k = \hbar k \lambda$ from the radiation field. This enables us to reinterpret the parameter λ as a quotient of two energies, i.e.,

$$\lambda = \frac{\bar{n}\hbar k}{\hbar k} = \frac{\text{mean energy transferred}}{\text{spacing of energies}}. \tag{4.5.12}$$

The function $f_N(\alpha)$ assumes its maximal value close to $\alpha = 1$. Figure 4.1 demonstrates that we may justly speak of a resonance behavior provided N is sufficiently large: as would be expected on general physical grounds, excitation of the ground state and transfer of energy occurs with appreciable probability only when the frequency ω of the external field is close to the internal frequency k of the target.

The maximum of the function $f_N(\alpha)$ grows quadratically with N, i.e., like $\frac{1}{2}N^2\pi^2$. At the same time, the width of the resonance curve in the vicinity of $\alpha = 1$ shrinks according to a $1/N$ law. This makes the area under the curve grow linearly:

$$\int_0^\infty d\alpha\, f_N(\alpha) = \pi^2 N. \tag{4.5.13}$$

Notice that N is our measure of duration of time. Consequently, if in a more realistic situation the radiation field $E(t)$ has a continuous spectrum with frequencies close to $\omega = k$, the above discussion would lead us to conclude that, roughly, energy is being transferred at a constant rate.

Let us compare our findings with what one would get from a classical analysis. The equation of motion (with respect to usual time) is

$$\ddot{x} + k^2 x = (q/m)\mathcal{E}\sin\omega t, \qquad \omega = 2\pi/T. \tag{4.5.14}$$

If the particle starts at rest, $x(0) = \dot{x}(0) = 0$ and thus

$$x(t) = \frac{(q/m)\mathcal{E}\omega}{\omega^2 - k^2}(k^{-1}\sin kt - \omega^{-1}\sin\omega t). \tag{4.5.15}$$

Consider the total energy of the oscillator, $H_{\mathrm{cl}}(t) = \frac{1}{2}m(\dot{x}^2 + k^2 x^2)$. After N periods of the incident sine wave,

$$H_{\mathrm{cl}}(NT) = \frac{q^2\mathcal{E}^2}{mk^2}f_N(\omega/k), \tag{4.5.16}$$

where $f_N(\alpha)$ coincides with the previously defined function. This shows that $H_{\mathrm{cl}}(NT)$ is in fact the mean energy $\bar{n}\hbar k$ obtained from the quantum mechanical treatment of the problem. As is often the case, mean values of quantum observables turn out to be classical expressions.

Though the foregoing discussion is nonperturbative, it gives no consistent account, as quantum electrodynamic does, of the interaction between charged particles and electromagnetic radiation. In a theory starting from the full set of Maxwell's equations, one finds that the influence of external radiation on the charge is marked by two processes: absorption and induced emission. Moreover, there is also the influence of the charge on the field known as *spontaneous emission*. The above treatment is incomplete from the physics point of view and affords but a glimpse of the absorption process.

5 The Linear-Coupling Theory of Bosons

If there be nothing new, but that which is
Hath been before, how are our brains beguil'd
Which, laboring for invention, bear amiss
The second burden of a former child!

W. Shakespeare

Spin and statistics of elementary particles are concepts that are already well known from courses on atomic physics and quantum mechanics. According to our present understanding, particles that have integer-valued spin obey Bose–Einstein statistics. They include for instance the photon, the pion, and the kaon. We refer to them as *bosons*. For completeness, one may also include pseudoparticles such as the *phonon*, which occurs in condensed matter, in the list of bosons. A special technique called *second quantization* is needed to handle situations where the particle number is not conserved during the course of time. Though the introduction of creation and annihilation operators for bosons provides the most elegant formulation of such problems, the ties to the harmonic oscillator formalism prompt us to seek an equivalent description in terms of path integrals [5.1]. Of course, there are certain limitations inherent in such an approach, as will soon become apparent.

5.1 Path Integrals for Bosons

A few introductory considerations will be needed to become familiar with the basic concepts and ideas. They concern the thermodynamics of bosons in general and pave the way towards an investigation of the polaron problem in Sect. 5.3. For simplicity, let us assume that the boson we are studying has merely a finite number of degrees of freedom, say N. Each degree is assigned some energy ϵ_n $(n = 1, \ldots, N)$, a creation operator a_n^*, and an annihilation operator a_n. We shall also assume that

$$0 < \epsilon_1 \leq \epsilon_2 \leq \cdots \leq \epsilon_N \ . \tag{5.1.1}$$

Bose–Einstein statistics is guaranteed by requiring that the canonical commutation relations be satisfied:

$$[a_n, a_m^*] = \delta_{nm} \ . \tag{5.1.2}$$

The Hamiltonian of a system of noninteracting bosons is written $H_{\mathrm{B}} = \sum \epsilon_n a_n^* a_n$. Note that the spectrum of the operator $a_n^* a_n$ (i.e., of the particle number associated with the nth degree of freedom) is the set $\{0, 1, 2, \ldots\}$.

As before, let $\beta^{-1} = k_{\mathrm{B}} T$. The so-called partition function of the free boson system is the sum of all Boltzmann weights:

$$Z_{\mathrm{B}} = \operatorname{tr} e^{-\beta H_{\mathrm{B}}} = \sum_{n_1=0}^{\infty} \cdots \sum_{n_N=0}^{\infty} e^{-\beta(n_1 \epsilon_1 + \cdots + n_N \epsilon_N)}$$

$$= \prod_{k=1}^{N} \sum_{n=0}^{\infty} e^{-\beta n \epsilon_k} = \prod_{k=1}^{N} \left(1 - e^{-\beta \epsilon_k}\right)^{-1}. \tag{5.1.3}$$

It is convenient to work with the free energy F_{B} obtained from the relation $Z_{\mathrm{B}} = e^{-\beta F_{\mathrm{B}}}$. Notice that this defines F_{B} as a function of β.

Formally, we are dealing with a harmonic oscillator in N dimensions. States of the system are nothing but complex wave functions of the oscillator coordinates $\xi = \{\xi_1, \ldots, \xi_N\} \in \mathbb{R}^N$, square-integrable with respect to the measure $d\xi = d\xi_1 \cdots d\xi_n$. The basic idea behind the oscillator picture is to use the following representation of the canonical commutation relations:

$$a_n = \sqrt{\frac{\epsilon_n}{2}} \xi_n + \sqrt{\frac{1}{2\epsilon_n}} \frac{d}{d\xi_n}$$

$$a_n^* = \sqrt{\frac{\epsilon_n}{2}} \xi_n - \sqrt{\frac{1}{2\epsilon_n}} \frac{d}{d\xi_n}. \tag{5.1.4}$$

After this change of description, the Hamiltonian becomes

$$H_{\mathrm{B}} = \sum_{n=0}^{N} \epsilon_n a_n^* a_n = \tfrac{1}{2} \sum_{n=0}^{N} \left(-\frac{d^2}{d\xi_n^2} + \epsilon_n^2 \xi_n^2 - \epsilon_n\right). \tag{5.1.5}$$

We will soon have occasion to study interactions of the bosons with a single Schrödinger particle[61], e.g., an electron, disregarding its spin. The electron is assigned coordinates $x \in \mathbb{R}^3$, a static potential $V(x)$ (e.g. the periodic potential in a crystal), and the energy $H_{\mathrm{el}} = -\tfrac{1}{2}\Delta + V$. For simplicity, the coupling is assumed to be linear in the boson operators, i.e., the total Hamiltonian we wish to study is of the form

$$H = H_{\mathrm{B}} + H_{\mathrm{el}} + \sum_n (a_n + a_n^*) u_n \tag{5.1.6}$$

---

[61] The interest in studying the interaction of nonrelativistic particles with a second-quantized scalar field has various origins. One is Fröhlich's model of the polaron [5.2], another is the interaction of an electron with its own radiation field. Then, new motivation came from the "dressing" concept in field theory which started another string of articles. This circle of ideas with the emphasis on mathematical questions originated from work of Nelson [5.3]. See also [5.4] for further details. A related system is the so-called spin-boson model which describes a spin (i.e., Pauli matrices) coupled to a scalar field [5.5–7].

with (real) functions $u_n(x)$, the choice of which depends on the physical problem at hand. In terms of the oscillator and electron coordinates, the interaction part of the Hamiltonian, $W = \sum(a_n + a_n^*)u_n$, is thus given by the function

$$W(\xi, x) = \sum_{n=1}^{N} \xi_n \sqrt{2\epsilon_n}\, u_n(x) \tag{5.1.7}$$

which is clearly linear in ξ.

5.1.1 The Partial Trace and Its Evaluation

The main problem of immediate physical interest is the evaluation of the free energy F for the combined system:

$$e^{-\beta F} = \operatorname{tr} e^{-\beta H} = \int dx \int d\xi \; \langle \xi, x | e^{-\beta H} | \xi, x \rangle. \tag{5.1.8}$$

For more information, one may also want to look at the *statistical operator* (sometimes called the *density operator*) of the electron:

$$\rho = \operatorname{tr}_B e^{-\beta H} / \operatorname{tr} e^{-\beta H}. \tag{5.1.9}$$

Here, tr_B designates the *partial trace* with respect to the bosonic states alone:

$$\langle x' | \operatorname{tr}_B e^{-\beta H} | x \rangle = \int d\xi \; \langle \xi, x' | e^{-\beta H} | \xi, x \rangle. \tag{5.1.10}$$

The transition amplitude, either written $\langle \xi', x' | e^{-\beta H} | \xi, x \rangle$ or $\langle \xi', x', \beta | \xi, x, 0 \rangle$, involves a twofold path integration: one integral over N-dimensional paths of the oscillator followed by another integral over three-dimensional paths of the electron. In Sect. 4.3 we have formulated a method of stepwise integration which will now be applied here. The easy step, i.e., the path integral pertaining to the oscillator, can be performed using our previous results on the driven harmonic oscillator. As the nth oscillator coordinate sees an "electric field" $E(t) = \sqrt{2\epsilon_n}\, u_n(\omega(t))$, we may use relation (4.4.16) several times to obtain a representation of the partial trace as a path integral pertaining to the electron degrees of freedom only:

$$\langle x' | \operatorname{tr}_B e^{-\beta H} | x \rangle = \int_{(x,0)}^{(x',\beta)} d\mu(\omega) \, \langle \beta | 0 \rangle_\omega \, \exp \left\{ -\int_0^\beta dt\, V(\omega(t)) \right\} \tag{5.1.11}$$

with the shorthand

$$\langle \beta | 0 \rangle_\omega = \int d\xi \; \langle \xi, \beta | \xi, 0 \rangle_\omega \,. \tag{5.1.12}$$

The path-dependent amplitude $\langle \xi, \beta | \xi, 0 \rangle_\omega$ conforms to the general scheme outlined in (4.3.3). In fact, the independence of the N driven harmonic oscillators makes it a product of N individual amplitudes:

$$\langle \xi', \beta | \xi, 0 \rangle_\omega = \prod_{n=1}^{N} f_n(\xi'_n, \xi_n; \omega), \tag{5.1.13}$$

where, for arbitrary real y and y',

$$
\begin{aligned}
f_n(y', y; \omega) = {}& \left[\frac{\epsilon_n/\pi}{1 - e^{-2\beta\epsilon_n}} \right]^{1/2} \exp\left\{ -\frac{\epsilon_n(y^2 + y'^2)}{2\tanh\beta\epsilon_n} + \frac{\epsilon_n yy'}{\sinh\beta\epsilon_n} \right. \\
& - \sqrt{2\epsilon_n} \int_0^\beta dt\, u_n(\omega(t)) \frac{y\sinh(\beta - t)\epsilon_n + y'\sinh t\epsilon_n}{\sinh\beta\epsilon_n} \\
& \left. + 2 \int_0^\beta dt \int_0^t dt'\, u_n(\omega(t)) u_n(\omega(t')) \frac{\sinh(\beta - t)\epsilon_n \sinh t'\epsilon_n}{\sinh\beta\epsilon_n} \right\}.
\end{aligned}
\tag{5.1.14}
$$

Equation (4.1.12) now becomes

$$\langle \beta | 0 \rangle_\omega = \prod_{n=1}^{N} \int_{-\infty}^{\infty} dy\, f_n(y, y; \omega), \tag{5.1.15}$$

and it remains to compute the Gaussian integral $\int dy\, f_n$. At first, the result looks a bit complicated. However, the following identity simplifies it drastically:

$$
\begin{aligned}
& \frac{2\sinh[(\beta - \max(t, t'))\epsilon_n]\, \sinh[\min(t, t')\epsilon_n]}{\sinh\beta\epsilon_n} \\
& + \frac{[\sinh(\beta - t)\epsilon_n + \sinh t\epsilon_n]\, [\sinh(\beta - t')\epsilon_n + \sinh t'\epsilon_n]}{\sinh\beta\epsilon_n\, (\cosh\beta\epsilon_n - 1)} \\
& = \frac{e^{-|t - t'|\epsilon_n} + e^{-(\beta - |t - t'|)\epsilon_n}}{1 - e^{-\beta\epsilon_n}}.
\end{aligned}
\tag{5.1.16}
$$

This way (5.1.15) becomes

$$\langle \beta | 0 \rangle_\omega = \exp\left\{ -\beta F_{\mathrm{B}} + \tfrac{1}{2} \int_0^\beta dt \int_0^\beta dt'\, G(\omega(t), t; \omega(t'), t') \right\}, \tag{5.1.17}$$

where

$$G(x, t; x', t') = \sum_{n=1}^{N} 2\beta\epsilon_n u_n(x) u_n(x')\, g(\beta^{-1}(t - t'), \beta\epsilon_n) \tag{5.1.18}$$

and

$$g(\tau,\gamma) = \frac{e^{-|\tau|\gamma} + e^{-(1-|\tau|)\gamma}}{2\gamma(1-e^{-\gamma})} = \frac{\cosh(|\tau|-1/2)\gamma}{2\gamma\sinh(\gamma/2)}. \tag{5.1.19}$$

Result (5.1.17) has to be inserted into (5.1.11). Now that the summation over the boson states has been carried out, no further simplification of the remaining electron path integral seems to be possible.

It is instructive to see that the function $g(\tau'-\tau,\gamma)$ may also be interpreted as the Green's function for some second-order differential operator. Let D be the operator on functions $f \in L^2(0,1)$ corresponding to the expression $-d^2/d\tau^2 + \gamma^2$ and the boundary condition $f(0) = f(1)$. The operator D is positive and its spectrum is discrete. To any eigenvalue $\lambda_k(\gamma) = (2\pi k)^2 + \gamma^2$ $(k \in \mathbb{Z})$ of D there corresponds the eigenfunction $e_k(\tau) = \exp(i2\pi k\tau)$, and, provided $\tau,\tau' \in [0,1]$,

$$\langle \tau'|D^{-1}|\tau\rangle = \sum_{k=-\infty}^{\infty} \frac{e_k(\tau')\bar{e}_k(\tau)}{(2\pi k)^2 + \gamma^2} = g(\tau'-\tau,\gamma). \tag{5.1.20}$$

Two further properties of the function g are worth mentioning. First, there exists a simple answer to the double integral:

$$\int_0^1 d\tau \int_0^1 d\tau' \, g(\tau - \tau',\gamma) = \gamma^{-2}. \tag{5.1.21}$$

Second, as is implied by (5.1.20), g has an obvious extension to a periodic function with respect to time satisfying $g(\tau + 1,\gamma) = g(\tau,\gamma)$, and, as γ becomes large,

$$\lim_{\gamma\to\infty} \gamma^2 g(\tau,\gamma) = \delta_*(\tau) \tag{5.1.22}$$

(convergence in the sense of distributions), where $\delta_*(\tau)$ designates the periodic Dirac function: $\delta_*(\tau) = \sum_{k=-\infty}^{\infty} \delta(\tau + k)$.

Exercise 1. For imaginary time, the function g as given by (5.1.19) furnishes the solution to the problem of determining the time correlation of a one-dimensional harmonic oscillator at a given temperature, that is, with $[a, a^*] = 1$ and $H = \epsilon a^* a$, one considers the statistical operator $\rho = e^{-\beta H}/\mathrm{tr}\,(e^{-\beta H})$. In the Heisenberg picture, one seeks the evolution of the position operator q and obtains

$$q_t = e^{itH} q e^{-itH} = \frac{1}{\sqrt{2\epsilon}}\left(ae^{-i\epsilon t} + a^* e^{i\epsilon t}\right).$$

There is a natural way to associate a time-correlation function with the state given by ρ. Prove that

$$\mathrm{tr}\,(\rho q_t q_{t'}) = \frac{e^{i(t'-t)\epsilon} + e^{-(\beta + i(t'-t))\epsilon}}{2\epsilon(1-e^{-\beta\epsilon})}.$$

5.2 A Random Potential for the Electron

The considerations that follow are motivated by a technique of M. Kac to *undo the square*, i.e., to write Gaussian functionals as expectations involving *linear* functionals (in the exponent). The implementation of this idea in the present context leads to the invention of random potentials. It is also a way to avoid multiline formulas.

In (5.1.17) we encountered a Gaussian functional (with regard to its dependence on $u_n(x)$, $n = 1, \dots, N$, as can be verified at a glance) that provides the answer to the problem of averaging the amplitude over the bosonic states. Since the functional in question involves a double time integral, no effective Schrdinger operator could be found for the electron that does the same thing, namely, reproduce the cumulative influence of bosonic environment – unless we apply the trick of undoing the square.

To this end, let us consider some N-dimensional Gaussian process Y_t with components Y_{tn}, $n = 1, \dots, N$, defined on the interval $0 \le t \le \beta$. We assume its mean to be zero and its covariance to be given by

$$E(Y_{tn}Y_{t'n'}) = \frac{e^{-|t-t'|\epsilon_n} + e^{-(\beta-|t-t'|)\epsilon_n}}{1 - e^{-\beta\epsilon_n}}\delta_{nn'} \,. \tag{5.2.1}$$

Apart from a factor, each component is thus a realization of the *oscillator bridge*[62]. In fact we are interested in another Gaussian process $Y(x, t)$, based on the former, that depends linearly on each function $u_n(x)$:

$$Y(x, t) = \sum_n u_n(x)Y_{tn} \qquad (x \in \mathrm{I\!R}^3). \tag{5.2.2}$$

Hence, $E\big(Y(x, t)Y(x', t')\big) = G(x, t; x', t')$ with the right-hand side given by (5.1.18). Finally, we define the *random potential* of the electron as the sum of two terms:

$$V_*(x, t) = V(x) + Y(x, t). \tag{5.2.3}$$

The first part, $V(x)$, coincides with the previously considered potential (neither dependent on t nor on β) as part of the Hamiltonian H_{el} referring to the electron; $V(x)$ is nonrandom, and hence deterministic, and unrelated to the bosonic background. The random part, $Y(x, t)$, has mean zero and fluctuates. It describes the effect of the interaction with the bosons at a given temperature. The following notation is simply a matter of convenience: we write

$$H_*(t) = -\tfrac{1}{2}\Delta + V_*(x, t) \tag{5.2.4}$$

for the random time-dependent Hamiltonian of the electron.

---

[62]See Exercise 2 at the end of Sect. 4.4.

The content of relations (5.1.11) and (5.1.17–19) can now be summarized as

$$\langle x'|\mathrm{tr}_B\, e^{-\beta H}|x\rangle = e^{-\beta F_{\mathrm{B}}} \int_{(x,0)}^{(x',\beta)} d\mu(\omega)\, \boldsymbol{E} \exp\left\{-\int_0^\beta dt\, V_*\big(\omega(t),t\big)\right\},$$

(5.2.5)

where the exponential is followed by an expectation $\boldsymbol{E}$ with respect to the Gaussian process $Y(x,t)$. So the solution to the original problem comes here as a sequence of operations applied to the random potential. Two of these operations, the path integral and the expectation, are linear and may be interchanged, allowing us to write

$$\mathrm{tr}_B\, e^{-\beta H} = e^{-\beta F_{\mathrm{B}}}\, \boldsymbol{E}\, T \exp\left\{-\int_0^\beta dt\, H_*(t)\right\}$$

(5.2.6)

with a time-ordered exponential followed by an expectation. Note that the partial trace defines $\mathrm{tr}_B\, e^{-\beta H}$ as an operator on $L^2(\mathbb{R}^3)$, the state space of the electron. The trace of this operator gives the partition function: $Z = \mathrm{tr}\left(\mathrm{tr}_{\mathrm{B}} e^{-\beta H}\right) = e^{-\beta F}$. We thus conclude that the free energy naturally splits into a bosonic and an electronic part, $F = F_{\mathrm{B}} + F_{\mathrm{el}}$, where F_{el} includes all effects of the interaction, and

$$e^{-\beta F_{\mathrm{el}}} = \boldsymbol{E}\, \mathrm{tr}\, T \exp\left\{-\int_0^\beta dt\, H_*(t)\right\}.$$

(5.2.7)

The trace on $L^2(\mathbb{R}^3)$ is now followed by an expectation. The right-hand side may be converted into a path integral when desired. One reason for conversion is that the path integral makes it easier to apply the Golden–Thompson–Symanzik inequality[63] to the trace:

$$e^{-\beta F_{\mathrm{el}}} \le (2\pi\beta)^{-3/2} \int dx\, e^{-\beta V(x)} \int_0^1 d\tau\, \boldsymbol{E}\left(e^{-\beta Y(x,\beta\tau)}\right)$$

$$= (2\pi\beta)^{-3/2} \int dx\, \exp\left\{-\beta V(x) + \frac{\beta^2}{2}\sum_n u_n(x)^2 \coth\frac{\beta\epsilon_n}{2}\right\}.$$

(5.2.8)

The latter integral is obtained by noting that

$$\boldsymbol{E}\big(Y(x,t)^2\big) = \sum_n u_n(x)^2 \coth(\beta\epsilon_n/2),$$

(5.2.9)

where the right-hand side is constant in t.

---

[63] Actually, we are using here the extended version of it. See the exercise at the end of Sect. 4.3.

Exercise 1. Use $E \exp\{-\int dt\, V_*\} \geq \exp\{-E \int dt\, V_*\}$ in (5.2.5) to prove that

$$e^{-\beta F_{\mathrm{el}}} \geq \mathrm{tr}\, e^{-\beta H_{\mathrm{el}}},$$

i.e., a (linear) coupling to bosons always lowers the free energy of an electron.

5.3 The Polaron Problem

If a slowly moving electron passes through a polar crystal, it will polarize its neighborhood owing to the accompanying Coulomb field[64], i.e., some force is exerted on both the band electrons and the ionic lattice. Only the latter interaction will be considered. Obviously, the perturbation is not static but follows the electron as it moves on. This is why electrons in crystal lattices are "dressed", that is, they are accompanied by vibrational modes of the lattice, and hence by *phonons*. An electron is called a *polaron* if it permanently drags along a cloud of (optical) phonons so that the cloud may also be considered part of the polaron[65]. The detailed structure of the interaction results from the classical picture, which says that the electron experiences an extra potential $-eV_1(x)$ caused by the (induced) charge density $\rho(x)$ which in turn follows from the polarization vector $P(x)$. More specifically, we have

$$\begin{aligned}
\Delta V_1(x) &= -\rho(x) \\
\rho(x) &= -\mathrm{div} P(x).
\end{aligned} \tag{5.3.1}$$

By applying the principles of second quantization, the polarization $P(x)$ will be treated as an *operator field* describing particles, i.e., phonons. We stress that the picture we are going to paint here relies to some extent on the linear response theory, and its validity should not be overestimated. From the vector character of the field it may be inferred that phonons carry a polarization and obey Bose statistics: there are three independent states for each momentum k. It should, however, be clear that only longitudinal modes enter the expression $\mathrm{div} P(x)$.

We take the three-dimensional lattice to be cubic with lattice spacing b. If we were to choose an infinitely extended lattice, the momentum k of a single phonon would vary continuously within the Brillouin zone $B_3 = [-\frac{\pi}{b}, \frac{\pi}{b}]^3$. It seems more practical to start from a finite lattice, say a cube Λ of volume L^3 such that $L \gg b$, and to adopt periodic boundary conditions (and hence make the cube look more like a torus). What we have accomplished this

---

[64]For sufficiently slow motion, magnetic effects may be neglected.
[65]Early ideas concerning the physics of the polaron are summarized in [5.8].

way is that the momenta accessible to phonons become discrete and finite in number.

We let Λ^* denote the set of those momenta $k \in B_3$, $k \neq 0$, whose components are multiples of $2\pi/L$. Each $k \in \Lambda^*$ is assigned a longitudinal polarization vector $e(k) = k/|k|$ which enters the plane wave decomposition of $P(x)$. From the obvious relation $ke(k) = |k|$ one infers that

$$\operatorname{div} P(x) \propto \sum_{k \in \Lambda^*} i|k|\big(a_k e^{ikx} - a_k^* e^{-ikx}\big), \tag{5.3.2}$$

where a_k^* and a_k are creation and annihilation operators for longitudinal phonons obeying canonical commutation relations: $[a_k, a_{k'}^*] = \delta_{kk'}$. Then ansatz (5.3.2) suggests writing

$$-eV_1(x) \propto \sum_{k \in \Lambda^*} i|k|^{-1}\big(a_k e^{ikx} - a_k^* e^{-ikx}\big). \tag{5.3.3}$$

We also need to specify the dispersion ϵ_k of a phonon, i.e., its frequency (or energy) as a function of the wave vector (or momentum). As for the optical branch of the dispersion and small $|k|$, $\epsilon_k = \epsilon =$ const is not too bad an approximation [5.9,10]. We shall then assume units to be such that $\epsilon = 1$. Polaron systems with general dispersion have also been discussed in the literature. Their properties are reviewed in [5.11] and [5.12]. An operator-theory approach may be found in [5.4].

The Hamiltonian is

$$H = H_{\text{el}} + \sum_{k \in \Lambda^*} a_k^* a_k + W, \tag{5.3.4}$$

where $H_{\text{el}} = -\tfrac{1}{2}\Delta + V$, and the interaction, varying with the electron coordinate x, is given by

$$\begin{aligned}
W(x) &= \lambda L^{-3/2} \sum_{k \in \Lambda^*} i|k|^{-1}\big(a_k e^{ikx} - a_k^* e^{-ikx}\big) \\
&= \lambda L^{-3/2} \sum_{k \in \Lambda^*} i|k|^{-1}(a_k - a_{-k}^*)e^{ikx}.
\end{aligned} \tag{5.3.5}$$

The coupling constant λ measures the polarizability of the crystal. Note that the Hamiltonian depends on the size of the lattice and the lattice spacing. Also, the external potential $V(x)$ constraining the motion of the electron has not yet been specified.

Interaction (5.3.4) is almost, but not quite, of the form chosen in (5.1.6). However, this can be cured by a change of variables:

$$\begin{aligned}
a_{k1} &= i(a_k + a_{-k})/\sqrt{2} \\
a_{k2} &= i(a_k - a_{-k})/\sqrt{2}.
\end{aligned} \tag{5.3.6}$$

Actually, since $a_{-k1} = a_{k1}$ and $a_{-k2} = -a_{k2}$, we are considering redundant variables. It is thus more appropriate to put k and $-k$ into one class and to replace the set Λ^* of momenta by the smaller set K of equivalence classes. Therefore, we may write the interaction as

$$W = \sum_{k \in K} \sum_{\sigma=1,2} (a_{k\sigma} + a_{k\sigma}^*) u_{k\sigma} \qquad (5.3.7)$$

with functions

$$u_{k\sigma}(x) = \lambda L^{-3/2} |k|^{-1} \sqrt{2} \begin{cases} \cos kx & \text{if } \sigma = 1 \\ \sin kx & \text{if } \sigma = 2. \end{cases} \qquad (5.3.8)$$

It may also be demonstrated that

$$\sum_{k \in \Lambda^*} a_k^* a_k = \sum_{k \in K} \sum_{\sigma=1,2} a_{k\sigma}^* a_{k\sigma}. \qquad (5.3.9)$$

Now that we have cast the Hamiltonian into the preferred form allowing us to apply the formalism of Sect. 5.1, the techniques developed therein prompt us to consider the sum

$$S(x, x') = \sum_{k \in K} \sum_{\sigma=1,2} u_{k\sigma}(x) u_{k\sigma}(x') , \qquad (5.3.10)$$

since our interest lies in

$$G(x, t; x', t') = S(x, x') \frac{e^{-|t-t'|} + e^{-(\beta-|t-t'|)}}{1 - e^{-\beta}} \qquad (5.3.11)$$

(recall that $\epsilon = 1$). We immediately see that

$$\begin{aligned} S(x, x') &= 2\lambda^2 L^{-3} \sum_{k \in K} |k|^{-2} \cos k(x - x') \\ &= \lambda^2 L^{-3} \sum_{k \in \Lambda^*} |k|^{-2} e^{ik(x-x')} , \end{aligned} \qquad (5.3.12)$$

and the resulting expression is translationally invariant, as it should be.

5.3.1 The Limit $L \to \infty$, $b \to 0$

We wish to perform two limits. First, $L \to \infty$ means that we pass to an infinite volume of the crystal; this is the so-called *thermodynamic limit*. It turns $S(x, x')$ into an integral over the Brillouin zone:

$$S(x, x') = \lambda^2 (2\pi)^{-3} \int_{B_3} dk\, |k|^{-2} e^{ik(x-x')}. \qquad (5.3.13)$$

Second, $b \to 0$ means that we are going to neglect the lattice structure altogether. This is the so-called *continuum limit*. It makes the Brillouin zone grow, $B_3 \to \mathrm{I\!R}^3$, yielding

$$S(x, x') = \lambda^2 (4\pi|x - x'|)^{-1}. \tag{5.3.14}$$

By choosing these limits we have considerably reduced the complexity of the original model. The limiting model, which is more tractable, relates to a simplified physical picture: it treats the crystal as an infinitely extended, homogeneous dielectric medium described solely by its polarizability, and hence by the constant λ, and by the phonon energy (put to unity).

Specific problems, which have to be taken care of, come up in connection with these limits. First, we have to accept and live with the singularity $S(x, x) = +\infty$. Second, we have to face the fact that the free energy of the bosons is an extensive thermodynamic variable: $F_B \propto L^3$. Consequently, we cannot claim convergence for $\langle x'|\mathrm{tr}_B\, e^{-\beta H}|x\rangle$, but only for $\langle x'|\rho|x\rangle$, where ρ is the statistical operator of the electron as defined by (5.1.9), and also for $F_{\mathrm{el}} = F - F_B$, the electronic free energy, granted certain conditions on the potential $V(x)$. Previous formulas, especially (5.1.11) and (5.1.17), together with (5.3.11) and (5.3.14), lead to the following path integral formula for ρ:

$$\langle x'|\rho|x\rangle = e^{\beta F_{\mathrm{el}}} \int d\mu_{x,0}^{x',\beta}(\omega) \exp\left\{ D(\omega) - \int_0^\beta dt\, V(\omega(t)) \right\}$$

$$D(\omega) = \frac{\lambda^2/2}{1 - e^{-\beta}} \int_0^\beta dt \int_0^\beta dt'\, \frac{e^{-|t-t'|} + e^{-(\beta-|t-t'|)}}{4\pi|\omega(t) - \omega(t')|} \; , \tag{5.3.15}$$

where

$$e^{-\beta F_{\mathrm{el}}} = \int dx \int d\mu_{x,0}^{x,\beta}(\omega) \exp\left\{ D(\omega) - \int_0^\beta dt\, V\big(\omega(t)\big) \right\}. \tag{5.3.16}$$

As concerns the electronic free energy, it would be natural to substitute

$$\omega(t) = x + \ell\bar{\omega}(\tau), \qquad t = \beta\tau, \quad \ell^2 = \beta$$
$$d\mu_{x,0}^{x,\beta}(\omega) = (2\pi\beta)^{-3/2} d\bar{\omega}, \tag{5.3.17}$$

thereby introducing paths $\bar{\omega}$ of the Brownian bridge. We may thus write

$$e^{-\beta F_{\mathrm{el}}} = \int d\bar{\omega}\, A(\bar{\omega}, V)\, e^{\beta B(\bar{\omega})}, \tag{5.3.18}$$

where

$$A(\bar{\omega}, V) = (2\pi\beta)^{-3/2} \int dx\, \exp\left\{ -\beta \int_0^1 d\tau\, V\big(x + \ell\bar{\omega}(\tau)\big) \right\} \tag{5.3.19}$$

and

$$B(\bar{\omega}) = \frac{\lambda^2 \ell/2}{1 - e^{-\beta}} \int_0^1 d\tau \int_0^1 d\tau'\, \frac{e^{-\beta|\tau-\tau'|} + e^{-\beta(1-|\tau-\tau'|)}}{4\pi|\bar{\omega}(\tau) - \bar{\omega}(\tau')|}. \tag{5.3.20}$$

Let us pause at this point for supplementary remarks. There is an alternative, however, equivalent, description borrowing notions and techniques from Sect. 5.2; in particular, we may take $V_*(x,t) = V(x) + Y(x,t)$ as the random potential of the polaron, where $Y(x,t)$ is a (generalized) Gaussian process with mean zero and covariance

$$E\big(Y(x,t)Y(x',t')\big) = \frac{\lambda^2}{1 - e^{-\beta}} \, \frac{e^{-|t-t'|} + e^{-(\beta - |t-t'|)}}{4\pi|x - x'|} \tag{5.3.21}$$

such that

$$\langle x'|\rho|x\rangle = e^{\beta F_{\mathrm{el}}} \int d\mu_{x,0}^{x',\beta}(\omega)\, E \exp\left\{ -\int_0^\beta dt\, V_*\big(\omega(t),t\big) \right\} \tag{5.3.22}$$

and

$$e^{-\beta F_{\mathrm{el}}} = \int dx \int d\mu_{x,0}^{x,\beta}(\omega)\, E \exp\left\{ -\int_0^\beta dt\, V_*\big(\omega(t),t\big) \right\}. \tag{5.3.23}$$

By definition, the process $Y(x,t)$ is spatially and temporally homogeneous. Noninvariance of the polaron model with respect to translations in x-space is entirely due to $V(x)$ being nonconstant.

The expression for $A(\bar\omega, V)$ as given by (5.3.19) is peculiar because its looks as if it were a classical phase space integral; in fact, setting $\hat H(p,q) = \frac{1}{2}p^2 + V(q)$ enables us to cast the expression into an integral over $\mathbb{R}^3 \times \mathbb{R}^3$:

$$A(\bar\omega, V) = (2\pi)^{-3} \int dpdq \, \exp\left\{ -\beta \int_0^1 d\tau\, \hat H\big(p, q + \ell\bar\omega(\tau)\big) \right\}. \tag{5.3.24}$$

Moreover, taking $\ell = 0$ we would find that $A(0, V)$ is a classical partition function.

An examination of (5.3.20) tells us that $0 \le B(\bar\omega) \le \infty$. The upper bound is reached for instance if $\bar\omega(\tau) = 0$. The conjecture, however, that the integral on the right-hand side of (5.3.18) fails to exist would be premature. Paths $\bar\omega$ giving large contributions to the integral have distances $|\bar\omega(\tau) - \bar\omega(\tau')|$ close to zero. If τ and τ' are well separated, this observation would lead us to study double points of Brownian motion. Though double points are abundant in one dimension, they are rare occurrences in three dimensions [5.13] and certainly no source of trouble. It remains to examine the question: how singular is $|\bar\omega(\tau) - \bar\omega(\tau')|^{-1}$ at $\tau = \tau'$? Or to put it differently: does the integral

$$\int_0^1 d\tau \int_0^1 d\tau' \, |\bar\omega(\tau) - \bar\omega(\tau')|^{-1} \tag{5.3.25}$$

exist for a typical path $\bar\omega$?

As follows directly from the definition of the Wiener process X_t, the random variables $X_t - X_{t'}$ and $(t - t')^{1/2} X_1$ $(t > t')$ are equally distributed. This simple fact accounts for most of the properties of Brownian paths. One assertion, roughly speaking, is that a typical path has to be Hölder continuous with index $0 < \alpha < \frac{1}{2}$. More precisely, for each $c > 0$, $T > 0$, and $0 < \alpha < \frac{1}{2}$, the event

$$\forall t \in [0,T] \; : \; \limsup_n n^\alpha |\omega(t + 1/n) - \omega(t)| < c \tag{5.3.26}$$

occurs with probability one.

The reason: it suffices to take $T = 1$. By choosing points $t_i = i/n$ $(i = 0, \ldots, n)$, the interval $[0,1]$ will be divided into subintervals of length $1/n$. The probability of the above event is the limit

$$W_\alpha = \lim_n \prod_{i=1}^n P\left(n^\alpha |X_{t_i} - X_{t_{i-1}}| < c\right)$$

$$= \lim_n P\left(|X_1| < r_n\right)^n \qquad r_n := cn^{-\alpha+1/2}.$$

(5.3.27)

Obviously, $r_n \to \infty$ for n large and $\alpha - \frac{1}{2} < 0$. Moreover,

$$1 - P\left(|X_1| < r_n\right) = O\left(\exp(-\tfrac{1}{2}r_n^2)\right)$$

(5.3.28)

so that $W_\alpha = 1$.

A complementary statement is: for $c > 0$, $T > 0$, and $\alpha > \frac{1}{2}$, the event (5.3.26) occurs with probability zero. The reason is that, for n large and $\alpha - \frac{1}{2} > 0$, we would get $r_n \to 0$ and $P\left(|X_1| < r_n\right) \to 0$, and hence $W_\alpha = 0$.

In view of these comments and the fact that $|\tau - \tau'|^{-1/2}$ is integrable, we may safely state that the free energy F_{el} is well-defined for all β.

As regards the regularity of Brownian motion, stronger results are available. There is, for instance, the *law of the iterated logarithm* [1.26], which states that almost surely

$$\limsup_n \frac{|\omega(t + 1/n) - \omega(t)|}{\sqrt{(2/n)\log\log n}} = 1.$$

(5.3.29)

5.3.2 The Free Energy of the Polaron

Having ignored the lattice structure of the crystal, it may only seem consistent that we neglect the periodic potential from the ions, too. After all, we want the excess electron free to wander and not to be constrained to any part of the medium. On the other hand, we are not allowed to choose $V(x) = 0$ from the beginning. For this would induce translational invariance of the relevant path integral, and quantities like $A(\bar\omega, V)$ would cease to exist, and, as a consequence of the delocalization of the electron, $e^{-\beta F_{\mathrm{el}}} = \infty$.

A judicious choice is the oscillator potential $V(x) = \frac{1}{2}k^2x^2$ with the limit $k \to 0$ taken at the end of the calculation[66]. From (5.3.19) and a little algebra,

$$A(\bar\omega, V) = \exp\left\{-\beta F_{\mathrm{osc}} - \left(\frac{\beta k}{2}\right)^2 \int_0^1 d\tau \int_0^1 d\tau' \, |\bar\omega(\tau) - \bar\omega(\tau')|^2\right\}, \quad (5.3.30)$$

where $F_{\mathrm{osc}} = 3\beta^{-1}\log(\beta k)$, which is the classical free energy of the oscillator. We stipulate its irrelevance here, take $F_{\mathrm{pol}} = F_{\mathrm{el}} - F_{\mathrm{osc}}$ as the free energy of the polaron, and perform the limit $k \to \infty$ on F_{pol}. This leads to the result

---

[66]The choice of $V(x) = \frac{1}{2}k^2x^2$ as the confining potential has been proposed by Spohn [5.14].

$$\exp(-\beta F_{\mathrm{pol}}) = \int d\bar{\omega} \, \exp\left(\beta B(\bar{\omega})\right) =$$

$$\int d\bar{\omega} \, \exp\left\{ \frac{\beta \lambda^2 \ell/2}{1 - e^{-\beta}} \int_0^1 d\tau \int_0^1 d\tau' \, \frac{e^{-\beta|\tau-\tau'|} + e^{-\beta(1-|\tau-\tau'|)}}{4\pi |\bar{\omega}(\tau) - \bar{\omega}(\tau')|} \right\}. \tag{5.3.31}$$

No explicit evaluation of this path integral seems to be possible. As $B(\bar{\omega}) \geq 0$, we have $F_{\mathrm{pol}} \leq 0$ for all β. At zero temperature ($\beta \to \infty$), the free energy F_{pol} becomes the ground state energy of the polaron, E_0.

If $V(x) = 0$, the polaron model is translationally invariant, i.e., the total Hamiltonian H we started from commutes with the operator of the total momentum,

$$P = -i\nabla + \sum_k k a_k^* a_k. \tag{5.3.32}$$

In this case, H and P can be simultaneously diagonalized. Let $E(p)$ denote the bottom of the energy spectrum for fixed total momentum p. Then, for small $|p|$,

$$E(p) = E_0 + \frac{p^2}{2m^*} \tag{5.3.33}$$

with m^* the *effective mass* of the polaron. As it turns out, the effective mass is strictly larger than the bare mass: $m^* > 1$. There are ways to define and estimate m^* in term of path integrals [5.15] which, however, are beyond our present scope.

5.3.3 Bounds on the Polaron Free Energy

Upper bounds of F_{pol} are abundant and easy to get. As usual, the key idea behind their derivation is a comparison with solvable path integrals[67]. An enormous repertoire of solvable path integrals is summarized in the formula:

$$\int d\bar{\omega} \, f\left(\bar{\omega}(\tau) - \bar{\omega}(\tau')\right) = \int dx \, K\left(x, s(1-s)\right) f(x), \tag{5.3.34}$$

where $s = |\tau - \tau'|$ and $\tau, \tau' \in [0,1]$. To demonstrate its validity we take $0 < \tau' < \tau < 1$, note that each path $\bar{\omega} \colon (0,0) \rightsquigarrow (0,1)$ decomposes naturally into three parts, and introduce $y' = \omega(\tau')$ and $y = \omega(\tau)$ as intermediate coordinates so that the above path integral becomes

$$\int dy' \int dy \, K(y' - 0, \tau' - 0) K(y - y', \tau - \tau') K(0 - y, 1 - \tau) f(y - y').$$

$$\tag{5.3.35}$$

---

[67]A path integral is termed 'solvable' if it can be reduced to some finite-dimensional integral.

By a change of variables putting $x = y - y'$ and $x' = \frac{1}{2}(y + y')$, we see that the integration with respect to x' can be performed leaving an integral with respect to x which can be brought into the form (5.3.34).

Without further ado, then, let us proceed to the elaboration of a three-dimensional case, where the x-integration in (5.3.34) can be performed, too:

$$\int \frac{d\bar{\omega}}{4\pi|\bar{\omega}(\tau) - \bar{\omega}(\tau')|} = \frac{(2\pi)^{-3/2}}{\sqrt{s(1-s)}}. \tag{5.3.36}$$

This integral is encountered when Jensen's inequality is applied to (5.3.31):

$$\int d\bar{\omega}\,\exp\left\{\beta B(\bar{\omega})\right\} \geq \exp\left\{\beta \int d\bar{\omega}\,B(\bar{\omega})\right\}. \tag{5.3.37}$$

Indeed, from definition (5.3.20)

$$\int d\bar{\omega}\,B(\bar{\omega}) = (2\pi)^{-3/2}(1 - e^{-\beta})^{-1}\lambda^2 \ell Q, \tag{5.3.38}$$

where

$$Q := \int_0^1 d\tau \int_0^\tau d\tau'\,\left(g(\tau - \tau') + g(1 - \tau + \tau')\right) \tag{5.3.39}$$
$$g(s) := [s(1-s)]^{-1/2}e^{-s\beta}.$$

By a change of variables putting $s = \tau - \tau'$ and $s' = \frac{1}{2}(\tau + \tau')$, the double time integral representing Q is traceable to the (modified) Bessel function I_0 of zero order:

$$\begin{aligned}
Q &= \int_0^1 ds\,(1-s)\big(g(s) + g(1-s)\big) \\
&= \int_0^{1/2} ds\,\big(g(s) + g(1-s)\big) \\
&= \int_0^1 ds\,g(s) \\
&= \pi e^{-\beta/2} I_0(\beta/2).
\end{aligned} \tag{5.3.40}$$

For historical reasons one puts $\lambda^2 = \sqrt{8}\,\pi\alpha$, calling α the Fröhlich parameter. Looking at previous relations such as (5.3.20), (5.3.31), and (5.3.38) we conclude that

$$F_{\text{pol}} \leq -\alpha\sqrt{\pi\beta}\,\frac{e^{-\beta/2} I_0(\beta/2)}{1 - e^{-\beta}}. \tag{5.3.41}$$

From $\lim_{x\to\infty} \sqrt{2\pi x}\,e^{-x} I_0(x) = 1$ it may be inferred that, as $\beta \to \infty$,

$$E_0 \leq -\alpha. \tag{5.3.42}$$

Expanding the ground energy E_0 into a power series with respect to α it is quickly realized that $E_0 = -\alpha + O(\alpha^2)$. So the upper bound (5.3.42) becomes almost exact in the weak coupling regime. To determine the next term of the power series requires more labor [5.9] and one gets the improved result $E = -\alpha - 0.0126\alpha^2 + O(\alpha^3)$ after a long-winded calculation. Feynman [5.10] also succeeded in deriving the following rigorous upper bound on the ground state energy:

$$E_0 \le -\alpha - \alpha^2/81. \tag{5.3.43}$$

He did so by starting from the variational principle for the free energy:

$$F_{\text{pol}} \le F_0 + \langle B_0 - B \rangle_0, \tag{5.3.44}$$

where $B(\bar{\omega})$ is given by (5.3.20), $B_0(\bar{\omega})$ denotes some arbitrary functional, and

$$e^{-\beta F_0} = \int d\bar{\omega}\, e^{\beta B_0(\bar{\omega})}$$
$$e^{-\beta F_0} \langle C \rangle_0 = \int d\bar{\omega}\, e^{\beta B_0(\bar{\omega})} C(\bar{\omega}), \tag{5.3.45}$$

where $C = B_0 - B$. In principle, (5.3.44) provides a very general tool of statistical mechanics and the details of the structure of B are irrelevant for the inequality to be valid. This is so since the proof relies on no more than the convexity of the exponential function and Jensen's inequality. As for the polaron, the following ansatz proved to be most convenient:

$$B_0(\bar{\omega}) = \tfrac{1}{2}c \int_0^1 d\tau \int_0^1 d\tau'\, |\bar{\omega}(\tau) - \bar{\omega}(\tau')|^2 e^{-w|\tau - \tau'|}. \tag{5.3.46}$$

The upper bound thus obtained depends on the free parameters c and w; at the end of the calculation they are chosen so as to obtain the best bound on F_{pol}.

Finally, the lack of respectable techniques (using path integrals) to place *lower* bounds on the polaron free energy is much more severe. Nevertheless, Lieb and Yamazaki [5.16], using methods from operator theory, were able to prove that

$$E \ge -3(p^2 - 1)(p^2 + 3)/4p^2 \tag{5.3.47}$$

with p the positive solution of $p^3(3p - 2\alpha) = 3$. Upper and lower bounds do not differ too much provided $0 \le \alpha \le 1$.

5.3.4 Pekar's Large-Coupling Result

Strictly speaking, the problem of rigorously proving the existence of the zero temperature limit

$$E(\lambda^2/2) = -\lim_{\beta \to \infty} \beta^{-1} \log \int d\bar{\omega}\, \exp\{\beta B(\bar{\omega})\} \tag{5.3.48}$$

calls for a proof patterned after the *theory of large deviations* of Donsker and Varadhan [5.17], and, in fact, this program has been carried out successfully [5.18]. We shall not go into this matter except to say that, along with their existence proof, the above authors confirmed Pecar's intriguing assertion [5.19] that, in the strong coupling limit,

$$\lim_{\lambda \to \infty} (\lambda^2/2)^{-2} E(\lambda^2/2) = c, \tag{5.3.49}$$

where

$$c = \inf \left\{ \tfrac{1}{2} \int dx \, |\mathrm{grad}\phi(x)|^2 - 2 \int dx \int dx' \, \frac{\phi(x)^2 \phi(x')^2}{|x - x'|} \right\}. \tag{5.3.50}$$

The infimum in (5.3.50) is taken with respect to all functions $\phi(x) : \mathbb{R}^3 \to \mathbb{R}$ satisfying $\int dx \, \phi(x)^2 = 1$. The task of determining the constant c has to be contrasted with the Ritz principle of ordinary quantum mechanics. There, the solution obeys the Schrödinger equation which is linear. But here, the solution $\phi(x)$ of (5.3.50) obeys some nonlinear Schrödinger equation known as the *Choquard equation*; its existence and uniqueness have been demonstrated by Lieb [5.20]. Also, there is numerical work [5.21] showing that $c = -0.038\,365\,1$.

Pekar's strong-coupling result and its connection with the variational principle (5.3.50) has many striking aspects[68]. One aspect is that the real function $\phi(x)$ may be considered an *order parameter* of the polaron model with the right-hand side of (5.3.50), i.e., Choquard's functional, taken as some kind of *effective free energy* (or Gibbs functional) for the polaron. Those who are familiar with the Ginzburg-Landau theory might find the analogy attractive (for additional information on 'effective energies' and 'effective actions' see the Sects. 8.5–6).

5.4 The Field Theory of the Polaron Model

The main questions raised in the foregoing section concerned the 'dressing' of an electron passing through a polar crystal, where we have disregarded all details of the lattice structure. But we may also wonder how the electron acts on the (transverse optical) phonon field. Evidently, one part of the story has not been told yet, namely, the part that deals with the field theory of the polaron model. We conjecture that definite answers will come from a field-theoretic investigation, in particular, answers to questions like:

1. What will be the mean charge density $\rho(x) = -\langle \mathrm{div} P(x) \rangle$ induced by the excess electron in the dielectric in thermal equilibrium?
2. How does the field $\mathrm{div} P(x)$, i.e., the longitudinal part of the polarization, fluctuate around its mean value?
3. What value do we get for the total (polarization) charge $q_{\mathrm{pol}} = \int dx \, \rho(x)$?

To start with, we consider a finite volume Λ ($|\Lambda| = L^3$), but shall neglect the lattice structure from the very beginning, i.e., we take $b = 0$. At a fixed

---

[68]A detailed analysis of the strong-coupling polaron theory may be found in an article by Gross [5.22].

time, say $t = 0$, we declare the second-quantized field of (optical transverse) phonons to be

$$\Phi(x) = L^{-3/2} \sum_{k \in \Lambda^*} i|k|\left(a_k e^{ikx} - a_k^* e^{-ikx}\right) \tag{5.4.1}$$

such that the relation $\mathrm{div}\, P(x) = \lambda \Phi(x)$ connects the field $\Phi(x)$ with the polarization field $P(x)$. Standard terminology calls $\Phi(x)$ a *scalar field*.

Next, we introduce an extra variable of the theory, $j(x)$, which is thought of as some real function used in integrals like

$$\Phi(j) = \int_\Lambda dx\, \Phi(x) j(x). \tag{5.4.2}$$

The objective now is to enlarge the Hamiltonian of the polaron model, H, by adding a source term to it, that is, we take

$$H(j) = H + \Phi(j) \tag{5.4.3}$$

as the new Hamiltonian. In principle, the source term could account for the emission or absorption of phonons due to an external force. But, as for (5.4.3), the source term has been introduced merely to probe the system and see the response of it. On a more formal level one may argue that all n-point functions of the phonon field are generated by the free energy $F(j)$ considered as a functional of the source function $j(x)$, and, in fact, this is all one needs to know. Finally, if so desired, one takes $j = 0$ at the end of the calculation.

Adopting the conventions and the notions of the preceding section, we arrive at the following alternative description of the source term:

$$\Phi(j) = \sum_{k \in K} \sum_{\sigma = 1,2} (a_{k\sigma} + a_{k\sigma}^*) j_{k\sigma} \tag{5.4.4}$$

with real coefficients $j_{k\sigma}$ given by

$$j_{k1} + i j_{k2} = \sqrt{2}|k| L^{-3/2} \int_\Lambda dx\, j(x) e^{ikx}. \tag{5.4.5}$$

We now take advantage of the fact that, by a canonical transformation $b_{k\sigma} = a_{k\sigma} + j_{k\sigma}$, the Hamiltonian $H(j)$ can be brought into a form which is similar to what we have been considering before (in Sect. 5.3). The effect of the transformation is seen from the identity

$$a_{k\sigma}^* a_{k\sigma} + (a_{k\sigma} + a_{k\sigma}^*)\left(u_{k\sigma}(x) + j_{k\sigma}\right) =$$

$$b_{k\sigma}^* b_{k\sigma} + (b_{k\sigma} + b_{k\sigma}^*) u_{k\sigma}(x) - 2 j_{k\sigma} u_{k\sigma}(x) - j_{k\sigma}^2 \tag{5.4.6}$$

with $u_{k\sigma}(x)$ given by (5.3.8). Notice also that

$$\sum_{k \in K} \sum_{\sigma=1,2} j_{k\sigma}^2 = \int_\Lambda dx \, \left(\mathrm{grad}\, j(x)\right)^2$$

$$\sum_{k \in K} \sum_{\sigma=1,2} j_{k\sigma} u_{k\sigma}(x) = \lambda j(x) \qquad (x \in \Lambda).$$

(5.4.7)

What has been accomplished is that the structure of the operator $H(j)$ follows the same pattern as for $j = 0$, i.e., to account for $j \neq 0$ in previous formulas we need only replace $a_{k\sigma}$ by $b_{k\sigma}$, $a_{k\sigma}^*$ by $b_{k\sigma}^*$, and the potential $V(x)$ by

$$V(x) - 2\lambda j(x) - \int_\Lambda dx \, \left(\mathrm{grad}\, j(x)\right)^2.$$

(5.4.8)

As partial traces with respect to the bosonic degrees of freedom such as $\mathrm{tr}_B \, e^{-\beta H(j)}$ remain unaffected by a canonical transformation, there is no need for starting another calculation. Denoting $F(j)$ the free energy of the total system, we define the electronic part of it as before by setting

$$F_{\mathrm{el}}(j) = \lim_{\Lambda \uparrow \mathbb{R}^3} \left\{ F(j) - F_B \right\}$$

(5.4.9)

and from (5.3.18) get the answer in terms of some path integral:

$$F_{\mathrm{el}}(j) = - \int dx \, \left(\mathrm{grad}\, j(x)\right)^2$$

$$- \beta^{-1} \log \int d\bar{\omega} \, A(\bar{\omega}, V - 2\lambda j) \exp\{-\beta B(\bar{\omega})\}$$

(5.4.10)

with functionals A and B given by (5.3.19–20).

There exists a formal expansion[69] of $F_{\mathrm{el}}(j)$ into powers of the source function j, the coefficients of which are said to be the *cumulants* of the field $\Phi(x)$. Only the first few terms of this expansion are of interest here:

$$F_{\mathrm{el}}(j) = F_{\mathrm{el}}(0) + \int dx \, \langle \Phi(x) \rangle j(x)$$

$$- \tfrac{1}{2}\beta \int dx \int dx' \, \langle \Phi(x); \Phi(x') \rangle j(x) j(x') + \cdots.$$

(5.4.11)

The interpretation is as follows. The constant term $F_{\mathrm{el}}(0)$ coincides with the free energy considered previously. The expectation value of the field in thermal equilibrium (with respect to H) is denoted $\langle \Phi(x) \rangle$. Fluctuations around the mean value are described (to lowest order) by the correlation function

$$\langle \Phi(x); \Phi(x') \rangle = \langle \Phi(x), \Phi(x') \rangle - \langle \Phi(x) \rangle \langle \Phi(x') \rangle,$$

(5.4.12)

where

---

[69]Compare (7.3.3).

$$\langle \Phi(x), \Phi(x') \rangle = \beta^{-1} \int_0^\beta ds \, \langle e^{sH} \Phi(x) e^{-sH} \Phi(x') \rangle \tag{5.4.13}$$

is the so-called *Duhamel two-point function*[70].

Adopting units such that the charge of an electron becomes -1, we derive the following expression for the density of the polarization charge:

$$\rho(x) := -\lambda \langle \Phi(x) \rangle = 2\lambda^2 \frac{\int d\bar{\omega} \, A_x(\bar{\omega}, V) \exp\{-\beta B(\bar{\omega})\}}{\int d\bar{\omega} \, A(\bar{\omega}, V) \exp\{-\beta B(\bar{\omega})\}} \geq 0. \tag{5.4.14}$$

In this formula, use has been made of the functional derivative

$$A_x(\bar{\omega}, V) := -\beta^{-1} \frac{\delta}{\delta V(x)} A(\bar{\omega}, V)$$

$$= (2\pi\beta)^{-3/2} \int_0^1 d\tau' \, \exp\left\{ -\beta \int_0^1 d\tau \, V\Big(x + \ell\big(\bar{\omega}(\tau) - \bar{\omega}(\tau')\big)\Big) \right\}. \tag{5.4.15}$$

Certain features of (5.4.14) are as expected: the density $\rho(x)$ is positive everywhere irrespective of the potential $V(x)$. For small coupling, the mean field $\langle \Phi(x) \rangle$ is of the order λ, so $\rho(x)$ is of the order λ^2. The fact that the mean field varies with x is entirely due to the presence of the potential. In regions where $V(x)$ assumes large values, the mean field $\langle \Phi(x) \rangle$ is seen to be exponentially small.

As is obvious from the definitions,

$$\int dx \, A_x(\bar{\omega}, V) = A(\bar{\omega}, V), \tag{5.4.16}$$

and hence $\int dx \, \langle \Phi(x) \rangle = -2\lambda$, and the total induced charge[71] is simply

$$q_{\text{pol}} = -\lambda \int dx \, \langle \Phi(x) \rangle = 2\lambda^2. \tag{5.4.17}$$

One striking aspect of this formula is that the result does not depend on our choice of $V(x)$. Another is that there cannot be quantum fluctuations of the integrated charge. To prove the latter assertion, let us consider the relevant operator, $Q_{\text{pol}} = -\lambda \int dx \, \Phi(x)$. The charge Q_{pol} does not fluctuate, i.e., it behaves like a classical observable, because $\int dx \, \Phi(x) = \int dx \, \langle \Phi(x) \rangle$. A formal proof of the latter property requires that we choose a constant source function $j(x)$ in (5.4.10) and watch what happens. Obviously, for $j = \text{const}$, $\Phi(j) = j \int dx \, \Phi(x)$ and

$$F_{\text{el}}(j) = F_{\text{el}}(0) - 2\lambda j \tag{5.4.18}$$

---

[70]See Exercise 1 at the end of this section.

[71]The positive charge, which is due to the polarization of the dielectric, screens the negative charge of the electron. The net charge $-1 + 2\lambda^2$ may be considered the charge of the polaron.

since $A(\bar{\omega}, V - 2\lambda j) = e^{2\beta\lambda j} A(\bar{\omega}, V)$. As the expansion of $F_{\mathrm{el}}(j)$ stops right after the linear term, the random variable $\int dx\, \Phi(x)$ has variance zero.

Suppose now hat we weaken the influence of the potential $V(x)$ by introducing a coupling constant which is ultimately set to zero. The effect would be that the electron's wave function spreads more and more: in a sense, the electron delocalizes and, as a consequence, one observes a "thinning" of the induced positive charge, i.e., $\rho(x)$ tends to zero everywhere though q_{pol} stays constant. On closer scrutiny, we shall find that, for $V(x) \to 0$ at large but finite volume Λ,

$$A_x(\bar{\omega}, 0)_\Lambda = (2\pi\beta)^{-3/2}, \tag{5.4.19}$$

and hence $A(\bar{\omega}, 0)_\Lambda = (2\pi\beta)^{-3/2}|\Lambda|$ and $\rho(x) = q_{\mathrm{pol}}/|\Lambda|$.

There is no doubt that the field-theoretic aspects of the polaron model deserve special attention and thorough study. As we have discovered, new kinds of path integrals come up in this context with great challenges to the researcher and with questions bearing on the physics of the electron–phonon system.

For more information and special features of path integrals related to the polaron see [5.23–33]. An excellent account of the quantum theory of large systems in given in [5.34].

Exercise 1.   Let H be the Hamiltonian of a thermodynamical system. For the sake of simplicity, assume that the energy spectrum is finite discrete so that we are dealing essentially with matrices. Given an observable A (a symmetric matrix), define

$$A(t) = e^{tH} A e^{-tH}, \qquad B(t) = e^{tH} e^{-t(H+A)}$$

so that $\dot{B}(t) = -A(t)B(t)$ and $B(0) = \mathbb{1}$, which is immediately turned into

$$B(t) = \mathbb{1} - \int_0^t ds\, A(s)B(s).$$

Formally, this integral equation may be solved by a recursive procedure, i.e., one writes $B(t) = \sum B_n(t)$, where

$$B_0(t) = \mathbb{1}, \qquad B_{n+1}(t) = - \int_0^t ds\, A(s)B_n(s).$$

In particular, $B_1(t) = -\int_0^t ds\, A(s)$ and $B_2(t) = \int_0^t ds \int_0^s ds'\, A(s)A(s')$. Consider now the Gibbs state that assigns the thermal expectation $\langle C \rangle = \mathrm{tr}\,(e^{-\beta H} C)/\mathrm{tr}\, e^{-\beta H}$ to any observable C. Prove that

$$\langle A(s) \rangle = \langle A \rangle, \qquad \langle A(s)A(s') \rangle = f(s - s'),$$

where f is some function satisfying $f(t) = f(\beta - t)$. Use these relations to prove that

$$\langle B_1(\beta)\rangle = -\beta\langle A\rangle, \qquad \langle B_2(\beta)\rangle = \tfrac{1}{2}\beta^2\langle A, A\rangle,$$

where $\langle A, C\rangle \equiv \beta^{-1}\int_0^\beta ds\,\langle A(s)C\rangle$ is the *Duhamel two-point function* (sometimes called the *Bogoliubov scalar product*). Observe next that

$$e^{-\beta F_\lambda} := \operatorname{tr} e^{-\beta(H+\lambda A)} = \operatorname{tr}\left(e^{-\beta H}\sum\lambda^n B_n(\beta)\right) = e^{-\beta F_0}\sum\lambda^n\langle B_n(\beta)\rangle,$$

and hence we have the expansion $F_\lambda = F_0 + \lambda\langle A\rangle - \tfrac{1}{2}\lambda^2\beta\langle A; A\rangle + O(\lambda^3)$, where $\langle A; A\rangle := \langle A, A\rangle - \langle A\rangle^2$. Under what conditions can this analysis be extended to circumstances, where the spectrum of the Hamiltonian is neither discrete nor bounded?

Exercise 2. With $\langle\cdot\rangle$ the same Gibbs state as in the preceding exercise and dropping the condition that A be symmetric, one can think of four reasonable ways to associate a "two-point function" to A:

$$a = \tfrac{1}{2}\langle A^*A + AA^*\rangle$$
$$b = \tfrac{1}{2}\langle A^*A - AA^*\rangle$$
$$c = \beta^{-1}\int_0^\beta ds\,\langle e^{sH}A^*e^{-sH}A\rangle$$
$$d = \tfrac{1}{4}\beta\langle[A^*, [H, A]]\rangle = \tfrac{1}{4}\beta\langle[[A^*, H], A]\rangle$$

Prove that all four expressions are real and, furthermore, that there exists a measure m on $\mathbb{R}$ such that

$$a = \int dm(x)\,\cosh x \qquad c = \int dm(x)\,x^{-1}\sinh x$$
$$b = \int dm(x)\,\sinh x \qquad d = \int dm(x)\,x\sinh x.$$

(Hint: apply Bochner's theorem to the function $f(t) = \langle e^{itH}A^*e^{-itH}A\rangle$.) Use this representation to prove the validity of the inequalities

$$b\coth(b/c) \le a \le \sqrt{dc}\coth\sqrt{d/c}.$$

Hint: the function $g(x) := x\coth x$ is convex, whereas $G(x) = \sqrt{x}\coth\sqrt{x}$ is concave. Use Jensen's inequality. Demonstrate that equality holds iff $[H, A] = EA$ for some real E. Further inequalities follow from the ones already given such as $b^2 \le dc$ and $b\coth(d/b) \le a$ (which strengthens Bogoliubov's inequality stating that $b^2 \le ad$). For further information on these inequalities, their history, and suggestions as to how they may be applied see [5.35–41].

6 Magnetic Fields

> The strength of our persuasions is no evidence at all of their
> own rectitude: crooked things may be as stiff and inflexible as
> straight, and men may be as positive and peremptory in error as
> in truth.
>
> *John Locke*

6.1 Heuristic Considerations

In the preceding chapters, we did not even mention interactions with external magnetic fields. This may have helped create the impression that, for some unknown reason, they are excluded from the path integral formalism. With great delay we shall come to this matter now, thus enlarging the repertoire of path integrals. We will be content with studying a *single* Schrödinger particle; hence we take $d = 3$ in previous formulas unless otherwise stated.

We start with some preliminaries. It is correct to say that quantum physics has failed to formulate magnetic interactions entirely in terms of the magnetic field $B(x)$, i.e., in a gauge-invariant fashion. Instead, the Hamiltonian can only be written in terms of the vector potential $A(x)$ from which the magnetic field derives: $B(x) = \mathrm{rot} A(x)$. However, a potential is never unique, a fact giving rise to the freedom of *choosing a gauge*. In fact, gauge transformations generally change both the potential and the wave function of a charged particle,

$$A \to A + \nabla f , \qquad \psi \to e^{if} \psi, \tag{6.1.1}$$

but leave invariant every directly observable physical quantity.

In the simplest of all situations, the Hamiltonian[1] is

$$H = \tfrac{1}{2}(i\nabla + A)^2. \tag{6.1.2}$$

We shall base our construction of the transition amplitude $\langle x'|e^{-sH}|x\rangle$ on *integrals of the vector potential along Brownian paths*. Such a notion arises from an attempt to generalize the well-known concept of a line integral, i.e., of integrating a vector field along some oriented rectifiable curve C in $\mathbb{R}^3$ from x to x':

---

[1] For brevity, the charge q is always included in the definition of $A(x)$. All vectors, e.g., x, A, etc. have three components, and scalar products, if necessary, will be emphasized by writing $x \cdot A$.

$$\mathcal{A}(C) = \int_C dx \cdot A(x). \qquad (6.1.3)$$

Under a gauge transformation (6.1.1),

$$\mathcal{A}(C) \to \mathcal{A}(C) + f(x') - f(x), \qquad (6.1.4)$$

where the gauge function f need not be differentiable. In essense, the key idea is to look upon the map $C \to \mathcal{A}(C)$ as the fundamental object (thereby avoiding the function $x \to A(x)$) in terms of which the time evolution and other physical processes should be discussed and formulated, i.e., if the particle propagates from x to x', then these should be the endpoints of the curve C. To this end, however, we need to be able to replace the curve C by a Brownian path ω in $\mathbb{R}^3$. In fixing the meaning of such a replacement we are faced with the difficulty that a Brownian path has no length, and hence is nonrectifiable: there is no a priori meaning that may be attributed to the expression $\mathcal{A}(\omega)$. When attempting to extend $\mathcal{A}(C)$, we shall adhere to the equality $\mathcal{A}(-C) = -\mathcal{A}(C)$, where $-C$ is C with the opposite orientation.

We begin by asserting that, for a "very short" path C from x to x', the formula

$$A(x, x') = (x' - x) \cdot (A(x') + A(x))/2 = -A(x', x) \qquad (6.1.5)$$

provides a reasonable approximation of $\mathcal{A}(C)$. Definition (6.1.5) in some sense corresponds to the trapezoidal rule of numerical quadrature. Other choices are possible, of course, such as

$$(x' - x) \cdot A\left(\frac{x' + x}{2}\right) \quad \text{or} \quad (x' - x) \cdot \int_0^1 d\tau \, A(x + (x' - x)\tau). \qquad (6.1.6)$$

The next step is to define a family of operators, Q_t $(t > 0)$, in terms of their integral kernels:

$$\langle x'|Q_t|x \rangle = K(x' - x, t)e^{iA(x,x')}, \qquad K(x' - x, t) = \langle x'|e^{t\Delta/2}|x \rangle. \qquad (6.1.7)$$

Note that these kernels are no longer real but complex, owing to the presence of a magnetic field. Also, two properties, the symmetry of K and the antisymmetry of $A(x, x')$, come together to make $\langle x'|Q_t|x \rangle$ a Hermitean kernel:

$$\langle x'|Q_t|x \rangle = \overline{\langle x|Q_t|x' \rangle}. \qquad (6.1.8)$$

Furthermore, it is important to realize that, for $t \to 0$, one factor, namely $K(x' - x, t)$, tends to $\delta(x' - x)$. Therefore, the main contribution to the kernel of Q_t comes from the vicinity of the diagonal $x = x'$ if t is sufficiently close to zero, and under such circumstances $A(x, x')$ is sufficiently close to any other version of $\mathcal{A}(C)$ with C running from x to x'.

The interesting fact about the above construction is that one recovers the Hamiltonian (6.1.2) as a limit:

$$\lim_{t \to 0} t^{-1}(Q_t - 1) = -H \tag{6.1.9}$$

(leaving aside domain questions). Taking this claim for granted, it by no means implies that H generates Q_t in the usual sense of the word: we are not allowed to write $Q_t = e^{-tH}$, since Q_t ($t \geq 0$) is not a semigroup of operators. Well, almost never.

Nevertheless, there is a standard procedure that turns a family into a semigroup without altering the infinitesimal structure:

$$e^{-tH} = \lim_{n \to \infty} (1 - \tfrac{t}{n} H)^n = \lim_{n \to \infty} (Q_{t/n})^n. \tag{6.1.10}$$

Taking this limit is reminiscent of the procedure used in the Lie–Trotter formula, and once more such a formula provides the key for constructing the path integral. For, if we concatenate the kernels of $Q_{t/n}$ n times to produce the kernel of $(Q_{t/n})^n$, the relevant phases will add up to an approximate version of the line integral along the Brownian path ω:

$$\begin{aligned}
\mathcal{A}_n(\omega) &= \sum_{k=1}^{n} A\big(\omega(\tfrac{k-1}{n}t), \omega(\tfrac{k}{n}t)\big) \\
&= \tfrac{1}{2} \sum_{k=1}^{n} \left[\omega(\tfrac{k}{n}t) - \omega(\tfrac{k-1}{n}t)\right] \cdot \left[A\big(\omega(\tfrac{k}{n}t)\big) + A\big(\omega(\tfrac{k-1}{n}t)\big)\right].
\end{aligned} \tag{6.1.11}$$

Suppose we find convincing arguments for the existence of

$$\mathcal{A}(\omega) = \lim_{n \to \infty} \mathcal{A}_n(\omega) \tag{6.1.12}$$

for a typical Brownian path $\omega \colon (x, 0) \leadsto (x', t)$. Then under a gauge transformation $A \to A + \nabla f$,

$$\mathcal{A}(\omega) \to \mathcal{A}(\omega) + f(x') - f(x) \tag{6.1.13}$$

and, from (6.1.10), we would get a Feynman–Kac like formula for the transition amplitude:

$$\langle x', t | x, 0 \rangle = \langle x' | e^{-tH} | x \rangle = \int d\mu_{x,0}^{x',t}(\omega) \, \exp\left\{i\mathcal{A}(\omega)\right\}. \tag{6.1.14}$$

As before, $d\mu(\omega)$ stands for the conditional Wiener measure. More importantly, the exponent of the integrand is a *linear functional* of the vector potential, whereas the Hamiltonian is quadratic in A. Notice, however, that, in the presence of magnetic fields, the integrand is no longer real but *complex*. Since it is a pure phase, it is certainly bounded by 1 in absolute value, and there can be no problem with regard to convergence of the path integral. By the same token, one obtains the bound

$$|\langle x', t | x, 0 \rangle| \leq \int d\mu_{x,0}^{x',t}(\omega) = K(x' - x, t). \tag{6.1.15}$$

The point of the path oriented approach is to prove (6.1.9) at the beginning and then to verify that (6.1.12) makes sense. The rest follows without much further ado. To prove (6.1.9), we need to calculate

$$\left[\nabla_x^2\, e^{iA(x,x')}\psi(x)\right]_{x'=x} = \left(\nabla - iA(x)\right)^2\psi(x) \tag{6.1.16}$$

for suitable wave functions ψ, which is easy. Next, to go from (6.1.15) to (6.1.9) we write

$$\begin{aligned}
[Q_t\psi](x') &= \int dx\, \langle x'|Q_t|x\rangle\psi(x) \\
&= \int dx\, \langle x'|e^{t\Delta/2}|x\rangle e^{iA(x,x')}\psi(x)
\end{aligned} \tag{6.1.17}$$

and $e^{t\Delta/2} = 1 + t\Delta/2 + O(t^2)$. Thus,

$$\begin{aligned}
\lim_{t\to\infty}\left[t^{-1}(Q_t - 1)\psi\right](x') &= \tfrac{1}{2}\int dx\, \delta(x - x')\nabla_x^2\, e^{iA(x,x')}\psi(x) \\
&= -\tfrac{1}{2}(i\nabla_{x'} + A(x'))^2\psi(x')
\end{aligned} \tag{6.1.18}$$

which proves the assertion.

6.2 Itô Integrals

The mathematical questions behind limit (6.1.12) get us into the subject of Itô integrals. This subject has been extensively treated in the literature (see [1.31], for instance) and attention has been drawn to it by exploring the solutions of stochastic differential equations. From this rich theory we shall present only a few essentials.

There is more than one way to give meaning to the integral

$$\int_0^t dX_s\cdot A(X_s, s) \tag{6.2.1}$$

if $A(x, s)$ is a vector-valued function on $\mathbb{R}^3 \times \mathbb{R}_+$ and with X_s the (three-dimensional) Wiener process. There are two notable approaches that carry names.

The most convenient and widely accepted approach, from the mathematics point of view, is due to Itô. Some physicists, however, quote an alternative notion from Stratonovitch. Either may be adopted in the present context without affecting the physical content of the resulting path integral. Since the former notion seems more popular among mathematical physicists and since the related integral calculus is more developed, Itô's idea will be reproduced here in some detail.

The basic strategy is, of course, to divide the fundamental interval $0 \leq s \leq t$ into smaller and smaller subintervals and to do this recursively: in the first step, two subintervals of length $t/2$ are created, and continuing the bisection algorithm until we reach the nth step, 2^n subintervals of length $t/2^n$ are obtained. As the recursion progresses, one evaluates $A_m := A(X_{s_m}, s_m)$ at $s_m = mt/2^n$ $(m = 0, 1, \ldots, 2^n - 1)$ and writes the Itô integral as the limit of telescopic sums:

$$\int_0^t dX_s \cdot A(X_s, s) = \lim_{n \to \infty} \sum_{m=0}^{2^n - 1} \Delta_m \cdot A_m, \qquad \Delta_m := X_{s_{m+1}} - X_{s_m}. \tag{6.2.2}$$

The important aspect of this definition is that the increment Δ_m points into the future, whereas the pivotal value A_m depends on X_{s_m}, i.e., on the present. Equivalently, Itô's convention may be stated as $dX_s = X_{s+ds} - X_s$. This fact clearly facilitates the analysis because it makes the random variables A_m and Δ_m statistically independent. It is by no means immaterial at what time $s \in [s_m, s_{m+1}]$ the integrand $A(X_s, s)$ is evaluated to form A_m: the error does not necessarily tend to zero, and standard integration theory (according to Riemann, Lebesgue, or Stieltjes) does not apply here.

Nevertheless, it may be shown that limit (6.2.2) exists[2] provided

$$\int_0^t ds\, \boldsymbol{E}\big(A(X_s, s)^2\big) < \infty. \tag{6.2.3}$$

More explicitly, this condition reads

$$\int_0^t ds\, (2\pi s)^{-3/2} \int dx\, e^{-x^2/2s} A(x, s)^2 < \infty \tag{6.2.4}$$

and is recognized to be relatively weak[3].

Itô integrals satisfy many remarkable rules and formulas.

1. Every Itô integral has mean zero: $\boldsymbol{E}\big(\int_0^t dX_s \cdot A(X_s, s)\big) = 0$. The reason: we have that $\boldsymbol{E}(\Delta_m \cdot A_m) = \boldsymbol{E}(\Delta_m) \cdot \boldsymbol{E}(A_m) = 0$ by the independence of Δ_m and A_m and by the identity $\boldsymbol{E}(\Delta_m) = 0$.

2. If wanted, one may compile a list of 'solvable' Itô integrals. However, such a list is coded in one fundamental formula[4]

---

[2] Convergence takes place with respect to the norm $\|R\| = \{\boldsymbol{E}(R^2)\}^{1/2}$, where R is any random variable with finite second moment.

[3] Though condition (6.2.3) is not expected to stir anybody's concern, it can become critical in certain cases. Suppose that $A(x) \sim |x|^{-1}$ as $|x| \to 0$. Then $\boldsymbol{E}(A(X_s)^2) \sim s^{-1}$ as $s \to 0$, and (6.2.4) is violated. However, this only means that the Itô integral cannot be evaluated at a Brownian path starting at the origin. There is no problem with paths starting at $x \neq 0$: for the evaluation to make sense we need only check that $\boldsymbol{E}(A(x + X_s)^2)$ is integrable at $s = 0$. Moral: when choosing the initial and final positions of a path, stay away from the singularities of the vector potential.

[4] See [1.22] for details and proof.

$$\int_0^t dX_s \cdot a \, e^{a \cdot X_s - a^2 s/2} = e^{a \cdot X_t - a^2 t/2} - 1, \tag{6.2.5}$$

where the vector $a \in \mathbb{R}^3$ is arbitrary. Expansion into powers of a yields

$$\int_0^t dX_{sk} = X_{tk}$$

$$\int_0^t (X_{sj} \, dX_{sk} + X_{sk} \, dX_{sj}) = X_{tj} X_{tk} - t\delta_{jk} \tag{6.2.6}$$

etc. $(j, k = 1, 2, 3)$. The latter formula may also be written as

$$X_{sj} \, dX_{sk} + X_{sk} \, dX_{sj} = d(X_{sj} X_{sk}) - ds \, \delta_{jk}. \tag{6.2.7}$$

3. A more sophisticated assertion is *Itô's lemma*:

$$dX_{sj} \, dX_{sk} = ds \, \delta_{jk} \tag{6.2.8}$$

$(j, k = 1, 2, 3)$. This reflects the fact that increments like $X_{s+ds} - X_s$ are of the order $ds^{1/2}$.

6.2.1 The Feynman–Kac–Itô Formula

Let us now state our desideratum: to extend the Feynman–Kac formula to magnetic interactions – even for time-dependent magnetic fields – in such a way that the earlier result for the transition amplitude without a magnetic field shall be contained in the formula as a special case. We therefore return to construction (6.1.11) and replace the path $\omega(s)$ by the Wiener process X_s. Unfortunately, we notice at once that the result differs from what one would get using the Itô procedure (6.2.2), i.e., it differs by the expression $F = \lim F_n$, where

$$F_n = \tfrac{1}{2} \sum_{m=0}^{2^n - 1} \Delta_m \cdot (A_{m+1} - A_m). \tag{6.2.9}$$

Introducing the first partial derivatives of the vector potential, $A_{j|k}(x, s) = (\partial/\partial x_k) A_j(x, s)$, however, we see that F_n tends to

$$F = \tfrac{1}{2} \int_0^t \sum_{j,k=1}^3 dX_{sj} \, dX_{sk} \, A_{j|k}(X_s, s). \tag{6.2.10}$$

By virtue of Itô's lemma (6.2.8),

$$F = \tfrac{1}{2} \int_0^t ds \sum_{k=1}^3 A_{k|k}(X_s, s) \equiv \tfrac{1}{2} \int_0^t ds \, \nabla \cdot A(X_s, s) \tag{6.2.11}$$

and we are faced with the fact that there is a finite additional term in the expression for $\mathcal{A}$ that extends (6.1.12) to time-dependent vector potentials:

$$\mathcal{A} = \int_s^{s'} (dX_t + \tfrac{1}{2}dt\nabla)\cdot A(X_t, t). \tag{6.2.12}$$

The term involving the divergence, $\nabla\cdot A$, is mainly there to guarantee the correct behavior of $\mathcal{A}$ under gauge transformations (and time reversal). In fact, ansatz (6.1.11) is at the heart of Stratonovitch's approach to the definition of a stochastic integral. If we were to follow his prescriptions, there would of course be no $\nabla\cdot A$ term in (6.2.12). The reader who feels uneasy about following Itô should realize that the distinction disappears as soon as the condition $\nabla\cdot A = 0$ is imposed: this is precisely what we shall assume from now on.

We come to state our conclusion: given a time-dependent Hamiltonian

$$H(t) = \tfrac{1}{2}(i\nabla + A(\cdot, t))^2 + V(\cdot, t), \tag{6.2.13}$$

with $V(x, t)$ a scalar and $A(x, t)$ a vector potential satisfying $\nabla\cdot A = 0$. Then, if we define the random variables

$$\mathcal{V} = \int_s^{s'} dt\, V(X_t, t), \qquad \mathcal{A} = \int_s^{s'} dX_t\cdot A(X_t, t), \tag{6.2.14}$$

the quantum mechanical transition amplitude is given by the *Feynman–Kac–Itô formula*:

$$\boxed{\langle x', s'|x, s\rangle = \int d\mu_{x,s}^{x',s'}(\omega) \, \exp\big\{ -\mathcal{V}(\omega) + i\mathcal{A}(\omega)\big\}.} \tag{6.2.15}$$

A few words of explanation are in order. In accordance with what has been said in Sect. 1.6.1 and with reference to $\mathcal{V}$ and $\mathcal{A}$, the term *random variable* is always understood to be a map $\Omega \to \mathbb{R}$, where Ω is the underlying space of sample paths. In accordance with this concept, we shall write $\mathcal{V}(\omega) = \int dt\, V\big(\omega(t), t\big)$ and $\mathcal{A}(\omega) = \int d\omega(t)\cdot A\big(\omega(t), t\big)$ if $\mathcal{V}$ and $\mathcal{A}$ are evaluated at $\omega \in \Omega$. Moreover, if $\partial A/\partial t = 0$, we suggestively write (by abuse of notation) $\mathcal{A}(\omega) = \int_\omega dx\cdot A(x)$.

The integrand in (6.2.15) is of the form $I = \exp(-\mathcal{V} + i\mathcal{A})$ with $\mathcal{V}$ and $\mathcal{A}$ real. Thus, $|I| = \exp(-\mathcal{V})$, which makes it easy for us to compare the full amplitude and that with vanishing vector potential:

$$\big|\langle x', s'|x, s\rangle\big| \le \langle x', s'|x, s\rangle_{A=0}. \tag{6.2.16}$$

In particular, if both potentials are constant with respect to time, taking $x' = x$, we have

$$\langle x|e^{-\beta H}|x\rangle \le \langle x|e^{-\beta H}|x\rangle_{A=0} \tag{6.2.17}$$

so that, after integration over x, (6.2.17) is turned into an inequality between free energies: $F(\beta) \geq F(\beta)_{A=0}$. Consequently, in the limit $\beta \to \infty$, we would get a comparison result concerning two ground state energies: $E \geq E_{A=0}$. In other words:

> *Turning on a magnetic field never lowers the ground state energy of a charged spinless particle (or system of such particles).*

For more information see [6.1–4].

6.2.2 The Semiclassical Approximation

Normally, paths have fixed endpoints. It is thus perfectly reasonable to bring the Itô integral into a new, more convenient form involving the Brownian bridge $\bar{X}_\tau$ only. This is easily done by the standard process of *rescaling* which has several benificial effects. One of them is that, from the resulting expression, we can read off the complete dependence on the coordinates of the endpoints without having to evaluate the integral explicitly. As regards the final formulation, however, one will certainly miss the elegance of the original integral. Nevertheless, the description is then based on dimensionless path and time variables which are best suited for a semiclassical analysis. In essence, what we have in mind is the replacement[5]

$$\begin{aligned}
X_t &\leftarrow x + (x' - x)\tau + \ell\bar{X}_\tau & t &= s + (s' - s)\tau \\
dX_t &\leftarrow (x' - x)d\tau + \ell d\bar{X}_\tau & dt &= (s' - s)d\tau
\end{aligned} \tag{6.2.18}$$

with ℓ a length parameter (see Sect. 3.1 for details) and $\tau \in [0, 1]$. To conform to Itô's rule we must define $d\bar{X}_\tau = \bar{X}_{\tau+d\tau} - \bar{X}_\tau$.

In terms of sample paths,

$$\begin{aligned}
\omega(t) &= x + (x' - x)\tau + \ell\bar{\omega}(\tau) \\
d\mu(\omega) &= K(x' - x, s' - s)d\bar{\omega},
\end{aligned} \tag{6.2.19}$$

where $\bar{\omega}$ denotes an *arbitrary* path of the Brownian bridge.

Knowing that the parameter ℓ sets the scale of quantum fluctuations, we call a power series, with respect to ℓ, of an amplitude a *semiclassical expansion* whenever such an expansion is possible. Formally, we write

$$\langle x', s' | x, s \rangle = K(x' - x, s' - s) \int d\bar{\omega} \, \exp \sum_{n=0}^{\infty} \ell^n I_n(\bar{\omega}) \tag{6.2.20}$$

and call

$$\langle x', s' | x, s \rangle_{\mathrm{sc}} = K(x' - x, s' - s) \exp I_0 \tag{6.2.21}$$

the *semiclassical approximation* of the amplitude $\langle x', s' | x, s \rangle$. Obviously,

---

[5] As random variables are maps – basically – , "replacement" here means *restriction to a subset of paths* (with fixed endpoints) and *change of variables*.

$$I_0 = -\mathcal{V}(\omega_0) + i\mathcal{A}(\omega_0) \tag{6.2.22}$$

with $\omega_0: (x, s) \rightsquigarrow (x', s')$ the linear (optical) path; hence,

$$I_0 = \int_0^1 d\tau \left\{ -(s' - s)V\big(x + (x' - x)\tau, s + (s' - s)\tau\big) \right. \tag{6.2.23}$$
$$\left. + i(x' - x)\cdot A\big(x + (x' - x)\tau, s + (s' - s)\tau\big) \right\}$$

which extends our previous findings (see (3.2.10)). Suppose $V = 0$ and $\partial A/\partial t = 0$. Then I_0 coincides with the ordinary line integral $i\int dx\cdot A(x)$ along the optical path.

Exercise 1.   To provide a correction to the semiclassical approximation (6.2.20) one needs to take the effect of $I_1(\bar{\omega})$ into account. Prove that

$$I_1(\bar{\omega}) = \int_0^1 f(\tau)\cdot d\bar{X}_\tau$$

and, with recourse to (3.1.25–27),

$$\int d\bar{\omega}\, e^{\ell I_1(\bar{\omega})} = \exp\left\{ \tfrac{1}{2}\ell^2 \mathrm{var}(f) \right\},$$

where $f(\tau)$ is an integral along the optical path up to time τ:

$$f(\tau) = \int_0^\tau d\tau' \left\{ -(s' - s)\,\mathrm{grad}V\big(x + (x' - x)\tau', s + (s' - s)\tau'\big) \right.$$
$$\left. - i(x' - x) \times \mathrm{rot}A\big(x + (x' - x)\tau, s + (s' - s)\tau\big) \right\}.$$

Exercise 2.   For constant electric and magnetic fields, E and B, one may choose a gauge such that $V(x) = -x\cdot E$ and $A(x) = -\tfrac{1}{2}x \times B$. Upon setting

$$G = (s' - s)E + i(x' - x) \times B$$

show that a translation in space changes the amplitude in the following way:

$$\langle x' + a, s'|x + a, s\rangle = \langle x', s'|x, s\rangle \exp\{a\cdot G\}.$$

Prove also that the semiclassical approximation yields

$$\langle x', s'|x, s\rangle_{\mathrm{sc}} = K(x' - x, s' - s) \exp\left\{ \tfrac{1}{2}(x' + x)\cdot G \right\}.$$

Moreover, $f(\tau) = -\tau G$, where f has been defined in the preceding exercise, and $\mathrm{var}(f) = \tfrac{1}{12}G^2$.

Exercise 3.   Increments of the Brownian bridge, though pointing into the future, depend on the present. Show for instance that, in one dimension and if $d\bar{X}_\tau = \bar{X}_{\tau + d\tau} - \bar{X}_\tau$,

$$\boldsymbol{E}(\bar{X}_\tau d\bar{X}_\tau) = -\tau d\tau \ ,$$

as opposed to $\boldsymbol{E}(X_t dX_t) = 0$.

6.3 The Constant Magnetic Field

It will certainly raise no objection when we propose to relax for a moment and contemplate the simplest of all problems that arise in the present context: a Schrödinger particle moving in a constant magnetic field B. The fascinating aspect of this problem is, as we shall see, that it reduces to studying the harmonic oscillator, for which we know the answer. The method is elegant but a little intricate.

We may hope to simplify the discussion by restricting it to two dimensions, i.e., to the plane orthogonal to the direction of the magnetic field. With $x = \{x_1, x_2\}$ an arbitrary point in the plane and

$$A(x) = \tfrac{1}{2}B\{-x_2, x_1\} \qquad (B > 0) \tag{6.3.1}$$

the vector potential, we treat the magnetic field B as if it were a scalar. In writing (6.3.1) we have chosen a particular gauge. In two dimensions, the exterior product (or *wedge product*) is simply a number:

$$a \wedge b = a_1 b_2 - a_2 b_1. \tag{6.3.2}$$

Keep in mind that, under a transformation $T \in O(2)$ of the plane (rotations + reflections), the exterior product behaves as a pseudoscalar: $(Ta) \wedge (Tb) = \det T\, a \wedge b$ with $\det T = \pm 1$.

In constructing the amplitude $\langle x', s' | x, s \rangle_B$ we need to replace x in $A(x)$ by $x + (x' - x)\tau + \ell \bar{X}_\tau$ with $\bar{X}_\tau$ the two-dimensional Brownian bridge and $\ell^2 = s' - s$. As $x \to A(x)$ is linear, the resulting expression splits into a deterministic and a nondeterministic part: $A\big(x + (x'-x)\tau\big) + \ell A(\bar{X}_\tau)$. The following shorthand proves to be convenient:

$$\nu = B(s' - s) \ , \qquad a = \{a_1, a_2\} = \ell B(x' - x). \tag{6.3.3}$$

We shall work rather with expectations $\boldsymbol{E}(\cdot)$ than with path integrals and represent the amplitude in the form

$$\langle x', s' | x, s \rangle_B \ = \ \langle x', s' | x, s \rangle_0 \, \boldsymbol{E}\big(\exp(iY) \big), \tag{6.3.4}$$

where the random variable Y is linear in the magnetic field B. According to the Feynman–Kac–Itô formula, Y decomposes as

$$Y = Y_1 + Y_2 + Y_3 + Y_4. \tag{6.3.5}$$

We will deal with each Y_i separately.

1. Y_1 is a conventional integral of some deterministic quantity:

$$Y_1 = \int_0^1 d\tau\,(x' - x)\cdot A\big(x + (x' - x)\tau\big) = \tfrac{1}{2}Bx \wedge x'. \qquad (6.3.6)$$

2. Y_2 is a conventionial integral of some stochastic quantity:

$$Y_2 = \ell \int_0^1 d\tau\,(x' - x)\cdot A(\bar X_\tau) = \tfrac{1}{2}\int_0^1 d\tau\,\bar X_\tau \wedge a. \qquad (6.3.7)$$

3. Y_3 is a stochastic integral of some ordinary function:

$$Y_3 = \ell \int_0^1 A\big(x + (x' - x)\tau\big)\cdot d\bar X_\tau. \qquad (6.3.8)$$

In fact, $Y_3 = Y_2$ (use partial integration and $\bar X_0 = \bar X_1 = 0$).

4. Y_4 is a genuine Itô integral:

$$Y_4 = \ell^2 \int_0^1 A(\bar X_\tau)\cdot d\bar X_\tau = \tfrac{1}{2}\nu \int_0^1 \bar X_\tau \wedge d\bar X_\tau. \qquad (6.3.9)$$

Substituting in (6.3.4) we get

$$E\big(\exp(iY)\big) = \exp(\tfrac{1}{2}iBx \wedge x')\,E\big(\exp(iZ)\big)$$

$$Z := Y_2 + Y_3 + Y_4 = \int_0^1 \bar X_\tau \wedge (a\,d\tau + \tfrac{1}{2}\nu d\bar X_\tau). \qquad (6.3.10)$$

Rotational invariance (in the plane) allows us to put $a_2 = 0$. Writing $\bar X_{\tau 1}$ and $\bar X_{\tau 2}$ for the components of $\bar X_\tau$, we find that

$$Z = \int_0^1 (a_1\tau + \nu \bar X_{\tau 1})d\bar X_{\tau 2}, \qquad (6.3.11)$$

where use has been made of

$$\int_0^1 d\tau\,\bar X_{\tau 2} = -\int_0^1 \tau d\bar X_{\tau 2}\ , \qquad \int_0^1 \bar X_{\tau 2}\,d\bar X_{\tau 1} = -\int_0^1 \bar X_{\tau 1}\,d\bar X_{\tau 2}.$$

$$(6.3.12)$$

As $\bar X_{\tau 1}$ and $\bar X_{\tau 2}$ are independent processes, we obtain $E(\cdot)$ in (6.3.10) by performing successively the expectations E_1 and E_2 with respect to $\bar X_{\tau 1}$ and $\bar X_{\tau 2}$, i.e., $E(\exp(iZ)) = E_1(E_2(\exp(iZ)))$. The first task is to determine the inner expectation, which is a known problem (see (3.1.28)):

$$E_2\big(\exp(iZ)\big) = \exp\big\{ -\tfrac{1}{2}\mathrm{var}(f)\big\}, \qquad f(\tau) := a_1\tau + \nu \bar X_{\tau 1}. \qquad (6.3.13)$$

The integral $\int_0^1 d\tau\, f(\tau)$ appears quadratically in var(f) making the evaluation of the outer expectation difficult. The trick to circumvent this difficulty is to *undo the square* by some extra Gaussian integral:

$$\boldsymbol{E}_2\big(\exp(iZ)\big) = (2\pi)^{-1/2} \int_{-\infty}^{\infty} dp\, e^{-p^2/2} \exp\left\{ -\int d\tau\, \big(pf(\tau) + \tfrac{1}{2}f(\tau)^2\big) \right\}.$$

$$(6.3.14)$$

The argument of the exponential on the right-hand side is now seen to be a single τ integral; it may also be rewritten so as to reveal its dependence on $\bar{X}_{\tau 1}$:

$$\tfrac{1}{2}p^2 + \int_0^1 d\tau\, \left(pf(\tau) + \tfrac{1}{2}f(\tau)^2\right) = \tfrac{1}{2}\int_0^1 d\tau\, (p + a_1\tau + \nu\bar{X}_{\tau 1})^2. \qquad (6.3.15)$$

The next step is to interchange the p-integration and the expectation $\boldsymbol{E}_1$ to get

$$\boldsymbol{E}\big(\exp(iZ)\big) = (2\pi)^{-1/2} \int_{-\infty}^{\infty} dp\, \boldsymbol{E}_1\left(\exp\left\{ -\tfrac{1}{2}\int_0^1 d\tau\, (p + a_1\tau + \nu\bar{X}_{\tau 1})^2 \right\}\right).$$

$$(6.3.16)$$

To determine the expectation $\boldsymbol{E}_1$ under the integral is a problem known to us from studying the one-dimensional harmonic oscillator with the frequency B, and to see the connection, we introduce the fictitious coordinates b and b' and quote Mehler's formula recalling the definitions $\nu = B(s' - s)$ and $\ell^2 = s' - s$:

$$\langle b', s'| b, s\rangle_{\mathrm{Me}} = K(b' - b, s' - s) \times$$

$$\boldsymbol{E}_1\left(\exp\left\{ -\tfrac{1}{2}(s' - s)B^2 \int_0^1 d\tau\, \left(b + (b' - b)\tau + \ell\bar{X}_{\tau 1}\right)^2 \right\}\right)$$

$$= \left[\frac{B}{2\pi \sinh\nu}\right]^{1/2} \exp\left\{ -\frac{B(b^2 + b'^2)}{2\tanh\nu} + \frac{Bbb'}{\sinh\nu} \right\}. \qquad (6.3.17)$$

The expectations $\boldsymbol{E}_1$ in (6.3.16) and (6.3.17) coincide if

$$b = (\ell B)^{-1}p\,, \qquad\qquad b' = (\ell B)^{-1}(a_1 + p). \qquad (6.3.18)$$

Now that all expectations have been performed, we are left with an integral with respect to p which poses no problem because it is Gaussian again:

$$\boldsymbol{E}(\exp(iZ)) = \left[\frac{\nu}{2\pi \sinh\nu}\right]^{1/2} \int_{-\infty}^{\infty} dp\, \exp\{-F(p)\} \qquad (6.3.19)$$

setting

$$F(p) = \frac{p^2 + (a_1 + p)^2}{2\nu \tanh \nu} - \frac{p(a_1 + p)}{\nu \sinh \nu} - \frac{a_1^2}{2\nu^2}$$
$$= (p + \tfrac{1}{2}a_1)^2 \frac{\tanh(\tfrac{1}{2}\nu)}{\nu} + \frac{a_1^2}{2\nu^2}\left(\tfrac{1}{2}\nu \coth(\tfrac{1}{2}\nu) - 1\right). \tag{6.3.20}$$

Summarizing, we state the result of the computation as

$$\boxed{\langle x', s' | x, s \rangle_B = \frac{B/2}{2\pi \sinh(\tfrac{1}{2}\nu)} \exp\left\{\frac{B}{2}\left(ix \wedge x' - \tfrac{1}{2}(x' - x)^2 \coth(\tfrac{1}{2}\nu)\right)\right\}}$$

$$\tag{6.3.21}$$

with $\nu = B(s' - s)$ and $s' > s$.

6.3.1 A Brief Discussion of the Result

As regards the final formula (6.3.21), there is only one place where the imaginary unit appears: in front of the exterior product $x \wedge x'$. As may be learned from the initial steps of above calculation, the imaginary contribution to the exponent in (6.3.21) comes from a line integral of the vector potential, the line being the optical path ω_0 from x to x':

$$\mathcal{A}(\omega_0) = \int_{\omega_0} dx \cdot A(x) = \tfrac{1}{2}Bx \wedge x'. \tag{6.3.22}$$

That this contribution is present under all circumstances has been shown before: see (6.3.22).

For parallel x and x', the phase disappears and the amplitude $\langle x', s' | x, s \rangle$ comes out real. Observe also that

$$\langle x', s' | x, s \rangle_B = \overline{\langle x, s' | x', s \rangle_B}, \tag{6.3.23}$$

which should not come as a surprise here since the Hamiltonian is self-adjoint.

To be able to analytically continue the amplitude to imaginary times we need to assume that $\nu \neq 0 \bmod 2\pi$. For $s = it$ and $s' = it'$, all hyperbolic functions in (6.3.21) become trigonometric functions. As a result, the argument of $\exp(\cdot)$ in (6.3.21) becomes purely imaginary and oscillates with the Larmor frequency $\tfrac{1}{2}B$.

Translations perpendicular to the magnetic field (i.e., translations in the plane) have nontrivial effects on the amplitude:

$$\langle x' + a, s' | x + a, s \rangle_B = e^{if(x')}\langle x', s' | x, s \rangle_B \, e^{-if(x)} \tag{6.3.24}$$

with $f(x) = \tfrac{1}{2}Bx \wedge a$. One way to interpret this behavior is to say that the system, when translated by a vector a, experiences a gauge transformation with the gauge function $f(x)$. This clearly shows that spatial and gauge symmetries are intimately connected.

If wanted, the restriction $B > 0$ can be dropped. A change of sign, $B \to -B$, then means reversal of the magnetic field. Any reflection in the plane, i.e., any $T \in O(2)$ with $\det T = -1$, reverses the field:

$$\langle Tx', s'|Tx, s\rangle_B = \langle x', s'|x, s\rangle_{-B}. \tag{6.3.25}$$

Exercise 1. What would be the effect of a Galilean transformation $x \to x - vt$, $x' \to x' - vt'$ on the amplitude $\langle x', it'|x, it\rangle_B$ apart from a time-dependent gauge transformation?

Exercise 2. Show that a Fourier transformation turns (6.3.21) into a formula whose structure is very similar:

$$(2\pi)^{-2} \int_{\mathbb{R}^2} dx \int_{\mathbb{R}^2} dx' \, e^{ip'x' - ipx} \langle x', s'|x, s\rangle_B$$

$$= \frac{2/B}{2\pi \sinh(\frac{1}{2}\nu)} \exp\left\{ \frac{2}{B}\left(ip \wedge p' - \tfrac{1}{2}(p' - p)^2 \coth(\tfrac{1}{2}\nu) \right) \right\}$$

$$(p, p' \in \mathbb{R}^2).$$

Exercise 3. For X_t the two-dimensional Wiener process starting at $x = 0$, define the "area below the curve" as the random variable

$$F_t = \tfrac{1}{2} \int_0^t X_s \wedge dX_s$$

and prove that

$$E\left(e^{i\alpha F_t}\right) = (\cosh \alpha t)^{-1/2} = (2\pi)^{-3/2} \int_{-\infty}^{\infty} dx \, |\Gamma(\tfrac{1}{4} + i\tfrac{x}{2})|^2 e^{i\alpha t x}$$

with $\Gamma(z)$ the gamma function for a complex argument. Hint: use (2.3.32) for $d = 1$. This by the way shows that the area per unit time, $t^{-1}F_t$, has a stationary distribution (which does not mean that $t^{-1}F_t$ is independent of t). What could be the connection of the above formula with the problem of a particle in a constant magnetic field?

6.4 Diamagnetism of Electrons in a Solid

To see at least one application of formula (6.3.21) we turn now to a problem in solid state physics and ask: how does the electronic part of the free energy change if a homogeneous magnetic field B is applied to a solid? Let

$$\dot{F}(B) = F_{\mathrm{orbit}}(B) + F_{\mathrm{spin}}(B) \tag{6.4.1}$$

be the free energy per particle (electron: charge e, mass m) at the temperature T and

$$\omega_L = \frac{eB}{2m} \qquad (6.4.2)$$

the Larmor frequency. Restricting (6.3.21) to $x' = x$ and setting

$$s' - s = \hbar\beta = \frac{\hbar}{k_B T} \, , \qquad \tfrac{1}{2}\nu = \frac{\hbar\omega_L}{k_B T} \, , \qquad (6.4.3)$$

we know that[6]

$$\frac{\exp\{-\beta F_{\text{orbit}}(B)\}}{\exp\{-\beta F_{\text{orbit}}(0)\}} = \frac{\langle x, s'|x, s\rangle_B}{\langle x, s'|x, s\rangle_0} = \frac{\nu/2}{\sinh(\nu/2)} \, . \qquad (6.4.4)$$

Conceptually, the resulting formula

$$F_{\text{orbit}}(B) = F_{\text{orbit}}(0) + \beta^{-1} \log \frac{\sinh(\nu/2)}{\nu/2} \qquad (6.4.5)$$

neither incorporates effects from the interaction of the electrons nor does it take into account the Dirac statistics. So its validity is limited.

We should not however neglect the spin of the electrons. Dirac theory says that a magnetic field experiences the magnetic moment

$$\mu = \frac{e\hbar}{2m} \, , \qquad (6.4.6)$$

leading to energy levels $\pm\mu B$. As for the spin part, the determination of the free energy is easy:

$$\exp\{-\beta F_{\text{spin}}(B)\} = 2\cosh(\nu/2) \, , \qquad \tfrac{1}{2}\nu = \frac{\mu B}{k_B T}. \qquad (6.4.7)$$

It seems now that we have two conflicting definitions of the parameter ν. However, the identity $\hbar\omega_L = \mu B$ shows that this is not the case. The spin part of the free energy,

$$F_{\text{spin}}(B) = -\beta^{-1} \log \Big(2\cosh(\nu/2) \Big), \qquad (6.4.8)$$

ought to be confronted with the orbital part (6.4.5). Both parts behave very differently. One obvious thing to observe is the difference in sign. But from the experimental point of view it would be more appropriate to look at the derivatives with respect to the magnetic field. The first derivative $F'(B)$

---

[6] To justify this formula we adhere to the independent particle model appropriate for a dilute electron gas. Moreover, the translational invariance of the model system requires that two intermediate steps have to be taken: introduction and elimination of a confining potential for the electron. As this has already been discussed in a similar case, i.e., for the polaron model in Sect. 5.3, we refrain here from repeating the arguments.

gives us the induced magnetic moment, whereas the second derivative $F''(B)$ tells us how a small change of B would influence the magnetic moment. Since

$$\frac{d^2}{dx^2}\log\frac{\sinh x}{x} = \frac{1}{x^2} - \frac{1}{\sinh^2 x} > 0 \quad (x \neq 0)$$

$$\frac{d^2}{dx^2}\log\cosh x = \frac{1}{\cosh^2 x} > 0, \tag{6.4.9}$$

we may now state the difference in behavior as

$$F''_{\text{spin}}(B) < 0 \qquad \text{(Pauli paramagnetism [6.5])}$$

$$F''_{\text{orbit}}(B) > 0 \qquad \text{(Landau diamagnetism [6.6])}. \tag{6.4.10}$$

Linear response theory focuses on the small B behavior and calls $\chi = F''(0)$ the *suceptibility* of the system. For a calculation of χ we need to expand two elementary functions:

$$\log\frac{\sinh x}{x} = \frac{1}{6}x^2 + O(x^4), \qquad \log\cosh x = \frac{1}{2}x^2 + O(x^4). \tag{6.4.11}$$

Accordingly, the susceptibility consists of two parts, χ_{orbit} and χ_{spin}, which satisfy

$$0 < \chi_{\text{orbit}} = -\tfrac{1}{3}\chi_{\text{spin}} \tag{6.4.12}$$

as can be inferred from expansions (6.4.11).

Our simple-minded calculation can in fact be repeated and relation (6.4.12) still holds for the case of a noninteracting Fermi gas to which a magnetic field is applied [6.7]. Such a calculation becomes necessary when the Pauli exclusion principle is expected to have a more serious influence. For strong magnetic fields, however, a separation of the induced magnetic moment into an orbit and a spin part is no longer possible.

Exercise 1. Standard moment inequalities for the discrete spectrum of a Schrödinger operator without magnetic field, as discussed in Sect. 3.4, have an obvious extension to situations where a magnetic field is present. For the Hamiltonian $H = \frac{1}{2}(i\nabla + A)^2 - V$ (not explicitly dependent on time and satisfying $V > 0$), consider the Birman–Schwinger operator

$$B_E = V^{1/2}\left(\tfrac{1}{2}(i\nabla + A)^2 - E\right)^{-1}V^{1/2} \qquad (E < 0).$$

Along the lines of Sect. 3.4 with the Feynman–Kac–Itô formula replacing the Feynman–Kac formula to account for the presence of the vector potential, show that

$$\operatorname{tr} F(B_E) = \int_0^\infty \frac{ds}{s}e^{Es}\int dx\int d\mu_{x,0}^{x,s}(\omega)\,e^{i\mathcal{A}(\omega)}\,f\left(\int_0^s dt\,V(\omega(t))\right),$$

where F and f are connected by relation (3.4.25) but are arbitrary otherwise. As the vector potential enters the path integral as a pure phase factor $e^{i\mathcal{A}}$, which is obviously bounded by 1, all estimates in Sect. 3.4 from (3.4.29) on remain true (with the same constants), regardless of the magnetic field that is applied to the system. This result is due to M. Birman (private communication by E. Lieb). See also [6.1] and [6.2].

6.5 Magnetic Flux Lines

By 1959 it had become evident that magnetic fields, though confined to some region in space, may have (quantum) effects on charged particles traversing the exterior of that region [6.8,9]. The somewhat paradoxical phenomenon had been predicted by Aharonov and Bohm and, later on, came to be known as the AB effect. It instantly received widespread attention because is has some bearing on the role of the vector potential in quantum physics. More importantly, the effect has no analogue in classical electrodynamics: the Lorentz force acts locally and no *action at a distance* with regard to fields and particles has ever been theoretically predicted or experimentally observed. Formally, the difference lies in the fact that quantum mechanics unavoidably relies on the Hamiltonian formalism rather than on some *equation of motion*. Therefore, the AB effect is deeply rooted in the fundaments of quantum mechanics and general acceptance by the physics community did not come immediately. Schulman [6.10] was first in stating that the AB effect has a topological origin and should be discussed using the Feynman path integral formalism.

A typical situation, perhaps the simplest with regard to the AB effect, arises when electrons travel freely in some region crossed by a single magnetic flux line (a flux tube with zero diameter but finite flux). For ease we shall assume that we are dealing with a straight line, say the x_3-axis. As is evident, the problem is essentially two-dimensional and suggests introducing polar coordinates:

$$x = \{x_1, x_2\} = \{r \cos \phi, r \sin \phi\}. \tag{6.5.1}$$

We take the vector potential to be

$$A(x) = \frac{\kappa}{r^2}\{-x_2, x_1\}, \tag{6.5.2}$$

which renders the magnetic field singular: $\mathrm{rot}A(x)$ is strictly confined to $x = 0$. It is fairly easy to realize that $dx \cdot A(x) = \kappa d\phi$ in polar coordinates, and thus the flux becomes

$$\Phi \equiv \oint dx \cdot A(x) = \kappa \int_0^{2\pi} d\phi = 2\pi\kappa \,, \tag{6.5.3}$$

where the line integral may be taken with respect to any closed curve around the origin which is oriented counterclockwise.

6.5.1 Winding Numbers

Brownian paths between given endpoints never take the shortest route, but almost certainly make a detour. It may therefore happen that they circulate around the origin an unspecified number of times and, as a result of this, produce a phase in the path integral. We have no difficulty recognizing that, if C is a planar curve starting at $x = \{r\cos\phi, r\sin\phi\}$, avoiding the origin, and ending at $x' = \{r'\cos\phi', r'\sin\phi'\}$, the *magnetic phase* is given by

$$\int_C dx \cdot A(x) = (\phi' - \phi + 2\pi n)\kappa \qquad (n \in \mathbb{Z}), \tag{6.5.4}$$

where n denotes the *winding number* of C. As should be clear from its name, the winding number indicates the number of times the curve circulates around $x = 0$. This is illustrated in Fig. 6.1. Clockwise and counterclockwise motions receive winding numbers that differ in sign. One may look upon n as some kind of *topological index* addressing the geometry of the punctured plane and of Brownian paths in it, an aspect not incorporated in and ignored by conventional quantum mechanics. The introduction of this index into the theory seems in marked contrast to studying genuine quantum numbers, i.e., eigenvalues of physical observables.

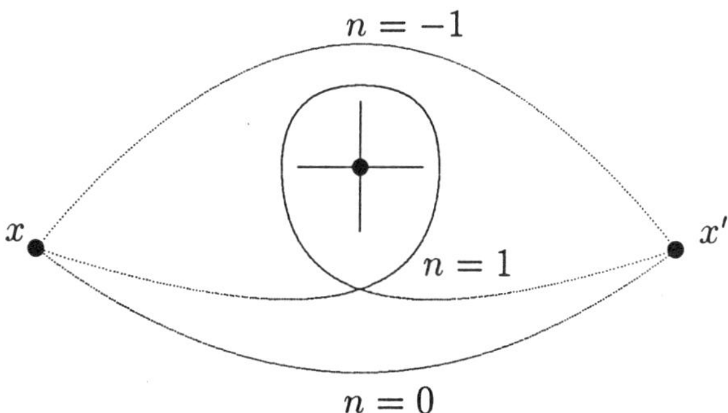

Fig. 6.1. Three different paths from x to x' with winding numbers -1, 0, and 1

We take $s > 0$ and call Ω the set of all planar Brownian paths $\omega : (x, 0) \leadsto (x', s)$ with x and x' fixed and not equal to the zero vector. The subset of those ω that pass the origin has zero measure and will henceforth be ignored. All other paths fall into homotopy classes: we call Ω_n the set of $\omega \in \Omega$ whose winding number (relative to the origin) is n. If $\mu(\Omega_n)$ is the conditional Wiener measure of the set Ω_n, we certainly have

$$\sum_{n=-\infty}^{\infty} \mu(\Omega_n) = \mu(\Omega) = k(x'-x,s) = (2\pi s)^{-1} \exp\{-(x'-x)^2/2s\}.$$

$$(6.5.5)$$

The Hamiltonian $H = \frac{1}{2}(i\nabla + A)^2$ offers a unique opportunity to calculate the numerical value of $\mu(\Omega_n)$. The main idea is to look upon $\langle x'|e^{-sH}|x\rangle$ as the generating function, i.e., utilizing the Feynman–Kac–Itô formula and (6.5.4), we find that

$$\begin{aligned}
\langle x'|e^{-sH}|x\rangle &= \int_\Omega d\mu(\omega)\, \exp\left\{i\int_\omega dx \cdot A(x)\right\} \\
&= \sum_n \mu(\Omega_n) \exp\left\{i(\phi'-\phi+2\pi n)\kappa\right\}.
\end{aligned}$$

$$(6.5.6)$$

As the partial amplitudes $\mu(\Omega_n)$ do not depend on the parameter κ, they may be regarded as the Fourier coefficients of the function

$$f(\kappa) := \exp\{i(\phi-\phi')\kappa\}\langle x'|e^{-sH}|x\rangle. \tag{6.5.7}$$

We are left with the question: is there an alternative route to compute $\langle x'|e^{-sH}|x\rangle$?

6.5.2 Spectral Decomposition

By far the simplest approach is to appeal to the spectral decomposition theorem from the theory of differential operators. In polar coordinates,

$$2H = -\frac{\partial^2}{\partial r^2} - \frac{1}{r}\frac{\partial}{\partial r} + \frac{(L-\kappa)^2}{r^2} \quad , \quad L = -i\frac{\partial}{\partial \phi}. \tag{6.5.8}$$

By the method of separation of coordinates, one obtains the solutions of $H\Phi = E\Phi$ (i.e., those that are regular at $r = 0$) as

$$\Phi_{E,m}(x) = \frac{1}{\sqrt{2\pi}}J_\nu(r\sqrt{2E})e^{im\phi} \quad , \qquad \nu = |m-\kappa| \tag{6.5.9}$$

with $E \in \mathbb{R}_+$ and $m \in \mathbb{Z}$ ($J_\nu(z) =$ Bessel function). As the spectrum of H is continuous, the (generalized) eigenfunctions $\Phi_{E,m}$ cannot be normalized in the L^2 sense. Instead, the normalizing factor has been chosen in (6.5.9) such that the relation of completeness reads

$$\int_0^\infty dE \sum_{m=-\infty}^{\infty} \Phi_{E,m}(x')\overline{\Phi_{E,m}(x)} = \delta(x'-x). \tag{6.5.10}$$

Having settled the question of normalization, we are ready to formulate the spectral decomposition and rewrite it in terms of the modified Bessel functions $I_\nu(z)$:

$$\langle x'|e^{-sH}|x\rangle = \int_0^\infty dE \sum_{m=-\infty}^\infty \Phi_{E,m}(x')e^{-sE}\,\overline{\Phi_{E,m}(x)}$$

$$= \frac{1}{2\pi} \sum_{m=-\infty}^\infty e^{im(\phi'-\phi)} \int_0^\infty dE\, e^{-sE} J_\nu(r'\sqrt{2E}) J_\nu(r\sqrt{2E})$$

$$= \frac{1}{2\pi s} \sum_{m=-\infty}^\infty e^{im(\phi'-\phi)} I_\nu\left(\frac{rr'}{s}\right) \exp\left(-\frac{r^2+r'^2}{2s}\right). \tag{6.5.11}$$

To simplify matters we introduce the function

$$g(t) = \frac{1}{2\pi s} e^{it(\phi'-\phi)} I_{|t|}\left(\frac{rr'}{s}\right) \exp\left(-\frac{r^2+r'^2}{2s}\right) \tag{6.5.12}$$

and obtain the following representation of f using definition (6.5.7) and our result (6.5.11):

$$f(\kappa) = \sum_{m=-\infty}^\infty g(m-\kappa) = \sum_{n=-\infty}^\infty \mu(\Omega_n)e^{i2\pi n\kappa}. \tag{6.5.13}$$

This is easily decoded to furnish the Fourier coefficients:

$$\mu(\Omega_n) = \int_0^1 d\kappa\, f(\kappa)e^{-i2\pi n\kappa}$$

$$= \sum_{m=-\infty}^\infty \int_0^1 d\kappa\, g(m-\kappa)e^{i2\pi n(m-\kappa)} \qquad (\text{since } e^{i2\pi nm} = 1)$$

$$= \int_{-\infty}^\infty dt\, g(t)e^{i2\pi nt}. \tag{6.5.14}$$

Upon introducing the function

$$I(z,u) = \int_{-\infty}^\infty dt\, I_{|t|}(z)e^{itu} = 2\int_0^\infty dt\, I_t(z)\cos tu, \tag{6.5.15}$$

we may state the final result as

$$\boxed{\mu(\Omega_n) = \frac{1}{2\pi s} \exp\left(-\frac{r^2+r'^2}{2s}\right) I\left(\frac{rr'}{s},\phi'-\phi+2\pi n\right).} \tag{6.5.16}$$

Certainly, $\mu(\Omega_n) \to 0$ as $n \to \infty$. The question is, *how fast?* Turning to (6.5.16) for help, we see that an immediate answer cannot be given. For to do so we need to investigate the behavior of the function $I(z,u)$ as $|u|$ becomes large. To this end, it will be helpful to recall a well known integral representation [6.11] of the modified Bessel functions:

$$I_t(z) = \frac{1}{\pi} \int_0^\pi d\theta\, \exp(z\cos\theta)\cos t\theta - \frac{\sin t\pi}{\pi} \int_0^\infty dv\, \exp(-z\cosh v - tv), \tag{6.5.17}$$

valid for $\mathrm{Re}\,z \geq 0$. Insertion into (6.5.15) yields a better description of the function I:

$$I(z,u) = \exp(z\cos u)\Theta(\pi - |u|) - \int_0^\infty dv\,\exp(-z\cosh v)j(u,v), \qquad (6.5.18)$$

where

$$j(u,v) = \frac{1}{\pi}\left\{\frac{\pi+u}{(\pi+u)^2+v^2} + \frac{\pi-u}{(\pi-u)^2+v^2}\right\}, \quad \Theta(u) = \begin{cases} 1 & u>0 \\ \frac{1}{2} & u=0 \\ 0 & u<0. \end{cases} \qquad (6.5.19)$$

Supposing that the optical path that joins x and x' avoids the origin, we know that there exists a unique $\bar{n} \in \mathbb{Z}$ satisfying

$$|\phi' - \phi + 2\pi\bar{n}| < \pi. \qquad (6.5.20)$$

It should be clear that $\bar{n}$ is the winding number of the optical path and it can only be -1, 0, or 1. With $u \equiv \phi' - \phi + 2\pi n$, we get the following correspondence:

$$\begin{aligned} n = \bar{n} &\Longleftrightarrow |u| < \pi &\Longleftrightarrow \inf_v j(u,v) = 0 \\ n \neq \bar{n} &\Longleftrightarrow |u| > \pi &\Longleftrightarrow \inf_v j(u,v) = 2(\pi^2 - u^2)^{-1}. \end{aligned} \qquad (6.5.21)$$

By estimating the integral in (6.5.18) one finds that, for $z > 0$,

$$0 \leq I(z,u) \leq \begin{cases} \exp(z\cos u) & \text{if } |u| < \pi \\ 2(u^2-\pi^2)^{-1}k_0(z) & \text{if } |u| > \pi, \end{cases} \qquad (6.5.22)$$

where

$$k_0(z) = \int_0^\infty dv\,\exp(-z\cosh v). \qquad (6.5.23)$$

One may also prove that $u^2 I(z,u) \to 2k_0(z)$ as $u \to \infty$, thereby showing that

$$\mu(\Omega_n) = O(n^{-2}) \qquad (6.5.24)$$

provided $rr' \neq 0$ and x, x', s are fixed.

6.5.3 Imaginary Times

The analytic continuation to the imaginary time axis ($s \to it$) turns the formerly real z into a purely imaginary variable and so, not surprisingly, $\mu(\Omega_n)$ becomes a complex number for each n. The connection with probability theory is then lost, but the winding numbers remain. This in particular means that the amplitude $\langle x'|e^{-itH}|x\rangle$ admits a decomposition with respect to n. As this amplitude serves to define the solution $\psi(x,it)$ of a Cauchy problem for the Schrödinger equation (e.g. with Cauchy data at $t = 0$), one infers that each $\psi(x,it)$ may also be decomposed,

$$\psi(x,it) = \sum_{n=-\infty}^{\infty} \psi_n(x,it), \qquad (6.5.25)$$

no matter what κ (or $2\pi\kappa$, the magnetic flux). Aside from considering such a decomposition to be of geometric interest in its own right, however, the components ψ_n should not be expected to satisfy the very same Schrödinger equation: they do not. Nor can they be regarded as belonging to orthogonal (Hilbert) subspaces. We must also refrain from attributing n to some hypothetical trajectory of the charged particle (e.g., an electron), since the claim that there is a classical trajectory cannot be substantiated in the context of quantum physics. In other words, we should not try to push beyond the confines of Brownian motion theory.

There is still another striking observation to be made which comes under the heading *flux quantization*. For, if κ happens to be an integer, $e^{i2\pi n\kappa} = 1$ for every $n \in \mathbb{Z}$. Therefore, the dependence on the winding number disappears from the magnetic phase (6.5.4) and all homotopy classes combine to yield the simple result

$$\langle x'|e^{-itH}|x\rangle = K(x' - x, it)\exp\{i(\phi' - \phi)\kappa\}. \tag{6.5.26}$$

Disregarding the overall magnetic phase for a moment, the quantum time evolution follows the pattern of a *free particle*. Moreover, the magnetic phase can be removed by a gauge transformation with the gauge function $f(x) = \kappa\phi$ (ϕ being the polar angle of x). Though $f(x)$ cannot be defined as a one-valued continuous function on the punctured plane, its exponential $e^{if(x)}$ can. There does not seem to be a trace of an interaction. Last not least, the overall magnetic phase vanishes if $\phi' = \phi$, a situation that arises if the Brownian particle performs a closed loop.

In physical units, the flux quantum is $\Phi_0 = 2\pi\hbar c/e = hc/e$. Magnetic fluxes are said to be quantized if they are integer multiples of Φ_0. Infinitely extended (or closed) strings carrying quantized fluxes are unobservable in principle. Those strings that possess one or two endpoints are observable in the sense that their endpoints have significance, whereas the string itself remains hidden. A flux line with one endpoint is called a *Dirac string*: it has one loose end extending to infinity. Such strings have been suggested by Dirac as part of a consistent theory of magnetic monopoles, which in turn were invented to account for the phenomenon of charge quantization[7].

Renewed interest in two-dimensional electronic systems has come from the discovery of the so-called *quantum Hall effect* by von Klitzing [6.20]: at low temperatures, the Hall conductance σ was found to be quantized, i.e., $2\pi\sigma$ turned out to be an integer (in units where $e = \hbar = 1$). It was quickly realized that the effect was of geometric origin [6.21] and needed an adiabatic invariant to be calculated [6.22]. In an idealized setting, one considers two flux lines piercing the plane, so that the determination of the

---

[7] For a review of the theory of monopoles see [6.12–13]. For an account of the Aharonov–Bohm effect (i.e., the interference pattern of electrons past a nonquantized flux) see [6.8] and [6.14–16]. For the relevance of flux quantization in superconductivity consult [6.17–19].

Hall conductance becomes an exercise in the quantum theory of electrons moving in a multiply connected space, in very much the same way as we would analyze the Aharonov–Bohm experiment. Unfortunately, a proper understanding of the quantum Hall effect on the basis of path integration and winding numbers has not yet been achieved.

Exercise 1. For $\cos u \neq -1$ and $\mathrm{Re}\, z$ large, the integral on the right-hand side of (6.5.18) gives a negligible contribution to $I(z, u)$. Turning to (6.5.16) and fixing n, demonstrate the validity of another asymptotic formula, i.e., for $rr' \gg s$, $\mu(\Omega_n) \approx K(x' - x, s)\delta_{n\bar{n}}$, provided the optical path from x to x' does not visit the origin. This kind of asymptotic estimate shows that, with increasing distance between the optical path and the flux line, all Brownian paths whose winding number n differs from that of the optical path become unimportant.

7 Euclidean Field Theory

Soon after its invention, quantum mechanics was superseded by the quantized theory of fields, which by now has reached a stage where the theory is firmly established and generally believed to provide a consistent and almost complete description of nature. It is designed so as to incorporate both the principles of quantum mechanics and the laws of special relativity. Still, despite its many successes, the theory is undoubtedly beset with great conceptual difficulties and paradoxes of infinity which prompted theoreticians to seek alternative formulations and to recast the theory over and over again until it became a practical language for describing the observed processes among elementary particles.

To complement our discussion of real-time quantum mechanics utilizing Brownian motion, the present chapter offers a brief introduction to the Euclidean continuum field theory, assuming that the serious student has already seen the Minkowski counterpart of it (i.e., the operator version). In accordance with our general strategy which is to try and use the successful methods of one subject in another, we wish to establish a connection with probabilistic concepts (rather than with operator theory). The present framework allows us to treat only Bose fields, while Fermi fields require a completely different mathematical setting, a discussion of which will be deferred till Chap. 10.

A correct formulation must take into account the finiteness of the velocity of light, c, giving spacetime its definite structure. Throughout the second half of this book, however, it will be assumed that $c = \hbar = 1$ unless specified otherwise.

7.1 What Is a Euclidean Field?

Having specified a system of coordinates in the Minkowski space M_4, one writes any vector in it as $x = \{x^0, x^1, x^2, x^3\} = \{x^0, \boldsymbol{x}\}$ with real components x^α and defines the Lorentz-invariant scalar product of x and y to be

$$(x, y) = x^0 y^0 - \boldsymbol{x} \cdot \boldsymbol{y}. \tag{7.1.1}$$

Another shorthand is $x \cdot y$ or xy (or even x^2 for xx) where there is no danger of confusion. Ever since its discovery, the pseudo-Euclidean structure of the Minkowski space proved to be a source of inspiration: physisists have always been fascinated by the fact that a simple "replacement" $ix^0 \to x^4$ converts the Minkowskian structure into a Euclidean structure and vice versa. Few puzzles have captured the fancy of students the way this one has.

To be little more specific as concerns the Euclidean space E_4, we would then consider vectors $x = \{x^1, x^2, x^3, x^4\} = \{\boldsymbol{x}, x^4\}$ and define the scalar product in E_4 as

$$xy = \boldsymbol{x} \cdot \boldsymbol{y} + x^4 y^4. \tag{7.1.2}$$

In fact, a formal replacement $ix^0 \to x^4$, $iy^0 \to y^4$ transforms (7.1.1) into (7.1.2) up to a minus sign. There lies an obvious advantage in using the Euclidean structure: we need no longer distinguish between covariant and contravariant coordinates of vectors and tensors. Yet, the transition from a real coordinate (x^4) to an imaginary coordinate (ix^0) is purely formal and tries to conceal the different geometries of the spaces M_4 and E_4. There is no sensible way to identify the two spaces, and this is certainly not what we plan to do.

The subject matter received new impulses from the theory of analytic continuation applied to n-point functions in field theory. The hypothesis then was that *time* should rather be treated as a complex variable, i.e., we ought to write $z = x^4 + ix^0$, at the same time strictly abandoning all previous "replacement tricks". The procedure is seen to conform to what has been suggested previously in this book with regard to ordinary quantum mechanics. Again, we must refrain from entering into philosophical discussions about the true concept of *time* except to say that there is more to *complex time* than just mathematical trickery, introduced solely to be able to do the pertinent field theoretic calculations (e.g., perform path integrations) in a convincing manner which otherwise would be plagued with ambiguities and infinities. For instance, there may be a context where it seems appropriate to call x^0 the physical time and another where x^4 accepts this role. Both variables are real and they are different.

We have, however, to exercise some care when denoting some spacetime point by x or y. For it is not a priori clear whether it is used to designate a point in M_4 or in E_4 or something which embodies a complex time. Fortunately, in most situations and formulas, ambiguities should not, perhaps cannot, arise since what is meant generally follows from the context.

We expect the reader to be familiar with the very basic elements of quantum field theory, for instance with the fact that states are described in the Heisenberg picture and that there exists a unique translationally invariant state, Ω, called the vacuum. One also requires that there be one-particle states belonging to some irreducible representation of the group of spacetime symmetries (i.e., the Poincaré group). In the simplest of all cases, one deals with a free neutral scalar field $A(x)$ satisfying the Klein–Gordon equation $(\Box + m^2)A(x) = 0$ so that[8]

$$A(x) = (2\pi)^{-3/2} \int \frac{d\boldsymbol{p}}{2\omega} \left\{ a(p)e^{-ipx} + a^\dagger(p)e^{ipx} \right\}. \qquad (7.1.3)$$

The technique of expressing field operators as Fourier integrals shows that, typically, positive and negative frequencies occur simultaneously. This very fact forbids, even on a formal level, to replace ix^0 by x^4. The reason: either the first part on the right-hand side of (7.1.3) becomes senseless (if $x^4 < 0$), or the second part becomes senseless (if $x^4 > 0$)[9]. What appears to be true for free fields is also true for general fields. There the argument (similar to the above but less conclusive) runs as follows. For $\Phi(x)$ an operator field and H the Hamiltonian,

$$\Phi(x^0, \boldsymbol{x}) = e^{ix^0 H}\Phi(0, \boldsymbol{x})e^{-ix^0 H}. \qquad (7.1.4)$$

A formal replacement $ix^0 \to x^4$ generates two factors, $e^{x^4 H}$ and $e^{-x^4 H}$. Depending on the sign of x^4, one of them cannot be defined everywhere on the Hilbert space because the energy, though positive, is unbounded above.

These remarks show that any attempt to formulate Euclidean field theory within the boundaries of operator algebra (with the same state space) is bound to fail. It is fairly obvious that we must give up the idea of introducing Euclidean fields as operators in any way similar to Minkowski fields and related to them via the process of analytic continuation.

The next sections will prepare the ground for a correct approach, and at the end, we shall be able to give a meaning to the Euclidean (Bose) field. It will then prove to be purely a matter of convenience and flexibility which of two languages is given preference:

(1) The Euclidean field is a random variable, or rather a *random field*. It thus generalizes suitably the notion of a random process in that 'time'

---

[8] Here, we adopt the following conventions: $p = \{\omega, \boldsymbol{p}\}$, $\omega = \sqrt{m^2 + \boldsymbol{p}^2}$, $px = \omega x^0 - \boldsymbol{px}$, $d\boldsymbol{p} = dp^1 dp^2 dp^3$. The operators $a(p)$ and $a^\dagger(p)$ satisfy the canonical commutation relations appropriate for a relativistic theory: $[a(p), a^\dagger(p')] = 2\omega\delta(\boldsymbol{p} - \boldsymbol{p}')$.

[9] Phrased differently: for most states ϕ, the expectation $(\phi, A(x)\phi)$ is not analytic in the time variable.

is replaced by the four-dimensional Euclidean continuum. We shall then speak of the *stochastic interpretation*.

(2) The Euclidean field is some kind of generalized "spin", i.e., a continuous analogue of the Ising spin, defined on the sites of a four-dimensional lattice (this is for the case of scalar fields only). The interaction is ferromagnetic and of the nearest-neighbor type. The continuum field (provided it exists) is recovered from the corresponding lattice version at a point of second order phase transition (a so-called critical point). Since this concepts borrows heavily from statistical mechanics, we shall then speak of the *statistical interpretation*.

At this point it should be noted that we have already seen the two different formalisms at work, namely, in the context of the Feynman–Kac formula. There, the method of path integration stood for the stochastic interpretation. On the other hand, a different view was presented in Sect. 2.7-8 with one-dimensional ferromagnetic spin chains playing a fundamental role.

Functional integrals for Euclidean fields will soon be introduced. In some sense, they resemble and generalize Wiener integrals. Unfortunately, the generalization entails that typical "paths" are no longer continuous. But as before, discretization of functional integrals always means *replacement by finite-dimensional integrals* or, by a change of language, it means *putting the field on a lattice*. The latter point of view is generally preferred since the study of a particular model then becomes an interprise within the territory of statistical mechanics.

7.2 The Euclidean Two-Point Function

Although Minkowski fields do not admit analytic continuations, their vacuum expectations do. Vacuum expectations are customarily taken of an n-fold product of field operators at n different spacetime points; they are the so-called n-point or Wightman functions of the theory. We assume the reader to be acquainted with the fact that the state space as well as the field operators may be reconstructed from the set of Wightman functions [7.1–3]. Indeed, these functions embody a complete description of the field theoretical model up to equivalence; all necessary information is coded in them, and this assertion continues to be correct when we replace every Minkowkian n-point function by its Euclidean counterpart, the n-point Schwinger function. More importantly, any Schwinger function, being real and symmetric with respect to permutations of its arguments, may be given a probabilistic interpretation, i.e., as the expectation of an n-fold product of Euclidean fields viewed as random variables. The underlying structure is entirely real commutative. The miracle of the Euclidean formulation therefore is that the necessity of dealing with noncommuting objects has disappeared from the scene.

To be more specific, let us consider some Minkowskian field $\Phi(x)$ and let

$$W(x - y) = (\Omega, \Phi(x)\Phi(y)\Omega) \tag{7.2.1}$$

be its two-point Wightman function[10], where Ω denotes the vacuum state. Owing to the translational invariance of the vacuum, it is merely a function of the difference $x - y$ as indicated. Roughly, it may also be regarded as some "matrix element" of the evolution operator e^{-itH} with $t = x^0 - y^0$ (use (7.1.2) and the invariance of vacuum). From $H \geq 0$ it may thus be inferred that $W(x)$ admits an analytic continuation with respect to x^0. Writing $S(\boldsymbol{x}, z)$ for the analytic function with $z = x^4 + ix^0$ restricted to the right halfplane $x^4 > 0$, we have that

$$W(x^0, \boldsymbol{x}) = \lim_{x^4 \downarrow 0} S(\boldsymbol{x}, x^4 + ix^0). \tag{7.2.2}$$

As a rule, the boundary values do not exist pointwise but rather in the sense of distributions. This fact makes Minkowski fields (even free fields) very singular objects: only smeared fields $\Phi(f)$, where f is from a suitable test function space, yield well-defined operators. By contrast, all points belonging to the semiaxis $z = x^4 > 0$ are in the domain of analyticity of S, and these are the points where the Euclidean two-point function is being evalated: $S(x) \equiv S(\boldsymbol{x}, x^4)$. Hence, there will be no singularity along the semiaxis except for the boundary point $x^4 = 0$.

It is instructive to study a prototype for which we take the free neutral scalar field $\Phi(x)$ of mass m from the previous section. The function W (in fact a distribution) is conveniently written as a Fourier integral:

$$W(x) \equiv \Delta_+(x; m) = \frac{1}{(2\pi)^3} \int \frac{d\boldsymbol{p}}{2\omega}\, e^{i(\boldsymbol{p}\boldsymbol{x} - \omega x^0)} \tag{7.2.3}$$

$(\omega = \sqrt{m^2 + \boldsymbol{p}^2})$. Therefore, in the Euclidean domain,

$$S(x; m) = \frac{1}{(2\pi)^3} \int \frac{d\boldsymbol{p}}{2\omega}\, e^{i\boldsymbol{p}\boldsymbol{x} - \omega x^4}. \tag{7.2.4}$$

This representation is valid only if $x^4 > 0$. It remains to be shown that one can actually convert the three-dimensional into a four-dimensional integral, thereby establishing the full $O(4)$ invariance of the Euclidean function $S(x; m)$. The argument is simple and relies on the integral formula

---

[10]Momentarily, we want to be sufficiently vague about the nature of this field, whether it is a scalar or a component of a vector or something else: Lorentz invariance is not our concern.

$$\frac{1}{2\omega}e^{-\omega x^4} = \frac{1}{2\pi}\int_{-\infty}^{\infty} dp^4 \,\frac{e^{ip^4 x^4}}{p^2 + m^2}, \qquad (x^4 > 0) \qquad (7.2.5)$$

with $p = \{p^1, p^2, p^3, p^4\} = \{\boldsymbol{p}, p^4\}$ the Euclidean momentum and $p^2 = \sum_\alpha (p^\alpha)^2$. From (7.2.4) and $px = \sum_\alpha p^\alpha x^\alpha$,

$$\boxed{S(x;m) = \frac{1}{(2\pi)^4}\int dp \,\frac{e^{ipx}}{p^2 + m^2}} \qquad (7.2.6)$$

$(dp = dp^1 dp^2 dp^3 dp^4)$. The function $S(x;m)$ may also be expressed in terms of the modified Bessel function K_1:

$$S(x;m) = \begin{cases} (2\pi)^{-2}m|x|^{-1}K_1(m|x|) & \text{if } m > 0 \\ (2\pi)^{-2}|x|^{-2} & \text{if } m = 0 \end{cases} \qquad |x| = \sqrt{x^2}. \quad (7.2.7)$$

What strikes the eye is the close relationship between the Schwinger function $S(x;m)$ from Euclidean field theory and the Feynman function from Minkowskian field theory:

$$\Delta_F(x;m) = \frac{1}{(2\pi)^4}\int dp \,\frac{e^{-ipx}}{p^2 - m^2 + i0} \qquad \begin{aligned} p^2 &= (p^0)^2 - \boldsymbol{p}^2 \\ px &= p^0 x^0 - \boldsymbol{p}\cdot\boldsymbol{x}. \end{aligned} \qquad (7.2.8)$$

Evidently, the transition from the Minkowskian to the Euclidean momentum turns p^2 into $-p^2$, and the Fourier transforms $\tilde{\Delta}_F(p;m)$ and $-\tilde{S}(p;m)$ are thus connected via an analytic continuation[11] with regard to the complex variable $z = p^0 + ip^4$. The two-point function is typical in this respect, and, as it is the main and only building block from which all remaining n-point functions of a free field are constructed, we have seen, at least from studying one example, the way in which Wightman und Schwinger functions are related[12]:

<table>
<tr><td>Wightman function
in x space</td><td>$\xleftrightarrow{\text{analyt. c.}}$</td><td>Schwinger function
in x space</td></tr>
<tr><td></td><td></td><td>$\updownarrow$ Fourier tr.</td></tr>
<tr><td>Feynman function
in p space</td><td>$\xleftrightarrow{\text{analyt. c.}}$</td><td>Schwinger function
in p space</td></tr>
</table>

---

[11]To go from the real to the imaginary axis in the complex plane is sometimes called a *Wick rotation*. In fact, upon writing $p^0 + ip^4 = re^{i\alpha}$ and $(p^0)^2 + i0 = \{p^0(1+i0)\}^2 = \{p^0 e^{i0}\}^2$, a Wick rotation is seen to change the angle α from $+0$ to $\pi/2$.

[12]Most textbooks "define" Feynman functions (or tau functions) in x space to be the vacuum expectations of *time-ordered products* which is much less convenient because the result is ambiguous. In order to change a Wightman function into a Feynman function and vice versa, we suggest instead chasing the function around the diagram shown.

It is worthwhile noting some properties of the Schwinger function of a free field:

1. Though, initially, the function $S(x; m)$ was defined only for $x^4 > 0$, the representation (7.2.6) extends it quite naturally to a symmetric function on the the entire x^4 axis, and hence, on E_4. We must however be aware that the extended function becomes singular at the origin, where it behaves like $1/x^2$. This fact should not worry us too much since the singularity is integrable with respect to the volume measure dx.

2. The function $S(x; m)$ is positive real, and the same is true for the Fourier transform $\tilde{S}(p; m) = (p^2 + m^2)^{-1}$.

3. The function is invariant with respect to the group O(4), i.e., the full group of four-dimensional rotations including the reflections. The group O(4) thus replaces the Lorentz group when dealing with Euclidean fields.

4. $S(x; m)$ may be viewed as the Green's function of the differential operator $-\Delta + m^2$ so that $(-\Delta + m^2)S(x; m) = \delta(x)$, where Δ is used to designate the four-dimensional Laplacian. Whereas the Klein–Gordon operator $\Box + m^2$ is a hyperbolic, its Euclidean counterpart $-\Delta + m^2$ is elliptic. This has many beneficial effects. One of them is that we no longer depend on $i0$-prescriptions when wishing to invert $-\Delta + m^2$. For the inversion is unique with integral kernel $\langle x'|(-\Delta + m^2)^{-1}|x\rangle = S(x' - x; m)$.

5. The function decays exponentially at large distances provided the mass m is nonzero:

$$S(x; m) \to \frac{m^2/2}{(2\pi m|x|)^{3/2}} \exp\{-m|x|\} \qquad |x| \to \infty.$$

In particular, the following integral exists: $\int dx\, S(x; m) = m^{-2}$.

Exercise 1. Find a way to prove identity (7.2.5) and extend it to all of $x^4 \in \mathbb{R}$. Hint: apply the method of residues to the integral.

Exercise 2. With $[a, a^*] = 1$ the canonical commutation relation for one (Bose) degree of freedom and $H = \omega a^* a$ the associated Hamiltonian, one defines the thermal average of an observable A to be the number $\langle A \rangle_\beta = \mathrm{tr}\,(e^{-\beta H} A)/\mathrm{tr}\,e^{-\beta H}$. Verify that

$$\langle a^* a \rangle_\beta = \frac{1}{e^{\beta\omega} - 1}\,, \qquad \langle a a^* \rangle_\beta = \frac{1}{1 - e^{-\beta\omega}}.$$

Exercise 3. (Fields at a finite temperature) The reader is now asked to take a deep breath, take one sheet of reversed computer paper, remember the way fields are quantized in a box, and use the result of the preceding exercise to show that, in the thermodynamic (large volume) limit, the thermal two-point function of a free neutral scalar field is of the form $\langle \Phi(x)\Phi(y)\rangle_\beta = W_\beta(x - y)$, where

$$W_\beta(x) = (2\pi)^{-3} \int \frac{d\boldsymbol{p}}{2\omega} \frac{e^{-ipx} + e^{-\beta\omega} e^{ipx}}{1 - e^{-\beta\omega}}$$

($\omega = \sqrt{m^2 + \boldsymbol{p}^2}$, $p = \{\omega, \boldsymbol{p}\}$). Demonstrate that W_β can be continued to a meromorphic function $S_\beta(\boldsymbol{x}, z)$ with respect to the variable $z = x^4 + ix^0$, such that W_β coincides with S_β along the imaginary axis. Where are the poles? Verify that the Schwinger function S_β approaches S as $\beta \to \infty$, where S is given by (7.2.6): thermal averages become vacuum expectations at zero temperature. Use the symmetry relation $S_\beta(\boldsymbol{x}, -x^4) = S_\beta(\boldsymbol{x}, x^4)$ to verify that

$$S_\beta(\boldsymbol{x}, x^4) = \sum_{n=-\infty}^{\infty} S(\boldsymbol{x}, x^4 + n\beta; m)$$

from which the periodicity property $S_\beta(\boldsymbol{x}, x^4 + \beta) = S_\beta(\boldsymbol{x}, x^4)$ follows. Why does the sum over the integers converge? How is S_β behaved at (spatial) infinity?

7.3 The Euclidean Free Field

7.3.1 The n-Point Functions

Within the Minkowskian framework, any neutral scalar field $\Phi(x)$ (not just a free field) may be specified by providing, for each n, the n-point Wightman function

$$W_n(x_1, \ldots, x_n) = (\Omega, \Phi(x_1) \cdots \Phi(x_n)\Omega). \tag{7.3.1}$$

The quantum character of the theory does not allow us to assume that $\Phi(x)$ and $\Phi(x')$ always commute. Therefore, Wightman functions cannot be symmetric with respect to permutations of their arguments. The lack of symmetry prevents us from constructing a generating functional from which all Wightman functions could be recovered.

There is still another way to specify a field, i.e., by providing all its Feynman functions (or tau functions):

$$\tau_n(x_1, \ldots, x_n) = (\Omega, T\Phi(x_1) \cdots \Phi(x_n)\Omega). \tag{7.3.2}$$

The symbol T is there to signify that the product of field operators has been time ordered. We shall not dwell on the kind of problems connected with time ordering [7.4–7] but assume that the Feynman functions exist.

From their naive definition, these functions are symmetric and thus admit the construction of a generating functional:

$$F(j) = (\Omega, T \exp\{i\Phi(j)\}\Omega) = 1 + \sum_{n=1}^{\infty} \frac{i^n}{n!}(\Omega, T\Phi(j)^n\Omega), \qquad (7.3.3)$$

where $\Phi(j) = \int dx\, \Phi(x)j(x)$ for suitable real "source" functions $j(x)$.

Again, the free field of mass m provides the simplest model in that the generating functional is Gaussian. To make it look a bit more interesting, we shall equip $\Phi(x)$ with some (real) nonzero vacuum expectation, c, the so-called *condensate*[13] so that the generating functional has only two major constituents which are the first and the second order tau functions:

$$\begin{aligned}
\tau_1(x) &= (\Omega, \Phi(x)\Omega) = c \\
\tau_2(x,y) &= (\Omega, T\Phi(x)\Phi(y)\Omega) = i\Delta_F(x - y; m).
\end{aligned} \qquad (7.3.4)$$

In fact,

$$\log F(j) = icI(j) - \tfrac{1}{2}i\Delta_F(j * j; m), \qquad (7.3.5)$$

where

$$\begin{aligned}
I(j) &= \int dx\, j(x) \\
\Delta_F(j * j; m) &= \int dx \int dy\, j(x)\Delta_F(x - y; m)j(y).
\end{aligned} \qquad (7.3.6)$$

The Gaussian character is revealed by the fact that the expansion of $\log F(j)$ with respect to j truncates after the second order term. One may further simplify the representation (7.3.5) by noting that the first order term can be eliminated by shifting the field: $\Phi(x) \to \Phi(x) - c$. We shall henceforth assume that $c = 0$ when discussing free scalar fields.

It is not hard to show that, provided $c = 0$, (7.3.5) is equivalent to the following recursive definition of the tau functions:

$$\begin{aligned}
\tau_1(x_1) &= 0 \\
\tau_2(x_1, x_2) &= i\Delta_F(x_1 - x_2; m) \\
\tau_n(x_1, \ldots, x_n) &= \sum_{k=1}^{n-1} \tau_{n-2}(x_1, \ldots, \hat{x}_k, \ldots, x_{n-1})i\Delta_F(x_k - x_n; m)
\end{aligned} \qquad (7.3.7)$$

(customarily, the hat ˆ on top of something means that the variable has been left out). This recursive scheme is in fact a unifying feature of free field theory. Not only does it apply to tau functions but also to Wightman functions (and other sets of functions):

---

[13]This assumption comes unmotivated at this point but provides a simple and also natural generalization. Translational invariance tells us that c cannot depend on x but is arbitrary otherwise.

$$W_1(x_1) = 0$$

$$W_2(x_1, x_2) = \Delta_+(x_1 - x_2; m) \tag{7.3.8}$$

$$W_n(x_1, \ldots, x_n) = \sum_{k=1}^{n-1} W_{n-2}(x_1, \ldots, \hat{x}_k, \ldots, x_{n-1})\Delta_+(x_k - x_n; m),$$

where $\Delta_+(x; m)$ has been defined in (7.2.3). A conclusion may immediately be drawn from the recursive scheme: since the function Δ_+ admits an analytic continuation, each W_n can be simultaneously be continued in all its time variables $x_1^0, \ldots, x_n^0$. This allows us to introduce the complex variables $z_k = x_k^4 + ix_k^0$ under the provision that we avoid coinciding arguments, i.e., $x_i - x_k = 0$ for no pair of indices. In real points of $\mathbb{C}^n$ (these are points where $z_k = x_k^4$, $k = 1, \ldots, n$), we obtain the nth-order Schwinger function of the Euclidean field. Formally,

$$S_n(x_1, \ldots, x_n) = W_n(x_1, \ldots, x_n)\Big|_{ix_k^0 \to x_k^4}. \tag{7.3.9}$$

As concerns the free field, these functions can also be constructed using the following recursive scheme:

$$S_1(x_1) = 0$$

$$S_2(x_1, x_2) = S(x_1 - x_2; m)$$

$$S_n(x_1, \ldots, x_n) = \sum_{k=1}^{n-1} S_{n-2}(x_1, \ldots, \hat{x}_k, \ldots, x_{n-1})S(x_k - x_n; m). \tag{7.3.10}$$

It then follows from the relation $S(-x; m) = S(x; m)$ that each Schwinger function comes out *symmetric* with respect to permutations of its arguments. This important property suggests looking for some generating functional $S\{f\}$ from which the set of Schwinger functions may be obtained. The obvious answer to this problem is

$$\log S\{f\} = -\tfrac{1}{2}S(f * f; m)$$

$$= -\tfrac{1}{2}\int dx \int dy \, f(x)S(x - y; m)f(y) \tag{7.3.11}$$

$$= -\tfrac{1}{2}(f, (-\Delta + m^2)^{-1}f)$$

so that

$$S\{f\} = 1 + \sum_{n=1}^{\infty} \frac{i^n}{n!} \int dx_1 \cdots \int dx_n \, S_n(x_1, \ldots, x_n)f(x_1) \cdots f(x_n). \tag{7.3.12}$$

With (7.3.12) and f taken from a suitable real test function space, $S\{f\}$ is called the *Schwinger functional* for the Euclidean field. So far, of course, we have seen but one example of a field. Nevertheless, this example deserves further study. A key feature is the property $S(f * f; m) \geq 0$ (and $= 0$ iff $f = 0$), which renders the Gaussian functional $S\{f\}$ nondegenerate. Also, not unexpectedly, the relationship between the free field formalism and the harmonic oscillator becomes manifest.

7.3.2 The Stochastic Interpretation

As vague as the concept of *Euclidean fields* is in general, it is very concrete for the free field, for which we write $\boldsymbol{\Phi}(x)$, not to be confused with $\Phi(x)$, the Minkowskian field. As the Schwinger functions are real symmetric, we may take them to be the expectations – in the sense of probability theory – of products $\boldsymbol{\Phi}(x_1)\cdots\boldsymbol{\Phi}(x_n)$, where the order of the factors does not matter. Each factor is some random variable and so is the product. Such expectations are then termed *correlation functions*, and it follows that the free Euclidean field is a Gaussian random variable given solely by its second-order correlation function (or covariance):

$$\boldsymbol{E}\big(\boldsymbol{\Phi}(x)\boldsymbol{\Phi}(y)\big) = S(x - y; m) = \langle x|(-\boldsymbol{\Delta} + m^2)^{-1}|y\rangle \qquad (7.3.13)$$

provided the field is *centered* so that $\boldsymbol{E}(\boldsymbol{\Phi}(x)) = 0$. Correlation functions such as

$$\boldsymbol{E}\big(\boldsymbol{\Phi}(x_1)\cdots\boldsymbol{\Phi}(x_n)\big) = S_n(x_1, \ldots, x_n) \qquad (7.3.14)$$

may be constructed recursively.

There is still one obstacle to the above interpretation: at coinciding points we encounter singularities (i.e., if $x_i = x_k$ for some pair i, k). The only reasonable way to modify the setting is to smear the Euclidean field with real test functions f, and, as the singularities are integrable, almost any function space would do. Hence, the idea is to take integrals $\boldsymbol{\Phi}(f) = \int dx\, \boldsymbol{\Phi}(x)f(x)$ as the basic random variables varying with f (instead of $\boldsymbol{\Phi}(x)$ varying with x) and to look upon the Schwinger functional as an expectation:

$$S\{f\} = \boldsymbol{E}\big(e^{i\boldsymbol{\Phi}(f)}\big). \qquad (7.3.15)$$

In a more abstract setting, $S\{f\}$ is said to be the characteristic functional of the *generalized stochastic process* $\boldsymbol{\Phi}(f)$ indexed by f, a member of some function space. As everything is determined by providing the characteristic functional, we ought to be able to compute, to all orders n, the probability distributions of the random process corresponding to the free field. The problem is not intriguing, but the result is nevertheless instructive. We will treat the simplest case first.

The Case $n = 1$. We think of $\boldsymbol{\Phi}(f)$ as taking random values in the reals. Replacing f by tf in the Schwinger functional and varying the real parameter t while keeping the function f fixed, we may write

$$\boldsymbol{E}\Big(\exp\{it\boldsymbol{\Phi}(f)\}\Big) = \exp\{-\tfrac{1}{2}at^2\} = \int d\mu(\alpha)\, \exp\{it\alpha\} \qquad (7.3.16)$$

with $a = (f, (-\boldsymbol{\Delta} + m^2)^{-1}f)$. This introduces μ, which depends on f, as a probability measure whose role is to predict the way the values of $\boldsymbol{\Phi}(f)$ will be distributed on the real line. If we were to study a complex scalar field (or, equivalently, two real fields), we would similarly get a distribution on $\mathbb{C}$ (or, equivalently, on $\mathbb{R}^2$). In the present situation,

$$d\mu(\alpha) = d\alpha \,(2\pi a)^{-1/2} \exp\{-(2a)^{-1}\alpha^2\} \tag{7.3.17}$$

with a the width of the Gaussian distribution.

The General Case. This time we write $f = \sum_{k=1}^n t_k f_k$ in $S\{f\}$ assuming that the test functions f_k are linearly independent:

$$\boldsymbol{E}\Big(\exp\{i \textstyle\sum_{k=1}^n t_k \boldsymbol{\Phi}(f_k)\}\Big) = \exp\left\{-\frac{1}{2} \sum_{j,k=1}^n t_j t_k a_{jk}\right\} \tag{7.3.18}$$

$$(f_j, (-\boldsymbol{\Delta} + m^2)^{-1} f_k) = a_{jk}.$$

The formula involves the $n \times n$ matrix $A = (a_{ik})$ which is positive symmetric and nonsingular. The reason: we know that $\sum t_j t_k a_{jk} = (f, (-\boldsymbol{\Delta} + m^2)^{-1} f) \geq 0$ which, if it vanishes, implies that $f = 0$. This in turn implies that $t_k = 0$ (for all k) because the system (f_k) is independent by assumption. Let the random variables $\boldsymbol{\Phi}(f_k)$ take values $\alpha_k \in \mathbb{R}$. Then the joint probability distribution for events $(\alpha_1, \ldots, \alpha_n)$ is Gaussian again and follows from

$$\boldsymbol{E}\Big(\exp\{i \textstyle\sum t_k \boldsymbol{\Phi}(f_k)\}\Big) = \int d\mu(\alpha_1, \ldots, \alpha_n) \, \exp\{i \textstyle\sum t_k \alpha_k\}. \tag{7.3.19}$$

This is solved for μ by applying a Fourier transformation:

$$d\mu(\alpha_1, \ldots, \alpha_n) = [\det(2\pi A)]^{-1/2} \exp\left\{-\frac{1}{2} \sum_{j,k=1}^n \alpha_j \alpha_k (A^{-1})_{jk}\right\} \prod_{k=1}^n d\alpha_k.$$

$$\tag{7.3.20}$$

The linear independence of (f_k) is thus seen to be crucial. For (7.3.20) would make no sense if $\det A = 0$. The discussion has made precise what is meant by the following statement:

The free Euclidean field $\boldsymbol{\Phi}(x)$ is a generalized Gaussian process on E_4, with mean zero and covariance $\boldsymbol{E}(\boldsymbol{\Phi}(x)\boldsymbol{\Phi}(y)) = \langle x|(-\boldsymbol{\Delta} + m^2)^{-1}|y\rangle$.

Conceptually, the above formulation generalizes the notion of a stochastic process in two ways:

(1) The process is no longer indexed by $t \in \mathbb{R}_+$ but by $x \in E_4$. Hence, time has been replaced by (Euclidean) spacetime.
(2) Though one cannot maintain that the field $\boldsymbol{\Phi}(x)$ itself makes sense as a random variable in the strict sense, the smeared field $\boldsymbol{\Phi}(f)$ is a bona fide random variable. The map $f \to \boldsymbol{\Phi}(f)$ is linear and said to be a *random functional* [7.8].

Naturally, a generalized random process is called *Gaussian* if, for any n, the joint distribution of the variables $\Phi(f_1), \ldots, \Phi(f_n)$ is Gaussian (as is the case for the free Euclidean field). Ordinary Gaussian processes that take values in $\mathbb{R}^n$ have $n \times n$ covariance matrices which are positive-definite. Quite analogously, for $\Phi(x)$ a generalized random process, the correlation functional

$$B(f,g) \equiv E\big(\Phi(f)\Phi(g)\big) - E\big(\Phi(f)\big)E\big(\Phi(g)\big) \tag{7.3.21}$$

is positive-definite bilinear. In case there exists an operator K such that $B(f,g) = (f, Kg) = \int dx\, f(x)(Kg)(x)$, it will be termed the *covariance operator* of the process. For the free field, we have $K = (-\Delta + m^2)^{-1}$. As a rule, the field is taken to be centered, i.e., we also have $E(\Phi(f)) = 0$ ('no condensate' condition).

7.4 Gaussian Functional Integrals

The ground is now prepared for the construction of a functional integral involving the Euclidean action and designed so as to yield the Schwinger functional of a free neutral scalar field (to begin with). Of course, no interesting physics is expected to come from such an integral representation, but it will prove to be the decisive step toward the quantization of other field theories which are of greater interest.

First of all, we need to select some "standard" function space, and we propose to take the Schwartz space[14] $\mathcal{S}$ for the sake of convenience. Let F be some n-dimensional subspace of $\mathcal{S}$ and F' be its dual[15]. Upon choosing a basis $(f_k)_{k=1,\ldots,n}$ in F, we know from algebra that there is a dual basis $(f_k^*)_{k=1,\ldots,n}$ in F' such that $f_j^*(f_k) = \delta_{jk}$. This entails the following: for $f \in F$ and $\phi \in F'$ we have $\phi(f) = \sum t_k \alpha_k$, where $f = \sum t_k f_k$ and $\phi = \sum \alpha_k f_k^*$; the real coefficients t_k and α_k are viewed as coordinates of the vectors f and ϕ respectively. The Fourier decomposition

$$E\Big(\exp\{i\sum t_k \Phi(f_k)\}\Big) = \int d\mu_{f_1,\ldots,f_n}(\alpha_1,\ldots,\alpha_n)\,\exp\{i\sum t_k \alpha_k\},$$

$$\tag{7.4.1}$$

as analyzed in the foregoing section, does the following: it generates, for each finite-dimensional subspace $F \subset \mathcal{S}$, a probability measure μ on the dual F' such that, if F is given the basis (f_k), the measure becomes $\mu_{f_1,\ldots,f_n}$, i.e., a concrete measure on $\mathbb{R}^n$. Obviously, changing the basis will change the concrete measure, but the abstract measure on F' is still the same.

Varying F will allow us to construct a measure on the space $\mathcal{S}'$, the dual of $\mathcal{S}$, and this is achieved in the following way. Given any measurable (Borel)

---

[14]This is the space of all C^∞-functions $f \colon E_4 \to \mathbb{R}$, such that f and all its derivatives decrease faster than any inverse power $|x|^{-n}$ as $|x| \to \infty$.

[15]The dual F' consists of all $\mathbb{R}$-linear functionals $\phi \colon F \to \mathbb{R}$.

set $A \subset \mathbb{R}^n$ and independent vectors $f_1, \ldots, f_n$, where n is arbitrary, we declare

$$\hat{A} = \left\{ \phi \in \mathcal{S}' \,\middle|\, (\phi(f_1), \ldots, \phi(f_n)) \in A \right\} \tag{7.4.2}$$

to be a measurable set in $\mathcal{S}'$ with same measure as A:

$$\int_{\hat{A}} d\mu(\phi) = \int_A d\mu_{f_1, \ldots, f_n}(\alpha_1, \ldots, \alpha_n). \tag{7.4.3}$$

Sets of the form (7.4.2) are called *cylinder sets* in $\mathcal{S}'$. The measure μ that has already been shown to make sense on cylinder sets can, by a standard procedure [7.8, Chap. IV], be extended to the so-called Borel algebra of $\mathcal{S}'$. This is the smallest σ-algebra containing every cylinder $\hat{A}$ with A some Borel set in $\mathbb{R}^n$.

The question is: what is the space $\mathcal{S}'$? Answer: it is no more and no less than the space of all continuous linear functionals [7.9] on the Schwartz space $\mathcal{S}$. Such a functional, say ϕ, is frequently termed a *distribution*[16], and one writes $\phi(f) = \int dx\, f(x)\phi(x)$, knowing that, in most cases, the generalized function $\phi(x)$ is not a function at all. No particular ϕ is given preference in our probabilistic approach. Rather, ϕ is chosen randomly which renders the typical ϕ "very discontinuous". Don't bet too much on the continuity of the map $f \to \phi(f)$: there are lots of jumps and singularities of $\phi(x)$ all over the place. Comparison with Brownian motion tells us that $\phi(x)$ is related to the Euclidean field $\boldsymbol{\Phi}(x)$ the same way the Brownian path $\omega(t)$ is related to the Wiener process X_t. So in order to emphazise the analogy, we will refer to ϕ as the *path of the Euclidean field*[17]. This also helps to avoid a conflict of terms if we were to call it a *distribution*. Consequently, we will refer to $\mathcal{S}'$ as the *path space*.

Let us be more concrete now and study Gaussian measures on $\mathcal{S}'$ since they are so intimately related to free fields. Also, they are the simplest. Take (f, Kf) to be any nondegenerate positive quadratic form on $\mathcal{S}$. Then there exists a Gaussian measure μ on $\mathcal{S}'$ such that

$$\int d\mu(\phi)\, \exp\{i\phi(f)\} = \exp\{-\tfrac{1}{2}(f, Kf)\} \tag{7.4.4}$$

which means roughly the following: upon choosing any subspace $F \subset \mathcal{S}$ with $n = \dim F < \infty$ and (f_k) a basis in F, the left-hand side of (7.4.4) can be written as some n-dimensional Gaussian integral with respect to the restricted set of coordinates $\alpha_k = \phi(f_k)$ of the path ϕ, i.e., for all $f \in \mathcal{S}$,

---

[16]Not to be confused with a probability distribution.

[17]Leaving aside any kind of mathematical sophistication, we may think of $\phi(\boldsymbol{x}, 0)$ as a classical field at a fixed time. As time progresses, $\phi(\boldsymbol{x}, x^4)$ exhibits how the field evolves, and hence describes the path of the field in some abstract space of infinite dimension.

$$\int d\mu(\phi)\,\exp\{i\phi(f)\} = \int d\mu_{f_1,\dots,f_n}(\alpha_1,\dots,\alpha_n)\,\exp\{i\sum t_k\alpha_k\}$$

$$= \exp\{-\tfrac{1}{2}\sum t_j t_k a_{jk}\}, \tag{7.4.5}$$

where $f = \sum t_k f_k$ and $a_{jk} = (f_j, Kf_k)$.

This offers the opportunity to write the Schwinger functional for the free field as a path integral (Feynman's *sum over histories*):

$$E(\exp\{i\boldsymbol{\Phi}(f)\}) = \int d\mu(\phi)\,\exp\{i\phi(f)\}$$

$$= \exp\{-\tfrac{1}{2}(f, (-\boldsymbol{\Delta} + m^2)^{-1}f)\}. \tag{7.4.6}$$

Setting $f = \sum t_k f_k$ in (7.4.6) and expanding into powers of $t_1, \dots, t_n$, we obtain

$$E(\boldsymbol{\Phi}(f_1)\cdots\boldsymbol{\Phi}(f_n)) = \int d\mu(\phi)\,\phi(f_1)\cdots\phi(f_n). \tag{7.4.7}$$

For fixed $f_1,\dots f_n$, the right-hand side turns out to be an ordinary n-dimensional integral. We also write

$$E(\boldsymbol{\Phi}(x_1)\cdots\boldsymbol{\Phi}(x_n)) = \int d\mu(\phi)\,\phi(x_1)\cdots\phi(x_n) \tag{7.4.8}$$

for the Schwinger functions. However, this is meant merely symbolically.

We may ask: is there a more concrete (less abstract) description of the measure μ on the path space other than writing down its Fourier transform (7.4.6)? To find such a description, one is tempted to extrapolate representation (7.3.20) to $n \to \infty$ writing

$$d\mu(\phi) = Z^{-1}\mathcal{D}\phi\,\exp\{-\tfrac{1}{2}(\phi, (-\boldsymbol{\Delta} + m^2)\phi)\}. \tag{7.4.9}$$

Again, the right-hand side is meant to be formal in that no a priori interpretation is claimed for its constituents. However, when considered on finite-dimensional dual pairs (F, F') with basis (f_k) and (f_k^*), they generate the following. Let $\phi = \sum \alpha_k f_k^*$. Then

1. $\mathcal{D}\phi$ stands for the Lebesgue measure $\prod d\alpha_k$.
2. Z stands for $[\det(2\pi A)]^{1/2}$, where A is a matrix with entries $a_{jk} = (f_j, (-\boldsymbol{\Delta} + m^2)^{-1}f_k)$.
3. $(\phi, (-\boldsymbol{\Delta} + m^2)\phi)$ stands for $\sum \alpha_j \alpha_k (A^{-1})_{jk}$.

As n, the dimension of F (or F'), tends to infinity, none of the mentioned expressions retains a separate meaning. In the limit, they need to come together in order to yield a meaningful expression, a fact which is a little irritating at first sight. However, we already know that there is a neat and simple characterization of the limit through the characteristic functional which prompts us to formulate a notational convention[18]:

With (f, Kf) a nondegenerate quadratic form on $\mathcal{S}$, the solution of

$$\int d\mu(\phi)\, e^{i\phi(f)} = e^{-\frac{1}{2}(f, Kf)}$$

(i.e., the solving measure on $\mathcal{S}'$) is formally written as

$$d\mu(\phi) = Z^{-1}\mathcal{D}\phi\, e^{-\frac{1}{2}(f, K^{-1}f)}.$$

Let's look again at the free scalar field where $K^{-1} = -\mathbf{\Delta} + m^2$. If $\phi(x)$ is sufficiently 'nice' (differentiable and decreasing at infinity), the scalar product $\frac{1}{2}(\phi, K^{-1}\phi)$ is quickly seen to be the *Euclidean action* of the classical field $\phi(x)$, because a partial integration brings the expression into its standard form:

$$\tfrac{1}{2}(\phi, K^{-1}\phi) = \tfrac{1}{2}\int dx\left[\sum_\alpha \{\partial_\alpha\phi(x)\}^2 + m^2\phi(x)^2\right], \tag{7.4.10}$$

where $\partial_\alpha = \partial/\partial x^\alpha$. This sheds some light on the role of the classical action $W(\phi)$ in quantum field theory: it defines a probability measure on the path space in very much the same way as the energy defines a Gibbs measure in statistical mechanics.

This fact is by now the starting point of every attempt to define a concrete field theory. Such an attempt proceeds in two steps: first, one writes down an action for it, and, second, one works hard to give a meaning to the implied Gibbs measure which usually rests on a tedious process of discretization and approximation. The simplest class of models one can think of is provided by the action

$$W(\phi) = \int dx\left[\tfrac{1}{2}\sum_\alpha \{\partial_\alpha\phi(x)\}^2 + U\big(\phi(x)\big)\right]. \tag{7.4.11}$$

The function $U\colon \mathbb{R} \to \mathbb{R}$ describes a local self-interaction of the neutral scalar field. We look upon $U(r)$ as a *potential* and we take it to be bounded below. The Euclidean n-point functions are obtained as path integrals:

$$E(\mathbf{\Phi}(x_1)\cdots\mathbf{\Phi}(x_n)) = \frac{\int \mathcal{D}\phi\, e^{-W(\phi)}\phi(x_1)\cdots\phi(x_n)}{\int \mathcal{D}\phi\, e^{-W(\phi)}}. \tag{7.4.12}$$

However, this is but a formal expression. Specific model are:

---

[18]Though the quadratic form defines a linear map $K\colon \mathcal{S} \to \mathcal{S}'$, when writing $(\phi, K^{-1}\phi)$, we do not mean to imply the existence of an inverse $K^{-1}\colon \mathcal{S}' \to \mathcal{S}$. As a rule, K^{-1} is an unbounded operator and (7.3.20) shows how to approximate such operators by matrices A^{-1} of increasing order.

- ϕ^4 *model* $\qquad\qquad U(r) = \frac{1}{2}m^2r^2 + \lambda r^4$ $\qquad\qquad$ $(\lambda > 0)$
- *Higgs model* $\qquad\qquad U(r) = -\frac{1}{2}\mu^2 r^2 + \lambda r^4$ $\qquad\qquad$ $(\lambda > 0)$
- *Sine-Gordon model* $\quad U(r) = \frac{1}{2}m^2r^2 + \lambda(\cos(\gamma r) - 1).$

Non-Gaussian measures on the path space $\mathcal{S}'$ are notoriously difficult to grasp. Some courageous people have taken up the challenge to construct some of them. Their main motive was to investigate the meaning of 'renormalization' outside perturbation theory and to prove the existence of nontrivial field theories. Numerous articles have been written on the construction of $P(\phi)_2$ fields: these are two-dimensional boson fields with polynomial self-interaction. Partial results exist in three dimensions, very few in four dimensions. They include the construction of Yukawa, sine-Gordon, and Higgs models. For an excellent account see [7.10]. In spite of this enormous amount of rigorous work on "constructive quantum field theory", there has not as yet emerged an elegant solution to the construction problem as a whole and in general. Nor can the final answer be given in closed form for any specific (nontrivial) model.

A related but different approach utilizes techniques from statistical mechanics, in particular, the familiar process of constructing Gibbs measures on infinite lattices by way of performing the thermodynamic limit. Then the second step, transition to the continuum, is the hardest. We will get a glimpse of this approach when we come to study lattice field theory in Chap. 8.

Let us address another question before going on. What could be the relation between field theory and quantum mechanics? Comparing the exposition of the Euclidean formalism with what has been said in Section 2.6, we come to the following conclusion:

Boson field theory in one dimension (time) reduces to quantum mechanics formulated in the Heisenberg picture. In particular, any quantum mechanical system with n degrees of freedom derives from some n-component scalar field model. The stochastic process X_s canonically associated with the Hamiltonian $H = H_0 + V$ corresponds to a Euclidean field $\Phi(x^4)$, self-interacting via the potential V, where $s = x^4$, and the nondifferentiable paths $\omega(s)$ of that process correspond to the paths $\phi(x^4)$ of the field. The ground state of the quantum mechanical system corresponds to the vacuum and the position operator $q_t = e^{itH}qe^{-itH}$ may be considered to be the Minkowskian field $\Phi(x^0)$, where $t = x^0$. The correlation functions for the process X_s coincide with the Schwinger functions of the Euclidean field. The harmonic oscillator and the free field correspond to each other.

Here, the dimension n, i.e., the number of degrees of freedom of a quantum mechanical system, should not be confused with the dimension d of spacetime. There is no relation between the two, and they may both be arbitrary. Experience has shown that, with an increase of d, it becomes ex-

ceedingly more difficult to construct nontrivial examples of Euclidean fields:
the case $d = 1$ is easy, while $d = 2$ already provides a formidable task.

Exercise 1.   The relation $\int d\mu(\alpha)\, e^{it\alpha} = 1$ characterizes the normalized
Dirac measure on $\mathbb{R}$ which is concentrated at $\alpha = 0$. So formally, $d\mu(\alpha) = \delta(\alpha)d\alpha$. Correspondingly, if F is a linear space and F' its dual, the relation

$$\int d\mu(\phi)\, e^{i\phi(f)} = 1 \qquad (f \in F, \phi \in F')$$

defines the Dirac measure on F', normalized and concentrated at the origin.
Let (f, Kf) be positive, however degenerate with $F = \{f \in \mathcal{S} \mid Kf = 0\}$
the kernel (or null space) of K. Show that (7.4.4) gives rise to a *singular
Gaussian measure* on $\mathcal{S}'$, that is to say, it behaves like a Dirac measure on
F' and like a ordinary Gaussian measure on

$$F^\perp = \{\phi \in \mathcal{S}' \mid \phi(f) = 0 \ \forall f \in F\}.$$

Exercise 2.   (Instantons in quantum mechanics) It is legal to replace E_4
by $\mathbb{R}$ (time) in order to test our understanding of field theory in a more
familiar context, i.e., ordinary quantum mechanics. Here is an example. For
$\phi(s)$ real, consider the classical action

$$W(\phi) = \int ds \left\{ \tfrac{1}{2}\dot{\phi}^2 + \lambda(\phi^2 - c^2)^2 \right\} \qquad (\lambda > 0).$$

Clearly, the absolute minimum ($W = 0$) is attained for $\phi(s) = \pm c$. Prove
that there exist also local minima at $W = (2/3)\lambda^{1/2}c^3$ for a continuum of
so-called *instanton solutions*:

$$\phi_a(s) = c\tanh\frac{s-a}{\tau}, \qquad \tau^{-1} = c(2\lambda)^{1/2} \qquad (a \in \mathbb{R})$$

(show that ϕ_a solves $\delta W/\delta\phi = 0$ and check the second derivative). Each
solution 'connects' the two absolute minima:

$$\lim_{s \to \pm\infty} \phi_a(s) = \pm c.$$

Guess what the contribution of these solutions is to the path integral, specifically to the two-point function $E\big(\Phi(s)\Phi(s')\big)$.

7.5 Basic Postulates

The point of the preceding discussion has been to call attention to the construction mechanism. An important question that comes next is: once constructed, does the probability measure μ or, equivalently, the Schwinger functional

$$S\{f\} = \int d\mu(\phi) \, \exp\{i\phi(f)\} \tag{7.5.1}$$

give rise to some interpretable field theory (in terms of physics)? Of course, we would like to know whether we can go on and construct the Minkowskian version of it, i.e., obtain the Wightman functions and finally be able to construct the Hilbert space, the field operators, and the energy-momentum operator (with a physically reasonable spectrum). Only those measures for which the answer is affirmative deserve further study.

Conditions on the Schwinger functional fall into two categories. One category lists all obvious properties that follow from representation (7.5.1) as it stands:

- $|S\{f\}| \le S\{0\} = 1$,
- $S\{f + tg\}$ is a continuous function of $t \in \mathrm{I\!R}$ for all $f, g \in \mathcal{S}$,
- the matrix with elements $a_{jk} = S\{f_j - f_k\}$ $(j, k = 1, \ldots, n)$ is positive definite (for arbitrary n and $f_k \in \mathcal{S}$).

For short, we call $S\{f\}$ a continuous normed functional of positive type. Let us comment on the last property (positivity), which appears to be less obvious. If

$$F(\phi) = \sum_{k=1}^{n} c_k \exp\{i\phi(f_k)\} \tag{7.5.2}$$

with $c_k \in \mathbb{C}$, we would find that $|F(\phi)|^2 = \sum_{j,k} c_j \bar{c}_k \exp\{i\phi(f_j - f_k)\}$, and hence

$$0 \le \int d\mu(\phi)|F(\phi)|^2 = \sum_{j,k} c_j \bar{c}_k S\{f_j - f_k\}. \tag{7.5.3}$$

Positivity of the measure μ and positivity of the characteristic functional S are corresponding properties. Either of them allows us to naturally associate the Hilbert space $\mathcal{E} := L^2(\mathcal{S}', d\mu)$ of square-integrable functions $F : \mathcal{S}' \to \mathbb{C}$ to the measure μ in such a way that the scalar product is

$$(F, G) = \int d\mu(\phi) \, \overline{F(\phi)} G(\phi). \tag{7.5.4}$$

Warning: this space is not yet the Hilbert space of physical states, but its introduction is an essential step towards the construction of the latter. A dense set of vectors is provided by (7.5.2).

The basic postulates belonging to the first category do not even guarantee the existence of the Schwinger functions to all orders. For their existence rests on whether the measure decays sufficiently fast at infinity so that it has moments to all orders. Supposing Schwinger functions do exist, their symmetry with respect to permutations will again be a consequence of ansatz (7.5.1) and therefore is regarded part of the basic properties.

The postulates of the second category leave much more room for discussion. Osterwalder and Schrader [7.11] were the first to write down a satisfactory list of "axioms" during 1973–75. Since then variants of this list have been discussed by B. Simon [7.12] and J. Glimm and A. Jaffe [7.10]. We will spend the rest of this section presenting and discussing a version which is close to that given by the latter authors. To do so we need to consider a slightly larger space of test functions: the space $\mathcal{S}^c$ of all complex $f = f_1 + i f_2$, where $f_1, f_2 \in \mathcal{S}$.

Analyticity. *The Schwinger functional $S\{f\}$ admits an extension to complex functions $f \in \mathcal{S}^c$ with the extension being analytic.* This postulate means roughly the following: for $f = \sum_{k=1}^{n} z_k f_k$ with $z_k \in \mathbb{C}$ and $f_k \in \mathcal{S}^c$, the mapping $(z_1, \ldots, z_n) \to S\{f\}$ becomes entire analytic in $\mathbb{C}^n$. It is thus guaranteed that $S\{f\}$ admits a power series expansion which converges everywhere and represents the functional. Phrased differently, the measure μ is "nicely behaved" at infinity. One may argue whether or not a weaker condition such as the C^∞ property of the map $t \to S\{tf\}$ (with $t \in \mathbb{R}$ and $f \in \mathcal{S}$) would do, but some sort of condition in this direction is absolutely necessary, as can be learned from studying misbehaved model systems. Consider for instance the model

$$\log S\{f\} = -c | \textstyle\int dx\, f(x)| - \tfrac{1}{2} S(f * f; m) \qquad (c > 0) \tag{7.5.5}$$

which does not look pathological at all at first glance. Nevertheless, the absolute value occurring in the term of the right-hand side spoils not only analyticity but also differentiability. The peculiar feature of this model is best illustrated by the well known fact in probability theory that $e^{-|s|}$ is the characteristic function of the Cauchy distribution:

$$e^{-|s|} = \frac{1}{\pi} \int_{-\infty}^{\infty} \frac{dt}{1 + t^2} e^{ist}. \tag{7.5.6}$$

Owing to the slow decrease of the distribution at infinity, not even the first moment can be formed.

Regularity. *There exist constants c_1 and c_2 such that, for all $f \in \mathcal{S}^c$,*

$$|S\{f\}| \le \exp \int dx \left(c_1 |f(x)| + c_2 |f(x)|^2 \right). \tag{7.5.7}$$

Moreover, the two-point function $\mathbf{E}(\boldsymbol{\Phi}(x)\boldsymbol{\Phi}(y))$ is assumed to be locally integrable. Bound (7.5.7) on the one hand expresses strong continuity and on the other hand limits the growth of the Schwinger functional for complex f. The supplementary postulate restricts the nature of the singularity of the two-point function at $x - y = 0$. In view of the remaining axioms, the regularity postulate seems rather ad hoc and, physically speaking, not even well founded. The next two postulates, however, are the main pillars on which field theory is built.

Invariance. *The functional $S\{f\}$ is invariant under all symmetries of the Euclidean space E_4.* In more detail: let the action of the Euclidean group on the test functions be described by

$$(a, R)f(x) = f\big(R^{-1}(x - a)\big) \qquad \big(a \in \mathbb{R}^4,\, R \in O(4)\big). \tag{7.5.8}$$

We then require that $S\{(a, R)f\} = S\{f\}$. Clearly, the symmetries of the Schwinger functional are the symmetries of the associated probability measure μ and also those of the Euclidean action. A unitary representation[19] $U_{a,R} : \mathcal{E} \to \mathcal{E}$ is then seen to be given by

$$U_{a,R}F(\phi) = \sum_{k=1}^{n} c_k \exp\big\{i\phi\big((a, R)f_k\big)\big\}, \tag{7.5.9}$$

where F is a special vector of the form (7.5.2) (extension by continuity). The constant function $F = 1$ plays the role of an invariant vector. It may be thought of as a preliminary version of the vacuum.

We stress the importance of two particular symmetries as elements of the Euclidean group:

- *Time reversal.* We write $\theta := (0, R)$, where $R(\boldsymbol{x}, x^4) = (\boldsymbol{x}, -x^4)$, and put $\Theta = U_\theta$. The unitary operator Θ is an involution: $\Theta^2 = 1$.
- *Time shift.* We write $U(t) := U_{a,1}$ where $a = (0, t)$ and obtain a (continuous) one-parameter unitary group: $U(0) = 1$, $U(t)^* = U(-t) = U(t)^{-1}$, $U(t)U(t') = U(t + t')$. Standard reasoning says that there exists a self-adjoint operator A such that $U(t) = \exp\{itA\}$. But A is not to be confused with the Hamiltonian.

It is easily verified that the following relation holds:

$$\Theta U(t) = U(-t)\Theta \tag{7.5.10}$$

which shows that the spectrum of A is symmetric about zero.

Reflection positivity. *The matrix with elements $a_{ik} = S\{f_i - \theta f_k\}$ ($i, k = 1, \ldots, n$, n arbitrary) is positive for all $f_k \in \mathcal{S}$ that vanish in the lower halfspace, i.e., if $f_k(x) = 0$ for $x^4 < 0$.* Test functions with this property form a subspace $\mathcal{S}_+$. In connection with the subspace, it is most easily seen that there is a corresponding subspace $\mathcal{E}_+$ of $\mathcal{E}$, which we define to be the closed complex-linear hull of all vectors of the form $F(\phi) = \exp\{i\phi(f)\}$ with

---

[19]Recall the definition $\mathcal{E} = L^2(\mathcal{S}', d\mu)$.

$f \in \mathcal{S}_+$. The main goal is that reflection positivity implies that $(\Theta F, G) \geq 0$ for all $F, G \in \mathcal{E}_+$ (for vectors of the form (7.5.2) to begin with, and then for all vectors by continuity). It is natural to introduce another scalar product in $\mathcal{E}_+$ by writing

$$\langle F, G \rangle := (\Theta F, G) \tag{7.5.11}$$

under the provision that we factor out the null space[20]

$$\mathcal{E}_0 := \{F \in \mathcal{E}_+ \mid \langle F, F \rangle = 0\}. \tag{7.5.12}$$

Do we get another Hilbert space? The answer is no, unless we add all Cauchy sequences to the quotient space, a process which is customarily called *completion*. Symbolically,

$$\mathcal{H} := (\mathcal{E}_+/\mathcal{E}_0)^-. \tag{7.5.13}$$

We anticipate that the Hilbert space $\mathcal{H}$ is in fact the physical state space. Still, this needs to be confirmed. Observe that the constant functional $F = 1$ is a member of $\mathcal{E}_+$ since $1 = \exp\{\phi(f)\}$ with $f = 0 \in \mathcal{S}_+$. We therefore anticipate the existence of a vacuum state in $\mathcal{H}$.

Two vectors F and G in $\mathcal{E}_+$ will give rise to the same vector in $\mathcal{H}$ if they differ by some vector in $\mathcal{E}_0$. It is thus reasonable to consider the equivalence class $F_\bullet$ associated with each $F \in \mathcal{E}_+$ and treat $F_\bullet$ as some vector in $\mathcal{H}$ so that the scalar product becomes

$$(F_\bullet, G_\bullet) = \langle F, G \rangle. \tag{7.5.14}$$

In particular, if $F = 1$, we propose to write Ω in place of $1_\bullet$ and call Ω the *vacuum*. Observe that Ω is properly normalized since μ is a probability measure: $\|\Omega\|^2 = \int d\mu = 1$.

7.5.1 The Hamiltonian

It is a remarkable fact that in order to obtain the Hamiltonian H of a quantum field theory we need not construct the Minkowskian version to begin with. Therefore, it fully suffices to stay with the Euclidean formulation to be able to answer questions concerning the mass spectrum of the theory. The key idea is to consider the semigroup of time shifts with $t \geq 0$. Then the following observation is basic to the construction: if the function $f(x, x^4)$ vanishes in the lower halfspace $x^4 < 0$, so does the function

$$f_t(x, x^4) = f(x, x^4 - t)$$

for all $t > 0$, i.e., the semigroup of time shifts maps the space $\mathcal{S}_+$ into itself. As a consequence, we have

---

[20]Notice that we are dealing with a linear subspace: $\langle F, F \rangle = 0$ iff $\langle F, G \rangle = 0$ for all $G \in \mathcal{E}_+$.

$$U(t) : \mathcal{E}_+ \to \mathcal{E}_+ \qquad (t \geq 0), \tag{7.5.15}$$

which makes $\mathcal{E}_+$ an invariant subspace of $\mathcal{E}$. What about $\mathcal{E}_0$, the most relevant subspace of $\mathcal{E}_+$? Is it invariant, too?

To find the answer to this question, we note that

$$\langle U(t)F, G \rangle = \langle F, U(t)G \rangle \qquad (F, G \in \mathcal{E}_+) \tag{7.5.16}$$

since $(\Theta U(t)F, G) = (U(-t)\Theta F, G) = (\Theta F, U(t)G)$. Then the argument runs as follows: $F \in \mathcal{E}_0$ implies that $\langle F, U(t)G \rangle = 0$, and hence $\langle U(t)F, G \rangle = 0$, and so $U(t)F$ is in $\mathcal{E}_0$ which means that $\mathcal{E}_0$ is invariant.

The invariance property, which is absolutely crucial in this reasoning, allows us to declare $U(t)$ a linear bounded operator acting on the quotient space $\mathcal{E}_+/\mathcal{E}_0$. A standard procedure using continuity extends the operator to all of $\mathcal{H}$, which is then denoted $U(t)_\bullet$. The construction is summarized by

$$U(t)_\bullet F_\bullet = (U(t)F)_\bullet \tag{7.5.17}$$

which is valid for $F \in \mathcal{E}_+$ and $t \geq 0$. The continuity of the map $t \to U(t)_\bullet$ is a minor technical detail and easily verified.

More importantly, equality (7.5.16) tells us that $U(t)_\bullet$ is in fact self-adjoint. In addition, $U(t)_\bullet = U(t/2)_\bullet^2$, implying $U(t)_\bullet \geq 0$. This is an easy lower bound. We wish to prove a bound, namely $U(t)_\bullet \leq 1$, which would guarantee that the semigroup is of the form

$$U(t)_\bullet = \exp(-tH) \qquad (t \geq 0, H \geq 0). \tag{7.5.18}$$

For a proof, we start from the obvious Cauchy–Schwarz bound (in $\mathcal{E}_+$):

$$\|U(t)_\bullet F_\bullet\|^2 = (\Theta U(t)F, U(t)F) \leq \|F\|^2. \tag{7.5.19}$$

We may also use the Cauchy–Schwarz inequality in $\mathcal{H}$:

$$\|U(t)_\bullet F_\bullet\|^2 = (F_\bullet, U(2t)_\bullet F_\bullet) \leq \|F_\bullet\| \|U(2t)_\bullet F_\bullet\|. \tag{7.5.20}$$

With the abbreviations

$$u(t) = \frac{\|U(t)_\bullet F_\bullet\|}{\|F_\bullet\|}, \qquad c = \frac{\|F\|}{\|F_\bullet\|} \qquad (F_\bullet \neq 0), \tag{7.5.21}$$

the bounds (7.5.19–20) assume the compact form

$$u(t) \leq c, \qquad u(t)^2 \leq u(2t) \qquad (t \geq 0). \tag{7.5.22}$$

Iterating these relations we would get

$$u(t)^{2^n} \leq u(2^n t) \leq c \qquad (n \geq 0). \tag{7.5.23}$$

From $\lim_n c^{2^{-n}} = 1$ one infers that $u(t) \leq 1$, and so $U(t)_\bullet \leq 1$.

Relation (7.5.18) provides an elegant characterization of the Hamiltonian H as an operator on $\mathcal{H}$. It should also be clear that the vacuum Ω is one (if not the only) ground state of H since $U(t)_\bullet \Omega = \Omega$ implies that $H\Omega = 0$. We have not included in our list some axiom that would allow us to deduce that the vacuum is unique, i.e., Ω may not be the only vector in $\mathcal{H}$ with zero energy.

7.5.2 The Free Field Revisited

At this point it seems natural to ask whether the free field is indeed reflection positive as it should, and also whether the construction of the Hilbert space can be understood on the basis of more traditional ideas and notions. As for the free field, it suffices to focus on the one-particle space. Everything else follows easily. So we expect reflection positivity to be solely a property of the two-point function. In fact, it may be formulated as

$$S(\bar{f} * \theta f; m) \geq 0 \qquad (f \in \mathcal{S}_+^c). \tag{7.5.24}$$

The correctness of this assertion can be seen as follows. To each $f \in \mathcal{S}_+^c$ we assign the function

$$f_\bullet(\boldsymbol{p}) = \int dx \, e^{i\boldsymbol{p}\boldsymbol{x} - \omega x^4} f(x) \tag{7.5.25}$$

($\omega = \sqrt{\boldsymbol{p}^2 + m^2}$) which is then thought of as a vector of some Hilbert space $\mathcal{H}_1$ with the scalar product

$$(f_\bullet, g_\bullet) = \frac{1}{(2\pi)^3} \int \frac{d\boldsymbol{p}}{2\omega} \, \overline{f_\bullet(\boldsymbol{p})} \, g_\bullet(\boldsymbol{p}). \tag{7.5.26}$$

Utilizing representation (7.2.4) of $S(x; m)$, one checks that

$$\|f_\bullet\|^2 = S(\bar{f} * \theta f; m) \qquad (f \in \mathcal{S}_+^c). \tag{7.5.27}$$

This confirms the validity of (7.5.24). Notice: by writing (7.5.25) we have side-stepped the purely algebraic process of forming a quotient space. Nevertheless, the quotient is there, though implicitly: let $\mathcal{S}_0^c$ be the kernel of the linear map $\mathcal{S}_+^c \to \mathcal{H}_1$ that takes functions f into $f_\bullet$. Hence, by completing the quotient $\mathcal{S}_+^c / \mathcal{S}_0^c$, we get a Hilbert space which is naturally identified with the one-particle space $\mathcal{H}_1$.

The action of the semigroup is as expected:

$$U(t)_\bullet f_\bullet(\boldsymbol{p}) = e^{-\omega t} f_\bullet(\boldsymbol{p}) \qquad (t \geq 0). \tag{7.5.28}$$

A formal proof starts from $U(t)f(\boldsymbol{x}, x^4) = f(\boldsymbol{x}, x^4 - t)$ and $U(t)_\bullet f_\bullet = (U(t)f)_\bullet$. Therefore, the Hamiltonian H, when restricted to the one-particle space, is nothing but *multiplication by* ω.

The free field also illustrates the general presence of another positivity property of a quite different nature corresponding to the positivity of the underlying probability measure. It gives rise to the construction of an unphysical Hilbert space $\mathcal{E}_1$, considerably larger than $\mathcal{H}_1$. Namely, for $f \in \mathcal{S}^c(E_4)$ arbitrary, we consider the Fourier transformed function

$$\tilde{f}(p) = (2\pi)^{-2} \int dx \, \exp\{ipx\} f(x) \tag{7.5.29}$$

as a member of some linear space with the scalar product

$$(\tilde{f}, \tilde{g}) = \int dp \, \frac{\overline{\tilde{f}(p)} \tilde{g}(p)}{p^2 + m^2} = S(\bar{f} * g; m). \tag{7.5.30}$$

Completion of this space yields

$$\mathcal{E}_1 = L^2(\mathbb{R}^4, d\nu), \qquad d\nu(p) = dp \, (p^2 + m^2)^{-1}. \tag{7.5.31}$$

Time shifts act unitarily: $U(t)\tilde{f}(p) = e^{ip^4 t}\tilde{f}(p)$ so that $U(t) = \exp\{itA\}$, where A is the (selfadjoint) generator. Thus, the action of A is simply *multiplication by p^4* with no restriction on p^4. It is striking to see that the operators A and H are in no way related. As predicted by the general theory, the spectrum of H is positive, whereas the spectrum of A is symmetric about the origin.

Another peculiar feature is the different role of time reversal as a symmetry of the two spaces. It acts unitarily on $\mathcal{E}_1$, but antiunitarily on $\mathcal{H}_1$:

$$\Theta \tilde{f}(\boldsymbol{p}, p^4) = \tilde{f}(\boldsymbol{p}, -p^4), \qquad Tf_\bullet(\boldsymbol{p}) = \overline{f_\bullet(-\boldsymbol{p})}. \tag{7.5.32}$$

Again, there is no relationship whatsoever between Θ and T. The concept of time reversal, which has many profound implications in Minkowskian field theory, is now seen not to rely on *Euclidean time reversal* as part of the group $O(4)$, but to be solely rooted in *complex conjugation*:

$$K : \mathcal{S}^c \to \mathcal{S}^c \,, \;\; f \mapsto \bar{f}. \tag{7.5.33}$$

For we have $T = K_\bullet$. What is true for the free field also holds in general with due changes: complex conjugation then means

$$K : \mathcal{E} \to \mathcal{E} \,, \;\; F \mapsto \bar{F}, \tag{7.5.34}$$

and it is easily verified that both $\mathcal{E}_+$ and $\mathcal{E}_0$ are left invariant under the map K which allows us to construct $T = K_\bullet$. By contrast, the reflection Θ does not leave the space $\mathcal{E}_+$ invariant and, therefore, does not lead, despite its importance within a different context, to some operator on $\mathcal{H}$ and hence to another physical symmetry.

Remark 1. For $f \in \mathcal{S}_+$ (so that f is real), the vector $f_\bullet \in \mathcal{H}_1$ as defined by (7.5.25) satisfies $f_\bullet = Tf_\bullet$. Clearly, vectors in $\mathcal{H}_1$ obeying this condition form a closed real-linear subspace $\mathcal{R}$. Moreover, any vector $u \in \mathcal{H}_1$ which is not in that subspace may be decomposed as $u_1 + iu_2$ such that u_1 and u_2 obey the above condition. Simply put $u_1 = \frac{1}{2}(u + Tu)$ and $u_2 = \frac{i}{2}(u - Tu)$. This fact is written $\mathcal{H}_1 = \mathcal{R} + i\mathcal{R}$ and it shows that, even for f restricted to $\mathcal{S}_+$ in (7.5.24-25), the state space of the free field is recovered with no restrictions by always taking the closed complex-linear hull of the restricted set of vectors.

Remark 2. For every function $f \in \mathcal{S}_+$ consider the vector $\hat{f} = F_\bullet \in \mathcal{H}$, where $F(\phi) = ce^{i\phi(f)}$ and c has been chosen such that $\|F_\bullet\| = 1$. Let the Schwinger functional S correspond to a free field with $\mathcal{H}_1$ the one-particle space. For $f, g \in \mathcal{S}_+$, the scalar product $(f_\bullet, g_\bullet)$ in $\mathcal{H}_1$ turns out to be real, and a short calculation reveals that, in $\mathcal{H}$,

$$(\hat{f}, \hat{g}) = \exp\left\{ (f_\bullet, g_\bullet) - \tfrac{1}{2}\|f_\bullet\|^2 - \tfrac{1}{2}\|g_\bullet\|^2 \right\},$$

which shows that $\hat{f}$ is a so-called *coherent state*. In fact, writing the canonical commutation relations in the form

$$[a(u), a^*(v)] = (u, v), \qquad a(u)\Omega = 0 \qquad (u, v \in \mathcal{H}_1),$$

we would find that $\hat{f}$ corresponds to the state $e^{-\frac{1}{2}\|f_\bullet\|^2} e^{a^*(f_\bullet)}\Omega$. So the construction of the physical state space has used the fact that the coherent states span the Fock space of a Bose particle. For more information about coherent states see [7.13] and [7.14].

Exercise 1. Extend the discussion of this section to the situation where S consists of vector-valued Schwartz functions $f \colon E_4 \to \mathbb{R}^n$. Equivalently, write $f = \{f_1, \ldots, f_n\}$ and $\Phi(f) = \sum_{i=1}^n \Phi_i(f_i)$. Argue why this provides a natural setting for the theory of the n-component scalar field.

Exercise 2. With S_1 and S_2 two Schwinger functionals, define a new functional S by

$$S\{f\} = S_1\{f_1\}S_2\{f_2\}, \qquad f = \{f_1, f_2\}.$$

Prove that if both S_1 and S_2 satisfy the basic postulates, so does S. What sense does it make to say that multiplication of Schwinger functionals corresponds to the addition of fields? Show also that, if $\mathcal{H}_1$ and $\mathcal{H}_2$ are the physical Hilbert spaces underlying S_1 and S_2 respectively, their (Hilbert) tensor product $\mathcal{H}_1 \otimes \mathcal{H}_2$ may naturally be taken as the physical space underlying S.

Exercise 3. With $f = \{f_1, f_2\}$ a two-component test function and

$$\sigma(f) = \left[\left(\int dx\, f_1(x) \right)^2 + \left(\int dx\, f_2(x) \right)^2 \right]^{1/2},$$

consider the Schwinger functional

$$S\{f\} = J_0\big(c\sigma(f)\big) \exp\left\{ -\tfrac{1}{2} \sum_{i=1,2} \Delta_+(f_i * f_i; m) \right\},$$

where $J_0(z)$ is the Bessel function of order zero and c a real number. Prove that S satisfies the basic postulates. Use the integral

$$\frac{1}{2\pi} \int_0^{2\pi} d\alpha \, e^{i(x\cos\alpha + y\sin\alpha)} = J_0\left(\sqrt{x^2 + y^2}\right)$$

to investigate the structure of the resulting field theory (of a complex scalar field). What is meant by the statement that the field has *long-range correlations*?

Exercise 4. Let $\boldsymbol{\Phi}(x)$ be the real Euclidean field of mass m as in Sect. 7.3.2 and let $F(\boldsymbol{\Phi})$ be some functional. Under suitable conditions on F, prove the following *integration by parts formula*:

$$\boldsymbol{E}\big(\boldsymbol{\Phi}(x)F(\boldsymbol{\Phi})\big) = \int dy \, S(x - y; m)\boldsymbol{E}\left(\frac{\delta F(\boldsymbol{\Phi})}{\delta\boldsymbol{\Phi}(y)}\right).$$

Moreover, if Wick powers $: \boldsymbol{\Phi}(f)^n :$ are defined by setting

$$\sum_{n=0}^{\infty} \frac{i^n}{n!} : \boldsymbol{\Phi}(f)^n : = \frac{\exp\left\{i\boldsymbol{\Phi}(f)\right\}}{\boldsymbol{E}\big(\exp\left\{i\boldsymbol{\Phi}(f)\right\}\big)} ,$$

it makes sense to think of $: \boldsymbol{\Phi}(x)^n :$ as another scalar field (the nth Wick power of $\boldsymbol{\Phi}(x)$). Prove that

$$\boldsymbol{E}\big(:\boldsymbol{\Phi}(x)^n: F(\boldsymbol{\Phi})\big) = \int dy \, S(x - y; m)\boldsymbol{E}\left(:\boldsymbol{\Phi}(x)^{n-1}: \frac{\delta F(\boldsymbol{\Phi})}{\delta\boldsymbol{\Phi}(y)}\right),$$

and $\delta:\boldsymbol{\Phi}(f)^n:/\delta\boldsymbol{\Phi}(y) = n:\boldsymbol{\Phi}(f)^{n-1}: f(y)$, and hence

$$\boldsymbol{E}\big(:\boldsymbol{\Phi}(x)^n::\boldsymbol{\Phi}(y)^m:\big) = \begin{cases} 0 & m \neq n \\ n! S(x - y; m)^n & m = n. \end{cases}$$

8 Field Theory on a Lattice

> There is no such thing at the microscopic level as space or time or spacetime continuum.
>
> *J.A. Wheeler*

8.1 The Lattice Version of the Scalar Field

Let us remind ourselves of the basic problem in Euclidean field theory, which is to give meaning to the Schwinger functions when the action is

$$W(\phi) = \int dx \left[\tfrac{1}{2} \sum_\alpha \left(\partial_\alpha \phi(x) \right)^2 + U(\phi(x)) \right].$$ (8.1.1)

As for the continuum, there seems to exist no direct approach to the problem outside perturbation theory, neither by analytical means to get closed-form expressions nor by high-speed computers to get numbers. So we might as well try to change the entire setting of the theory. The most promising structure of spacetime, in which the action $W(\phi)$ would instantly make sense for all ϕ and which gives rise to a bona fide Gibbsian measure $d\mu(\phi)$ without renormalization, is that of a finite lattice so that path integrals become finite-dimensional integrals (though the dimension could be enormous). To some people this might seem a rather strange way of looking at things and also a rather roundabout way to construct a continuum theory and to explore its properties. However, it is the safest. There may perhaps be also some deeper reason for not being able to finally approach the continuum for some of the more prominent models of elementary particle physics, mainly owing to quantum gravity limiting the applicability of these models at very small distances, where new parameters (such as the Planck length) might dictate the true behavior.

As for terminology, we let $\mathbb{Z}_N \equiv \mathbb{Z}/(N\mathbb{Z})$ denote the integers modulo N, and introduction of a lattice always means that we replace the Euclidean space E_4 by the four-dimensional lattice $(\mathbb{Z}_N)^4$ of period N (in all directions for simplicity). So, as a rule, the lattice spacing is taken to be unity. A lattice spacing a different from 1 may be introduced later on via some scaling transformation. We would then write $(a\mathbb{Z}_N)^4$ for the lattice. It should be clear that a lattice of this type cannot be embedded into the Euclidean four-space but could conveniently be thought of as lying on a four-torus. From the view point of functional analysis, the introduction of a periodic lattice means that we have decided to work with periodic boundary conditions. The reason for this decision is well known: when giving up Euclidean symmetry, we wish to save as much as possible from it.

The four-lattice of period N has N^4 sites, which we denote x, y, and so on. Each site x has coordinates $x^i \in \{0, 1, \dots, N-1\}$ $(i = 1, \dots, 4)$. Functions on the lattice can easily be summed. For instance, the sum $\sum_x a^4 f(ax)$ always exists and approximates the integral $\int dx\, f(x)$ provided N is sufficiently large and a is sufficiently small. The Euclidean action of a scalar field on a lattice (with $a = 1$) assumes the form

$$W(\phi) = \sum_x \left[\tfrac{1}{2} \sum_i \left(\partial_i \phi(x) \right)^2 + U\left(\phi(x) \right) \right]. \tag{8.1.2}$$

We have to agree upon which of the lattice versions of the gradient should be adopted. Among the various options we propose to take the forward difference:

$$[\partial_i \phi](x) := \phi(x + e_i) - \phi(x). \tag{8.1.3}$$

Here, $x + e_i$ stands for a neighboring lattice site obtained by going one step in the ith direction.

To begin with, all quantities carry no physical dimension whatsoever. This in particular applies to points x in spacetime, the mass m, the field $\Phi(x)$, and all kinds of coupling constants. It is only after the introduction of Planck's constant $\hbar$, the velocity of light c, and the lattice spacing a (as the unit of length) that all those quantities acquire their proper physical dimension.

Within the stochastic interpretation, the Euclidean field $\Phi(x)$ on a lattice would still be thought of as some random variable taking values in $\mathbb{R}$, whereas the path $\phi(x)$ describes a particular event: it simply assigns a real number to each lattice site with no restriction. In view of this, the path space may be taken to be $\mathbb{R}^{N^4}$. Any path integral thus becomes equivalent to some trustworthy N^4-dimensional integral, a fact which is comforting on the one hand but frustrating on the other when it comes to actual computations: even for a modest choice, say $N = 5$, there is little hope of being able to perform a 625-dimensional integral unless it is Gaussian.

It is plain that there are important advantages to the lattice formulation: fields need no longer be smeared with test functions since there is no distinction any more between smooth and singular functions, or between ordinary and generalized functions. "Differentiability" loses its meaning. So the formula

$$E\left(\Phi(x_1) \cdots \Phi(x_n) \right) = \frac{\int \mathcal{D}\phi\, e^{-W(\phi)} \phi(x_1) \cdots \phi(x_n)}{\int \mathcal{D}\phi\, e^{-W(\phi)}} \tag{8.1.4}$$

poses no problem. The Lebesgue measure is well-defined,

$$\mathcal{D}\phi = \prod_x d\phi(x), \tag{8.1.5}$$

since there are only finitely many points on the lattice. Yet, a minor problem remains. Can we be sure that the integral on the right-hand side of (8.1.4) converges? If it does, the theory is said to be stable. A mild condition guaranteeing stability is the following:

> *The potential $U(r)$ of the Euclidean action (8.1.2) admits a lower bound of the form $U(r) > -c + \mu^2 r^2$ with suitable constants c and $\mu^2 > 0$.*

Recall that a similar condition was needed to give meaning to the Feynman–Kac formula in quantum mechanics. In the present situation, the condition is somewhat stronger. The potential has to be bounded from below by some parabola, and strict positivity of μ^2 is indispensable to avoid problems owing to the presence of zero-momentum modes.

Take the ϕ^4 theory, for instance. Then the stability condition implies that we may not deliberately alter the sign of the coupling constant without running the risk of losing stability, though this is of no consequence in perturbation theory. This gives us cause to think that no closed-form result nor statement deserves to be termed 'nontrivial' if the sign of the coupling plays no essential role in it.

How would we proceed to extract information about the continuum version (if it exists) of the model field? That turns out to be the hardest part of the construction since the continuum field can only be reached in three nontrivial steps:

1. *Thermodynamic limit.* We let the period of the lattice, N, tend to infinity, hoping that the Schwinger functions converge. The lattice is still there and we write $\mathbb{Z}^4$ for it. As yet, none of the mathematical terms has acquired a physical dimension.

2. *Scaling.* The lattice spacing a is introduced as a new variable and a unit of length. It may be helpful to think of a as being of the order of 10^{-13} cm and below. Scaling replaces the unit lattice $\mathbb{Z}^4$ by $(a\mathbb{Z})^4$, which is then considered as being embedded in the Euclidean continuum. Scaling also affects most parameters of the theory (masses, coupling constants, etc.), which now acquire their physical dimension appropriate for the third step. As a result, we obtain some a-dependent lattice model.

3. *Continuum limit.* With dependence on a suitably adjusted and the parameters of the theory properly tuned, the Schwinger functional $S\{\phi\}$ approaches a limit as the lattice constant a tends to zero. During this process, the correlation length λ (inverse mass) is kept at a fixed value, i.e., its physical value. Since it does not make any difference whether $a \to 0$ in a scaled theory or $\lambda \to \infty$ in an unscaled theory, it is thus apparent that the lattice model (with $a = 1$) tends to one of its critical points (of second-order phase transition). This is so because statistical

mechanics tells us that it is only at a critical point that the correlation length may diverge. Critical behavior is a rare occurrance in parameter space: this is why we have to tune the parameters of the theory. In a sense, the continuum field theory explores the properties of the associated four-dimensional lattice model at and in the vicinity of the critical point.

These three procedures have to be followed precisely in the above order and act as a substitute for the more traditional renormalization techniques known from perturbation theory.

8.2 The Euclidean Propagator on the Lattice

8.2.1 The Fourier Representation

The symmetry of the periodic 4-lattice with regard to discrete translations is enough to make Fourier transformation a useful tool. It naturally leads to the momentum p varying over a discrete set, the *dual lattice*. As for our case, it looks very much like the original lattice, except that the spacing becomes $2\pi/N$. So $p = \{p_1, \ldots, p_4\} \in (\frac{2\pi}{N}\mathbb{Z}_N)^4$, and as before, we put $px = \sum_{k=1}^{4} p_k x^k$. Plain waves are written

$$f_p(x) = N^{-2} e^{ipx}. \tag{8.2.1}$$

As is easily verified, they form a complete orthonormal set of vectors with respect to the scalar product $(f, g) = \sum_x \overline{f(x)} g(x)$. The lattice gradient and its adjoint are

$$\begin{aligned}
\partial_k f(x) &= f(x + e_k) - f(x) \\
\partial_k^* f(x) &= f(x - e_k) - f(x).
\end{aligned} \tag{8.2.2}$$

We take the lattice Laplacian to be $-\Delta = \sum_{k=1}^{4} \partial_k^* \partial_k$. From these definitions and $pe_k = p_k$, one deduces that the plain waves form a common set of eigenvectors:

$$\begin{aligned}
\partial_k f_p(x) &= (e^{ip_k} - 1) f_p(x) \\
\partial_k^* f_p(x) &= (e^{-ip_k} - 1) f_p(x) \\
-\Delta f_p(x) &= \sum_{k=1}^{4} |e^{ip_k} - 1|^2 f_p(x).
\end{aligned} \tag{8.2.3}$$

Putting

$$E_p = \sum_{k=1}^{4} 2(1 - \cos p_k), \tag{8.2.4}$$

one could also write $-\boldsymbol{\Delta} f_p(x) = E_p f_p(x)$ and then state the following:

On the lattice, the spectrum of the operator $-\boldsymbol{\Delta}$ is purely discrete and falls into the interval $[0, 16]$.

Knowing the spectral decomposition of the operator $-\boldsymbol{\Delta}$ makes it easy for us to spectrally decompose any operator $F(-\boldsymbol{\Delta})$ with $F(t)$ a complex-valued function on the real interval $0 \le t \le 16$:

$$F(-\boldsymbol{\Delta}) f_p(x) = F(E_p) f_p(x). \tag{8.2.5}$$

To see an application of this formula, we propose to take $F(t) = \log(t+m^2)$. Returning to the position space representation, we get

$$\left[\log(-\boldsymbol{\Delta} + m^2)\right]_{xy} = N^{-4} \sum_p e^{ip(x-y)} \log(E_p + m^2) \tag{8.2.6}$$

which gives

$$\operatorname{tr} \log(-\boldsymbol{\Delta} + m^2) = \sum_x \left[\log(-\boldsymbol{\Delta} + m^2)\right]_{xx} = \sum_p \log(E_p + m^2). \tag{8.2.7}$$

This way we have calculated our first path integral

$$Z \equiv \int \mathcal{D}\phi\, e^{-W_0(\phi)} = \int \mathcal{D}\phi\, \exp\{-\tfrac{1}{2}(\phi, (-\boldsymbol{\Delta} + m^2)\phi)\}$$
$$= \left[\det\left(\frac{-\boldsymbol{\Delta} + m^2}{2\pi}\right)\right]^{-1/2} = \exp\left\{-\tfrac{1}{2}\sum_p \log\frac{E_p + m^2}{2\pi}\right\}, \tag{8.2.8}$$

where $W_0(\phi) = \sum_x \tfrac{1}{2}\left(\partial_i \phi(x)\right)^2$ and use has been made of the identity $\det = \exp \operatorname{tr} \log$.

As for another application of (8.2.5), choose $F(t) = (t + m^2)^{-1}$ and get a formula for the Euclidean lattice propagator of a free particle:

$$E\left(\boldsymbol{\Phi}(x)\boldsymbol{\Phi}(y)\right)_N \equiv Z^{-1} \int \mathcal{D}\phi\, e^{-W_0(\phi)} \phi(x)\phi(y)$$
$$= S_N(x - y; m) = N^{-4} \sum_p \frac{e^{ip(x-y)}}{E_p + m^2}. \tag{8.2.9}$$

Once more the result indicates that, on the lattice, E_p assumes the role normally played by p^2, the square of the momentum.

The limit $N \to \infty$ poses no problem with regard to the above formulas. The effect would simply be that every sum over the momentum p turns into an integral over the Brillouin zone $B_4 = [-\pi, \pi]^4$ (see also the remarks at the end of Sect. 8.3):

$$S_\infty(x; m) = \frac{1}{|B_4|} \int_{B_4} \frac{dp\, e^{ipx}}{E_p + m^2} \quad , \qquad |B_4| = (2\pi)^4. \tag{8.2.10}$$

Utilizing the integral representation

$$\frac{1}{E_p + m^2} = \int_0^\infty ds \, \exp(-sm^2) \prod_{i=1}^4 \exp\{-2s(1 - \cos p_i)\}, \tag{8.2.11}$$

the right-hand side of (8.2.10) becomes a one-dimensional integral:

$$S_\infty(x; m) = \int_0^\infty ds \, \exp(-sm^2) \prod_{i=1}^4 e^{-2s} I_{x_i}(2s) \tag{8.2.12}$$

($I_n(z)$ = modified Bessel function). The familiar continuum version is reached letting $a \to 0$ after scaling:

$$S(x; m) = \lim_{a \to 0} a^{-2} S_\infty(a^{-1}x; am) \qquad (x \neq 0). \tag{8.2.13}$$

In words: with $\boldsymbol{\Phi}(x; m)$ the free Euclidean field of mass m on the infinite lattice, the scaled field $\boldsymbol{\Phi}^a(x; m) := a^{-1}\boldsymbol{\Phi}(a^{-1}x; am)$, where $x \in (a\mathbb{Z})^4$, tends to the continuum field of mass m as $a \to 0$. The fact that we have excluded $x = 0$ in (8.2.13) indicates that, generally speaking, convergence occurs for functionals (i.e., smeared fields) only.

To study the two-point correlation function (i.e., the Euclidean propagator) in the presence of an interaction is one of the main tasks for the theoretician, and computer programs have been written to solve the problem numerically. Then formula (8.2.9) is no longer correct, though expected to be asymptotically valid when $|x - y|$ grows. The large distance behavior is dictated solely by m, the lowest mass being found among those states that carry the quantum numbers of the field $\boldsymbol{\Phi}$. Unfortunately, the distance $|x - y|$ cannot be made arbitrarily large on a finite lattice needed to experience the fall-off, which, for the continuum, would be of the form $e^{-m|x-y|}$. So the question arises: what replaces the exponential law when the lattice is finite?

The answer is supposed to come from inspecting (8.2.9), and it may be complicated. A reasonable and simple result follows if we project onto $p = 0$ keeping $p_4 \neq 0$, a procedure we will describe in some detail. From $\sum_x N^{-3} e^{ipx} = \delta_{p,0}$ it follows that

$$\sum_x S_N(x; m) = N^{-1} \sum_{p_4} \frac{\exp\{ip_4 x^4\}}{m^2 + 2(1 - \cos p_4)} \, . \tag{8.2.14}$$

The right-hand side can now be evaluated in closed form:

Let the modified mass parameter M be determined by $\sinh(M/2) = m/2$. *Then, for* $0 \leq x^4 < N$,

$$\sum_{x^1, x^2, x^3} S_N(x; m) = C_N \cosh\{M(x^4 - N/2)\}, \tag{8.2.15}$$

where $C_N^{-1} = 2 \sinh M \sinh(MN/2)$.

To prove (8.2.15) we put $f(\sigma) = [m^2 + 2(1 - \cos \sigma)]^{-1}$ and expand:

$$f(\sigma) = \sum_{n=-\infty}^{\infty} c_n e^{-in\sigma}. \tag{8.2.16}$$

The Fourier coefficients are

$$c_n = \frac{1}{2\pi} \int_{-\pi}^{\pi} e^{in\sigma} f(\sigma) d\sigma = \frac{1}{2\pi} \int_0^{\pi} \frac{\cos n\sigma \, d\sigma}{a - \cos \sigma}$$
$$= \frac{1}{2\sqrt{a^2-1}} \left(a - \sqrt{a^2-1} \right)^{|n|}, \tag{8.2.17}$$

where $a = 1 + \frac{1}{2}m^2 > 1$. To simplify the result we set $a = \cosh M$. Another way to express the relation between M and m is $\sinh(M/2) = m/2$. This gives

$$c_n = \frac{e^{-M|n|}}{2\sinh M}, \tag{8.2.18}$$

and we may write

$$\sum_{\mathbf{x}} S_N(x, m) = \frac{1}{N} \sum_{p_4} e^{ip_4 x^4} f(p_4) = \sum_{n=-\infty}^{\infty} c_n \delta_N(n' - n), \tag{8.2.19}$$

where $x_4 = n'$, $p_4 = 2\pi k/N$, and

$$\delta_N(n) = \frac{1}{N} \sum_{k=0}^{N-1} e^{i2\pi kn/N} = \begin{cases} 1 & \text{if } n = 0 \bmod N \\ 0 & \text{otherwise.} \end{cases} \tag{8.2.20}$$

Hence,

$$\sum_{n=-\infty}^{\infty} c_n \delta_N(n' - n) = \sum_{j=-\infty}^{\infty} c_{n'+jN}. \tag{8.2.21}$$

We also observe that, for $0 \le n' < N$,

$$|n' + jN| = \begin{cases} n' + jN & \text{if } j \ge 0 \\ N - n' + kN & \text{if } k \ge 0, \quad j + k + 1 = 0 \end{cases} \tag{8.2.22}$$

and so

$$\sum_{j=-\infty}^{\infty} e^{-M|n'+jN|} = \frac{e^{-Mn'} + e^{-M(N-n')}}{1 - e^{-MN}} = \frac{\cosh\{M(n' - N/2)\}}{\sinh(MN/2)}, \tag{8.2.23}$$

completing the proof.

The problem we posed seems to be solved and the answer is that, on a finite lattice, the cosh law of (8.2.15) replaces the exponential decay of correlations as seen in infinitely extended systems provided there is a gap in the spectrum. As for the period-N lattice, the maximal distance of two points on the time axis can at most be $N/2$, and our result shows that, for the maximal distance, correlations become minimal. Moreover, the parameter M that enters (8.2.15) does not coincide with the parameter m of (8.2.9), and what is even more surprising, the difference between the two values does not go away if we let N tend to infinity. The reason for not getting the identity $M = m$ is the fact that the lattice is still present. Suppose now that we scale the theory by changing the lattice constant from 1 to a. Then the equation that relates the two masses M and m becomes $\sinh(\frac{1}{2}aM) = \frac{1}{2}am$. If we finally let a tend to zero, we would indeed get the identity $M = m$ as requested. The conclusion to be drawn from this discussion is that, on a lattice, there is certainly more than one way to introduce "the mass" as a descriptive term, and any such concept may differ in a systematic fashion from the actual mass of the continuum theory.

The simple example considered above also throws some light on the numerics and serves to clarify to what extent a computer simulation should be relied on for solving a continuum problem on a finite lattice. Take $M = 1/5$, for instance, so that $m \approx M$. Figure 8.1 depicts the decay of correlations, i.e., the spatially averaged function $S_N(x; m)$ in the domain $0 \leq x^4 \leq 12$, as the lattice size is varied between 12^4 und 120^4, and where, for convenience, a normalization has been adopted such that the value 1 is always reached at $x^4 = 0$.

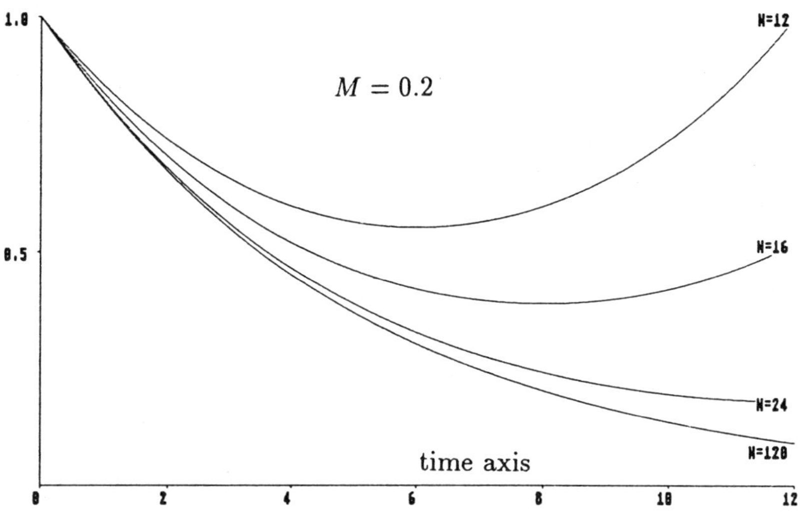

Fig. 8.1. The decay of correlations along the time axis over a distance of 12 lattice sites. The curves shown represent cosh functions characteristic of a lattice model where the "mass" is $M = 0.2$ and the period is $N = 12$, 16, 24, and 120 respectively.

Notice that, in our model, the distance of five lattice sites corresponds to the Compton wavelength $\lambda = m^{-1}$ of the particle (in units of the lattice spacing). In order to ensure that the N-lattice correlation is reasonable close to its large-N limit (say, within 5%) in the domain $0 \leq x^4 \leq 5$, we need to choose $N = 24$ or larger which says that the linear dimension of the lattice ought to be at least 4 to 5 Compton wavelengths!

8.2.2 Random Paths on a Lattice

Since the decay of correlations constitutes one of the main issues both in statistical mechanics and in field theory, we propose now to tackle this question from a different angle with a new technique employing random paths on an infinite lattice. The starting point is a rather clever description of the lattice Laplacian:

$$\Delta = \sum_{i=1}^{4}(S_i + S_i^{-1} - 2)$$

$$[S_i f]_{x'} \equiv \sum_{x}(S_i)_{x'x} f_x = f_{x'-e_i}.$$

$$(8.2.24)$$

Think of S_i as a unit step in the direction i and of S_i^{-1} as a step in the opposite direction. At the same time, S_i and S_i^{-1} are operators affecting any function we might consider on the lattice.

Recall next the principles of the random walk on a hypercubic lattice. Fix some initial site x on the lattice. Any sequence of unit vectors, $\pm e_{i_1}, \pm e_{i_2}, \ldots, \pm e_{i_n}$, is found to describe a path ω of length $|\omega| = n$ leading to the final position x', where

$$x' = (\cdots((x \pm e_{i_1}) \pm e_{i_2}) \pm \cdots \pm e_{i_n}).$$

$$(8.2.25)$$

Translations constitute an Abelian group and thus bracketing is not needed in (8.2.25). Of course, permuting the translation vectors e_i generally alters the path but leaves the endpoint x' invariant. It thus makes sense to think of the altered paths as belonging to some equivalence class denoted $[\omega]$. To each individual path ω we associate the sequence of operators, $S_{i_n}^{(-1)}, \cdots, S_{i_2}^{(-1)}, S_{i_1}^{(-1)}$, where we agree upon writing S_i^{-1} in place of S_i in case the translation vector e_i has a minus sign in front. From the sequence we pass to the product $S_\omega = S_{i_n}^{(-1)} \cdots S_{i_2}^{(-1)} S_{i_1}^{(-1)}$, which is a class function: the operator S_ω depends on $[\omega]$ only. It acts on functions defined on the infinite lattice and may be viewed as an infinite matrix with elements

$$(S_\omega)_{x'x} = \begin{cases} 1 & \text{if } \omega: x \rightsquigarrow x' \\ 0 & \text{otherwise.} \end{cases}$$

$$(8.2.26)$$

But the main combinatorial fact implied by (8.2.26) is that the total number of paths $\omega: x \rightsquigarrow x'$ of length n can be expressed algebraically:

$$\sum_{\substack{\omega:x\rightsquigarrow x' \\ |\omega|=n}} 1 = \sum_{\substack{\omega:x\rightsquigarrow x' \\ |\omega|=n}} (S_\omega)_{x'x} = \left(\sum_i(S_i + S_i^{-1})\right)^n_{x'x}.$$

$$(8.2.27)$$

Now, the formula (8.2.24) for the lattice Laplacian immediately springs to mind, and we may use both (8.2.24) and (8.2.27) when the lattice propagator is expanded into a geometric series to get:

$$(-\Delta + m^2)^{-1}_{x'x} = \sum_{n=0}^{\infty} (m^2 + 8)^{-n-1} \left(\sum_i (S_i + S_i^{-1}) \right)^n_{x'x}$$

$$= \sum_{\omega:x \leadsto x'} \lambda^{|\omega|+1}, \qquad \lambda = (m^2 + 8)^{-1}. \tag{8.2.28}$$

What has been achieved then is that correlations become expressible as path sums in such a way that very long paths acquire exponentially small weights. All terms in the sum are positive, and the dominant terms are those for which $|\omega|$ is smallest, corresponding to the shortest route from x to x', which however may not be unique.

This raises two questions:

1. Does the expansion (8.2.28) converge?
2. What is the asymptotic behavior of the sum when the distance $|x'-x| = \min_{\omega:x \leadsto x'} |\omega|$ becomes large?

The answer to the first question is rather obvious, while the second question is rather "deep". The key observation is that the total number of paths of length n, starting at x and ending anywhere, is $N(n) = 8^n$ (generally $(2d)^n$ in d dimensions by simple combinatorial reasoning). Since $8\lambda < 1$ provided $m^2 > 0$,

$$0 < (-\Delta + m^2)^{-1}_{x'x} < \sum_{n=|x'-x|}^{\infty} 8^n \lambda^{n+1} = m^{-2} e^{-\mu|x'-x|}, \tag{8.2.29}$$

where $\mu = \log(1+m^2/8)$. The crude estimate obtained from combinatorics is not as satisfactory as result (8.2.15). Though it recaptures the exponential fall-off, some strange constant μ appears in a place where we expect to encounter M (from $\sinh(M/2) = m/2$). It is not hard to show that $M > \mu$.

The random walk on a lattice as a technique was introduced into field theory by Symanzik [8.1] in 1969. It was much later (about 1983) that it started to develop into a powerful tool (see, for instance, [8.2, Chap. 21]). The technique has also been applied to classical spin systems [8.3]. It has led to inequalities for the four-point Schwinger function and thus played a decisive role in arguments that showed the triviality of the ϕ^4 theory in dimensions $d \geq 4$.

Exercise 1.   Random paths may also be considered on a finite lattice of period N. Prove that

$$N^{-4}\mathrm{tr} \log(-\Delta + m^2) = -\sum |\omega|^{-1} \lambda^{|\omega|} \log \lambda,$$

where $\lambda = (m^2 + 8)^{-1}$ and the sum extends over all path $\omega:0 \leadsto 0$ on the lattice such that $|\omega| \geq 2$. Show that both sides approach a limit as $N \to \infty$. Hint: use (8.2.7).

8.3 The Variational Principle

The purpose of this section is to give special focus to another concept, *entropy*, and to stimulate discussion about its role in field theory. Such discussion will allow us to tackle questions such as: can classical field theory be recovered from quantum field theory by letting $\hbar \to 0$? It also provides another approach to the problem of characterizing Gibbs measures in general. Above all, we wish to emphasize the relevance of the subject and its ties to statistical mechanics. The central theme will be the formulation of a variational principle, which we will do in the spirit of the familiar Gibbs principle characterizing the (Helmholtz) free energy in thermodynamics.

8.3.1 The Case of a Discrete Configuration Space

We suggest leaving field theory for a moment and studying the example of a finite state space for pedagogical reasons. The reader who wishes to visualize abstract sets is advised to think of the state space as describing, for instance, all possible spin configurations of an Ising model defined on a finite, however, large, lattice. As for this section, the actual interpretation of the states enumerated $1, 2, \ldots, n$ is absolutely irrelevant.

Instead of talking about some probability measure μ on field configurations, we are now dealing with a finite probability distribution $p = (p_1, \ldots, p_n)$ such that $0 \le p_i \le 1$ and $\sum_i p_i = 1$. The idea is not to look at a particular p but at the convex set formed by all distributions. To each distribution p one assigns a number called its *entropy*:

$$S(p) = -\sum_{i=1}^{n} p_i \log p_i, \tag{8.3.1}$$

where it is agreed that $p_i \log p_i = 0$ if $p_i = 0$. Notice that in thermodynamics one is inclined to follow a slightly different convention. There, the entropy is taken to be $k_B S(p)$ with k_B the Boltzmann constant. We have chosen not to follow this tradition since k_B plays no role outside thermodynamics.

Obviously, $S(p) \ge 0$. The entropy is maximal if and only if $p_i = n^{-1}$:

$$\sup_p S(p) = S(\tfrac{1}{n}, \ldots, \tfrac{1}{n}) = \log n. \tag{8.3.2}$$

We go on to generalize, arguing that we ought to be able to compare two distributions p and $\alpha = (\alpha_1, \ldots, \alpha_n)$, and call

$$S(p|\alpha) = -\sum_i p_i \log(p_i/\alpha_i) \tag{8.3.3}$$

the *relative entropy of p with respect to* α, allowing for $-\infty$ as a possible value of $S(p|\alpha)$. As it turns out, the absolute entropy $S(p)$ provides a particular example of the relative entropy arising from $\alpha_i = n^{-1}$:

$$S(p \mid \tfrac{1}{n}, \dots, \tfrac{1}{n}) = S(p) - \log n. \tag{8.3.4}$$

Up to an irrelevant shift by a constant, both entropies are the same. We assert that

$S(p|\alpha) \le 0$ *always, and* $S(p|\alpha) = 0$ *iff* $p = \alpha$.

For a proof, it suffices to assume that $\alpha_i > 0$ for all i. The function $f(u) = u \log u$ is convex on the halfaxis $u > 0$ since $f''(u) = u^{-1} > 0$. By virtue of Jensen's inequality (see Appendix C),

$$f\left(\sum_i \alpha_i u_i\right) \le \sum_i \alpha_i f(u_i). \tag{8.3.5}$$

Setting $u_i = p_i/\alpha_i$ we have that $\sum_i \alpha_i u_i = \sum_i p_i = 1$, and from $f(1) = 0$,

$$0 \le \sum_i p_i \log(p_i/\alpha_i) \tag{8.3.6}$$

as required. Moreover, the function $f(u)$ is strictly convex ($f''(u) > 0$). Therefore, equality occurs in (8.3.5) precisely for coinciding u_i's, and hence if $u_i = q$ for some q. Then, $p_i = q\alpha_i$ and so $1 = \sum_i p_i = q \sum_i \alpha_i = q$, i.e., $\alpha_i = p_i$, which completes the proof.

Suppose now that $w_1, \dots, w_n$ are arbitrary real numbers. The distribution α, where

$$\alpha_i = z^{-1} e^{-w_i}, \qquad z = \sum_i e^{-w_i}, \tag{8.3.7}$$

mimics the definition of a Gibbsian ensemble, and in this situation,

$$\sum_i p_i w_i - S(p) = -S(p|\alpha) - \log z. \tag{8.3.8}$$

The assertion concerning $S(p|\alpha)$ proved a minute ago yields

The Variational Principle. *With real coefficients* w_i,

$$\inf_p \left\{ \sum_i w_i p_i - S(p) \right\} = -log z, \tag{8.3.9}$$

where the infimum is reached only for $p = \alpha$; α *and* z *are given by* (8.3.7).

For our purposes it is best to think of the number w_i as the value of some observable W with respect to the state i, e.g., of the energy in units of $k_B T$, or of the Euclidean action in units of $\hbar$.

Example. We consider the effect of a magnetic field B on the energy levels

$$E_m = -m\mu B, \qquad m = -j, -j+1, \dots, j \quad (2j \in \mathbb{N}) \tag{8.3.10}$$

of a single spin j when μ is the magnetic moment and look for a distribution $p = (p_{-j}, \ldots, p_j)$ which maximizes the entropy $S(p)$ subject to the condition $\sum_m p_m E_m = E$. The solution is found by the Lagrange multiplier method which means that we first solve a related problem,

$$\beta \sum_m p_m E_m - S(p) = minimum. \tag{8.3.11}$$

and then solve for the multiplier β. By the variational principle, we get a unique solution p as a function of the multiplier: $p_m = z^{-1} \exp(-\beta E_m)$. From

$$z = \sum_{m=-j}^{j} \exp(-\beta E_m) = \frac{\sinh((j + \frac{1}{2})\beta\mu B)}{\sinh(\frac{1}{2}\beta\mu B)} \tag{8.3.12}$$

it follows that

$$-\sum_m p_m E_m = \frac{\partial}{\partial\beta} \log z = \mu B F_j(\beta\mu B) \tag{8.3.13}$$

with $F_j(x) = (j + \frac{1}{2}) \coth(j + \frac{1}{2})x - \frac{1}{2} \coth \frac{1}{2}x$ the Brillouin function. Notice that F_j maps the real line onto the open interval $(-j, j)$. Monotonicity guarantees the existence of the inverse map F_j^{-1}. To solve for β we invoke the condition $\mu B F_j(\beta\mu B) = -E$. If $-j\mu B < E < j\mu B$, we thus get

$$\beta = \frac{1}{\mu B} F_j^{-1}\left(-\frac{E}{\mu B}\right). \tag{8.3.14}$$

To this we add the remark that the Lagrange multipier β may also be viewed as the inverse temperature and that the magnetization M comes out as expected:

$$M = \beta^{-1} \frac{\partial}{\partial B} \log z = \mu F_j(\beta\mu B) = -E/B. \tag{8.3.15}$$

The derivation of (8.3.14) gives some idea of how the variational principle is applied to real systems.

8.3.2 The Deterministic Limit

Classical mechanics is fully deterministic, while statistical mechanics is not. Most strikingly, in both disciplines we encounter variational principles. So the question arises: what makes these principles differ? Why does a system like the one considered above behave statistically, others, however, deterministically? Answer: it is the presence of entropy in the variational principle that causes measured values to fluctuate. To see more closely the transition from statistics to determinism, we scale the entropy term in (8.3.9) by some parameter and then let the parameter go to zero[21]. In the limit, we arrive at the problem

$$\sum_i p_i w_i = minimum, \qquad 0 \leq p_i \leq 1, \qquad \sum_i p_i = 1. \tag{8.3.16}$$

To solve it, we first observe that $\min_p \sum_i p_i w_i = \min_i w_i$. If among the w_i there is precisely one element whose value is smallest, say w_1, the solution of (8.3.16) will be unique:

$$p_i = \begin{cases} 1 & \text{if } i = 1 \\ 0 & \text{otherwise.} \end{cases} \tag{8.3.17}$$

Since all probabilities are either zero or one, we have thus shown that the system behaves deterministically as soon as the entropy term is dropped in (8.3.9).

The degenerate case, where there is more than one minimal element among the numbers w_i, is a nuisance. As the solutions of (8.3.16) will then form a convex set of nontrivial probability distributions, determinism cannot be claimed for such a system.

8.3.3 Continuous Configuration Space

Returning to the definition of entropy, we take a continuous configuration space as our next choice. Such a choice accounts for what we need to assume in field theory, even though we work with a lattice. We shall model after the prototype only, the neutral scalar field. So the Euclidean action $W(\phi)$ is taken to be of the form (8.1.2). The treatment of more complicated (Bose) fields follows a similar pattern.

The new feature of our discussion is the explicit introduction of Planck's constant $\hbar$ as a varying parameter. We adhere to the idea that the action $W(\phi)$ is a purely "classical" expression in the sense that it does not involve $\hbar$. This essentially means that each (Bose) quantum field theory has a classical counterpart which quantization chooses as its basis and the starting point. We also think of the configuration space, on which the field ϕ is allowed to vary, as some classical construct, where $\hbar$ plays no role. Hence, it is only the measure μ where $\hbar$ enters:

$$d\mu(\phi) = Z^{-1}\mathcal{D}\phi \, \exp\{-\hbar^{-1}W(\phi)\}$$
$$Z = \int \mathcal{D}\phi \, \exp\{-\hbar^{-1}W(\phi)\}. \tag{8.3.18}$$

Needless to say, the measure is well defined for any value of $\hbar$ as long as it is positive and the lattice is finite.

---

[21] In thermodynamics we are likely to invoke the Gibbs variational principle. Here, it is the temperature T that multiplies the entropy, and the deterministic limit is reached at $T = 0$.

We now turn to the entropy definition which is owing to Boltzmann, Gibbs, and Shannon.

For μ a probability measure such that $d\mu(\phi) = \mathcal{D}\phi\, p(\phi)$ with $p(\phi) \geq 0$, the entropy is

$$S(\mu) = -\int \mathcal{D}\phi\, p(\phi) \log p(\phi). \tag{8.3.19}$$

Implicit in this definition is the convention that $r \log r = 0$ if $r = 0$. Notice that there is no maximal value of $S(\mu)$ as we vary μ. For, if there were a maximal value, it would correspond to $p(\phi) =$ const, which does not lead to a normalizable measure. So the range of the entropy is the entire real axis.

To compare two probability measures μ and ν, we define their relative entropy:

$$S(\mu|\nu) = -\int d\mu(\phi)\, \log g(\phi) \quad , \qquad d\mu(\phi) = d\nu(\phi)\, g(\phi). \tag{8.3.20}$$

For short, $S(\mu|\nu) = -\int d\mu \, \log(d\mu/d\nu)$. Assuming that $g(\phi) > 0$, we may state the following:

$S(\mu|\nu) \leq 0$ always, and $S(\mu|\nu) = 0$ iff $\mu \doteq \nu$ (equality everywhere except on sets of measure zero).

The proof, which naturally relies on the convexity of $u \log u$, follows the same lines as in the discrete case and is thus omitted.

Take, for instance, $d\mu(\phi) = \mathcal{D}\phi\, p(\phi)$ and ν a Gibbs measure:

$$d\nu(\phi) = Z^{-1} \mathcal{D}\phi \, \exp\{-\hbar^{-1} W(\phi)\}$$

$$Z = \int \mathcal{D}\phi \, \exp\{-\hbar^{-1} W(\phi)\}, \tag{8.3.21}$$

where it is assumed that $W(\phi)$ satisfies the stability condition of Sect. 8.1. It is not hard to see that

$$\int d\mu(\phi)\, W(\phi) - \hbar S(\mu) = -\hbar S(\mu|\nu) - \hbar \log Z. \tag{8.3.22}$$

This way we encounter a hitherto unknown path integral,

$$\langle \mu, W \rangle = \int d\mu(\phi)\, W(\phi) \tag{8.3.23}$$

which will be referred to as the *mean action*. The above statement about the relative entropy can be recast to read:

The Variational Principle. *With $W(\phi)$ some stable action,*

$$\inf_{\mu} \{ \langle \mu, W \rangle - \hbar S(\mu) \} = -\hbar \log Z. \tag{8.3.24}$$

The infimum with respect to all probability measures is reached only if $\mu \doteq \nu$, where ν and Z are given by (8.3.21).

It is characteristic of the Gibbs measure that it solves a variational problem.

8.3.4 The Classical Limit

When $\hbar \to 0$, quantum field theory, whether Minkowkian or Euclidean, will change its character: it approaches classical field theory. To see the details of the transition, we turn to the variational formulation (8.3.24), and almost instantly we discover *Hamilton's principle of least action*:

$$\langle \mu, W \rangle = minimum. \qquad (8.3.25)$$

Actually, it is *not* the familiar form of the principle we normally encounter in classical mechanics and, moreover, it is the *Euclidean* version of that principle. The trick to find the solution μ of (8.3.25) is to look for some ϕ_0, the "classical field" that minimizes $W(\phi)$. Suppose such an element in configuration space can be found and is unique. Then the Dirac measure μ_0 supported by ϕ_0 solves (8.3.25) and no other solution exists:

$$\langle \mu, W \rangle = W(\phi_0) = \min_{\phi \in \Omega} W(\phi). \qquad (8.3.26)$$

Imagine now what a Dirac measure would do to the Schwinger functions:

$$\lim_{\hbar \to 0} E\big(\Phi(x_1) \ldots \Phi(x_n)\big) = \phi_0(x_1) \ldots \phi_0(x_n). \qquad (8.3.27)$$

We discover that the classical limit is trivial in the sense that correlations disappear (i.e., n-point functions factorize). Also, we need only know the classical field $\phi_0(x)$ furnished by Hamilton's principle of least action. The classical field is among the solutions of the Euler–Lagrange equations, or field equations, associated with the action[22].

Normally, a complete description of classical field theory does not involve probabilistic terms. It may happen though that there is more than one solution to the problem $W(\phi) = minimum$. This indicates a degeneracy and thereupon the measure μ solving (8.3.26) cannot be unique; it may not even be true that the measure appears as a finite combination of Dirac measures. Then the discussion of the classical limit is more delicate since the probabilistic character stays with the system, though it is expected to disappear when $\hbar \to 0$ in most cases.

Example. Suppose that the action is given by

$$W(\phi) = \sum_x \left\{ \tfrac{1}{2} \sum_i \big(\partial_i \phi(x)\big)^2 + \lambda \big(\phi(x) - c\big)^4 \right\} \qquad (\lambda > 0,\ \phi \text{ real}).$$

$$(8.3.28)$$

---

[22]Classical Euclidean field theory is in many ways different from classical Minkowskian field theory. First of all, Euclidean fields satisfy elliptic, Minkowskian fields hyperbolic, differential equations. Second, Euclidean fields minimize the underlying action, whereas Minkowskian fields make the action only stationary.

Then its minimum is zero, and the constant field, $\phi(x) = c$, is the unique solution of the problem $W(\phi) = 0$: it represents the classical field. The Dirac measure supported by this solution is said to be the equilibrium state, or ground state:

$$d\mu(\phi) = \prod_x d\phi(x)\,\delta\big(\phi(x) - c\big). \tag{8.3.29}$$

Consider next the model

$$W(\phi) = \sum_x \left\{ \tfrac{1}{2} \sum_i \big(\partial_i \phi(x)\big)^2 + \lambda\big(\phi(x)^2 - c^2\big)^2 \right\} \qquad (\lambda > 0,\ \phi\ \text{real}). \tag{8.3.30}$$

In this case there exist two classical solutions, $\phi(x) = \pm c$, which break the $\mathbb{Z}_2$-gauge symmetry of the model: $W(\phi) = W(-\phi)$. Finally,

$$W(\phi) = \sum_x \left\{ \tfrac{1}{2} \sum_i |\partial_i \phi(x)|^2 + \lambda\big(|\phi(x)|^2 - c^2\big)^2 \right\} \qquad (\lambda > 0,\ \phi\ \text{complex}) \tag{8.3.31}$$

admits a continuum of classical solutions, $\phi(x) = e^{i\alpha}c$, which break the $U(1)$-gauge symmetry: $W(\phi) = W(e^{i\alpha}\phi)$ $(0 \le \alpha < 2\pi)$.

8.3.5 Fluctuations Around the Classical Solution

How can quantum fields in any way come close to classical fields? As we have added $\hbar$ to our set of control parameters, we may now start from the classical situation, thus reversing the limiting procedure $\hbar \to 0$. As the parameter $\hbar$ is turned on, the (Euclidean) quantum field starts to fluctuate owing to the presence of the entropy term in the variational principle. According to this view, it is entropy that makes $\phi(x)$ a random variable.

It has been demonstrated that there may be an abundance of classical equilibrium states, or ground states. By contrast, the quantum equilibrium state, or Gibbs measure (and hence the vacuum) is always unique, provided we work with a finite lattice and $\hbar > 0$. It should be pointed out, however, that uniqueness may nevertheless be lost in the thermodynamic limit $(N \to \infty)$. We expect this to happen if the infinite system undergoes a phase transition. Such a transition occurs if some control parameter like $\hbar$ is being varied. So it appears that, for an infinite lattice, the feature of nonuniqueness (or degeneracy) is no longer a privilege of the classical limit $(\hbar = 0)$, but extends into the quantum domain $\hbar > 0$. At some critical value $\hbar_c$ of Planck's constant, the degeneracy presumably disappears leaving a unique equilibrium state for $\hbar > \hbar_c$. Experience tells us that, as a rule, ground state degeneracy is accompanied by the spontaneous breakdown of some internal symmetry.

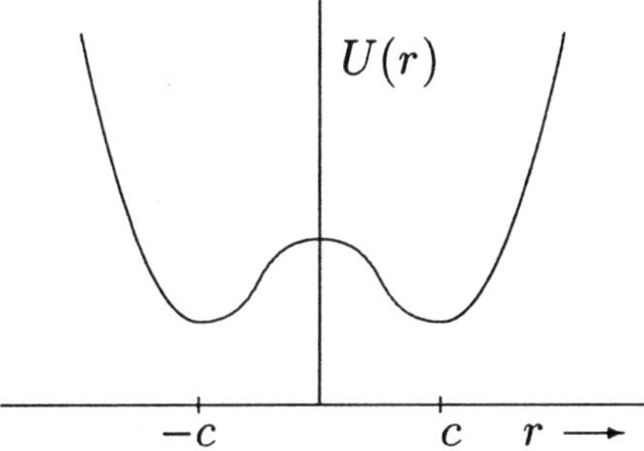

Fig. 8.2. A double-well potential $U(r)$ with two minima giving rise to a spontaneous breaking of the $\mathbb{Z}_2$-gauge symmetry.

To name but one example: the potential $U(r) = \lambda(r^2 - c^2)^2$ of (8.3.30) possesses two minima (see Fig. 8.2). For sufficiently small $\hbar$, there ought to be two equilibrium states (and hence two vacuum states) for the system. The quantum field thus fluctuates around one of the minima. To describe such fluctuations it may not be too big a mistake if we were to choose a parabola approximating $U(r)$ in the vicinity of the minimum. Such an approximation inevitably leads to a free field of mass $m = |c|\sqrt{8\lambda}$. The free field will lose its significance when $\hbar$ increases: quantum fluctuations tend to suppress symmetry breaking.

To repeat, the variational principle of Euclidean field theory parallels the Gibbs principle of statistical mechnics. Let us therefore briefly compare the two. The latter says: a many-particle system is in thermal equilibrium if $\omega(H) - \beta^{-1}S(\omega)$ assumes its minimal value, where $\beta^{-1} = k_BT$. This way we get $F = U - TS$, the free energy. The minimum is searched for by varying the state ω with finite entropy $S(\omega)$, subject to the condition that it be a legitimate state of the system. We let $\omega(H)$ denote the expectation value of the Hamiltonian, either classical or quantum mechanical. With ω_0 the minimizing state, the expectation $U \equiv \omega_0(H)$ is said to be the *internal energy* of the system and $S = k_BS(\omega_0)$ its entropy.

Against this background, comparison seems easy: quantum field theory simply replaces $\omega(H)$ by $\langle \mu, W \rangle$, the mean action, and β^{-1} by $\hbar$, the Planck constant. Free energy and internal energy do exist for fields but carry no particular names. Adhering to the thermodynamic terminology, we may for instance talk about the *free energy of a quantum field*:

$$F = \inf_{\mu}[\langle \mu, W \rangle - \hbar S(\mu)] = -\hbar \log Z. \tag{8.3.32}$$

As is well known from statistical mechanics, this defines F as an extensive state variable since it grows like N^4. If we let N tend to infinity, the limit $f = \lim N^{-4}F$ stays well-defined and f is said to be the free energy per site.

Example. Consider the free neutral scalar field of mass m on a lattice of size N^4. By (8.2.8), the free energy is

$$F = \frac{\hbar}{2} \sum_p \log \frac{E_p + m^2}{2\pi\hbar}$$

$$E_p = \sum_{k=1}^{4} 2(1 - \cos p_k), \qquad p \in \left(\tfrac{2\pi}{N} \mathbb{Z}_N\right)^4. \tag{8.3.33}$$

Recall that we started out by saying that each component p_k of the momentum belongs to the intervall $[0, 2\pi]$. Taking advantage of the periodicity of the trigonometric functions involved, we may also take the interval $[-\pi, \pi]$, which proves more convenient. As N increases, the Brillouin zone $B_4 = [-\pi, \pi]^4$ becomes densely filled by the admissible momenta. In the limit, $\left(\tfrac{2\pi}{N}\right)^4 \sum_p \cdots$ tends to $\int dp \cdots$ so that

$$f = \lim_{N \to \infty} N^{-4} F = \frac{\hbar/2}{|B_4|} \int_{B_4} dp \, \log \frac{E_p + m^2}{2\pi\hbar}. \tag{8.3.34}$$

The free energy per site, f, turns out to be a well-defined integral over the Brillouin zone depending on $\hbar$ and m.

Exercise 1.   Calculate the "internal energy" U and the entropy S of a free field such that $F = U - \hbar S$. Show that U and S are extensive variables and find out what happens in the thermodynamic limit, i.e., determine

$$u = \lim_{N \to \infty} N^{-4} U, \qquad s = \lim_{N \to \infty} N^{-4} S.$$

8.4 The Effective Action

Any probability measure μ on the configuration space of a field $\boldsymbol{\Phi}(x)$ gives rise to an expectation (or mean) value

$$\phi'(x) = \boldsymbol{E}\big(\boldsymbol{\Phi}(x)\big) = \langle \mu, \boldsymbol{\Phi}(x) \rangle = \int d\mu(\phi) \, \phi(x) \tag{8.4.1}$$

for which we also write $\phi' = \langle \mu, \boldsymbol{\Phi} \rangle$, and any function ϕ' obtained this way is again an element of the configuration space. Conversely, with ϕ' an arbitrary function, there is always some probability measure μ such that (8.4.1) holds (take a Dirac measure for instance).

Stretching our physical intuition and terminology a bit, we may think of μ as describing a *quantum state* and of $\langle \mu, \boldsymbol{\Phi} \rangle = \phi'$ as being the *classical field* connected with that state[23]. This way we get a more intuitive picture of the role of the measure μ. Suppose the quantum field $\boldsymbol{\Phi}(x)$ satisfies a linear field equation. Then so does the classical field ϕ'. This argument in particular applies to free fields. Interacting fields are different: there is no indication whatsoever that quantum fields obey any kind of field equations (linear or nonlinear). This makes expectations like $\langle \mu, \boldsymbol{\Phi} \rangle$ look suspect and leaves many open questions concerning their interpretation as *classical fields*. Never mind. Let us pose the following problem:

If μ is the Gibbs measure obtained from the action W and $\phi = \langle \mu, \boldsymbol{\Phi} \rangle$, are there field equations for ϕ?

This problem is clearly ill posed as it stands. But what we really have in mind are two things. First, there ought to exist an *effective action* such that ϕ obeys Hamilton's principle of least action, $W_{\mathrm{eff}}(\phi) = minimum$. Hence, the field equations eluded to are in fact the Euler–Lagrange equations derived from the effective action. Second, the construction ought to be universal: W_{eff}, though depending on W, makes no reference to the state μ.

It is easy to see that there exists a solution to the problem. For, if we start from the variational principle $\langle \mu, W \rangle - \hbar S(\mu) = minimum$, we propose to construct the solution in a two-step process:

1. For arbitrary but fixed ϕ, find the constrained infimum

$$W_{\mathrm{eff}}(\phi) = \inf_{\mu} \big\{ \langle \mu, W \rangle - \hbar S(\mu) \mid \langle \mu, \boldsymbol{\Phi} \rangle = \phi \big\}. \tag{8.4.2}$$

   This step defines the effective action.
2. Next, vary ϕ and search for the solution of $W_{\mathrm{eff}}(\phi) = minimum$. This step suspends the condition of constraint, $\langle \mu, \boldsymbol{\Phi} \rangle = \phi$, on the right-hand side of (8.4.2) and takes us back to the problem $\langle \mu, W \rangle - \hbar S(\mu) = minimum$.

In passing, we note that the infimum of $W_{\mathrm{eff}}(\phi)$ is nothing but the free energy F. To keep track of what is happening here, we point out that the measure μ, the solution to the problem (8.4.2), depends on ϕ in some complicated unknown fashion. It coincides with the Gibbs measure obtained from the action W as soon as ϕ minimizes the effective action.

Any constrained variational problem is conveniently solved by the Lagrange multiplier method. As for our case, we have to introduce a (real) multiplier $j(x)$ *at each site x of the lattice* and then study the related problem

---

[23]Statements of this type have the flavor of a *correspondence principle*. The idea works well in other contexts like electromagnetic radiation, where the quantum state ω describes photons, while the expectation $f_{\mu\nu}(x) = \omega(F_{\mu\nu}(x))$ is a classical solution of Maxwell's equations. Of course, such a solution $f_{\mu\nu}$ by no means determines ω. It should also be noted that, taken as a general idea, the correspondence principle would never work in the case of Fermi fields.

$$\langle \mu, W \rangle - \hbar S(\mu) - \sum_x j(x)\langle \mu, \boldsymbol{\Phi}(x)\rangle = minimum \tag{8.4.3}$$

(j fixed, μ varying). Actually, the new problem has the same form as in (8.3.24), except that $W(\phi)$ appears now to be replaced by $W(\phi) - \sum_x j(x)\phi(x)$. This observation allows us to immediately write down the solution as a Gibbs measure:

$$d\mu_j(\phi) = \mathcal{D}\phi\, Z\{j\}^{-1} \exp \hbar^{-1}\left[\sum_x j(x)\phi(x) - W(\phi)\right]$$

$$Z\{j\} = \int \mathcal{D}\phi \exp \hbar^{-1}\left[\sum_x j(x)\phi(x) - W(\phi)\right] \tag{8.4.4}$$

$$minimum = -\hbar \log Z\{j\}.$$

The new feature of the measure μ_j is that it yields a family of field theories indexed by j, with n-point functions

$$\boldsymbol{E}_j(\boldsymbol{\Phi}(x_1)\cdots\boldsymbol{\Phi}(x_n)) = \int d\mu_j(\phi)\,\phi(x_1)\cdots\phi(x_n). \tag{8.4.5}$$

This is not entirely new to the practitioner in the field who has worked with formulas such as (8.4.4) off and on, calling $j(x)$ an "external source" and $Z\{j\}$ a "generating functional" with applications for varied purposes. He or she also views j and ϕ as *conjugate variables*. In statistical physics, it would be more appropriate to think of $j(x)$ as some inhomogeneous external (magnetic, electric etc.) field. There the key issue is whether the system responds *linearly* to some external force which is of no concern here.

As for us, $j(x)$ is some auxiliary variable. It is ultimately eliminated by solving $\boldsymbol{E}_j(\boldsymbol{\Phi}(x)) = \phi(x)$ for j. Before doing so, let us consider the functional

$$W_{\text{eff}}^*(j) = \hbar \log Z\{j\}\,, \tag{8.4.6}$$

which serves to generate the cumulants of the field. In particular,

$$\frac{\partial W_{\text{eff}}^*(j)}{\partial j(x)} = Z\{j\}^{-1}\hbar\frac{\partial Z\{j\}}{\partial j(x)} = \boldsymbol{E}_j(\boldsymbol{\Phi}(x))\,, \tag{8.4.7}$$

and, hence, the equation of constraint, $\boldsymbol{E}_j(\boldsymbol{\Phi}(x)) = \phi(x)$, is equivalent to

$$\frac{\partial W_{\text{eff}}^*(j)}{\partial j(x)} = \phi(x), \tag{8.4.8}$$

which is seen to solve the variational problem

$$\sum_x j(x)\phi(x) - W_{\text{eff}}^*(j) = maximum \tag{8.4.9}$$

(ϕ fixed, j varying). The reason: first, condition (8.4.8) leads to some extremum of the the left-hand side of (8.4.9), and second, the matrix K formed by the second derivatives of $W_{\text{eff}}^*(j)$ is positive:

$$\hbar K_{xx'} = \hbar^2 \frac{\partial^2 \log Z\{j\}}{\partial j(x)\partial j(x')} = E_j(\boldsymbol{\Phi}(x)\boldsymbol{\Phi}(x')) - E_j(\boldsymbol{\Phi}(x))E_j(\boldsymbol{\Phi}(x')). \tag{8.4.10}$$

The essence of the argument is that $\hbar K$ coincides with the covariance of the field $\boldsymbol{\Phi}$ with respect to the measure μ_j. Indeed, this shows that $W_{\text{eff}}^*(j)$ is a *convex* functional. We now come to state our main proposition:

Let $W(\phi)$ be the Euclidean action, $W_{\text{eff}}(\phi)$ the corresponding effective action, and $W_{\text{eff}}^*(j)$ the generating functional of the cumulants. Then W_{eff} and W_{eff}^* are convex functionals and related to each other by Legendre transformations:

$$W_{\text{eff}}(\phi) = \sup_j \left(\sum_x j(x)\phi(x) - W_{\text{eff}}^*(j)\right) \tag{8.4.11}$$

$$W_{\text{eff}}^*(j) = \sup_\phi \left(\sum_x j(x)\phi(x) - W_{\text{eff}}(\phi)\right). \tag{8.4.12}$$

To prove this, we note that, for arbitrary ϕ and j,

$$\begin{aligned}
\sum_x j(x)\phi(x) - W_{\text{eff}}^*(j) &= \sum_x j(x)\phi(x) - \hbar \log Z\{j\} \\
&= \inf_\mu \left[\langle \mu, W\rangle - \hbar S(\mu) + \sum_x j(x)\{\phi(x) - \langle\mu, \boldsymbol{\Phi}(x)\rangle\}\right] \\
&\leq \inf_\mu \left[\langle\mu, W\rangle - \hbar S(\mu) \mid \langle\mu, \boldsymbol{\Phi}\rangle = \phi\right] \\
&= W_{\text{eff}}(\phi). \tag{8.4.13}
\end{aligned}$$

The inequality arises because the absolute infimum never lies above the constrained infimum. However, the inequality turns into an equality provided j has been chosen such that $E_j(\boldsymbol{\Phi}(x)) = \phi(x)$. Then

$$W_{\text{eff}}(\phi) = \langle\mu_j, W\rangle - \hbar S(\mu_j) \tag{8.4.14}$$

and therefore (8.4.11) holds. The validity of (8.4.12) follows in a similar fashion, and the convexity of W_{eff}^* has already been shown. It remains to prove the convexity of W_{eff}. Let $\phi = \alpha\phi_1 + (1-\alpha)\phi_2$, where $0 < \alpha < 1$. For j arbitrary,

$$W_{\text{eff}}(\phi_i) \geq \sum_x j(x)\phi_i(x) - W_{\text{eff}}^*(j) \qquad (i = 1, 2), \tag{8.4.15}$$

and thus

$$\alpha W_{\text{eff}}(\phi_1) + (1-\alpha)W_{\text{eff}}(\phi_2) \geq \sum_x j(x)\phi(x) - W_{\text{eff}}^*(j). \tag{8.4.16}$$

Taking the supremum with respect to j, we get

$$\alpha W_{\text{eff}}(\phi_1) + (1-\alpha)W_{\text{eff}}(\phi_2), \geq W_{\text{eff}}(\phi) \tag{8.4.17}$$

which proves convexity.

What has been discovered here is that field theory has three equally important levels of description, where each level is represented by its own "action". The transition from one level of description to another is by no means easy but is nevertheless well defined:

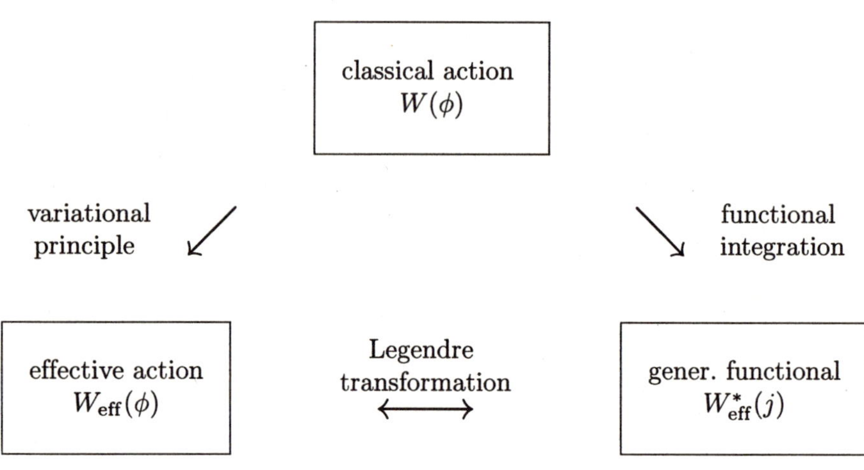

Among other things the diagram suggests that functional integration (according to a widespread opinion, the most difficult computational technique ever devised by physicists) can be evaded by another two-step instruction: first, solve a variational problem, and second, apply a Legendre transformation.

Example. The free field may again serve as an illustration though it brings us no enlightenment on its structure. It is the only case we know of where one gets W_{eff} and W_{eff}^* in closed form. The action W defined by

$$W(\phi) = \tfrac{1}{2}(\phi, (-\Delta + m^2)\phi) := \tfrac{1}{2}\sum_x \left[\sum_i \{\partial_i\phi(x)\}^2 + m^2\phi(x)^2\right] \quad (8.4.18)$$

poses no problem and yields, after some manipulations,

$$\begin{aligned}
W_{\mathrm{eff}}^*(j) &= \tfrac{1}{2}(j, (-\Delta + m^2)^{-1}j) - F \\
W_{\mathrm{eff}}(\phi) &= \tfrac{1}{2}(\phi, (-\Delta + m^2)\phi) + F,
\end{aligned} \quad (8.4.19)$$

where the constant F is the free energy of the free field; it may be ignored. The simple relation $W_{\mathrm{eff}}(\phi) = W(\phi) + F$ is characteristic of the Gaussian model.

Exercise 1. For f some extended (allowing for $+\infty$ as a possible value) real function on the the real line, prove that the Legendre transform

$$f^*(s) = \sup_r \{rs - f(r)\}$$

is a convex extended real function. Moreover, the double transform $f^{**}(r)$ coincides with the so-called *convex closure* of $f(r)$: this is the largest convex function smaller than f. Hence, $f = f^{**}$ if the function f is convex.

Exercise 2. In the classical limit, the effective action is expected to approach the classical action. This is "almost" the case and, after some minor modification, can indeed be turned into a theorem. Show that

$$\lim_{\hbar \to 0} W_{\text{eff}}(\phi) = W^{**}(\phi),$$

where W^{**} is the convex closure of the classical action. Hint: prove first that

$$\lim_{\hbar \to 0} W^*_{\text{eff}}(j) = W^*(j),$$

where $W^*(j)$ is the Legendre-transformed classical action.

8.5 The Effective Potential

It is conceivable that most students, upon reflecting on the content of the preceding section, will ask the same question: are there reasons to believe, or can one even prove, that the effective action assumes the form

$$W_{\text{eff}}(\phi) = \sum_x \left\{ \tfrac{1}{2} \sum_i \left(\partial_i \phi(x)\right)^2 + U_{\text{eff}}\left(\phi(x)\right) \right\} \tag{8.5.1}$$

with $U_{\text{eff}}(r)$ some unknown potential? Actually this is too much asking. A moment's reflection will tell us that it would be wise to change (8.5.1) to

$$W_{\text{eff}}(\phi) = \sum_x \left\{ U_{\text{eff}}\left(\phi(x)\right) + \tfrac{1}{2} Z\left(\phi(x)\right) \sum_i \left(\partial_i \phi(x)\right)^2 + \cdots \right\}, \tag{8.5.2}$$

where $U_{\text{eff}}(r)$ is bounded below, $Z(r) > 0$, and the dots "$\cdots$" indicate further terms containing higher powers of the gradient and "higher-order derivatives" of $\phi(x)$ (whatever that means on a lattice). With this modification, the ansatz (8.5.2) looks more promising. However, we have to admit that the implied expansion of the effective action is not very systematic. Nor are we able to offer any proof for it.

On closer inspection, we find a way to perform at least the *first* step of the expansion (8.5.2):

The effective potential U_{eff} is defined to be the effective action per site for constant $\phi(x)$:

$$U_{\text{eff}}(r) = N^{-4} W_{\text{eff}}(\phi), \qquad \phi(x) = r. \tag{8.5.3}$$

The reason this definition works is that, by definition, all terms in (8.5.2) other than the first one are assumed to contain derivatives of the field which vanish if $phi(x)$ is constant.

8.5.1 Spontaneous Breakdown of Symmetry

Let us take for granted that expansion (8.5.2) makes sense and that the higher terms represented by the dots can be ignored in a first approximation. As remarked earlier, to find the average field $\phi = \boldsymbol{E}(\boldsymbol{\Phi})$, we must solve $W_{\text{eff}}(\phi) = minimum$ for ϕ. This can only be achieved if two things happen:

1. $\partial_i \phi(x) = 0$, i.e., $\phi(x) = c = const$,
2. $U_{\text{eff}}(r)$ is minimal for $r = c$.

Since the solution comes out translationally invariant, there can be no spontaneous breakdown of the translational symmetry (on a finite lattice). The problem of finding the minimum of the effective action is thus reduced to the problem of finding the minimum of the effective potential.

An interesting situation arises if the theory is symmetric with respect to a global $\mathbb{Z}_2$-gauge transformation: $\phi \rightarrow -\phi$. Then W, U, W_{eff}, U_{eff}, and many other quantities will reflect this symmetry. Suppose now that U is some double-well potential, so symmetry breaking occurs at the classical level ($\hbar = 0$). Since[24] $\lim_{\hbar \to 0} U_{\text{eff}}(r) = U^{**}(r)$, it seems feasible at first sight that symmetry breaking continues to be present provided $\hbar$ is sufficiently small. Looking closer, we realize that, at least on a finite lattice, this cannot be true for two reasons:

1. The effective potential is a convex function, the reason being that it is the restriction of a convex functional (the effective action) to a constant field.

2. For a finite lattice and as long as $\hbar > 0$, maps like

$$z \mapsto W_{\text{eff}}^*(zj) \qquad \text{and} \qquad z \mapsto W_{\text{eff}}(z\phi)$$

   turn out to be (entire) analytic functions of $z \in \mathbb{C}$. No particular property of the classical potential is capable of ruining analyticity. By (8.5.3), $U_{\text{eff}}(r)$ is analytic in r.

The three conditions, convexity, analyticity, and symmetry, imply that the minimum of $U_{\text{eff}}(r)$ occurs at $r = 0$ and is unique (unless the potential is constant; this case is ignored). Conclusion:

For a finite lattice and $\hbar > 0$, there is no phase transition, nor a spontaneous breakdown of gauge symmetries, nor an anomalous average $\boldsymbol{E}(\boldsymbol{\Phi}(x)) \neq 0$.

---

[24]See Exercise 2 from the foregoing section. $U^{**}(r)$ is the convex closure of the classical potential. If $U(r)$ looks like in Fig. 8.2, the graph of $U^{**}(r)$ will come out as in Fig. 8.3.

The argument fails for an infinitely extended lattice: analyticity is lost if $N \to \infty$. This loss brings about the chance to observe symmetry breaking. There is only one way, specific to the present situation, in which the effective potential on an infinite lattice allows for more than one equilibrium point r. The situation is depicted in Fig. 8.3.

Briefly stated, symmetry breaking always rests on the property that the effective potential remains constant within some symmetric interval $[-c, c]$ since this is the only way in which a convex symmetric function acquires a nonunique minimum and the field an anomalous average. Such a function is never analytic (unless it is constant everywhere). To the endpoints of the constant interval there correspond equilibrium states, i.e., measures μ_- and μ_+ such that $\langle \mu_\pm, \boldsymbol{\Phi}(x) \rangle = \pm c$. Besides, any value between $-c$ and c may arise as an anomalous average. For, if $0 < \alpha < 1$ and $\mu = \alpha\mu_- + (1 - \alpha)\mu_+$, we have $\langle \mu, \boldsymbol{\Phi}(x) \rangle = \alpha(-c) + (1 - \alpha)c$. Among those equilibrium states there is one for which the average is in fact zero. Simply take $\alpha = \frac{1}{2}$.

The general lesson to be drawn is that equilibrium states form some convex set: if μ_1 and μ_2 are two such states, the convex combination $\mu = \alpha\mu_1 + (1 - \alpha)\mu_2$, where $0 < \alpha < 1$, is again an equilibrium state. A state is said to be *extremal* within a convex set if it cannot be written as a nontrivial convex combination of other states from that set. The above discussion has shown that extremal equilibria have extremal anomalous averages ($\pm c$) and every equilibrium state can be uniquely decomposed into extremal states (by the way, a convex set with this property is called a *simplex*).

8.5.2 Order Parameters

As $\hbar$ increases, quantum fluctuations become large, thus inhibiting the occurrence of nonsymmetric equilibrium states, i.e., symmetry is restored as we pass some critical value of $\hbar$. The terms "phase" and "phase transition" used in thermodynamics refer to the different behavior on either side of the critical point. Below the critical point, when symmetry is destroyed, a new parameter is created, called the "order parameter", which thermodynamic potentials depend upon and which is needed for a complete characterization of the (macroscopic) state. To conform with the standard terminology of statistical mechanics, we would call c, r, and more generally, $\phi(x) = \boldsymbol{E}(\boldsymbol{\Phi}(x))$ *order parameters* of the field theory and, moreover, call functionals like W_{eff} and U_{eff} *thermodynamic potentials*. It also conforms to the general picture that the equilibrium value of the order parameter is found by solving $W_{\text{eff}}(\phi) = minimum$ (though it may not render the solution unique).

The most salient example of an order parameter is provided by the magnetization (a vector) of a ferromagnet in three dimensions when the energy is assumed to be fully O(3)-symmetric. Above the Curie point, where there is no order parameter, the equilibrium state is indeed rotationally invariant. Below, rotational symmetry is destroyed. To stabilize the anomalous value of the order parameter (i.e., to create a nonvanishing magnetization), one

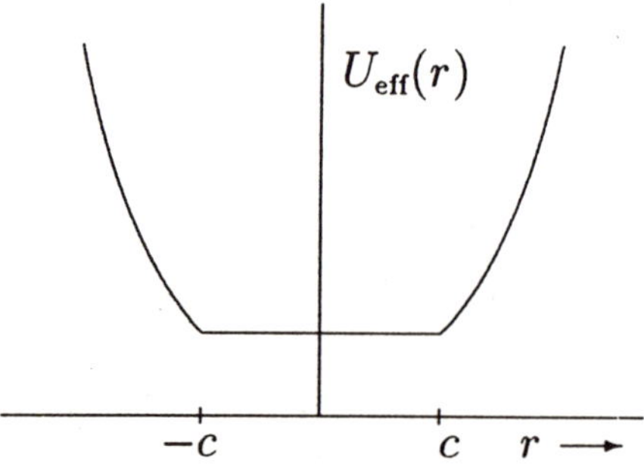

Fig. 8.3. An effective potential on an infinite lattice which, though symmetric, gives rise to many nonsymmetric equilibrium solutions

applies a magnetic field to the system. When the external field is slowly switched off, one observes that the magnetization stays nonzero. Naturally, it is the component m of the magnetization parallel to the applied field that survives, and m is seen to always fall into the same fixed interval $[-M, M]$ so that each value of that interval is accessible to the experimenter. As the Curie point is approached, M (and thus m) tends to zero in a continuous fashion to stay zero beyond that point, a behavior which is typical of a second-order phase transition.

8.6 The Ginzburg–Landau Equations

To put the subject of the effective action in a proper perspective, let us present a broad outline of the most prominent and time-honored effective theory, which has become known as the *Ginzburg–Landau approach* to superconductivity. That work may be said to mark the beginning of the *effective theory* as a serious discipline. Though the underlying microscopic structure is that of an interacting electron gas at low temperatures, one has chosen to deal only with macroscopic variables as is characteristic of any thermodynamic theory. The heart of the matter is the response of a superconductor to an external magnetic field and the theoretical ansatz to explain it. By now the *Landau approach* has acquired a status where it may be applied to any second-order phase transition as an effective theory valid in the vicinity of the critical point.

As for the superconductor in an external field, experience has shown that such a system does not admit an equilibrium state with "normal" properties, i.e., where an internal homogeneous magnetic field could coexist with a superconducting phase. Specifically, the response is always one of the following three possibilities [8.4]:

1. The magnetic field is destroyed (Meissner–Ochsenfeld effect).

2. The superconducting state is destroyed.

3. An *inhomogeneous* superconducting state is created.

The third option defines the type II superconductor. Assuming that there exists a complex order parameter $\phi(\boldsymbol{x})$ (a deterministic variable with no fluctuations) that characterizes the superconducting equilibrium state, Ginzburg and Landau [8.5] wrote down a set of equations for $\phi(\boldsymbol{x})$ and the electromagnetic vector potential $\boldsymbol{A}(\boldsymbol{x})$ with $\boldsymbol{x} \in E_3$ and no dependence on time:

$$-\frac{1}{4m}(\nabla + 2ie\boldsymbol{A})^2\phi = \Lambda\phi - 2\lambda|\phi|^2\phi$$

$$\nabla \times (\nabla \times \boldsymbol{A}) = \frac{ie}{2m}(\bar{\phi}\nabla\phi - \phi\nabla\bar{\phi}) - \frac{2e^2}{m}|\phi|^2\boldsymbol{A}.$$

$$(8.6.1)$$

In fact, these are the Euler-Lagrange equations for the variational problem

$$W_{\text{eff}}(\phi, \boldsymbol{A}) = minimum \tag{8.6.2}$$

where W_{eff}, the "Landau free energy", is a functional of remarkable structure:

$$W_{\text{eff}} = \int dx \left\{ \tfrac{1}{2}(\nabla \times \mathbf{A})^2 + \tfrac{1}{4m}|(\nabla + 2ie\mathbf{A})\phi|^2 - \Lambda|\phi|^2 + \lambda|\phi|^4 \right\} \tag{8.6.3}$$

$(dx = dx_1 dx_2 dx_3)$. Notice that the functional is bounded below. So the theory is stable. Let us spell out in more detail how things should be interpreted:

- The Ginzburg–Landau functional provides a macroscopic description of Cooper pairs. These are weakly bound states of two electrons with opposite helicity and momenta in the vicinity of the Fermi surface. In fact, Cooper pairs act like bosons of mass $2m$ and charge $-2e$. Hence, $\phi = 0$ means that there are no Cooper pairs at the macroscopic level and the superconducting phase is absent.

- Apart from some irrelevant constant, W_{eff} is the free energy of the thermodynamic system when the temperature T is close enough to its critical value T_c. W_{eff} is obtained as the solution of a constrained Gibbs variational principle. All details of the quantum mechanical many-particle Hamiltonian are hidden in two fundamental parameters, λ and Λ. These depend analytically on T and, more importantly, Λ changes sign at T_c, i.e., $(T_c - T)\Lambda(T) \geq 0$ for all T.

- The complex character of the order parameter indicates that ϕ cannot be obtained through classical considerations. Nor can ϕ be observed in any experiment[25]. To make contact with the underlying microscopic theory, one takes the constraints to be $\phi(\boldsymbol{x}) = \langle \psi_\uparrow(\boldsymbol{x})\psi_\downarrow(\boldsymbol{x}) \rangle$

---

[25]It is the *phase* of the order parameter that remains unobservable, while $|\phi|^2$ is related to the density of Cooper pairs. However, the Josephson effect provides an interesting possibility to measure the *difference of phases* associated with two superconductors separated by a thin dielectric.

and $\boldsymbol{A}(\boldsymbol{x}) = \langle\boldsymbol{A}_{\mathrm{op}}(\boldsymbol{x})\rangle$, where $\langle\cdot\rangle$ means expectation with respect to a microstate. The nonrelativistic electron field is $\psi = \psi_\downarrow + \psi_\uparrow$ (the arrows refer to the helicity states) and the photon field is $\boldsymbol{A}_{\mathrm{op}}$.

- Global $U(1)$-gauge transformations of the Fermi field, $\psi(\boldsymbol{x}) \to e^{i\alpha}\psi(\boldsymbol{x})$, give rise to gauge transformations of the order parameter, and gauge symmetry of the underlying Hamiltonian reflects itself in the gauge symmetry of the Ginzburg-Landau functional. Thus, $\phi \neq 0$ entails a spontaneous breakdown of the $U(1)$ symmetry.
- The Ginzburg–Landau functional involves the potential function

$$f(r) = -\Lambda r^2 + \lambda r^4. \tag{8.6.4}$$

This fails in one important aspect: $f(r)$ is not a convex function for any T below T_c (since $\Lambda > 0$). However, the defect can be cured by taking the convex closure, i.e., by the replacement $f \to f^{**}$, where

$$f^{**}(r) = \begin{cases} -\Lambda r^2 + \lambda r^4 & \text{if } |r| > a \\ -\lambda a^4 & \text{if } |r| \leq a, \end{cases} \qquad a = \sqrt{\frac{\Lambda}{2\lambda}}. \tag{8.6.5}$$

Such a replacement leaves the physics unchanged and, moreover, f and f^{**} coincide if $T > T_c$ ($\Lambda < 0$).

For $T < T_c$, there is a critical B field above which normal conductance is observed. This critical field is found approximately if the cubic term in (8.6.1) is dropped and if $\nabla \times \boldsymbol{A}$ is assumed to be constant:

$$-\frac{1}{4m}(\nabla + ie\boldsymbol{B}\times\boldsymbol{x})^2\phi(\boldsymbol{x}) = \Lambda\phi(\boldsymbol{x}). \tag{8.6.6}$$

If we require the solution $\phi(\boldsymbol{x})$ to be bounded in the Euclidean three-space, we are formally dealing with the Schrödinger equation for a particle in constant field. This then leads to the well-known Landau levels

$$\Lambda = \frac{eB}{m}(n + \tfrac{1}{2}) + \frac{p^2}{4m} \qquad (B = |\boldsymbol{B}|, n = 0, 1, 2, \ldots), \tag{8.6.7}$$

where p is the projection of the momentum onto the axis of the magnetic field; this part of the momentum is not quantized. From (8.6.7) we infer the inequality

$$\Lambda(T) \geq \frac{eB}{2m}, \tag{8.6.8}$$

which determines the domain of the (T, B)-plane, where a superconducting phase coexists with a magnetic field. A closer look at the equations reveals that such a state is always inhomogeneous. To summarize, $\phi = 0$ is the only solution of (8.6.6) if the magnetic fields exceeds $B_c = 2(m/e)\Lambda(T)$, given the temperature T (below T_c).

For Λ the bottom of the spectrum and the magnetic field oriented along the x_3-axis, the solution of (8.6.6) becomes

$$\phi_0(\boldsymbol{x}) = (x_1 - ix_2)C \exp\{-\tfrac{1}{2}eB(x_1^2 + x_2^2)\}, \qquad C \in \mathbb{C}. \tag{8.6.9}$$

One aspect is that ϕ_0 is also an eigenfunction of the angular momentum about the x_3-axis, another that ϕ_0 is peaked at $x_1 = x_2 = 0$. The solution describes a *quantized vortex* which is essentially localized within a tube of diameter

$$d = \sqrt{\frac{\hbar c}{eB}}. \tag{8.6.10}$$

Outside the (thin) vortex tube the order parameter is negligibly small. As a result, superconductivity is confined to the vicinity of a line. There is a remarkable parallel between vortices in superconductors and vortices in a superfluids like liquid He^4.

In the presence of a constant magnetic field[26], translations in space are implemented in a very special way:

$$\phi(\boldsymbol{x}) \to \phi^{\boldsymbol{a}}(\boldsymbol{x}) \equiv \phi(\boldsymbol{x} - \boldsymbol{a}) \exp\{ie\boldsymbol{B}\cdot(\boldsymbol{a}\times\boldsymbol{x})\} \qquad (\boldsymbol{a} \in \mathbb{R}^3). \tag{8.6.11}$$

From this definition it follows that translations form a symmetry group for the problem (8.6.6). This group when applied to ϕ_0 provides an abundance of further solutions. Specifically, the solution $\phi_0^{\boldsymbol{a}}$ corresponds to another vortex whose center appears translated by the vector $\boldsymbol{a}$. Moreover, we may form superpositions to get even more general solutions. Consider for instance a lattice $\mathcal{G}$ in the (x_1, x_2)-plane, which is the plane orthogonal to the magnetic field. Then

$$\phi^{\mathcal{G}}(\boldsymbol{x}) = \sum_{\mathbf{a}\in\mathcal{G}} \phi_0(\boldsymbol{x} - \boldsymbol{a}) \exp\{ie(\boldsymbol{B}\cdot(\boldsymbol{a}\times\boldsymbol{x})\} \tag{8.6.12}$$

defines a solution of the linearized Ginzburg–Landau equation (8.6.6). Abrikosov [8.6] was able to demonstrate that, within the subset of functions given by (8.6.12), the functional W_{eff} is minimized if $\mathcal{G}$ is a triangular lattice. This amounts to a close-packing of equal spheres in two dimensions. Luckily, the theoretical prediction has been experimentally confirmed.

This concludes the brief account of results obtained from an effective theory concerning the superconductor. Though we have not really fathomed the implications of the Ginzburg–Landau equations, we have at least got an idea of the wealth of concrete physical results that usually follow from an effective-action approach. Frequently, one proceeds phenomenologically, i.e., the detailed structure of the effective action is not always justifiable by appeal to first principles, and parameters are adjusted to fit the data.

Field theory, as we have argued, is very much like four-dimensional Ginzburg–Landau theory, except that there is a severe lack of reliable candidates for effective actions W_{eff} in elementary particle physics (say for the

---

[26]Compare the discussion in Sect. 6.3, especially (6.3.25).

"standard theory" of electro-weak interactions when the condition of constraint is $\phi(x) = \boldsymbol{E}(\boldsymbol{q}(x)\bar{\boldsymbol{q}}(x))$ with $\boldsymbol{q}(x)$ the quark field).

Exercise 1.   Assume that the superconductor occupies the domain $\Omega \subset E_3$ of finite volume $|\Omega|$. Show that the Ginzburg–Landau functional admits a strict lower bound

$$W_{\text{eff}}(\phi, \boldsymbol{A}) \geq |\Omega| \, (\Lambda/\lambda)^2 \,.$$

With no external magnetic field, $\nabla \times \boldsymbol{A}$ vanishes at the boundary $\partial\Omega$, and the lower bound is reached for $\boldsymbol{A} = 0$ and $\phi(x) = e^{i\alpha}a$ (provided $T < T_c$), where a is given by (8.6.5) and α is arbitrary. This defines the *homogeneous superconducting phase*. Show also that ϕ vanishes like $(T_c - T)^{1/2}$ at T_c.

Exercise 2.   Show that the Ginzburg–Landau functional is invariant under gauge transformations:

$$\boldsymbol{A}(\boldsymbol{x}) \to \boldsymbol{A}(\boldsymbol{x}) + \nabla f(\boldsymbol{x}), \qquad \phi(\boldsymbol{x}) \to \phi(\boldsymbol{x}) \exp\big\{ -i2ef(\boldsymbol{x})\big\}.$$

If $\boldsymbol{A}(\boldsymbol{x}) = (2e(x_1^2 + x_2^2))^{-1}\{-x_2, x_1, 0\}$, there is a magnetic vortex line in the superconductor carrying precisely one flux quantum $\Phi_0 = \pi/e$. Prove that, because of the quantization, the vector potential $\boldsymbol{A}$ can be "gauged away".

8.7 The Mean-Field Approximation

Let us recall the prescription to construct the effective action in lattice field theory. For $W(\phi) = \sum_x \{\frac{1}{2} \sum_i (\partial_i \phi(x))^2 + U(\phi(x))\}$ the classical action, one considers the mean action $\langle \mu, W \rangle$ and the entropy $S(\mu)$ associated with any probability measure μ, the 'trial state', and obtains $W_{\text{eff}}(\phi)$ as the infimum of $\langle \mu, W \rangle - \hbar S(\mu)$ subject to the condition of constraint, $\langle \mu, \boldsymbol{\Phi} \rangle = \phi$. From the standpoint of practical calculations, it might be more convenient if, and one would expect only small changes in the effective action when, the infimum is searched for within a sufficiently large subset of measures. A highly successful method, the so-called *mean-field approximation*, uses product measures. These are measures of the type

$$d\mu(\phi) = \prod_x d\nu_x\big(\phi(x)\big), \tag{8.7.1}$$

where each lattice site x has been equipped with some probability measure ν_x on $\mathbb{R}$ (generally on $\mathbb{R}^n$ if ϕ has n components). Product measures ignore the geometry of the lattice to a large extent and fail to correctly delineate the correlations of the field in neighboring sites[27]. The product measure is said to be translationally invariant if $d\nu_x(s) = d\nu(s)$ for all x. Here and in what follows, the real variable s stands for $\phi(x)$ when x is fixed.

Let $\mathcal{P}$ denote the set of product measures (8.7.1). When trying to achieve equilibrium in $\mathcal{P}$, we generally get a larger effective action:

$$W_{\mathrm{MF}}(\phi) := \inf_{\mu \in \mathcal{P}} \left\{ \langle \mu, W \rangle - \hbar S(\mu) \mid \langle \mu, \boldsymbol{\Phi} \rangle = \phi \right\} \geq W_{\mathrm{eff}}(\phi). \tag{8.7.2}$$

We must not expect to be able to prove the convexity[28] of the functional $W_{\mathrm{MF}}(\phi)$. An immediate finding is that the entropy of a product measure turns out to be a sum of individual entropies $S(\nu_x)$ particular to the site x:

$$S(\mu) = \sum_x S(\nu_x) \equiv -\sum_x \int d\nu_x(s) \, \log(d\nu_x/ds). \tag{8.7.3}$$

A second finding is what may be called *decoupling of constraints*: the equation $\langle \mu, \boldsymbol{\Phi} \rangle = \phi$ becomes

$$\int d\nu_x(s)s = \phi(x). \tag{8.7.4}$$

We finally realize that the mean action also comes out as a sum over the lattice sites:

$$\langle \mu, W \rangle = \sum_x \left\{ \tfrac{1}{2} \sum_i \left(\partial_i \phi(x) \right)^2 + \int d\nu_x(s) u\big(s, \phi(x)\big) \right\} \tag{8.7.5}$$
$$u(s, r) = 4(s - r)^2 + U(s) \qquad (s, r \in \mathbb{R})$$

with $U(s)$ the potential that entered the classical action W. Let us summarize:

The mean-field method approximates the effective action from above by

$$W_{\mathrm{MF}}(\phi) = \sum_x \left\{ \tfrac{1}{2} \sum_i \left(\partial_i \phi(x) \right)^2 + U_{\mathrm{MF}}\big(\phi(x)\big) \right\}, \tag{8.7.6}$$

where the mean-field effective potential is

$$U_{\mathrm{MF}}(r) = \inf_{\nu} \left\{ \int d\nu(s) \, u(s, r) - \hbar S(\nu) \mid \int d\nu(s) \, s = r \right\}. \tag{8.7.7}$$

---

[27]Transcribed into the vernacular of statistical mechanics, the mean field method replaces the pair interaction between the atoms of a large system by the average interaction of each individual atom with some background field, the so-called "mean field", produced collectively and self-consistently by all the atoms. Self-consistency means that the product state is optimal as regards the value of the free energy.

[28]The reason: if μ_1 and μ_2 are two product measures, their convex combination $\alpha\mu_1 + (1 - \alpha)\mu_2$ $(0 < \alpha < 1)$ is almost never in $\mathcal{P}$. Therefore, product measures do not form a convex set, which would be indispensable for the convexity of W_{MF}.

The infimum is with respect to all probability measures ν on $\mathbb{R}$.

To determine $W_{\mathrm{MF}}(\phi)$ it suffices to compute the potential $U_{\mathrm{MF}}(r)$. It is startling that the latter problem makes no reference to the lattice structure. Instead we are asked to solve, once and for all, the same problem for each lattice site x, which is to find the constrained infimum with respect to the trial measure ν. We are allowed to suppress the index x, writing ν in place of ν_x, since formulation (8.7.7) reveals that the initial reference to the lattice site is lost (as long as U does not *explicitly* depend on x). As a consequence of the translational invariance of the problem we started from, the solving product measure comes out translationally invariant, too. More explicitly, we have

$$d\nu(s) = \frac{ds \, \exp\{-\hbar^{-1}u(s,t)\}}{\int ds \, \exp\{-\hbar^{-1}u(s,t)\}} \, , \tag{8.7.8}$$

where t is fixed by the condition of constraint, $\int d\nu(s)s = r$.

8.7.1 The Curie–Weiss Approximation of the Ising Model

Frequently one encounters mean-field theory in solid state physics, where it first showed the extraordinary fertility of its ideas. To illustrate its use in that field, we take the nonisotropic Ising model with a ferromagnetic nearest-neighbor interaction on a d-dimensional hypercubic (periodic) lattice of volume N^d:

$$H(\sigma) = -\sum_x \sum_i J_i \sigma_x \sigma_{x+e_i} \qquad (J_i > 0, \sigma_x = \pm 1). \tag{8.7.9}$$

As should be clear from the preceding discussion, H now assumes the role of the action, while σ_x stands for the field. A *configuration* σ assigns either $+1$ or -1 to σ_x for each x of the lattice, and a *state* p assigns a probability $p(\sigma)$ to each configuration σ. There is also a mean energy $\langle p, H \rangle = \sum_\sigma p(\sigma)H(\sigma)$ and an entropy $S(p) = -\sum_\sigma p(\sigma) \log p(\sigma)$. The basic problem is the determination of the free energy as a function of the temperature T:

$$F = \inf_p \{\langle p, H \rangle - \beta^{-1} S(p)\}, \qquad \beta^{-1} = k_B T. \tag{8.7.10}$$

When solved for p, the Gibbs variational principle leads at once to the solution

$$p(\sigma) = \exp\{-\beta H(\sigma)\} / \sum_\sigma \exp\{-\beta H(\sigma)\}. \tag{8.7.11}$$

The effective potential dependent on the magnetization m (such that $-1 \leq m \leq +1$) is constructed in a similar fashion:

$$N^d U_{\mathrm{eff}}(m) = \inf_p \{\langle p, H \rangle - \beta^{-1} S(p) \mid \langle p, \sigma_x \rangle = m\}. \tag{8.7.12}$$

So far, everything is precise but nothing is explicit. Let us thus turn to mean-field theory to get approximate however explicit expressions.

A few things are quickly realized. First, a product state p has the characteristic property that, if A is any subset of lattice points,

$$\left\langle p, \prod_{x \in A} \sigma_x \right\rangle = \prod_{x \in A} \langle p, \sigma_x \rangle, \tag{8.7.13}$$

and any such product is completely determined by providing every expectation value $m_x := \langle p, \sigma_x \rangle$. Second, knowing m_x for all x means knowing the product state:

$$p(\sigma) = \prod_x \tfrac{1}{2}(1 + m_x \sigma_x). \tag{8.7.14}$$

The latter property is special to Ising variables. The entropy of a product state assumes the form $S(p) = \sum_\sigma s(m_x)$, where

$$s(m) = -\frac{1+m}{2} \log \frac{1+m}{2} - \frac{1-m}{2} \log \frac{1-m}{2} \tag{8.7.15}$$

$(-1 \leq m \leq 1)$. Consider the set $\mathcal{P}$ of product states. Then the mean-field effective potential is given by

$$N^d U_{\mathrm{MF}}(m) = \inf_{p \in \mathcal{P}} \left\{ \langle p, H \rangle - \beta^{-1} S(p) \mid \langle p, \sigma_x \rangle = m \right\}. \tag{8.7.16}$$

The solution is trivial since there is only one state in $\mathcal{P}$ satisfying the conditions of constraint. This is the state $p(\sigma) = \prod_x \tfrac{1}{2}(1 + m\sigma_x)$ for which

$$U_{\mathrm{MF}}(m) = -Jm^2 - \beta^{-1} s(m), \qquad J := \sum_i J_i. \tag{8.7.17}$$

As expected, the potential is symmetric: $U_{\mathrm{MF}}(-m) = U_{\mathrm{MF}}(m)$. To discuss further aspects, we need to know the first and second derivatives:

$$\begin{aligned} U'_{\mathrm{MF}}(m) &= -2Jm + \beta^{-1} \tfrac{1}{2} \log \frac{1+m}{1-m} \\ U''_{\mathrm{MF}}(m) &= -2J + \beta^{-1} \frac{1}{1-m^2} \, . \end{aligned} \tag{8.7.18}$$

Though manifestly approximate, the theory suggests that we distinguish two situations:

1. $2\beta J \leq 1$. The potential is convex and its minimum at $m = 0$ is unique. The system is in its paramagnetic phase.
2. $2\beta J > 1$. The potential is not convex and there are two minima. The two magnetizations $\pm m_0$ that correspond to these minima are the solutions of the transcendental equation

$$m = \tanh(2m\beta J) \qquad (m \neq 0). \tag{8.7.19}$$

The system is in its ferromagnetic phase.

This exceedingly encouraging result shows that the direction is right. However, the only *rigorous* statement we can offer is that U_{MF} approximates U_{eff} from above and so does U_{MF}^{**}, the convex closure:

$$U_{\mathrm{eff}} \leq U_{\mathrm{MF}}^{**} \leq U_{\mathrm{MF}}. \tag{8.7.20}$$

In the paramagnetic phase, there is no difference between U_{MF} and its convex closure. In the ferromagnetic phase, however, there is. In fact, $U_{\mathrm{MF}}^{**}(m)$ behaves like a constant on the interval $-m_0 \leq m \leq m_0$.

We conclude by adding a few additional remarks.

Remark 1. The *mean field* from which the method got its name is hidden in the product state. It is a constant magnetic field B caused simultaneously by all spins of the lattice and is tied to the magnetization via the relation $m = \tanh(\beta B)$. The connection may be seen as follows. When $H(\sigma)$ is replaced by the approximate energy $\bar{H}(\sigma) = -\sum_x B\sigma_x$, it implies that the approximate Gibbs state is the desired product state:

$$\frac{\exp\{-\beta\bar{H}(\sigma)\}}{\sum_\sigma \exp\{-\beta\bar{H}(\sigma)\}} = \prod_x \frac{\exp(\beta B\sigma_x)}{2\cosh(\beta B)} = \prod_x \tfrac{1}{2}(1 + m\sigma_x) . \tag{8.7.21}$$

The mean field vanishes as soon as the system is in the paramagnetic phase.

Remark 2. Applying an inhomogeneous magnetic field to the Ising system changes the Hamiltonian and the mean-field free energy in a characteristic manner:

$$H_B(\sigma) = H(\sigma) - \sum_x B_x\sigma_x$$

$$F_{\mathrm{MF}}(B) = \inf_{p\in\mathcal{P}} \left\{ \langle p, H_B \rangle - \beta^{-1}S(p) \right\}. \tag{8.7.22}$$

As a rule, equilibrium can no longer be attained for any translationally invariant state. To get a better understanding of the correlations in an Ising system, we look upon $F_{\mathrm{MF}}(B)$ as a generating functional for the cumulants of the spin variables. For instance, on the paramagnetic side,

$$\langle\sigma_x\rangle_{\mathrm{MF}} = \frac{\partial F_{\mathrm{MF}}}{\partial B_x}(0) = 0,$$

$$\langle\sigma_x\sigma_y\rangle_{\mathrm{MF}} = \beta^{-1}\frac{\partial^2 F_{\mathrm{MF}}}{\partial B_x\partial B_y}(0) \xrightarrow{N\to\infty} \frac{1}{|B_d|}\int_{B_d} \frac{dp\, e^{ip(x-y)}}{1 - 2\beta\sum_i J_i\cos p_i}. \tag{8.7.23}$$

Based on this result, we get the magnetic susceptibility as

$$\chi_{\mathrm{MF}} = \frac{1}{N^d} \sum_{x,y} \langle \sigma_x \sigma_y \rangle_{\mathrm{MF}}$$

$$= \sum_{x} \langle \sigma_x \sigma_0 \rangle_{\mathrm{MF}} \xrightarrow{N \to \infty} \frac{1}{1 - 2\beta J}.$$

(8.7.24)

The right-hand side diverges as we approach the critical point $\beta_c^{-1} = 2J$. Warning: the mean value $\langle \cdot \rangle_{\mathrm{MF}}$ is purely formal in the sense that there is no guaranty that it can be reproduced by some probability distribution. After all, it is only an approximation to a state. Quite surprisingly, the mean-field theory of the Ising model has demonstrated that, even though one is dealing with product states, there are not only short-range but also long-range mean-field correlations.

Remark 3. Experience has shown that the mean-field method works well in high dimensions but fails if $d = 1$ or 2. To see how badly it fails for the two-dimensional Ising model, we appeal to the Onsager solution which among many other things has shown us [8.7] that the critical manifold in the (J_1, J_2)-plane is given by the equation

$$\sinh(2\beta J_1) \sinh(2\beta J_2) = 1,$$

(8.7.25)

while the mean-field method predicts it to be given by $2\beta(J_1 + J_2) = 1$.

8.7.2 The Ising Spin Limit of the Neutral Scalar Field

The four-dimensional Ising model may also be viewed as a special limiting case of a scalar lattice field theory. To see the details of that limit, we take $J_i = J > 0$ $(i = 1, \ldots, 4)$ and $\sigma_x = J^{-1/2} \phi(x)$ so that

$$H(\sigma) = -J \sum_{x} \sum_{i} \sigma_x \sigma_{x+e_i} = J \sum_{x} \tfrac{1}{2} \sum_{i} (\sigma_{x+e_i} - \sigma_x)^2 + \mathrm{const}$$

$$= \sum_{x} \tfrac{1}{2} \sum_{i} \{\partial_i \phi(x)\}^2 + \mathrm{const} \equiv W_0(\phi) + \mathrm{const}.$$

(8.7.26)

The restriction of the neutral scalar field to the values $\phi(x) = \pm J^{1/2}$ (irrespective of x) can be achieved through the introduction of a Dirac function into the measure:

$$d\mu(\phi) = Z^{-1} \exp\{-\beta W_0(\phi)\} \prod_{x} \delta(\phi(x)^2 - J) d\phi(x)$$

$$Z = \int \exp\{-\beta W_0(\phi)\} \prod_{x} \delta(\phi(x)^2 - J) d\phi(x).$$

(8.7.27)

With this definition, expectations acquire the desired form:

$$\int d\mu(\phi)\, f(\phi) = \frac{\sum_{\sigma} e^{-\beta H(\sigma)} f(J^{1/2}\sigma)}{\sum_{\sigma} e^{-\beta H(\sigma)}}\ , \qquad (8.7.28)$$

where f may be any (reasonable) functional of the field. Properly normed, the Gaussian function $\exp\{-\lambda(\phi(x)^2 - J)^2\}$ will tend to $\delta(\phi(x)^2 - J)$ as $\lambda \to \infty$. Let us therefore introduce the Euclidean action

$$W_\lambda(\phi) = \sum_x \left\{ \tfrac{1}{2} \sum_i \left(\partial_i \phi(x)\right)^2 + \lambda\left(\phi(x)^2 - J\right)^2 \right\} \qquad (8.7.29)$$

and use it to define the Gibbsian measure

$$d\mu_\lambda(\phi) = Z_\lambda^{-1} \mathcal{D}\phi \, \exp\left\{ -\hbar^{-1} W_\lambda(\phi) \right\}$$
$$Z_\lambda = \int \mathcal{D}\phi \exp\left\{ -\hbar^{-1} W_\lambda(\phi) \right\}. \qquad (8.7.30)$$

Then μ_λ tends to μ (convergence in the sense of expectations) provided $\hbar$ is being identified with β in the above formulas. Ising systems are easier to analyze than field theories. The lesson is that any statement about the four-dimensional Ising system translates into some asymptotic statement about the self-interacting neutral scalar field.

8.8 The Gaussian Approximation

Though the mean-field method provides the most versatile approximation scheme, it is good to look for an alternative. We shall now focus on the *Gaussian approximation* which for some unknown reason has received little attention in the literature.

We take a lattice of size N^4 as before. But instead of considering trial states in the set $\mathcal{P}$ of product measures, we search for the optimal trial state in the set $\mathcal{K}$ of Gaussian measures:

$$d\mu(\phi) = Z^{-1} \mathcal{D}\phi \, \exp\{-\tfrac{1}{2}(\phi, K^{-1}\phi)\}, \qquad (8.8.1)$$

where the covariance operator K is arbitrary, i.e., the variational problem $\langle \mu, W \rangle - \hbar S(\mu) = minimum$ is to be solved for $\mu \in \mathcal{K}$. If a solution exists, it will look very much like a free field and will be called a *Gaussian approximation*.

8.8.1 A Case Study

Consider the prototype, the neutral scalar field with the action

$$W(\phi) = \sum_x \left\{ \tfrac{1}{2} \sum_i \left(\partial_i \phi(x)\right)^2 + \lambda \phi(x)^4 \right\}. \qquad (8.8.2)$$

Though W does not contain a mass term, it is generally conjectured that the interaction can create a nonzero mass (for the lowest excitation above the vacuum), hence a gap in the spectrum. But no one can possibly have any inkling of the gap generation mechanism if he or she seeks advice from perturbation theory, which is doomed to failure. Let us examine what the Gaussian approximation can do for us.

Needless to say that it suffices to restrict attention to translationally invariant Gaussian measures μ, which means that the covariance operator of that measure, if written as a matrix, assumes the form

$$K_{xy} = N^{-4} \sum_p e^{ip(x-y)} k_p \qquad (k_p > 0). \tag{8.8.3}$$

After some algebra we find that $\langle \mu, \phi(x)^4 \rangle = 3\langle \mu, \phi(x)^2 \rangle^2 = 3(K_{xx})^2$. Therefore,

$$N^{-4} \langle \mu, W \rangle = \tfrac{1}{2} N^{-4} \sum_p E_p k_p + 3\lambda \left(N^{-4} \sum_p k_p \right)^2, \tag{8.8.4}$$

where $E_p = 2\sum_i (1-\cos p_i)$. Also, $S(\mu) = \tfrac{1}{2}\langle \mu, (\phi, K^{-1}\phi) \rangle + \log \det(2\pi K)^{1/2}$, i.e.,

$$N^{-4} S(\mu) = \tfrac{1}{2}\left\{ 1 + N^{-4} \sum_p \log(2\pi k_p) \right\}. \tag{8.8.5}$$

This allows us to determine the free energy per site:

$$\begin{aligned}
f &= \lim_{N\to\infty} N^{-4}\{\langle \mu, W \rangle - \hbar S(\mu)\} \\
&= \frac{1}{2|B_4|} \int_{B_4} dp\, E_p k_p + 3\lambda \left(\frac{1}{|B_4|} \int_{B_4} dp\, k_p \right)^2 \\
&\quad - \frac{\hbar}{2}\left(1 + \frac{1}{|B_4|} \int_{B_4} dp\, \log(2\pi k_p) \right)
\end{aligned} \tag{8.8.6}$$

($B_4 = [-\pi, \pi]^4$). The minimum is assumed if the unknown function k_p satisfies the condition $\delta f / \delta k_p = 0$ (checking positivity of the second derivative):

$$E_p + \frac{12\lambda}{|B_4|} \int_{B_4} dp\, k_p - \hbar k_p^{-1} = 0. \tag{8.8.7}$$

The solution is of the form $k_p = \hbar(E_p + m^2)^{-1}$, where m^2 is implicitly given by

$$m^2 = 12\frac{\lambda\hbar}{|B_4|} \int_{B_4} dp\,(E_p + m^2)^{-1}. \tag{8.8.8}$$

The behavior for small λ is different from that for large λ:

$$\lambda \to 0: \quad m^2 \to 12c\lambda\hbar \quad , \qquad c = \frac{1}{|B_4|} \int_{B_4} dp\, E_p^{-1} \approx 0.155$$

$$\lambda \to \infty: \quad m^2 \to (12\lambda\hbar)^{1/2}.$$

$$(8.8.9)$$

We have learned that there exists a free field that approximates best. The Gaussian approximation, so to speak, has revealed the presence of free bosons of mass $m > 0$ in the system and thus predicts that there should be a gap in the spectrum. It is still utterly unclear, however, how close the prediction comes to the truth.

9 The Quantization of Gauge Theories

> Dans quelque pays inconnu
> Écorcher la terre revêche
> Et pousser une lourde bêche
> Sous notre pied sanglant et nu?
>
> *Charles Baudelaire*

Modern field theory mainly focuses on gauge theories, be they Abelian or non-Abelian, and we have come to appreciate *gauge invariance* as a unifying principle of particle physics with far reaching consequences. In the past we witnessed a period of bursting creativity when new ideas were developed in a rather hasty spirit. But it seems that we have now entered another period in which there is an urge to consolidate past achievements. The origin of these developments lies in the formulation of quantum electrodynamics (Abelian) and of the Yang-Mills theory (non-Abelian) [9.1].

What makes gauge theories so special is that they exhibit features normally not encountered in "ordinary" field theories. A common characteristic of all gauge theories is the necessity of dealing with *vector fields* and to cope with a large group of local symmetry transformations, the gauge group. This chapter is devoted to vector fields and local gauge invariance. The study of this topic brings us closer to the path integrals of present-day particle physics.

It is advisable to start with the prototype of a vector field, i.e., the electromagnetic potential for which there is a Minkowskian and a Euclidean version. In the next section we will restrict our attention to some aspects of the Euclidean formulation.

9.1 The Euclidean Version of Maxwell Theory

Before indulging in quantized $U(1)$-gauge theory, we shall briefly recapitulate what can be said about the nonquantized version from the variational-principle perspective. This will ultimately prepare the ground for the path integral formulation of gauge theories in general.

9.1.1 The Classical Situation ($\hbar = 0$)

When writing $A_k(x)$ for the Euclidean vector potential we must emphasize two things. First, the index k runs from 1 to 4 and $x = \{\boldsymbol{x}, x^4\}$. Second, each component A_k is *real* and there is no relationsship between the Euclidean and the Minkowskian vector potential at the classical level: they simply refer to different spaces and field equations[29]. Local gauge transformations are defined as usual,

$$A_k \rightarrow A_k + \partial_k f, \tag{9.1.1}$$

and potentials that differ by a gradient are said to be equivalent. The Euclidean field strength or Maxwell field is introduced as the antisymmetric tensor formed by the first derivatives: $F_{k\ell} = \partial_\ell A_k - \partial_k A_\ell$. Formally, it may be decomposed into an *electric* and an *magnetic* part (not to be confused with the Minkowskian electric and magnetic field):

$$\begin{aligned}
\boldsymbol{E} &= \{F_{14}, F_{24}, F_{34}\} = \{F_{23}^*, F_{31}^*, F_{12}^*\} \\
\boldsymbol{B} &= \{F_{23}, F_{31}, F_{12}\} = \{F_{14}^*, F_{24}^*, F_{34}^*\}
\end{aligned} \tag{9.1.2}$$

with F^* the dual tensor of F. Notice that $F^{**} = F$ in the Euclidean domain. If $\boldsymbol{E} = \boldsymbol{B}$ or, equivalently, $F^* = F$, the tensor F is said to be *self-dual*. On the other hand, if $\boldsymbol{E} = -\boldsymbol{B}$, equivalently, $F^* = -F$, the tensor F is called *anti-self-dual*. An arbitrary Maxwell field admits a unique decomposition into a self-dual and an anti-self-dual part: $F = \frac{1}{2}(F + F^*) + \frac{1}{2}(F - F^*)$.

By definition, the Euclidean Maxwell field transforms under the rotations of the Euclidean four-space as a tensor does. Generally, tensors are reducible and may uniquely be decomposed into their irreducible components. Antisymmetry of the Maxwell field tensor implies that we are dealing with the reducible representation $(1,0) \oplus (0,1)$ of the group $SO(4)$. The selfdual and the antiselfdual part of F correspond to the irreducible representations $(1,0)$ and $(0,1)$. The implied decomposition of F is thus $SO(4)$-invariant. It is not $O(4)$-invariant, however, since a reflection $I\colon E_4 \rightarrow E_4$ (satisfying $\det I = -1$) will interchange the selfdual and the antiselfdual parts.

In passing, we mention that the universal covering of the group $SO(4)$ is the product group $G = SU(2) \otimes SU(2)$ (both Lie groups, G and $SO(4)$, have identical Lie algebras). This makes life easy as far as representation theory goes. First, every finite-dimensional representation is equivalent to a unitary representation and thus is *completely reducible*, i.e., it may be written as a direct sum of irreducible representations. Second, each irreducible ray representation of $SO(4)$ is of the form (j, k), where j and k are "spins" (and hence half-integers) referring to irreducible representations of the first or second $SU(2)$ respectively of the covering group G. We get a spinor or a vector representation depending on whether $2j + 2k$ is odd or even. Its dimension is $(2j + 1)(2k + 1)$.

Euclidean fields are naturally classified according to their (j, k)-contents. Most surprisingly, this classification scheme continues to makes sense when the transition is made from the Euclidean space to Minkowski space. It has been known for a long time and referred to as the *Weyl trick*: the unitary representation (j, k) becomes, after analytic continuation, the nonunitary representation $\mathcal{D}^{jk}$ of the Lorentz group.

---

[29]Remember, analytic continuation is of no avail to us in the classical limit of a field theory.

There are some doubts as to whether the electromagnetic potential A_k should be regarded as a *vector field* at all, the reason being that $SO(4)$-transformations (or Lorentz transformations) and gauge transformations appear hopelessly intertwined. For instance, the Coulomb gauge condition $\nabla \cdot \boldsymbol{A} = 0$ can be maintained in any system of reference utilizing the freedom of performing a gauge transformation. At the quantized level, strange things happen, too: frequently, the two-point function of the second-quantized vector potential is not a tensor.

We write the Euclidean action for the potential A in the presence of an external source as

$$W(A) = \int dx \left\{ \tfrac{1}{4} F^{k\ell}(x) F_{k\ell}(x) - j_k(x) A^k(x) \right\}. \tag{9.1.3}$$

Since it is at most bilinear in A, nothing spectacular is expected to come from such an ansatz. A property we might want to check is the convexity of the functional $W(A)$. We will get an affirmative aswer a little later. But the really urgent question is this: does $W(A)$ admit a lower bound? If the answer is *yes*, the theory would be stable and it would make sense to pass to the next question: where is the minimum? To see that the answer depends on the current j and might be *no*, concentrate on

$$F^2 \equiv F^{ik} F_{ik} = 2(\boldsymbol{E}^2 + \boldsymbol{B}^2) \geq 0 \tag{9.1.4}$$

and observe that $F^2 = 0$ iff $F = 0$ which means that $A^k = \partial^k f$ for some gauge function f. Then integration by parts yields

$$W(A) = \int dx \, f(x) \partial_k j^k(x) \tag{9.1.5}$$

thereby demonstrating that $W(A)$ is not bounded below (f is arbitrary) unless the condition $\partial_k j^k(x) = 0$ is satisfied. We conclude that we do not get a satisfactory theory even at the classical level if the vector potential is coupled to a *nonconserved* current. For a *conserved* current, however, one finds that

$$\begin{aligned}
W(A) &= \tfrac{1}{2}(A, \mathcal{C}A) - (A, j) \\
&= \tfrac{1}{2}\left(A + \boldsymbol{\Delta}^{-1}j, \mathcal{C}(A + \boldsymbol{\Delta}^{-1}j)\right) - \tfrac{1}{2}\left(j, (-\boldsymbol{\Delta})^{-1}j\right) \\
&\geq -\tfrac{1}{2}\left(j, (-\boldsymbol{\Delta})^{-1}j\right),
\end{aligned} \tag{9.1.6}$$

where[30] $\mathcal{C} = \partial \otimes \partial - \boldsymbol{\Delta}$. For convenience we have employed the notation $(A, j) = \int dx \, A^k(x) j_k(x)$. Use has also been made of the fact that current conservation implies the relation $\mathcal{C}(-\boldsymbol{\Delta})^{-1}j = j$.

Taking for granted that the current j decreases sufficiently fast at spatial infinity, (9.1.6) shows that we do get a lower bound for the classical action. Moreover, as soon as current conservation is invoked, gauge invariance of

---

[30] As before, $\boldsymbol{\Delta}$ stands for the four-dimensional Laplacian. The operator $\mathcal{C}$ acts on vector fields. It may be helpful to think of $\mathcal{C}$ as a matrix whose elements are differential operators: $\mathcal{C}_{k\ell} = \partial_k \partial_\ell - \delta_{k\ell} \boldsymbol{\Delta}$.

the action becomes manifest. It may justly be said that stability implies current conservation which in turn implies gauge invariance.

The freedom to choose a gauge tells us that the variational problem $W(A) = minimum$ has many solutions,

$$A_k = (-\varDelta)^{-1} j_k + \partial_k f, \tag{9.1.7}$$

in harmony with the fact that (9.1.7) is the most general solution[31] of the field equation $\partial^k F_{k\ell} = j_\ell$ (equivalently, $\mathcal{C}A = j$). In particular, we would get $F_{k\ell} = 0$ for the case where there is no external current.

From the mathematical view point, nonuniqueness of solution (9.1.7) is caused by the fact that the operator $\mathcal{C}$, though positive, cannot be inverted. The kernel of the operator $\mathcal{C}$, i.e., the set of solutions of $\mathcal{C}A = 0$ consists of "pure gauges", $A = \partial f$. Any two solution of $\mathcal{C}A = j$ differ by some element from the kernel of $\mathcal{C}$, and hence by a gauge transformation.

It should be clear however that the minimum of the action does not depend on our choice of the gauge:

$$\inf_A \left\{ \tfrac{1}{2}(A, \mathcal{C}A) - (A, j) \right\} = -\tfrac{1}{2} \left(j, (-\varDelta)^{-1} j \right)$$

$$= -\tfrac{1}{2}(2\pi)^{-2} \int dx \int dy \, \frac{j^k(x) j_k(y)}{|x - y|^2} \, . \tag{9.1.8}$$

Recall that $(2\pi)^{-2} |x-y|^{-2} = S(x-y; 0)$ is the two-point Schwinger function of a massless scalar field.

So far we have been vague about the restrictions on j and A. But result (9.1.8) convinces us that currents should be viewed as members of the space

$$\mathcal{J} = \left\{ j : E_4 \to E_4 \mid \left(j, (-\varDelta)^{-1} j \right) < \infty \right\}, \tag{9.1.9}$$

while potentials A belong to the dual space

$$\mathcal{J}^* = \left\{ A : E_4 \to E_4 \mid \left(A, (-\varDelta)A \right) < \infty \right\}. \tag{9.1.10}$$

Concerved currents are members of the subspace

$$\mathcal{J}_0 = \left\{ j \in \mathcal{J} \mid \partial_k j^k = 0 \right\}. \tag{9.1.11}$$

Though the discussion has shown that there exist, for $j \in \mathcal{J}_0$, many equilibrium solutions $A \in \mathcal{J}^*$, the solution would still be unique when interpreted as an element of the dual $\mathcal{J}_0^*$. This is so since $\mathcal{J}_0^*$ consists of equivalence classes, i.e., two potentials A and A' from $\mathcal{J}^*$ are considered equivalent if $(A, j) = (A', j)$ for all $j \in \mathcal{J}_0$, implying that $A' = A + \partial f$ with f some gauge function. Let $\mathcal{G}$ denote the group of local gauge transformations induced by gauge functions f such that $\partial f \in \mathcal{J}^*$. Then $\mathcal{J}_0^*$ is nothing but the space of $\mathcal{G}$-orbits in $\mathcal{J}^*$ and we write

---

[31]There is no room for electromagnetic waves in classical Euclidean Maxwell theory, i.e., for nontrivial solutions of the homogeneous differential equation $\partial^k F_{k\ell} = 0$.

$$\mathcal{J}_0^* = \mathcal{J}^*/\mathcal{G}. \tag{9.1.12}$$

As for our model, $\mathcal{J}_0^*$ provides the correct configuration space containing the *relevant degrees of freedom* only. The larger space $\mathcal{J}^*$ has redundant degrees of freedom which are *indispensable for a manifest local formulation of the theory.*

9.1.2 Gauge Fixing

The distinction of relevant and redundant and the orbit description of the relevant degrees of freedom is central to any gauge theory. Yet, orbits are not easy to deal with when it comes to path integrals over orbit spaces. A simple notion that pervades various mathematical disciplines is that of a *section*. It enters whenever one seeks *representatives* for the orbits in question, either locally or globally[32]. A section may be visualized as a manifold transversal to the orbits and intersecting each orbit at precisely one point. In the present context, choosing a section for the orbits under the gauge group is referred to as *fixing the gauge*, and the section is referred to as the *gauge manifold*.

As a rule, the gauge manifold is given in terms of a defining equation satisfied by any potential that belongs to the manifold. Such an equation is called a *gauge condition*. A popular way to fix the gauge is to choose the Lorentz condition $\partial_k A^k = 0$. The implied section of the gauge orbits is called the *Lorentz manifold*. In a way, the Lorentz condition splits up into a continuum of constraints: for each $x \in E_4$ there is one equation. There is however a way to get around this. We write $\int dx\,(\partial_k A^k)^2 = 0$ for the condition of constraint when turning to the constrained variational problem. Following the usual strategy we introduce a Lagrange multiplier $\lambda > 0$ and study the related problem $W(A; \lambda) = minimum$, where

$$W(A; \lambda) = \int dx \left\{ \tfrac{1}{4} F^{k\ell} F_{k\ell} + \tfrac{1}{2}\lambda(\partial_k A^k)^2 - j_k A^k \right\}. \tag{9.1.13}$$

Let us emphasize what distinguishes the present situation from the previous one:

- $\lambda = 0$: Though the action is a convex functional of A and bounded below, it is constant in certain directions of the configuration space owing to gauge invariance: $W(A + \partial f; 0) = W(A; 0)$. The minimum is never unique.
- $\lambda > 0$: The action is *strictly* convex in that

$$W\big(\alpha A + (1 - \alpha)A'; \lambda\big) < \alpha W(A; \lambda) + (1 - \alpha)W(A'; \lambda) \tag{9.1.14}$$

---

[32]The most common notion is this. Let X and A be topological spaces, where the map $p\colon A \to X$ is continuous and onto. For Y a subset of X, a section over Y (and hence a *local section* for p) is a continuous map $s\colon Y \to A$ with $p \circ s\colon Y \to Y$ the identity. If $Y = X$, s is said to be a *global* section. There may be obstructions to the existence of global sections.

if $A \neq A'$ and $0 < \alpha < 1$. Gauge invariance is violated and the minimum is unique.

In the situation $\lambda > 0$ when gauge invariance is lost, it is no longer necessary to require that the current j be conserved to maintain stability. This is not quite so academic as it looks since arbitrary currents will be employed in generating functionals later on.

It may be illuminating to repeat the simple calculation that gives the minimum of the action, this time assuming that $\lambda > 0$:

$$\begin{aligned}
W(A;\lambda) &= \tfrac{1}{2}(A, C_\lambda A) - (A, j) \\
&= \tfrac{1}{2}\big(A - C_\lambda^{-1} j, C_\lambda (A - C_\lambda^{-1} j)\big) - \tfrac{1}{2}(j, C_\lambda^{-1} j),
\end{aligned} \tag{9.1.15}$$

where $C_\lambda = (1 - \lambda)\partial \otimes \partial - \boldsymbol{\Delta} > 0$ so that

$$C_\lambda^{-1} = (1 - \lambda^{-1})\boldsymbol{\Delta}^{-2} \partial \otimes \partial - \boldsymbol{\Delta}^{-1}. \tag{9.1.16}$$

From representation (9.1.15) it is obvious that

$$\inf_{A \in \mathcal{J}^*} W(A;\lambda) = -\tfrac{1}{2}(j, C_\lambda^{-1} j) \qquad (j \in \mathcal{J}). \tag{9.1.17}$$

The minimum is unique and attained for $A = C_\lambda^{-1} j$. Uniqueness is thus related to the fact that the operator C_λ can be inverted when $\lambda > 0$. As for conserved currents, we may write the solution in the simpler form $A = (-\boldsymbol{\Delta})^{-1} j$ and get

$$\inf_{A \in \mathcal{J}^*} W(A;\lambda) = -\tfrac{1}{2}\big(j, (-\boldsymbol{\Delta})^{-1} j\big) \qquad (j \in \mathcal{J}_0) \tag{9.1.18}$$

with no dependence on λ. This explains why we can skip the second step of our calculation: there is no need to determine the value for λ from the Lorentz condition $\partial_k A^k = 0$, nor would such an attempt be successful, since the Lorentz condition is automatically satisfied. The conclusion is that the extra parameter λ has no physical significance as long as current conservation is adhered to.

Therefore, the principal effect of demanding $\lambda > 0$ is that it *fixes the gauge*, i.e., from the set of gauge-equivalent solutions, $A = (-\boldsymbol{\Delta})^{-1} j + \partial f$, it picks the solution $A = (-\boldsymbol{\Delta})^{-1} j$ as the representative of the gauge orbit. The choice is such that the Lorentz condition is satisfied. The extra term in the action, $\tfrac{1}{2}\lambda \int dx\, (\partial_k A^k)^2$, which did the trick is called the *gauge fixing term*.

9.1.3 The Quantized Situation ($\hbar > 0$)

Let us now look at what happens if Planck's constant is turned on. We expect to arrive at some rudimentary version of quantum electrodynamics, i.e., a theory of photons coupled to an external current. Hamilton's principle of least action changes to the Gibbs variational principle,

$$\langle \mu, W \rangle - \hbar S(\mu) = minimum, \tag{9.1.19}$$

to be solved for μ a probability measure on $\mathcal{J}^*$. Here the action is assumed to include the gauge fixing term, while the other essential term containing the current j will be treated separately. Thus,

$$W(A; \lambda) = \tfrac{1}{2}(A, \mathcal{C}_\lambda A) = \int dx \left\{ \tfrac{1}{4} F^{k\ell} F_{k\ell} + \tfrac{1}{2} \lambda (\partial_k A^k)^2 \right\} \tag{9.1.20}$$

($\lambda > 0$). Not unexpectedly, gauge fixing has the beneficial effect that the solving (Gaussian) Gibbs measure μ_λ is both unique and well defined. So the ensuing path integral is well defined.

A convenient way to characterize a measure is to write down an explicit formula for the Fourier transform, i.e., to provide its characteristic functional. By analogy to the situation of the free scalar field, we get, in the present context,

$$\int d\mu_\lambda(A) \, \exp\{i(A, j)\} = \exp\left\{ -\tfrac{1}{2}\hbar(j, \mathcal{C}_\lambda^{-1} j) \right\} \qquad (j \in \mathcal{J}) \tag{9.1.21}$$

with $\hbar \mathcal{C}_\lambda^{-1}$ the covariance operator. Another way is to drop the imaginary unit in the exponent (the resulting formula is modeled after the two-sided Laplace transform). This changes (9.1.21) to what we refer to as the *real version* of that formula:

$$\int d\mu_\lambda(A) \, \exp\{(A, j)\} = \exp\left\{ \tfrac{1}{2}\hbar(j, \mathcal{C}_\lambda^{-1} j) \right\}. \tag{9.1.22}$$

Still another way is to pass to an expression emphasizing the Gaussian character of the measure:

$$d\mu_\lambda(A) = Z^{-1} \mathcal{D}A \, \exp\left\{ -\hbar^{-1} W(A; \lambda) \right\} \qquad (A \in \mathcal{J} \tag{9.1.23}$$

But this is merely formal and serves as a substitute of the rigorous definition (9.1.21).

All reference to λ disappears from the left-hand side of (9.1.21) or (9.1.22) if $j \in \mathcal{J}_0$, i.e., if the current is conserved, implying that the induced measure on $\mathcal{J}_0^*$ (the space of gauge orbits) *does not depend on* λ. This is, of course, a desired effect. But beware: viewed as a measure on $\mathcal{J}^*$, μ_λ varies with the parameter λ. To see what λ does to the measure, consider the orbit ${}^f\!A = A + \partial f$, where A has been fixed in such a way that $\partial_k A^k = 0$. Concentrating on the map

$$f \mapsto \exp(-\hbar^{-1}W(^fA; \lambda), \tag{9.1.24}$$

we would experience a Gaussian behavior as we move along the gauge orbit. The Gaussian is centered at $f = 0$ and its width is proportial to $\lambda^{-1/2}$. As λ is increased, the width will shrink and the dominant contribution to the measure μ_λ will come from the vicinity of the manifold given by the Lorentz condition. It is only in the limit that μ_∞ acts like a Dirac measure on each $\mathcal{G}$-orbit in $\mathcal{J}^*$. The measure is then seen to be supported by the Lorentz manifold.

It is straightforward to get explicit formulas for the n-point Schwinger functions of the theory. Owing to the Gaussian character of the field, it suffices to consider the two-point function:

$$\int d\mu_\lambda(A)\, A_k(x)A_\ell(y) = \hbar\langle x|[\mathcal{C}_\lambda^{-1}]_{k\ell}|y\rangle$$

$$= \frac{\hbar}{(2\pi)^4} \int \frac{dp}{p^2}\left\{(\lambda^{-1} - 1)\frac{p_k p_\ell}{p^2} + \delta_{k\ell}\right\} e^{ip(x-y)}. \tag{9.1.25}$$

As is evident from this formula, Schwinger functions involving the potential will always depend on λ, and they are not gauge invariant. However, when passing to the field strength, we lose any influence of the unphysical parameter λ on expectation values:

$$\int d\mu_\lambda(A)\, F_{jk}(x)F_{\ell m}(y) = \frac{\hbar}{(2\pi)^4} \int \frac{dp}{p^2}\, c_{jk,\ell m}(p)\, e^{ip(x-y)}, \tag{9.1.26}$$

where

$$c_{jk,\ell m}(p) = \delta_{k\ell}p_j p_m + \delta_{jm}p_k p_\ell - \delta_{j\ell}p_k p_m - \delta_{km}p_j p_\ell. \tag{9.1.27}$$

We have discovered that all quantities that are gauge invariant at the classical level remain unaffected after quantization by the presence of gauge-fixing terms in the action.

There are two special choices for λ that carry names:

- $\lambda = 1$: *Feynman gauge*. This is marked by the fact that the two-point function becomes proportional to $\delta_{k\ell}$.
- $\lambda^{-1} = 0$: *Landau gauge*. This is marked by the fact that the matrix with elements $p_{k\ell} = \delta_{k\ell} - p^{-2}p_k p_\ell$ enters (9.1.25). It represents a projector. The operator $\mathcal{C}_\lambda^{-1}$ exists, but $\mathcal{C}_\lambda$ does not. The support of the measure μ_∞ is the Lorentz manifold.

Exercise 1.   Adopt $\hbar = 1$ throughout. To justify the common terminology which refers to electrodynamics as a $U(1)$-gauge theory, show that, if $\mathcal{G}$ is a (differentiable, trivial) $U(1)$-bundle over E_4, a gauge transformation is given by

$$^uA_k(x) = A_k(x) - iu^{-1}(x)\partial_k u(x) \qquad (u \in \mathcal{G}),$$

then the previously used gauge function f is obtained from the relation $u(x) = e^{if(x)}$. What would be a 'gauge transformation' of objects like

$$P(A; \omega) = \exp\left\{ i\int_\omega dx^k A_k(x) \right\},$$

where ω is some path in E_4 running from x to x' ? Prove that (formally)

$$\int d\mu_\lambda(A)\, P(A; \omega) = \exp\left\{ -\tfrac{1}{2} \int_\omega dx^k \int_\omega dy^\ell\, \langle x|[C_\lambda^{-1}]_{k\ell}|y\rangle \right\}.$$

Hint: extend (9.1.21) to more singular currents like $j^k(y) = \int_\omega dx^k \delta(x - y)$. Show also that, if ω is a loop, the result will not depend on λ. Does the above double integral exist? If not, find a way to regularize it.

Exercise 2.   Sometimes it is good to know that, along a gauge orbit $^fA = A + \partial f$, there is precisely one potential satisfying the Lorentz condition $\partial_k A^k = 0$. To this end suppose that the potential A is twice differentiable and there exists $c > 0$ such that

$$|\partial_k A^k(x)| < c|x|^{-2-\epsilon} \qquad (\epsilon > 0).$$

Prove that there is a unique gauge function f, twice differentiable and tending to zero at infinity, such that $\partial_k(A^k - \partial^k f) = 0$. Hint: existence follows from setting

$$f(x) = -\frac{1}{(2\pi)^2} \int dx'\, \frac{\partial_k A^k(x')}{|x' - x|^2}\ .$$

Uniqueness follows from standard theorems about harmonic functions, i.e., about solutions of $\Delta h(x) = 0$.

9.2 Non-Abelian Gauge Theories: Preliminaries

Having mastered the Euclidean Maxwell theory, we are ready for a more complex structure. General evidence indicates that non-Abelian gauge theory provides the appropriate framework for the description of the strong and weak interactions between elementary particles. Gauge transformations are then considered to be functions on the Euclidean four-space taking values in some (compact) Lie group G other than $U(1)$. In general, two different gauge transformations will no longer commute. The theory is thus said to be *non-Abelian* [9.2–6].

To be more specific about the model and to ease the discussion, let us consider the $SU(n)$ gauge theory without matter fields (no coupling to Dirac fields or Higgs fields). The subject is also known under the name *Yang–Mills field theory*. It can be shown that the Lie algebra $\mathbf{su}(n)$ consists of all anti-Hermitean traceless $n \times n$ matrices. These matrices are considered to be members of some real-linear space of dimension n^2-1 for which there exists a basis $-it_a$ ($a = 1, \ldots, n^2-1$) such that the matrices t_a are Hermitean traceless and satisfy the orthogonality relation

$$\tfrac{1}{2}\operatorname{tr} t_a t_b = \delta_{ab}. \tag{9.2.1}$$

This clearly extends the definition of the Pauli matrices (for $SU(2)$) to $SU(n)$. The Euclidean gauge field $A_k(x)$ takes values in $\mathbf{su}(n)$ and therefore admits a decomposition with respect to the basis chosen:

$$A_k(x) = -i \sum_a A_k^a(x) t_a. \tag{9.2.2}$$

The 'coefficients' $A_k^a(x)$ should be viewed as ordinary real vector fields. For most purposes, however, it is far more convenient to deal with the Lie-algebra-valued field $A_k(x)$ directly.

Physically speaking, the non-Abelianness of the gauge group manifests itself in the fact that the vector field *interacts with itself*. The strength of this self-interaction is controlled by some coupling constant g which in some sense generalizes the concept of *charge*. Typically it appears in front of Lie brackets. However, unless one is interested in the perturbation expansion, it is advisable to eliminate g from the field equations via a scaling transformation, i.e., $A_k(x)$ is replaced by $g^{-1}A_k(x)$. As a result of this replacement, the factor g^{-2} appears in front of the action to join Planck's constant in the path integral. The interesting feature of quantized Yang–Mills theory is that $\hbar g^2$ becomes the only essential parameter. As the coupling g gets small, the classical limit is approached.

Calling $D_k \equiv \partial_k - A_k$ the *covariant derivative*, we find a number of different ways to write the field strength designed to take values in the Lie algebra:

$$\begin{aligned} F_{k\ell} &= [D_k, D_\ell] = D_\ell A_k - D_k A_\ell \\ &= A_{k|\ell} - A_{\ell|k} + [A_k, A_\ell]. \end{aligned} \tag{9.2.3}$$

The vertical line $|$ in front of an index indicates a derivative. It is used mainly to shorten formulas. The Euclidean action, though expressible in terms of the field strength alone, is considered a functional of the gauge field A:

$$W(A) = -\frac{1}{8g^2} \int dx \operatorname{tr}\left(F_{k\ell} F^{k\ell}\right). \tag{9.2.4}$$

By (9.2.1) and the decomposition $F_{k\ell} = -i \sum_a F_{k\ell}^a t_a$, one gets

$$W(A) = \frac{1}{4g^2} \int dx \sum_{ak\ell} (F_{k\ell}^a)^2. \tag{9.2.5}$$

Let us turn to the topic of gauge transformations designed to leave the action (9.2.4) invariant. Let $\mathcal{G}$ stand for the group of functions[33]

$$u : E_4 \to SU(n) , \quad x \mapsto u(x) , \tag{9.2.6}$$

and let $u \in \mathcal{G}$ act on the gauge field in the following manner:

$$A_k \ \to \ {}^u\!A_k := uA_k u^{-1} + u_{|k} u^{-1}. \tag{9.2.7}$$

To justify the definition, observe that, for every $x \in E_4$ and index k, the matrix $u_{|k} u^{-1}$ is in the Lie algebra $\mathbf{su}(n)$. It is straightforward to see that prescription (9.2.7) transforms $F_{k\ell}$ into $uF_{k\ell}u^{-1}$. So the action (9.2.4) stays invariant.

The classical variational problem $W(A) = minimum$ is solved trivially by $F_{k\ell} = 0$, as in the Abelian case. In terms of the gauge field it means that $A_k = u_{|k} u^{-1}$ for some $u \in \mathcal{G}$. Such field configurations are said to be *pure gauge*.

The equilibrium condition $F_{k\ell} = 0$ is to be distinguished from the *Yang–Mills equation*

$$[D^k, F_{k\ell}] \equiv \partial^k F_{k\ell} - [A^k, F_{k\ell}] = 0, \tag{9.2.8}$$

whose role is characterize those configurations for which the action becomes stationary. Surprisingly enough, there exist local minima of the action above zero in the non-Abelian case, i.e., solutions of the Yang–Mills equations that are either self-dual or anti-self-dual. These minima correspond to the so-called *instanton solutions* of (9.2.8) which have been extensively studied for the the group $SU(2)$ [9.7].

The existence of local minima is related to another aspect of $SU(2)$ gauge theory: the existence of a *topological charge* $n \in \mathbb{Z}$. To appreciate this aspect, let us confine our attention to the set Ω of finite-action configurations by demanding that $F_{k\ell}(x) = o(|x|^{-2})$. It follows that any $A \in \Omega$ must be *asymptotically pure gauge*, i.e., there exists $r > 0$ depending on A such that, for any x satisfying $|x| > r$,

$$A_k(x) = u_{|k}(x)u(x)^{-1} + o(|x|^{-1}), \tag{9.2.9}$$

where $u(x)$ is regular for $|x| > r$. The crux of the matter is that, in general, u does not admit a continuous extension to all of E_4. Typically, one encounters a point singularity somewhere in the ball $|x| \leq r$. This question brings us to the subject of homotopy. Consider some three-sphere $|x| = R$ where $R > r$. Restricting $u(x)$ to this sphere gives a map $S_3 \to SU(2)$. Such maps fall into homotopy classes. Namely, two such maps belong to the same class if they are continuous deformations of one another. These classes are considered to be elements of $\pi_3(SU(2))$, the third homotopy group of $SU(2)$. By continuity, the homotopy class of u is independent of R provided $R > r$, and, by (9.2.9), may be used to characterize the asymptotic behavior of $A_k(x)$.

The fact that $SU(2)$ is a group plays no role in homotopy theory. It is only the topology that counts. From the topology point of view, the group $SU(2)$ and the sphere S_3 can be identified, and mathematicians have shown that $\pi_3(S_3) = \mathbb{Z}$, which accounts for the occurrence of a topological charge $n \in \mathbb{Z}$. One consequence is that, as a topological space, Ω is in fact the union of components Ω_n (separated from each other like distinct

---

[33]Here we will be content with demanding differentiability. Further restrictions concerning the behavior at infinity are soon to follow. See (9.3.6).

islands). Another is that path integrals become sums over n of integrals over Ω_n: each component may be attributed a separate measure μ_n. Since gauge orbits are connected, they cannot jump between components and hence $n(^uA) = n(A)$. Most strikingly, the index $n \in \mathbb{Z}$ of $A \in \Omega$ may be computed by an integral [9.8]:

$$n(A) = -\frac{1}{16\pi^2} \int dx \operatorname{tr}(F^*_{k\ell} F^{k\ell}). \tag{9.2.10}$$

The minimum of $W(A)$ is zero and attained if $A_k = u_{|k} u^{-1}$ for all $x \in E_4$. As $u(x)$ is continuous, it is homotopically equivalent to a constant function (simply take the family $u(x;t) = u(tx)$, $0 \leq t \leq 1$) and $n(A) = 0$. Thus the absolute minima lie in Ω_0.

From (9.2.10) and $0 \leq -\operatorname{tr}\{(F \pm F^*)^2\}$ one infers that $W(A) \geq 8\pi^2 g^{-2}|n(A)|$ with equality iff $F = \epsilon F^*$ where $\epsilon = \operatorname{sgn}\{n(A)\}$. Therefore, in Ω_n the minimal value of the action is $8\pi^2 g^{-2}|n|$. Despite these appealing features of classical Yang–Mills theory, it seems difficult to judge whether instantons play a decisive role in the quantized version of it.

9.3 The Faddeev–Popov Quantization

We begin our discussion of the quantization procedure by reminding ourselves what goes wrong even on a formal level when writing

$$d\mu(A) = Z^{-1}\mathcal{D}A \exp\left\{-\hbar^{-1}W(A)\right\} \tag{9.3.1}$$

for the probability measure. The reasons for misbehavior are the same as in the Maxwell theory: gauge invariance forces the action to stay constant along a $\mathcal{G}$-orbit and, in a sense, each orbit by itself has infinite volume. The next thing to try is to explicitly break gauge invariance by introducing a suitable gauge fixing term into the action, where 'suitable' means that path integrals of gauge invariant quantities ought to come out independent of the parameter that multiplies the gauge fixing term. Experience has shown that this goal is not so easily reached as in the Abelian case.

The following procedure is purely heuristic and follows a suggestion by Faddeev and Popov [9.9]. The strategy is to formally introduce a δ-function into the path integral whose chief aim is to restrict the integration to the gauge manifold, a section of the $\mathcal{G}$-orbits given by the gauge condition

$$f(A) = C. \tag{9.3.2}$$

Implicit in this condition is the assumption that f stands for some function of $A_k(x)$ and its first derivatives, $A_{k|\ell}(x)$, which take values in the Lie algebra $\mathbf{su}(n)$, and that $C(x)$ is another $\mathbf{su}(n)$-valued function independent of A. Possible candidates for f include:

$$\begin{aligned}
f(A) &= \partial_k A^k & &\text{Landau gauge,}\\
f(A) &= \sum_{k=1}^{3} \partial_k A^k & &\text{Coulomb gauge,}\\
f(A) &= \partial_k A^k + [B_k, A^k] & &\text{Background gauge (B some vector field),}\\
f(A) &= n_k A^k & &\text{axial gauge (n some vector).}
\end{aligned}$$

Infinitesimal gauge transformations $a\colon E_4 \to \mathbf{su}(n)$ are obtained from one-parameter families $u(s) \in \mathcal{G}$ satisfying $u(0) = \mathbb{1}$:

$$u(s) - \mathbb{1} = sa + O(s^2) \qquad (s \in \mathbb{R}) \tag{9.3.3}$$

(with dependence on x suppressed). Whenever defined, the scalar product of two infinitesimal gauge transformations is written

$$(a, b) = -\tfrac{1}{2} \int dx \, \mathrm{tr}\left(a(x)b(x)\right) \tag{9.3.4}$$

so that $\|a\|^2 \equiv (a, a) > 0$ unless $a = 0$ almost everywhere. This gives

$$\mathcal{H} = \{\, a\colon E_4 \to \mathbf{su}(n) \mid \|a\| < \infty \,\} \tag{9.3.5}$$

the structure of a real Hilbert space. It seems natural to go on and put restrictions on the gauge group:

$$\mathcal{G} = \{\, u\colon E_4 \to SU(n) \mid u_{|k}u^{-1} \in \mathcal{H}, \; k = 1, \dots, 4 \,\}. \tag{9.3.6}$$

Let us now consider infinitesimal transformations of the vector potential:

$$^{u(s)}A_k - A_k = s d_k(A)a + O(s^2). \tag{9.3.7}$$

It is not hard to show that $d_k(A)$ is the following (unbounded) operator on $\mathcal{H}$:

$$d_k(A)a = [D_k, a] = a_{|k} + [a, A_k]. \tag{9.3.8}$$

We do not bother to explore the domain of definition except to say that, for any reasonable choice of the domain, we would find the relation $(a, d_k(A)b) = -(d_k(A)a, b)$, i.e., the operator $d_k(A)$ is *antisymmetric*.

Next we consider infinitesimal transformations of the function $f(A)$ appearing in gauge condition (9.3.2):

$$f(^{u(s)}A) - f(A) = -s\mathcal{M}(A)a + O(s^2). \tag{9.3.9}$$

This introduces yet another operator $\mathcal{M}(A)$ on $\mathcal{H}$:

$$\mathcal{M}(A) \;=\; -\partial^k d_k(A) \qquad \text{Landau gauge}$$

$$\mathcal{M}(A) \;=\; -\sum_{i=1}^{3} \partial^k d_k(A) \quad \text{Coulomb gauge}$$

$$\mathcal{M}(A) \;=\; -d^k(B) d_k(A) \qquad \text{background gauge}$$

$$\mathcal{M}(A) \;=\; -n^k d_k(A) \qquad \text{axial gauge.}$$

To exploit gauge invariance one adds the following hypothesis:

There exists an invariant[34] measure (Haar measure) $\mathcal{D}u$ on $\mathcal{G}$. Formally, the volume of the gauge group is written $|\mathcal{G}| = \int \mathcal{D}u$, although this is an infinite constant.

Consider the integral[35]

$$\Delta(A, C) := \int \mathcal{D}u \, \delta\big(f({}^{u}\!A) - C\big), \tag{9.3.10}$$

which defines $\Delta(A, C)$ as a *functional* of A and C with no dependence on x. By definition, the functional is gauge invariant: $\Delta({}^{u}\!A, C) = \Delta(A, C)$ for all $u \in \mathcal{G}$.

In order to be able to "evaluate" $\Delta(A, C)$ when $f(A) = C$ we need to require that the gauge manifold be a section:

For fixed A, $u = \mathbb{1}$ is the only solution of the equation $f({}^{u}\!A) = f(A)$.

It is not a-priori obvious that this condition is fulfilled for any of the cases considered above. Until we know better, it remains a conjecture. But, if it were true, we would get (at least formally)

$$\Delta\big(A, f(A)\big) = |\det \mathcal{M}(A)|^{-1}. \tag{9.3.11}$$

$\det \mathcal{M}(A)$ is said to be the *Fadeev–Popov (FP) determinant*. Again it is a formal object.

Formula (9.3.11) rests on certain properties of the Dirac functional. To explain these, let us consider the one-dimensional case first. Take $U \subset \mathbb{R}$ a neighborhood of $x = 0$. For any bijection $\phi : U \to U$ satisfying $\phi(0) = 0$ and $\phi(x) = mx + O(x^2)$, $m \neq 0$,

$$\int_{U} dx \, \delta\big(\phi(x)\big) = |m|^{-1}. \tag{9.3.12}$$

Armed with this result, one is able to prove that for $U \subset \mathbb{R}^n$ a neighborhood of $x = 0$ and any bijection $\phi : U \to U$ such that $\phi(0) = 0$ and $\phi(x) = Mx + O(x^2)$, $\det M \neq 0$,

$$\int_{U} dx \, \delta\big(\phi(x)\big) = |\det M|^{-1}. \tag{9.3.13}$$

There are situations where an extension of (9.3.13) to infinite-dimensional spaces (like $\mathcal{H}$) makes sense. Granted this, (9.3.11) would follow from (9.3.10) using (9.3.9).

---

[34]Invariance of the measure means that $\int \mathcal{D}u \, h(uv) = \int \mathcal{D}u \, h(u)$ for all $v \in \mathcal{G}$. Standard proofs of the existence of a Haar measure fail: $\mathcal{G}$ is not locally compact.
[35]The integral involves a Dirac functional which is formal and thought of as an infinite-dimensional analogue of a δ-function.

9.3.1 Division by $|\mathcal{G}|$

The considerations that follow are aimed at the elimination of the infinite volume of the gauge group from the numerator and the denominator in formulas like

$$E(h) \;=\; \frac{\int \mathcal{D}A \, \exp(-\hbar^{-1}W(A))h(A)}{\int \mathcal{D}A \, \exp(-\hbar^{-1}W(A))} \;, \tag{9.3.14}$$

where $h(A)$ is some gauge invariant functional: $h(^{u}\!A) = h(A)$. The technique we apply is to introduce the number 1 written as

$$1 = \Delta(A,C)^{-1} \int \mathcal{D}u \, \delta\big(C - f(^{u}\!A)\big) \tag{9.3.15}$$

into the path integral and to change freely the order of integration:

$$E(h) \;=\; \frac{\int \mathcal{D}u \int \mathcal{D}A \, \Delta(A,C)^{-1}\delta(C - f(^{u}\!A)) \exp(-\hbar^{-1}W(A))h(A)}{\int \mathcal{D}u \int \mathcal{D}A \, \Delta(A,C)^{-1}\delta(C - f(^{u}\!A)) \exp(-\hbar^{-1}W(A))} \;. \tag{9.3.16}$$

Gauge invariance is claimed for $\mathcal{D}A$. So

$$E(h) = \frac{\int \mathcal{D}u \int \mathcal{D}A \, \Delta(^{u^{-1}}\!A,C)^{-1}\delta(C-f(A)) \exp(-\hbar^{-1}W(^{u^{-1}}\!A))h(^{u^{-1}}\!A)}{\int \mathcal{D}u \int \mathcal{D}A \, \Delta(^{u^{-1}}\!A,C)^{-1}\delta(C-f(A)) \exp(-\hbar^{-1}W(^{u^{-1}}\!A))} \;. \tag{9.3.17}$$

Recall that the functionals $\Delta(A,C)$, $W(A)$, and $h(A)$ are gauge invariant. Consequently, the variable u disappears from the integrand, relation (9.3.17) becomes

$$E(h) \;=\; \frac{|\mathcal{G}| \int \mathcal{D}A \, \Delta(A,f(A))^{-1}\delta(C - f(A)) \exp(-\hbar^{-1}W(A))h(A)}{|\mathcal{G}| \int \mathcal{D}A \, \Delta(A,f(A))^{-1}\delta(C - f(A)) \exp(-\hbar^{-1}W(A))} \tag{9.3.18}$$

and the infinite group volume cancels. We may now use (9.3.11) to find

$$E(h) \int \mathcal{D}A \, |\det \mathcal{M}| \, \delta(C - f(A))e^{-\hbar^{-1}W(A)} =$$

$$\int \mathcal{D}A \, |\det \mathcal{M}| \, \delta(C - f(A))e^{-\hbar^{-1}W(A)}h(A). \tag{9.3.19}$$

This is not yet the desired result, but we are close to it. What annoys us most is the appearance of a Dirac functional in (9.3.19) pretending that $E(h)$ depends on $C \in \mathcal{H}$, while the original formula (9.3.14) makes no reference to C. However, the philosophy has always been that, for gauge-invariant $h(A)$, the expectation $E(h)$ ought to come out *independent of the gauge condition.* We therefore take for granted that $\boldsymbol{E}(h)$ does not depend in C and see what this assumption will lead us to.

There exists a bona fide Gaussian measure on $\mathcal{H}$ with Fourier transform

$$\int d\nu(C)\,\exp\left\{i(C,a)\right\} = \exp\left\{-\frac{2\hbar g^2}{\lambda}\|a\|^2\right\} \qquad (a \in \mathcal{H},\ \lambda > 0).$$

$$(9.3.20)$$

Integrating both sides of (9.3.19) with respect to ν, utilizing the fact that $\boldsymbol{E}(h)$ does not depend on C, and changing the order of integration, we get

$$\boldsymbol{E}(h)\int \mathcal{D}A\,|\det\mathcal{M}|\exp\left\{-\frac{1}{\hbar}W(A) - \frac{\lambda}{4\hbar g^2}\|f(A)\|^2\right\} =$$
$$\int \mathcal{D}A\,|\det\mathcal{M}|\exp\left\{-\frac{1}{\hbar}W(A) - \frac{\lambda}{4\hbar g^2}\|f(A)\|^2\right\}h(A).$$

$$(9.3.21)$$

Our goal has been achieved: a gauge fixing term has been created to supplement the action. We define

$$W(A;\lambda) = -\frac{1}{8g^2}\int dx\,\mathrm{tr}\left\{F_{k\ell}F^{k\ell} + \lambda f(A)^2\right\}$$

$$(9.3.22)$$

and may thus write

$$\boldsymbol{E}(h) = \frac{\int \mathcal{D}A\,|\det\mathcal{M}|\exp\{-\hbar^{-1}W(A;\lambda)\}h(A)}{\int \mathcal{D}A\,|\det\mathcal{M}|\exp\{-\hbar^{-1}W(A;\lambda)\}}.$$

$$(9.3.23)$$

Gauge invariance of the action has been destroyed, giving most expectations an extra dependence on the parameter λ. But, if everything works well, $\boldsymbol{E}(h)$ does not vary with λ granted the condition $h(A) = h(^uA)$.

One obvious pecularity remains. There is an extra factor, the FP determinant, in the path integral which does not cancel out. Though our "derivation" of formula (9.3.23) has many soft spots, it at least motivates the occurrence of the factor $|\det\mathcal{M}(A)|$ in the integrand, which we claim is characteristic of non-Abelian theories. Why is there no determinant in *Abelian* gauge theories? The answer is as simple as this: since $\mathcal{M} = -\mathbb{1}\Delta$ independent of A in Abelian theories, the determinant drops out of (9.3.23).

Is there a sensible way to get rid of the FP determinant in non-Abelian theories? One way is to choose the axial gauge condition $n^k A_k = C$ in (9.3.19). Then $\mathcal{M} = -\mathbb{1}n^k\partial_k$ independent of A and $\det\mathcal{M}$ cancels out. Another advantage of the axial gauge is that the gauge manifold can be shown to intersect each $\mathcal{G}$-orbit at precisely one point [9.10]. However, the serious disadvantage of the axial gauge is that it destroys the Euclidean invariance[36] of the theory.

---

[36]Euclidean invariance is lost with certainty at the level of the potential A. What is not so clear is whether it is also lost at the level of the field strength F.

9.3.2 Faddeev–Popov Ghosts

It is widely believed that the Landau gauge condition $\partial_k A^k = C$ should be given preference. Let us now examine this case in some detail. Here we find that:

$$(a, \mathcal{M}b) - (\mathcal{M}a, b) = ([a, C], b) \qquad (a, b \in \mathcal{H}) \tag{9.3.24}$$

when $\partial_k A^k = C$. Unless $C = 0$, the operator $\mathcal{M}(A)$ cannot be symmetric. If this is what we want, we had better invoke the Lorentz condition $\partial_k A^k = 0$ in (9.3.19). As for the path integral of (9.3.23), we need to pass to the limit $\lambda \to \infty$ in order to recover the Lorentz condition. Suppose this has been done. Can we prove that $\mathcal{M}(A)$ is a positive operator so as to guarantee that $|\det \mathcal{M}| = \det \mathcal{M}$ at least formally? This question, as it turns out, is very delicate. It may be checked at the point $A = 0$ of the configuration space, where we get an affirmative answer because $\mathcal{M}(0) = -\mathbb{1}\Delta$. One has to work harder to prove that $\mathcal{M}(A) > 0$ continues to hold in a neighborhood of $A = 0$ (see the remarks at the end of this section). Ultimately one assumes that the correct theory is obtained by stripping off the absolute sign from the FP determinant in (9.3.23).

Physically speaking, the FP determinant introduces some extra self-interaction of the gauge field. The most direct attempt to treat this interaction perturbatively starts from

$$\det \mathcal{M} = \det(-\mathbb{1}\Delta) \det(1 + \mathcal{K})$$
$$\mathcal{K}a = -\Delta^{-1}\partial^k[A_k, a] \qquad (a \in \mathcal{H})$$
$$\log \det(1 + \mathcal{K}) = \operatorname{tr} \log(1 + \mathcal{K})$$
$$= \operatorname{tr}\left(\mathcal{K} - \tfrac{1}{2}\mathcal{K}^2 + \cdots\right). \tag{9.3.25}$$

The constant $\det(-\mathbb{1}\Delta)$ cancels out in (9.3.23) and the action $W(A; \lambda)$ receives extra terms from (9.3.25) proportional to $\operatorname{tr} \mathcal{K}^n$ $(n = 1, 2, \ldots)$. These describe *nonlocal* interactions (action at a distance) due to Δ^{-1} in the definition of the operator $\mathcal{K}$. Regarding nonlocality as an undesirable feature of the theory, we reject this approach and look for an alternative.

The most promising attempt to formulate the theory such that locality and Euclidean symmetry are manifest is to enlarge the configuration space even further and introduce redundant variables, i.e., fictitious $\mathbf{su}(n)$-valued scalar fields $\eta(x)$ and $\bar{\eta}(x)$, so-called *ghost fields*, interacting with the gauge field via the operator $\mathcal{M}(A)$:

$$\det(\hbar^{-1}\mathcal{M}) = \int \mathcal{D}\eta \mathcal{D}\bar{\eta} \, \exp\{-\hbar^{-1}(\bar{\eta}, \mathcal{M}\eta)\}$$
$$(\bar{\eta}, \mathcal{M}\eta) = -\tfrac{1}{2}\int dx \operatorname{tr}\left\{\bar{\eta}(x)(\mathcal{M}\eta)(x)\right\} \tag{9.3.26}$$
$$= -\tfrac{1}{2}\int dx \operatorname{tr}\left\{\bar{\eta}^{|k}(x)\eta_{\|k}(x)\right\}.$$

Apart from the ordinary derivative $\bar{\eta}^{|k}$, we introduced the *covariant deriva-tive* via

$$\eta_{\|k}(x) = [D_k, \eta](x) = \partial_k \eta(x) + [\eta(x), A_k(x)]. \tag{9.3.27}$$

In order that representation (9.3.26) be correct, it is necessary to assume that the new fields $\eta(x)$ and $\bar{\eta}(x)$ are *Grassmann variables* (this is the subject of Chap. 10, and the reader is asked to look there for further information). What is actually meant here is this: decompose $\eta = -i \sum_a \eta_a t_a$ and $\bar{\eta} = i \sum_a \bar{\eta}_a t_a$ with respect to the basis $-it_a$ of the Lie algebra $\mathbf{su}(n)$. Then the components η_a and $\bar{\eta}_a$ are generators of some Grassmann algebra.

The final form of the action including the gauge fixing term and ghost part is

$$\boxed{W(A, \eta, \bar{\eta}; \lambda) = -\tfrac{1}{2} \int dx\, \mathrm{tr}\left\{ (2g)^{-2}\big[F_{k\ell} F^{k\ell} + \lambda(A^k_{|k})^2\big] + \bar{\eta}^{|k}\eta_{\|k} \right\}.}$$

$$(9.3.28)$$

Locality has been forcibly restored at the expense of having introduced even more redundant variables than would seem necessary form a naive point of view. It seems that we have to accept and live with oddly behaved particles, Faddeev–Popov ghosts[37], for which we cannot claim any existence in reality. For some this may seem to be too big an atrocity. However, it has only been recently that we have come to understand and appreciate the central importance and specific role of ghost variables in quantizing gauge systems. At a time when it was widely believed that ghosts are nothing but an artifact of the Faddeev–Popov quantization, Becchi, Rouet, and Stora [9.11] on the one hand and Tyutin [9.12-13], on the other, discovered a new kind of symmetry of action (9.3.28), the BRST symmetry, which mixes ghosts with other fields. They advocated the view that ghosts are really necessary to maintain unitarity (of the scattering operator) and for a local description of the full set of symmetries and are hence indispensable. So ghosts can never be completely divorced from the gauge bosons, and the triple $(A, \eta, \bar{\eta})$ ought to be viewed as a single object with intrinsic geometrical significance. It was realized later on [9.14–15] that the BRST argument is even more fundamental than was thought in the beginning. In fact, it can be discovered in quantum mechanics, even classical mechanics, when constrained systems are envisaged, and it relates to the geometry of phase

---

[37]It is clear beyond doubt that Faddeev–Popov ghosts should be fermions with spin zero and no mass, thus violating the spin-statistics theorem. Moreover, there is a multiplet of n^2-1 ghosts since η transforms according to the adjoint representation of the group $SU(n)$, and there are antighosts, and all of them inaccessible to observation. In the Feynman diagrammar, ghosts are confined to contributing to virtual processes only, thus avoiding conflict with the spin-statistics theorem. In other words, physical states have ghost number zero.

space: the BRST discovery opened the door to a new wonderful world where ghosts and nonghosts are united.

Gribov ambiguity. As it turned out, one of our hypotheses is untenable [9.16]: the Lorentz manifold $\partial_k A^k = 0$ *does not provide a global section* of the $\mathcal{G}$-orbits though it is a local section (and the same would be true for the Landau condition $\partial_k A^k = C$ or the Coulomb condition $\nabla \cdot \boldsymbol{A} = C$). This can be seen as follows. The two equations

$$\partial^k A_k = 0 \quad , \quad \partial^k(u A_k u^{-1} + u_{|k} u^{-1}) = 0 \tag{9.3.29}$$

together lead to the nonlinear differential equation

$$\partial^k a_k + [a_k, {}^{u}\!A^k] = 0 \tag{9.3.30}$$

for the function $a_k := u_{|k} u^{-1} \in \mathcal{H}$. Even for the simple ansatz $a_k = \partial_k \alpha$ and under the assumptions (1) α small and (2) A_k sufficiently large, there exists a nontrivial solution of (9.3.30) (where we would have expected $\alpha = 0$ to be the only solution).

If two potentials A and ${}^{u}\!A$ satisfy

$$\partial^k A_k = \partial^k({}^{u}\!A_k), \tag{9.3.31}$$

they are said to be *Gribov copies* of one another. As a rule, Gribov copies are widely separated. It may happen, however, that two Gribov copies A and ${}^{u}\!A$ are so close to each other that $u \in \mathcal{G}$ is infinitesimal, i.e., $u(x) = \mathbb{1} + s a(x) + O(s^2)$ with $|s| \ll 1$: the $\mathcal{G}$-orbit is tangential to the Lorentz manifold. Then we are at the boundary of some uniqueness domain and (9.3.31) becomes equivalent to

$$\mathcal{M}(A)a = 0 \qquad (a \in \mathcal{H}, a \neq 0). \tag{9.3.32}$$

This can be put into a form that resembles the (time-independent) Schrödinger equation with A acting as the potential energy. The problem then is to decide whether zero is an eigenvalue. Starting our search at $A = 0$, we learn from $\mathcal{M}(0) = -\boldsymbol{\Delta}$ that there is no such eigenvalue. As A increases, we get one, two, three or more solutions ("zero energy bound states") at certain values for A. These values mark the boundary between different domains of the configuration space such that any two Gribov copies belong to different domains. The boundary between the 'innermost domain' containing $A = 0$ and the next domain is called the *Gribov horizon* . In principle the horizon is computable from (9.3.32): find A such that $\mathcal{M}(A)$ becomes singular (noninvertible) with a one-dimensional kernel (null space). For further properties concerning the Gribov horizon consult [9.17,18]. It is by no means clear, however, to what extent the existence of Gribov copies influences the quantization of Yang–Mills theories or whether it invalidates the Faddeev–Popov scheme.

Exercise 1. The Maxwell field is known to obey a homogeneous field equation. By analogy, the field strength in non-Abelian gauge theory obeys the differential equation (also known as the Bianchi identity in differential geometry)

$$[D_k, F_{\ell m}] + [D_m, F_{k\ell}] + [D_\ell, F_{mk}] = 0.$$

Prove that this identity is a consequence of the way in which the field F has been defined.

Exercise 2. Use the scalar product $(A, B) = -\frac{1}{2} \int dx \, \mathrm{tr}\left(A_k(x) B^k(x)\right)$ to introduce a distance between two gauge fields A and B and prove that the distance stays constant along the $\mathcal{G}$-orbits in the sense that

$$\|{}^{u}\!A - {}^{v}\!B\| = \|A - B\|.$$

Exercise 3. Let $\mathcal{A}$ stand for the linear space of gauge fields A satisfying $\|A\| < \infty$ (notation from the previous exercise). For fixed $A \in \mathcal{A}$, let $S_A : \mathcal{G} \to \mathbb{R}$ be the functional $S_A(u) = \|{}^{u}\!A\|^2$. Prove two things:

1. S_A is stationary at u iff ${}^{u}\!A$ satisfies the Lorentz condition $\partial_k({}^{u}\!A)^k = 0$.
2. If $\partial_k({}^{u}\!A)^k = 0$ and $\mathcal{M}({}^{u}\!A)$ is positive, S_A has a local minimum at u.

Exercise 4. With $S_0 = S_{A=0}$ (notation from the previous exercise) show that $S_0(u) < \infty$ iff $u \in \mathcal{G}$ (see definition (9.3.6)). Prove also that S_0 is stationary at u iff $u(x)$ is constant. This in particular implies that $u_{|k}u^{-1} = 0$, and hence there is no Gribov copy of $A = 0$ in $\mathcal{A}$.

Exercise 5. The Faddeev–Popov action (9.3.28) contains a linear coupling of the gauge field A to the ghosts. The relevant term of the integrand is of the form

$$\tfrac{1}{2}\mathrm{tr}\left(\bar{\eta}^{|k}[\eta, A_k]\right) = \tfrac{1}{2}\mathrm{tr}\left(j^k A_k\right)$$

with $j^k(x)$ the 'ghost current'. Write $\eta = -i\sum \eta_a t_a$, $\bar{\eta} = i\sum \bar{\eta}_a t_a$, $j^k = -i\sum j_a^k t_a$, and $A_a^k = -i\sum A_a^k t_a$. Assume that $\eta_a\bar{\eta}_b = -\bar{\eta}_b\eta_a$ and

$$\tfrac{1}{2}\mathrm{tr}\left([t_a, t_b]t_c\right) = if_{abc}$$

with f_{abc} the structure constants of the Lie algebra $\mathbf{su}(n)$. Find an explicit expression for the components j_a^k. What sense does it make to write $j^k = \bar{\eta}^{|k}\eta + \eta\bar{\eta}^{|k}$?

9.4 Gauge Theories on a Lattice

The continuum formulation of gauge theories characterizes the gauge field as a function taking values in some Lie algebra $\mathbf{g}$ when G is the underlying group[38]. It is not easy to maintain this characterization when we now replace the Euclidean four-space by the finite lattice $\Lambda = (\mathbb{Z}_N)^4$. For we have to face the fact that, if u is some gauge transformation, $u_{|k}(x)u(x)^{-1}$ ceases to represent some element in the Lie algebra, once the derivative has been replaced by the difference: $u_{|k}(x) \to u(x + e_k) - u(x)$. Wilson [9.19] has put forth an elegant method that gets around this difficulty.

---

[38]It goes without saying that G is assumed to be some compact Lie group of complex $n \times n$ matrices. Modulo discrete factors, every compact Lie group can be thought of as a product of *simple* Lie groups and $U(1)$ factors.

The method requires only consideration of group-valued functions. We no longer deal with the gauge potential $A_k(x)$ directly but rather with a finite system of dynamical variables $U_{xk} \in G$ related to it. Note that the pair of indices, xk, stands for a *link* of the lattice. A link is thought of as connecting two neighboring sites. It is *oriented* in that it leads from the initial site x to $x+e_k$ and it has a length called the lattice spacing. Normally, the lattice spacing is taken to be unity.

By a local gauge transformation we mean a change of the dynamical variables which pays attention to the endpoints of every link xk:

$$U_{xk} \rightarrow {}^u U_{xk} := u(x + e_k)U_{xk}u(x)^{-1} \qquad u\colon \Lambda \rightarrow G. \tag{9.4.1}$$

There is no need to impose any restriction on $u \in \mathcal{G}$ and, most importantly, as a direct product of compact groups, the gauge group is automatically compact:

$$\mathcal{G} = \prod_x G \qquad \text{(the product extends over the sites).} \tag{9.4.2}$$

One beneficial effect of compactness is that the gauge group comes equipped with a normalized Haar measure which is the product measure obtained from the representation (9.4.2),

$$\mathcal{D}u = \prod_x d\nu\big((u(x))\big), \qquad \int_{\mathcal{G}} \mathcal{D}u = 1, \tag{9.4.3}$$

where ν stands for the Haar measure of the group G. So, by construction, the gauge group has 'finite volume'. Moreover, the configuration space Ω underlying the link variables U_{xk}, as a direct product of compact spaces, is a compact space, too:

$$\Omega = \prod_{xk} G \qquad \text{(the product extends over the links).} \tag{9.4.4}$$

This explains why we refer to Wilson's approach as the *compact formulation* of gauge field theory and also why there are theories such as compact quantum electrodynamics (QED) and compact quantum chromodynamics (QCD) etc. The formulation continues to be compact when N grows and also when the lattice becomes infinitely extended, i.e., in the thermodynamic limit. Compactness, however, is inevitably lost in the continuum limit of lattice gauge theory.

The choice of compact formulations as principal theme implies a shift of emphasis from the Lie algebra **g** to the Lie group G and, to some extent a neglect of tools which proved to be important in continuum field theory. So the question arises: given some compact formulation on an infinite lattice, how are we to recover the noncompact formulation, i.e., the continuum version of the theory? It is only after scaling that we are ready to make the connection. Scaling introduces a lattice constant a. Let us write the dynamical variables for the scaled version as

$$U_{xk} = 1\!\!1 + A_k(x)a + O(a^2), \tag{9.4.5}$$

where $x \in (a\mathbb{Z})^4$ and $A_k(x) \in \mathbf{g}$. In the limit as a tends to zero, the gauge field $A_k(x)$ exhibits the correct behavior under gauge transformations: for $u : E_4 \to G$ differentiable and ${}^u\!A_k = uA_ku^{-1} + u_{|k}u^{-1}$,

$$u(x + ae_k)U_{xk}u(x)^{-1} = 1\!\!1 + {}^u\!A_k(x)a + O(a^2). \tag{9.4.6}$$

To prove this, simply expand the left-hand side around $a = 0$.

Conversely, starting from the noncompact continuum formulation, we may construct the associated compact variables on a lattice by applying the exponential map. Quite generally, when C is some oriented curve in the Euclidean continuum, we may define the line integral with respect to C and take the exponential[39] of it:

$$U_C = \exp \int_C dx^k A_k(x). \tag{9.4.7}$$

Lattices contain elementary curves of length a, the lattice spacing. They have been termed *links*. Suppose that C coincides with a link, i.e., with the straight line from x to $x + ae_k$. Then we write $U_C = U_{xk}$, and once we have made this intuitive identification, the result would be a lattice approximation of the continuum theory we started from.

As for noncompact QED, the line integral of A, by virtue of Stokes's theorem, transforms to a surface integral if C happens to be the boundary of an (oriented) surface Q. This is indicated by writing $C = \partial Q$, and Stokes's theorem says that

$$\int_{\partial Q} dx^k A_k(x) = \iint_Q dx^k \wedge dx^\ell F_{k\ell}(x). \tag{9.4.8}$$

Non-Abelian theories are different, but there is an analogous construction called the *path-ordered exponential* or, for short, the *P-exponential*:

$$U_C = \mathrm{P}\exp \int_C dx^k A_k(x). \tag{9.4.9}$$

---

[39]For $\hbar = c = 1$, the gauge field has physical dimension $length^{-1}$. Thus line integrals carry no dimension. If $\hbar \neq 1$, we are forced to write $U_C = \exp\{\hbar^{-1} \int_C dx^k A_k(x)\}$. Notice also that the gauge field has been normalized in such a way that it incorporates the coupling constant. This convention is, however, uncommon in $U(1)$ gauge theory. In addition, all fields are taken to be real in a $U(1)$ theory. So, in principle, the considerations of the present section are applicable to QED, but only after the following replacements have been performed:

$$A \to -ieA, \qquad F \to -ieF, \qquad \tfrac{1}{2}\mathrm{tr} \to 1, \qquad g^2 \to e^2 = 4\pi\alpha$$

The P-exponential is defined to be limit of a sequence of path-ordered products:

$$P \exp \int_C dx^k\, A_k(x) = \lim_{n \to \infty} U_{nn} \cdots U_{n2} U_{n1}$$

$$U_{ni} = \exp \int_{C_{ni}} dx^k\, A_k(x) \qquad\qquad (9.4.10)$$

$$\lim_{n \to \infty} \max_{1 \le i \le n} |C_{ni}| = 0, \qquad C = C_{nn} \circ \cdots \circ C_{n2} \circ C_{n1}.$$

In other words, the curve C has been subdivided into small pieces C_{ni}, starting with C_{n1}, in such a way that their lengths $|C_{ni}|$ tend to zero as $n \to \infty$. In QED and for Q a square of side a,

$$U_{\partial Q} = \mathbb{1} + \iint_Q dx^k \wedge dx^\ell\, F_{k\ell}(x) + O(a^3) \qquad (a \to 0) \qquad (9.4.11)$$

as can be inferred from (9.4.8–9). Path ordering is totally irrelevant in the Abelian case, but becomes very crucial for the analysis of non-Abelian theories.

Path-ordered exponentials are the result a "parallel displacement" with the gauge field $A_k(x)$ defining an "affine connection" for the trivial bundle $E_4 \times \mathbb{C}^n$ whenever we assume the matrix group G to act on vectors $\xi \in \mathbb{C}^n$. Given some vector ξ attached to the point x, an infinitesimal parallel displacement yields the vector $\xi + d\xi$ at the point $x + dx$, where

$$d\xi = dx^k A_k \xi. \qquad\qquad (9.4.12)$$

If a point x moves along some curve C, it carries the vector ξ along whose infinitesimal changes are described by (9.4.12). Suppose that initially the vector was ξ_{i}. Then, at the end of the movement, we get the displaced vector ξ_{f} and (9.4.12) may be integrated formally to yield the following formula for *finite displacements*:

$$\xi_{\mathrm{f}} = U_C \xi_{\mathrm{i}}, \qquad\qquad (9.4.13)$$

where U_C, the parallel transporter, is given by (9.4.9). There are four conditions that characterize finite displacements:

1. The group element U_C depends differentiably on the endpoints of the curve C.
2. $U_{-C} = U_C^{-1}$ with $-C$ the reversed curve.
3. $U_{C'} U_C = U_{C' \circ C}$ when the endpoint of C coincides with the starting point of C' and $C' \circ C$ stands for the concatenation of the two curves.
4. A local gauge transformation changes U_C to ${}^u U_C = u(x_{\mathrm{f}}) U_C u(x_{\mathrm{i}})^{-1}$, where x_{i} and x_{f} denote the endpoints of C.

Generally the result of a finite parallel transport will depend on the particular curve, not merely on the endpoints of it. Therefore, if ξ is subjected to a parallel displacement around a loop, i.e., if $C = \partial Q$, it may not be true that $\xi_{\mathrm{i}} = \xi_{\mathrm{f}}$. Such a property is called *nonintegrability* and relates to the presence of 'curvature'. To arrive at a local measure of the curvature, we take $U_{\partial Q}$ and contract the surface Q to the infinitesimal neighborhood of a point. It is seen from (9.4.11) and (9.4.16) below that the field strength $F_{k\ell}(x)$ characterizes the infinitesimal behavior of $U_{\partial Q}$ and hence provides a local measure of the curvature. The formula (9.2.3) connecting the field strength F with the potential

A parallels the formula that connects the Riemann curvature tensor with the Christoffel symbols in Riemannian geometry, and gauge transformations in Yang–Mills theory, parallel arbitrary coordinate transformations in general relativity leaving the geometry of spacetime unchanged. Thus, it should not come as a surprise that, from the differential geometry point of view, gravity falls into the category of gauge theories, the only difference being that, in gravity, gauge transformation are *external* (affect the reference frames at the points x), while, in Yang–Mills theory, they are *internal* (leave the coordinatization of spacetime unaffected). There is an interesting harmony among internal and external gauge theories in that the classical action (no matter fields) serves as a global measure of the *total curvature* in the system, and classical equilibrium is reached if there is no curvature at all. Therefore, *classical* field theory says that it is *matter* that generates curvature. By contrast, the *quantized* field allows for zero-point fluctuations of the curvature even in the absence of matter.

Equation (9.4.11) helps to make a "good" guess of what corresponds to the field strength in the compact formulation of a gauge theory. The idea is to consider *plaquettes* of the unit lattice Λ. These are oriented elementary squares bounded by four links as shown in Fig. 9.1.

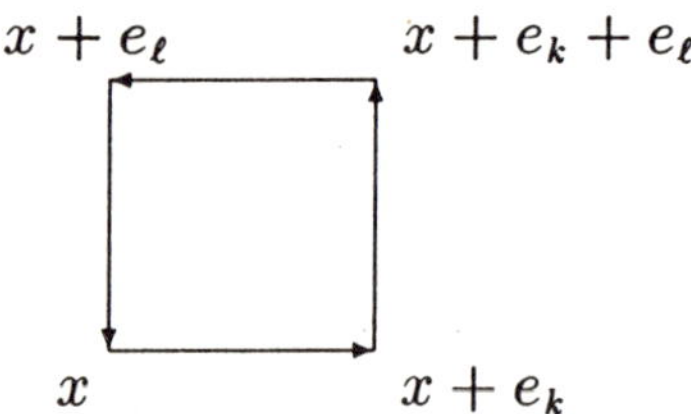

Fig. 9.1. A plaquette of the lattice with the base point x

Any plaquette has a base point x (Fig. 9.1: the lower left corner) and two directions, e_k and e_ℓ. It may therefore be specified by a triple, $p = xk\ell$, where $1 \leq k < \ell \leq 4$. The boundary is traversed as shown in Fig. 9.1. In doing so, two of the links are obviously traversed in the opposite direction, taking into account that links are given a natural orientation. The rule is then that this is taken care of by replacing U by U^{-1} for any such link within the path-ordered product:

$$U_{\partial p} := U_{x,\ell}^{-1} U_{x+e_\ell,k}^{-1} U_{x+e_k,\ell} U_{x,k}. \tag{9.4.14}$$

By construction, this quantity has a simple behavior under gauge transformations:

$$^u U_{\partial p} = u(x) U_{\partial p} u(x)^{-1} \tag{9.4.15}$$

with x the base point of the plaquette p. The relation to the field strength $F_{k\ell}(x)$ can be read off from the following:

After scaling when the lattice spacing is a, a small-a expansion yields

$$U_{\partial p} = \mathbb{1} - a^2 F_{k\ell}(x) + O(a^3), \qquad p = \text{plaquette}(xk\ell), \tag{9.4.16}$$

with $F_{k\ell} = A_{k|\ell} - A_{\ell|k} + [A_k, A_\ell]$ and $U_{x,k} = \exp\{aA_k(x)\}$, granted the differentiability of $A_k(x)$.

One has to work a little harder for the proof of this statement, but it is nevertheless straightforward. The Euclidean action appropriate for the lattice formulation is

$$W_0(U) = \frac{1}{4g^2} \sum_p \operatorname{tr}\left(\mathbb{1} - U_{\partial p}\right)^*\left(\mathbb{1} - U_{\partial p}\right)$$

$$= \frac{1}{2g^2} \sum_p \operatorname{tr}\left(\mathbb{1} - \tfrac{1}{2}(U_{\partial p} + U_{\partial p}^*)\right) \geq 0 \tag{9.4.17}$$

The summation extends over all plaquettes and $\sum_p$ ought to be interpreted as $\sum_x \sum_{k<\ell}$. As the lattice spacing tends to zero, we obtain the approximation we·aimed at:

$$W_0(U) \approx -\frac{1}{8g^2} \sum_x a^4 \operatorname{tr}\left(F_{k\ell}(x)F^{k\ell}(x)\right) \tag{9.4.18}$$

For we know that $\sum_x a^4 (\cdots)$ tends to $\int dx\,(\cdots)$.

Path integrals are with respect to the Gibbs measure

$$d\mu(U) = Z^{-1}\mathcal{D}U \exp\left\{-W_0(U)\right\}, \tag{9.4.19}$$

where $\mathcal{D}U = \prod_{xk} d\nu(U_{xk})$ (the product is over the links). Schematically, we write expectations values as

$$E(f) = \int d\mu(U)\,f(U). \tag{9.4.20}$$

On first thought, it is puzzling that the compact formulation, which uses matrix-valued link variables, successfully avoids the introduction of gauge-fixing terms and ghost fields. Nor is gauge invariance ever broken at any stage of the quantization procedure. On second thought, it is clear that the simplifying features of the Wilson approach can all be traced and attributed to the compactness of the gauge group on the lattice. It should also be noted that the absence of gauge fixing terms renders expectation values like $E(f)$ gauge invariant even if f itself is not invariant, that is to say, if $f(^uU) \neq f(U)$ for some $u \in \mathcal{G}$. Gauge invariance of the underlying measure has many peculiar effects. One is that it forces some expectation values of gauge-dependent objects to come out trivial. Take the group $SU(n)$ for example and consider a path C on the lattice leading from the point x to the point x'. If $x \neq x'$,

$$\int d\mu(U)\,U_C = 0 \tag{9.4.21}$$

because, if some $n \times n$ matrix A satisfies $A = u'Au^{-1}$ for all $u, u' \in SU(n)$, then $A = 0$.

At present, there is nothing that could throw some light on the relation between the Faddeev–Popov and the Wilson quantization procedures. Do they refer to different theories with no connection whatsoever? What could possibly be the role of the gauge-fixing parameter λ in a compact setting of the theory? The following answer seems very suggestive: λ is some order parameter necessary to distinguish different vacuum states that arise from the Wilson model in the thermodynamic limit. But such an answer would imply spontaneous breaking of gauge symmetry and thus contradicts Elitzur's theorem [9.20]. Nevertheless, the answer could still be correct, provided one takes the limit $a \to 0$, too. So the ultimate solution to the problem is expected to come from performing the continuum limit since it is in this step that the gauge group $\mathcal{G}$ becomes noncompact. For more serious attempts in this direction see [9.21–23].

9.5 Wegner–Wilson Loops

9.5.1 The Static Approximation in Minkowskian Field Theory

The tools so far developed already enable us to characterize, in a gauge-invariant manner, the force acting between massive particles and is mediated by the gauge field. Since we do not have the complete theory including dynamical fermions at our disposal, we must be content with discussing the static limit when masses become very large. To motivate the subsequent analysis, we go back to Minkowskian QED and consider two charged particles of mass M interacting via the exchange of a photon. Most textbooks on field theory treat this problem extensively and show that each particle exerts a Coulomb force on the other particle when $M \to \infty$. This is so because a heavy particle loses on dynamical degrees of freedom in the static limit, which simplifies matters considerably[40]. The Feynman propagator of the vector potential takes the form

$$(\Omega, TA_\mu(x)A_\nu(x')\Omega) = -g_{\mu\nu}D_F(x - x')$$
$$+ \text{ gauge-dependent terms,} \tag{9.5.1}$$

where $g_{\mu\nu} = \mathrm{diag}(1, -1, -1, -1)$ and

$$D_F(x) = \frac{1}{(2\pi)^4} \int dp \, \frac{e^{-ipx}}{p^2 + i\epsilon}. \tag{9.5.2}$$

The presence of a heavy particle at rest is felt by the electromagnetic field through the external conserved current

---

[40]In the Feynman diagrammar, the limit $M \to \infty$ has the effect that there are no contributions coming from diagrams which contain the heavy particle in internal lines. This explains what is meant by the phenomenon of *decoupling*. It is typical of the Abelian case that, if all charged particles decouple from the photon in the above way, the gauge potential is seen to be quantized as a *free field*.

$$j^0 = q\delta^3(\boldsymbol{x}) , \quad j^1 = j^2 = j^3 = 0. \tag{9.5.3}$$

Moreover, whatever the current, it acts as a source of a classical potential:

$$A_\mu^{\text{class}}(x) = (\Omega, TA_\mu(x)A(j)\Omega). \tag{9.5.4}$$

From (9.5.3) and (9.5.4),

$$A_0^{\text{class}} = qV(r), \quad A_i^{\text{class}} = 0, \ i = 1, 2, 3, \qquad r = |\boldsymbol{x}|, \tag{9.5.5}$$

where

$$V(r) = -\int_{-\infty}^{\infty} dx^0 \, D_F(x) = \frac{1}{4\pi r} . \tag{9.5.6}$$

As for the $SU(n)$ gauge theory in the presence of matter fields, one would like to argue in a similar fashion. However, even in the static limit, when matter and gauge bosons decouple, the gauge potential does not behave like a free field owing to the non-Abelianness of the gauge group. Nevertheless, some of the above formulas are still correct. The presence of a heavy particle is felt as an external current $j^\mu(x) \in \mathbf{su}(n)$, which in turn produces a classical potential $A_\mu^{\text{class}}(x) = (\Omega, TA_\mu(x)A(j)\Omega) \in \mathbf{su}(n)$. But there is as yet no way to predict the precise shape of such a potential.

We are used to thinking that correlations decay:

$$(\Omega, TA_\mu(x)A_\nu(x')\Omega) \to 0 \qquad |\boldsymbol{x} - \boldsymbol{x}'| \to \infty. \tag{9.5.7}$$

Such decay may be fast (exponentially) or slow (algebraically) depending on whether there is a spectral gap or not. In such a situation, two massive particles, if bound by their mutual potential, could be separated and ultimately freed by raising the total energy. This is what we observe in QED. There are, however, reasons to believe that QCD, whose gauge structure is based on the group $SU(3)$ (the so-called color group), behaves differently: the mutual potential of a quark–antiquark pair will *increase* with the separation of the pair[41]. This phenomenon which is strongly suggestive and also supported by experiment is referred to as *quark confinement*.

Let us now turn to the Euclidean formulation and see whether we get changes. It should not surprise us that, in the static limit, the potential pays no attention to the metric involved, be it Euclidean or Minkowkian. As for the case of QED, the potential $V(r) = (4\pi r)^{-1}$ is seen also to follow from the Schwinger function of the free photon field:

---

[41]This is correct only as long as *vacuum polarization* does not screen the color charge. That is, when distance between the quark-antiquark pair is sufficiently large, the energy stored in the surrounding gauge field will suffice to produce two or more *real* $q\bar{q}$-pairs. Hence, by adding energy, a color-neutral system will go over into a new hadronic state consisting of several color-neutral hadrons. Still, for heavy quarks, all effects coming from vacuum polarization may be neglected.

$$V(r) = \int_{-\infty}^{\infty} dx^4 \, S(x) \qquad r = |\boldsymbol{x}|, \tag{9.5.8}$$

$$\boldsymbol{E}(A_k(x)A_\ell(x')) = \delta_{k\ell} S(x - x') + \textit{gauge-dependent terms.} \tag{9.5.9}$$

The reason for this behavior may be phrased as follows: the Fourier transform $\tilde{S}(\boldsymbol{p}, p^4)$ results from an analytic continuation of $-\tilde{D}_F(p^0, \boldsymbol{p})$. A subsequent time integration projects onto $p^0 = p^4 = 0$. But then $\tilde{S}(\boldsymbol{p}, 0) = -\tilde{D}_F(0, \boldsymbol{p})$.

How does the integral in (9.5.8) arise in the Euclidean context? Naturally, it arises in very much the same way as in the Minkowskian case. First, an arbitrary conserved (Euclidean) current j^k acts as a source of a (Euclidean) potential:

$$A_k^{\text{class}}(x) = \boldsymbol{E}(A_k(x)A(j)) = \frac{1}{4\pi^2} \int dx' \, \frac{j_k(x')}{(x - x')^2}. \tag{9.5.10}$$

For reasons we have already mentioned, however, we are interested in $j^1 = j^2 = j^3 = 0, j^4 = q\delta^3(\boldsymbol{x})$. This way we recover the electrostatic potential of a point charge q.

9.5.2 Loop Variables in Euclidean QED

We have argued that the electromagnetic field responds to a charge whose current is j by producing a classical potential. This potential in turn is felt by a second charge whose current is j'. The mutual energy E is symmetric with respect to the interchange of j and j' and may be calculated as follows:

$$E = \lim_{T \to \infty} \frac{1}{4\pi^2 T} \int_{x_4 = -T/2}^{x_4 = T/2} dx \int_{x_4' = -T/2}^{x_4' = T/2} dx' \, \frac{j^k(x) j_k'(x')}{(x - x')^2} \,. \tag{9.5.11}$$

Notice that E is obtained in a two-step process: first, one determines the mutual action of the two currents between times $-T/2$ and $T/2$, and, second, one interprets *energy* as *action per time* in the limit $T \to \infty$. When the charges are separated by a distance R, then $E \to 0$ as $R \to \infty$.

To elucidate the ties of formula (9.5.11) to the fundamental loop variables

$$A_C = \int_C dx^k \, A_k(x), \tag{9.5.12}$$

we consider the expectation value

$$\boldsymbol{E}(A_C A_C) = \int_C dx^k \int_C dx_k' \, S(x - x') \tag{9.5.13}$$

for C a rectangle in Euclidean spacetime whose sides are T and R (*Wegner–Wilson loop* [9.19,24]) as depicted in Fig. 9.2.

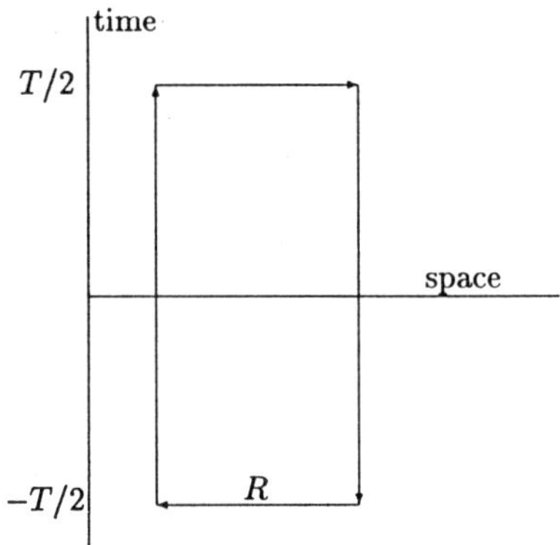

Fig. 9.2. A rectangular loop in Euclidean spacetime

Comparison with (9.5.11) shows that $\frac{1}{2}\boldsymbol{E}(A_C A_C)$ represents the finite-time action of two pointlike charges 1 and -1 at rest and separated by the distance R plus the selfaction of each charge. Owing to the infinite self-energy of pointlike charges, the integral on the right-hand side of (9.5.13) is singular in a continuum theory. It can, however, be regularized by the replacement $S(x) \to S_a(x)$, where[42]

$$S_a(x) = \begin{cases} (2\pi x)^{-2} & \text{if } x^2 > a^2 \\ 0 & \text{otherwise,} \end{cases} \tag{9.5.14}$$

to obtain $\boldsymbol{E}_a(A_C A_C)$, which is finite if $a > 0$. The integration is now elementary and yields a number of terms:

$$\begin{aligned} 2\pi^2 \boldsymbol{E}_a(A_C A_C) = {} & \frac{T}{a} - \log\frac{T}{a} + \log\left(1 + \frac{T^2}{R^2}\right) - 2\frac{T}{R}\arctan\frac{T}{R} \\ & + \frac{R}{a} - \log\frac{R}{a} + \log\left(1 + \frac{R^2}{T^2}\right) - 2\frac{R}{T}\arctan\frac{R}{T} \end{aligned} \tag{9.5.15}$$

$(0 \le \arctan q \le \pi/2)$. As has been suggested before, the potential by which two particles of opposite charge attract each other should be the result of the following limit:

$$V_a(R) := \lim_{T \to \infty} \frac{1}{2T} \boldsymbol{E}_a(A_C A_C). \tag{9.5.16}$$

Indeed, from (9.5.15) we infer[43] that

---

[42] The procedure of introducing the regulator a should be viewed as an attempt to cure the bad ultraviolet behavior of QED. Notice, however, that the proposed replacement destroys *positivity*: $S(x)$ has a positive definite Fourier transform, while $S_a(x)$ has not. As a consequence, there is no guaranty that $\boldsymbol{E}_a(A_C A_C) \ge 0$. A more appropriate procedure would be to introduce a lattice. Then the lattice spacing a would play the role of the ultraviolet regulator.

[43] To maintain positivity, assume that $\pi a < R$. This in particular implies that, in QED, we should not trust (9.5.17) when R becomes small.

$$V_a(R) = \frac{1}{4\pi^2 a} - \frac{1}{4\pi R} \, , \qquad\qquad (9.5.17)$$

which is quite satisfactory since the constant $(4\pi^2 a)^{-1}$ is unobservable and may be ignored.

Having justified the use of loop integrals (9.5.12) in the problem of determining the static forces, we find it desirable to go on and rewrite everything in terms of the even more fundamental loop variables, members of the underlying group $U(1)$:

$$U_C = \exp\{-ieA_C\} = \exp\{-ie \int_C dx^k \, A_k(x)\}. \qquad\qquad (9.5.18)$$

The application of U_C is immediate, once we realize that A_C is a Gaussian variable and thus

$$\boldsymbol{E}(U_C) = \exp\{-\tfrac{1}{2}e^2 \boldsymbol{E}(A_C A_C)\}. \qquad\qquad (9.5.19)$$

This just means that, after regularization, we will get the asymptotic formula

$$\boldsymbol{E}_a(U_C) = \exp\{-e^2 V_a(R)T\} \qquad T \to \infty. \qquad\qquad (9.5.20)$$

9.5.3 Area Law or Perimeter Law?

The asymptotic behavior of $\boldsymbol{E}(U_C)$ may be examined in a more systematic manner, treating space and time on the same footing. Take for instance some loop C and subject it to a dilatation that affects all coordinates simultaneously. As for the rectangular loop, it would mean that R and T grow, while R/T stays fixed. It is characteristic of QED that, asymptotically,

$$\boldsymbol{E}_a(U_C) = \exp\{-\sigma|C|\}, \qquad |C| = 2R + 2T \, , \quad R, T \to \infty, \qquad (9.5.21)$$

where $\sigma = e^2/(8\pi^2 a)$ as can be inferred from (9.5.19) and (9.5.15). This remarkable result is pleasantly simple and suggests that it is generally valid: C is now taken to be any loop with $|C|$ its perimeter. The asymptotic behavior expressed by (9.5.21) is called the *perimeter law*. The constant σ which appears in the exponent deals with the self-energy of two charges, e and $-e$, since it is selfenergy that remains when two point charges are separated:

$$\lim_{R \to \infty} \tfrac{1}{2}(-e)eV_a(R) = -\frac{e^2}{8\pi^2 a}. \qquad\qquad (9.5.22)$$

We leave QED and address the same problem in the context of Yang–Mills theory. For $A_k(x)$ the gauge field, loop integrals like $\int_C dx^k A_k$ are singular objects unless some ultraviolet cut-off is employed. It is therefore wise to always work with a lattice. If the lattice spacing is a and C is an rectangular (R,T)-loop, we expect $\boldsymbol{E}(U_C)$ to be a function of R/a and T/a. So we might as well take $a = 1$.

Recall that the unitary matrix U_C has been defined as a path-ordered product. In the present situation, it is regarded as a product over links $c_1, \ldots, c_n$, traversed by the rectangular (R,T)-path C (such that R and T are integers):

$$U_C = U_{c_n} \cdots U_{c_2} U_{c_1} \, , \qquad U_{c_i} = \begin{cases} U_{xk} & \text{if } c_i : x \rightsquigarrow x + e_k \\ U_{xk}^{-1} & \text{if } c_i : x + e_k \rightsquigarrow x. \end{cases} \qquad (9.5.23)$$

The fact that C is the boundary of some rectangle Q is written $C = \partial Q$. This notation, however, is slightly misleading because U_C also depends on x, the site where the path starts (and ends). This has to do with the other fact that U_C depends on the gauge. Both properties are particular to non-Abelian gauge theories. Nevertheless, these dependencies disappear if we decide to take the trace of U_C.

Two different asymptotic laws suggest themselves:

$$\boldsymbol{E}(\operatorname{tr} U_{\partial G}) \to \begin{cases} \exp\{-\sigma|\partial G|\} & \text{perimeter law (deconfinement)} \\ \exp\{-\kappa|G|\} & \text{area law (confinement)} \end{cases} \qquad (9.5.24)$$

$(\sigma > 0, \kappa > 0)$. Though these options are favored by their "naturalness", other options cannot be excluded as far as we know. It should also be clear that the correctness of the conjecture can only be tested on a truly infinite lattice.

If the perimeter law were correct, isolated massive colored particles – call them *quarks* – would be seen in nature. Let us demonstrate that the area law accounts for the empirical fact that isolated quarks do not exist. For Q a (R,T)-rectangle, we have, on the one hand,

$$\boldsymbol{E}(\operatorname{tr} U_{\partial Q}) \to \exp\{-V(R)T\}, \qquad T \to \infty \qquad (9.5.25)$$

and, on the other hand,

$$\boldsymbol{E}(\operatorname{tr} U_{\partial Q}) \to \exp\{-\kappa RT\}, \qquad R, T \to \infty. \qquad (9.5.26)$$

Therefore,

$$V(R) = \kappa R \qquad R \to \infty. \qquad (9.5.27)$$

As a consequence of the area law, we have found that quarks and antiquarks attract each other with a constant force κ. It is generally believed that this situation arises in QCD, the theory of strong interaction, with color group $SU(3)$.

9.6 The $SU(n)$ Higgs Model

Interesting physics is expected to come only from adding more structure, more field variables, and more particles (quarks, leptons, Higgs bosons). The simplest model that extends Yang–Mills theory and is still within the limits of boson field theory is the $SU(n)$ Higgs model with some bearing on the Glashow–Salam–Weinberg theory of electroweak interactions. It concerns the interaction of a Higgs particle with the gauge field. The action is conveniently written in terms of the lattice variables:

$$W(U, \phi) = W_0(U) + \tfrac{1}{2} \sum_{xk} |D_{xk}\phi(x)|^2 + \lambda \sum_x \left(|\phi(x)|^2 - \kappa^2\right)^2 \qquad (9.6.1)$$

for $\lambda > 0$. The first term on the right-hand side designates the Wilson action (9.5.17) and has been defined as a sum over plaquettes, while the second term is a sum over links. In writing (9.2.1) we have associated a covariant derivative with each link:

$$D_{xk}\phi(x) = \phi(x + e_k) - U_{xk}\phi(x). \qquad (9.6.2)$$

To justify this strange looking ansatz for D_{xk}, we apply the process of scaling which turns the expression $\phi(x + e_k) - U_{xk}\phi(x)$ on $\mathbb{Z}^4$ into

$$a\phi(x + ae_k) - \exp\big(aA_k(x)\big)a\phi(x) = a^2 D_k(x)\phi(x) + O(a^3) \qquad (9.6.3)$$

on $(a\mathbb{Z})^4$. The operator $D_k(x) \equiv \partial_k - A_k(x)$ is recognized to be the standard covariant derivative characteristic of a continuum gauge theory. The third term in (9.6.1), called the Higgs potential, is a sum over sites and represents the selfinteraction of the Higgs field. The significance of the parameter κ is that, in the classical limit ($\hbar \to 0$), there are equilibrium solutions satisfying $|\phi(x)|^2 = \kappa^2$.

Though the Higgs field ϕ behaves like a *scalar* with respect to *external* transformations (i.e., rotations of the lattice), it is still a *vector* with respect to 'internal' transformations. It follows from (9.6.2) that the Higgs field transforms according to the natural representation[44] of the group $SU(n)$, which means that ϕ has n complex ($2n$ real) components:

$$\begin{pmatrix} \phi_1 + i\phi_2 \\ \phi_3 + i\phi_4 \\ \vdots \\ \phi_{2n-1} + i\phi_{2n} \end{pmatrix}. \qquad (9.6.4)$$

Under a gauge transformation, $\phi(x) \to {}^u\phi(x) = u(x)\phi(x)$. This relation emphasizes the role of the complex components. We may, however, write

---

[44]We carefully avoid the term "fundamental representation", which leads to confusion. For instance, the group $SU(3)$ has two inequivalent fundamental representation but only one natural representation.

$|\phi|^2 = \sum_{\alpha=1}^{2n} \phi_\alpha^2$, thus emphasizing the role of the real components and the $O(2n)$ symmetry of the Higgs potential.

Non-Abelianness is of central importance in any Higgs model. One may nevertheless ask: what kind of model would emerge from the present scheme if the group $SU(n)$ were replaced by the Abelian group $U(1)$? A moments reflection tells us that the result would simply be a lattice version of the four-dimensional Ginzburg–Landau field theory. In a way, the unquantized Higgs model can be viewed as being a non-Abelian generalization of the Ginzburg–Landau model.

Quantization proceeds along the familiar lines, and path integrals are taken with respect to the measure

$$d\mu(\phi, U) = Z^{-1} \mathcal{D}\phi \, \mathcal{D}U \, \exp\left\{ - \hbar^{-1} W(\phi, U) \right\}, \tag{9.6.5}$$

where $\mathcal{D}\phi = \prod_{\alpha=1}^{2n} \prod_x d\phi_\alpha(x)$ (Lebesgue measure on $\mathbb{R}^m$, $m = 2nN^4$) and $\mathcal{D}U$ is the Haar measure of the gauge group $\mathcal{G}$.

We treat the quantized Higgs field as a random vector $\boldsymbol{\Phi}(x)$. By a symmetry argument and for a finite lattice, one obtains

$$\boldsymbol{E}\big(\boldsymbol{\Phi}(x)\big) \equiv \int d\mu(\phi, U) \, \phi(x) = 0 \tag{9.6.6}$$

even though ansatz (9.6.1) suggests that there is an anomalous average (i.e., a "condensate"). This however addresses the question whether there is a spontaneous breaking of the symmetry $\phi \to -\phi$ in the infinitely extended system since previous analysis has shown that such a breakdown can only be observed in the thermodynamic limit. Hence, to stabilize the anomalous average we need to introduce an external source j conjugate to ϕ (so j has $2n$ real components) and to study the partition function

$$Z\{j\} = \int \mathcal{D}\phi \int \mathcal{D}U \, \exp\left\{ \hbar^{-1} \left[\sum_x j(x)\phi(x) - W(U, \phi) \right] \right\}. \tag{9.6.7}$$

The next step is to define $W_{\text{eff}}^*(j) = \hbar \log Z\{j\}$ and to perform a Legendre transformation followed by the thermodynamic limit to obtain the *effective* $O(2n)$-symmetric potential:

$$U_{\text{eff}}(r) = \lim_{N \to \infty} N^{-4} \sup_j \left(\sum_x j(x) r - W_{\text{eff}}^*(j) \right) \qquad (r \in \mathbb{R}^{2n}). \tag{9.6.8}$$

This provides a convex function of $2n$ variables whose minimum may or may not be unique. To decide which case is realized in the system and how nonuniqueness depends on the parameters $\hbar$, g, λ, and κ is an ambitious and formidable task. There are, however, ways to put lower and upper limits on the effective potential. A first result in this direction may be phrased as follows:

The effective potential of the $SU(n)$ Higgs field admits bounds

$$w_2(r) \leq U_{\text{eff}}(r) \leq w_1(r), \tag{9.6.9}$$

where $r \in \mathbb{R}^{2n}$. The functions $w_i(r)$ are convex and $O(2n)$-symmetric. Each of them is obtained via a Legendre transformation:

$$w_i(r) = \sup_s \big(sr - w_i^*(s)\big) \qquad (s \in \mathbb{R}^{2n}, i = 1, 2)$$

$$w_i^*(s) = \hbar \log \int_{\mathbb{R}^{2n}} dr \, \exp\left\{ \hbar^{-1}\big(sr - V_i(r)\big) \right\} \tag{9.6.10}$$

with

$$V_1(r) = 3ng^{-2} + 4r^2 + \lambda(r^2 - \kappa^2)^2$$
$$V_2(r) = \lambda(r^2 - \kappa^2)^2. \tag{9.6.11}$$

The upper bound in (9.6.9) follows from applying Jensen's inequality to the inner integral of (9.6.8) and from the integral

$$\int \mathcal{D}U \, W(U, \phi) = \sum_x V_1\big(\phi(x)\big), \tag{9.6.12}$$

while the lower bound results from the inequality $W(U, \phi) \geq \sum_x V_2(\phi(x))$.

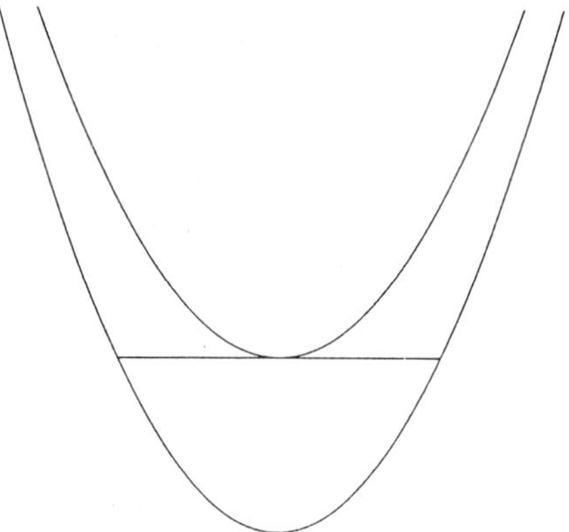

Fig. 9.3. A sketch of the lower and upper bounds of the effective potential as convex functions of $|r|$ and a plateau of maximal width between them

Suppose now that the minimum of the effective potential is not unique, i.e., $U_{\text{eff}}(r)$ stays constant within a ball $|r| \leq c$ with $c > 0$. Let c be the largest radius with that property. So c is the maximal value for the "condensate". The existence of convex lower and upper bounds for $U_{\text{eff}}(r)$ indicates that c cannot be arbitrarily large:

Let a be the radius of the ball $\{ r \in \mathbb{R}^{2n} \mid w_2(r) \leq A \}$ with $A = \inf\{ w_1(r) \mid r \in \mathbb{R}^{2n} \}$ and $w_i(r)$ given by (9.6.10). Then $c \leq a$.

The details of the argument can be read off from Fig. 9.3; a formal proof is left to the reader. The lesson to be drawn is this: while it frequently seems that the effective potential is not a very tractable object, simple expressions for lower and upper bounds may emerge that are meaningful even in the thermodynamic limit.

Exercise 1.   Use symmetry arguments to verify that, for A as defined above, one has $A = w_1(0) = -w_1^*(0)$ and thus

$$e^{-\hbar^{-1}A} = \int_{\mathbb{R}^{2n}} dr\, e^{-\hbar^{-1}V_1(r)}.$$

Think of A as a function of the coupling g. Show that $A(g) = A(0) + 3ng^{-2}$.

Exercise 2.   For ν the Haar measure on $SU(n)$, show that $\int d\nu(u)\, u = 0$. Hint: use invariance. From (9.4.3), $\int \mathcal{D}U\, U_{xk} = 0$ and $\int \mathcal{D}U\, U_{\partial p} = 0$, where xk is a link and p is a plaquette. Use this result to prove the integral formula (9.6.12). Observe that, on a periodic four-dimensional lattice, there are four links and six plaquettes attached to each site x.

Exercise 3.   What are the classical equilibrium configurations of the Higgs model? This is to say: when is $W(U, \phi)$ minimal? Naturally, the action takes zero as its minimal value.

10 Fermions

Clearly this is a subject in which common sense will have to guide
the passage between the Scylla of mathematical Talmudism and
the Charibdis of mathematical nonsense.

Jeremy Bernstein

This chapter deals with fermions which, according to common terminology,
are particles obeying Fermi–Dirac statistics. By the spin-statistics theorem
[7.2,3] we are assured that fermions possess half-integer spin. Moreover, it
seems a well established fact that the most fundamental fermions such as
quarks and leptons have spin $\frac{1}{2}$ and are thus adequately described by the
Dirac theory.

A major part of the Dirac formalism is purely algebraic. It concerns
the spinor calculus, the properties of γ-matrices, and symmetry transformations. Another part, very different from the first, deals with the rules
of quantization based on anticommutation relations and the construction
of the fermionic Fock space. The systematic technique of Feynman is then
available for the evaluation of the scattering amplitudes and has become
a central theme and thus occupies a large portion of any textbook on the
subject (see for instance [7.18]). There is now an ample supply of explicit
calculations proving the practical usefulness of the Dirac formalism.

It remains to show how the second-quantized Dirac theory fits into the
framework of path integration. To this end two things need to be done: (1)
adapt the spinor calculus to the Euclidean metric and (2) reinterpret spinor
fields (normally operators on Fock space) as elements of some Grassmann
algebra.

10.1 The Dirac Field in Minkowski Space

We start out by reviewing the basic elements of the Dirac formalism in
Minkowski space and take $\hbar = c = 1$ everywhere in our formulas. Consider
some particle of mass $m > 0$. The assertion is that there exists an antiparticle of the same mass and opposite charge. Both particle and antiparticle are
described by a common spinor field ψ with components ψ_a $(a = 1, \ldots, 4)$,
and, if interactions are absent, the equations are

$$(i\gamma^\mu \partial_\mu - m)\psi(x) = 0$$

$$\gamma^\mu\gamma^\nu + \gamma^\nu\gamma^\mu = 2g^{\mu\nu}\,\mathbb{1}, \qquad g^{\mu\nu} = \mathrm{diag}(1, -1, -1, -1). \tag{10.1.1}$$

Though the particular choice of the γ-matrices is irrelevant for most purposes, it proves convenient to work with the representation

$$\gamma^0 = \begin{pmatrix} 0 & \mathbb{1} \\ \mathbb{1} & 0 \end{pmatrix}, \qquad \gamma^k = \begin{pmatrix} 0 & -\sigma_k \\ \sigma_k & 0 \end{pmatrix} \qquad (k = 1, 2, 3) \tag{10.1.2}$$

(σ_k = Pauli matrices) and to write $\partial\!\!\!/ = \gamma^\mu \partial_\mu$. To fully specify the second-quantized free Dirac field, it suffices to write down its two-point Wightman functions:

$$\begin{aligned}
(\Omega, \psi(x)\bar\psi(y)\Omega) &= (i\partial\!\!\!/ + m)\Delta_+(x - y; m) \\
(\Omega, \psi(x)\psi(y)^T \Omega) &= 0 \qquad (\Omega = \text{vacuum}).
\end{aligned} \tag{10.1.3}$$

It is common practice to interpret $\bar\psi\psi$ as a scalar but $\psi\bar\psi$ as a 4×4 matrix.

Before discussing further technical aspects of the Dirac theory, we should emphasize a point of methodological importance. Pauli's nonrelativistic theory of the electron uses two-component spinors, while Dirac needs four-spinors. As Weyl and others have emphasized, two components are sufficient *in principle* even in the relativistic situation where we wish to include the antiparticle. The issue is clearly related to the existence of left- and right-handed spinors (which, as we know, play different roles in weak interactions) and can be traced to the fact that there is a γ^5 matrix. Phrased more formally, Dirac spinors are reducible under Lorentz transformations[45]. In fact the decomposition into irreducible left- and right-handed two-spinors (Weyl spinors) is accomplished by

$$\psi = \begin{pmatrix} \xi \\ \chi \end{pmatrix}. \tag{10.1.4}$$

Adopting the Weyl philosophy we would restrict attention to ξ (or χ) and write

$$\begin{aligned}
(\Box + m^2)\xi(x) &= 0 \\
(\Omega, \xi(x)\xi^*(y)\Omega) &= i\sigma^\mu \partial_\mu \Delta_+(x - y)
\end{aligned} \tag{10.1.5}$$

where $(\sigma_\mu) = \{\mathbb{1}, \sigma_1, \sigma_2, \sigma_3\}$. All this is to say that, in the Dirac formalism, one deliberately introduces redundant field variables. The extension to a four-component formalism will certainly be appreciated when it comes to studying interactions. However, even without interactions there is sufficient reason for redundancy: space reflection or *parity transformation* acts linearly on four-spinors and does not involve derivatives:

---

[45] From the point of view of representation theory, Dirac spinors transform according to the direct-sum represention $\mathcal{D}^{\frac{1}{2},0} \oplus \mathcal{D}^{0,\frac{1}{2}}$ of the group $L_+^\uparrow$ consisting of all proper orthochronous Lorentz transformations [7.2]. The projectors associated with the irreducible subspaces are $\frac{1}{2}(\mathbb{1} \pm \gamma^5)$. As we pass to the Euclidean formulation, Dirac spinors will then transform according to the unitary representation $(\frac{1}{2}, 0) \oplus (0, \frac{1}{2})$ of the group $SO(4)$. For zero mass (i.e., neutrinos) left- and right-handed spinors decouple.

$$\psi(x) \to \psi'(x) = \pm\gamma^0\psi(x^0, -\boldsymbol{x}) \tag{10.1.6}$$

(the sign in front is the intrinsic parity of the field). Thus a parity transformation is seen to interchange the two-spinors ξ and χ. If we were to eliminate χ by making use of the Dirac equation, the resulting parity transformation of the remaining field ξ would still be linear but no longer look simple because it involves derivatives.

Let us now look at another operation, *charge conjugation*:

$$\psi^c(x) = C\bar\psi(x)^T, \qquad \bar\psi = \psi^*\gamma^0, \tag{10.1.7}$$

with C a unitary 4×4 matrix satisfying

$$(\gamma^\mu C)^T = \gamma^\mu C, \qquad C^T = C^* = C^{-1} = -C. \tag{10.1.8}$$

These properties guarantee that charge conjugation is a symmetry of the free Dirac equation (though it is not a symmetry of the world), and $\psi^{cc} = \psi$. With regard to representation (10.1.2),

$$C = i\gamma^2\gamma^0 = \begin{pmatrix} -\epsilon & 0 \\ 0 & \epsilon \end{pmatrix}, \qquad \epsilon = \begin{pmatrix} 0 & 1 \\ -1 & 0 \end{pmatrix}. \tag{10.1.9}$$

To see the effect of charge conjugation even more clearly, we pass to the two-component formalism, writing

$$\psi^c = \begin{pmatrix} \xi^c \\ \chi^c \end{pmatrix} \tag{10.1.10}$$

so that

$$\begin{aligned} \xi^c(x) &= -\epsilon\chi^*(x)^T \\ \chi^c(x) &= \epsilon\xi^*(x)^T. \end{aligned} \tag{10.1.11}$$

It is instructive to see that *left* and *right* are interchanged also under a charge conjugation. However, even with the redundancy present in the Dirac formalism, charge conjugation is not a linear operation. So why not extend the formalism by introducing even more redundant field variables? Such an extension may not be necessary in all circumstances, but it can be convenient in some. Let us therefore consider *bispinors* having eight components:

$$\Psi = \begin{pmatrix} \psi \\ \psi^c \end{pmatrix} \quad \text{or} \quad \Psi_a = \begin{cases} \psi_a & \text{if } a = 1,\dots,4 \\ \psi^c_{a-4} & \text{if } a = 5,\dots,8. \end{cases} \tag{10.1.12}$$

It is clear from the former definition of charge conjugation that

$$\Psi^c = \boldsymbol{C}\Psi, \qquad \boldsymbol{C} = \begin{pmatrix} 0 & \mathbb{1} \\ \mathbb{1} & 0 \end{pmatrix} \tag{10.1.13}$$

with $\boldsymbol{C}$ an 8×8 matrix. The significance of the eight-component formalism will come to light more clearly in the Euclidean context.

For now, let us investigate the consequences of the use of bispinors in vacuum expectations. As a first step, one verifies two simple relations:

$$(\Omega, \psi(x)\psi^c(y)^T \Omega) = (\Omega, \psi^c(x)\psi(y)^T \Omega)$$
$$(\Omega, \psi^c(x)\psi^c(y)^T \Omega) = (\Omega, \psi(x)\psi(y)^T \Omega).$$

$$(10.1.14)$$

From these relations and $\psi^c(y)^T = -\bar{\psi}(y)C$, we infer that

$$(\Omega, \Psi(x)_a \Psi(y)_b \Omega) = W_2(x, y)_{ab},$$

$$(10.1.15)$$

where (interpreting $W_2(x, y)$ as some 8×8 matrix)

$$W_2(x, y) = \begin{pmatrix} 0 & -(i\partial\!\!\!/ + m)C \\ -(i\partial\!\!\!/ + m)C & 0 \end{pmatrix} \Delta_+(x - y; m).$$

$$(10.1.16)$$

More generally, we wish to consider n-point Wightman functions

$$W_n(x_1, \ldots, x_n)_{a_1 \cdots a_n} = (\Omega, \Psi_{a_1}(x_1) \cdots \Psi_{a_n}(x_n)\Omega)$$

$$(10.1.17)$$

for which there exists a recursive scheme, since it is a free Fermi field:

$$W_n(x_1, \ldots, x_n)_{a_1 \cdots a_n} =$$

$$(10.1.18)$$

$$\sum_{i=1}^{n-1} (-1)^{i+1} W_{n-2}(x_1, \ldots, \hat{x}_i, \ldots, x_{n-1})_{a_1 \cdots \hat{a}_i \cdots a_{n-1}} W_2(x_i, x_n)_{a_i a_n}$$

$(n = 4, 6, 8, \ldots)$. It is characteristic of the Fermi case that each contribution to the above sum enters with the sign $(-1)^{i+1}$, which would be absent in the Bose case. The extra sign results from invoking anticommutation relations.

Exercise. In an attempt to eliminate the undesirable vacuum expectation value of the electromagnetic current $\bar{\psi}\gamma^\mu\psi$, Jauch and Rohrlich [10.1] proposed a slightly altered definition, namely

$$j^\mu = \tfrac{1}{2}(\bar{\psi}\gamma^\mu\psi - \bar{\psi}^c\gamma^\mu\psi^c).$$

Explain why this is in fact an expression bilinear in the bispinor Ψ.

10.2 The Euclidean Dirac Field

When contrasting the Euclidean version with the Minkowskian version of the Dirac theory, it seems necessary to carefully distinguish the two sets of γ matrices involved. However, since from now on we shall deal exclusively with γ matrices that relate to the Euclidean metric, we may use the same notation without danger of confusion, i.e., we define matrices

$$\gamma^k = \begin{pmatrix} 0 & i\sigma_k \\ -i\sigma_k & 0 \end{pmatrix} \quad (k = 1, 2, 3), \qquad \gamma^4 = \begin{pmatrix} 0 & \mathbb{1} \\ \mathbb{1} & 0 \end{pmatrix}. \tag{10.2.1}$$

They are Hermitean and satisfy[46]

$$\gamma^k \gamma^\ell + \gamma^\ell \gamma^k = 2\delta^{k\ell}\mathbb{1} \qquad (k, \ell = 1, \dots, 4). \tag{10.2.2}$$

Let us continue to write $\partial\!\!\!/ = \gamma^k \partial_k$ (with the sum extending over $k = 1, \dots, 4$) and $\Psi(x)$ for the Euclidean bispinor with $x \in E_4$. As has been shown in Sect. 7.2, the two-point function $\Delta_+(x - y; m)$ admits an analytic continuation in the difference variable. In Euclidean points $x, y \in E_4$, the resulting (Schwinger) function is $S(x - y; m)$. In view of (10.1.15), this motivates the introduction of a Euclidean two-point function (formally obtained by a replacement $ix^0 \to x^4$, $iy^0 \to y^4$) for the bispinor $\Psi(x)$:

$$S_2(x, y) = \begin{pmatrix} 0 & (\partial\!\!\!/ - m)C \\ (\partial\!\!\!/ - m)C & 0 \end{pmatrix} S(x - y; m). \tag{10.2.3}$$

Notice that $(\gamma^k C)^T = \gamma^k C$ continues to hold for the Euclidean γ matrices. Hence, there is no need for a new charge conjugation matrix C. It satisfies $C^T = C^{-1} = -C$ as before. From $S(x - y, m) = S(y - x, m)$ and $((\partial\!\!\!/ - m)C)^T = (\partial\!\!\!/ + m)C$ one infers that the matrix function $S_2(x, y)$ is *antisymmetric*:

$$S_2(x, y)_{ab} = -S_2(y, x)_{ba} . \tag{10.2.4}$$

It is obvious that, if we recursively define (for $n = 4, 6, 8, \dots$)

$$S_n(x_1, \dots, x_n)_{a_1 \cdots a_n} = \tag{10.2.5}$$

$$\sum_{i=1}^{n-1} (-1)^{i+1} S_{n-2}(x_1, \dots, \hat{x}_i, \dots, x_{n-1})_{a_1 \cdots \hat{a}_i \cdots a_{n-1}} S_2(x_i, x_n)_{a_i a_n},$$

the Euclidean n-point functions S_n will inherit the property of being antisymmetric from S_2. More precisely, we find that

$$S_n(x_1, \dots, x_n)_{a_1 \cdots a_n} = \operatorname{sgn}\pi \, S_n(x_{\pi(1)}, \dots, x_{\pi(n)})_{a_{\pi(1)} \cdots a_{\pi(n)}} \tag{10.2.6}$$

with π an arbitrary permutation of $1, 2, \dots, n$. This property is fundamental to constructing the free Euclidean Dirac field. For it is reasonable to expect that antisymmetry of all n-point functions is the result of the field variables being antisymmetric themselves:

$$\Psi_a(x)\Psi_b(y) = -\Psi_b(y)\Psi_a(x). \tag{10.2.7}$$

---

[46]Mathematically speaking, the Euclidean γ matrices are the generators of some Clifford algebra over the real-linear space E_4.

Notice that this relation is assumed to hold for *all* $x, y \in E_4$. Looking upon (10.2.7) from the algebraic point of view, it is permissible to speak of the variables $\Psi_a(x)$ as elements of some Grassmann algebra.

Though recursion formula (10.2.5) applies to free fields only, relation (10.2.6) and thus (10.2.7) will be adopted as basic principles. We therefore postulate the following:

Euclidean Fermi fields have antisymmetric Schwinger functions, Bose fields symmetric Schwinger functions, to all orders.

In the preceding discussion we have been sloppy about singularities at coinciding points. Therefore, in a more formal approach, 'smearing' should be applied to the Schwinger functions from the very start. This implies smearing of the field $\Psi(x)$, a technique which, as it turns out, is desirable for another reason: in constructing generating functionals, we would like to deal with *commuting quantities* as arguments of ordinary functions (like the exponential function). But how can integrals of the form

$$\Psi(\eta) = \int dx\, \Psi_a(x)\eta^a(x), \tag{10.2.8}$$

with $\eta^a(x)$ certain source functions, be commuting objects? To achieve this goal we boldly write

$$\eta^a(x)\eta^b(y) + \eta^b(y)\eta^a(x) = 0$$
$$\eta^a(x)\Psi_b(y) + \Psi_b(y)\eta^a(x) = 0 \tag{10.2.9}$$
$$\Psi_a(x)\Psi_b(y) + \Psi_b(y)\Psi_a(x) = 0$$

$(a, b = 1, \ldots, 8)$, i.e., we wish to interpret the expressions $\Psi_a(x)$ and $\eta^a(x)$, indexed by x and a, as members of a *common* Grassmann algebra. Fields and sources are viewed as "conjugate" or "dual" to each other, a fact that will soon be reflected by the abstract algebraic framework (where we shall write η_a^* in place of Ψ_a).

Before indulging in Grassmann theory, we shall proceed heuristically to see what relations (10.2.9) could do for us. One property we should always be aware of is *anticommutativity* in the sense that

$$\int dx\, \Psi_a(x)\eta^a(x) = -\int dx\, \eta^a(x)\Psi_a(x), \tag{10.2.10}$$

but *commutativity* in the sense that $\Psi(\eta)$ commutes with "everything", i.e., with every element of the Grassmann algebra. Whereas $\Psi_a(x)$ and $\eta^a(x)$ are *odd* objects, $\Psi(\eta)$ is an *even* object in a sense to be made precise. Evenness implies that a quantity behaves more or less like a classical observable would do, though it is not yet a *number*. In order to get numbers, two further steps appear to be necessary. The first and most decisive step in this direction will always be to take expectations $\langle \cdot \rangle$. In particular, we would write the moments as

$$\langle \Psi(\eta)^{2n} \rangle = (-1)^n \int dx_1 \cdots \int dx_{2n} \, \langle \Psi_{a_1}(x_1) \cdots \Psi_{a_{2n}}(x_{2n}) \rangle \times$$
$$\eta^{a_1}(x_1) \cdots \eta^{a_{2n}}(x_{2n})$$
$$= (-1)^n \int dx_1 \cdots \int dx_{2n} \, S_{2n}(x_1, \ldots, x_{2n})_{a_1 \cdots a_{2n}} \times$$
$$\eta^{a_1}(x_1) \cdots \eta^{a_{2n}}(x_{2n})$$

$$\langle \Psi(\eta)^{2n+1} \rangle = 0. \tag{10.2.11}$$

A little reflection serves to convince ourselves that the nth moment $\langle \Psi(\eta)^n \rangle$ completely determines the Euclidean n-point function S_n, whereby the property of antisymmetry, as expressed by (10.2.6), is automatically guaranteed by the Grassmann property $\eta^a(x)\eta^b(y) = -\eta^b(y)\eta^a(x)$. In a way, taking expectations eliminates the field variables $\Psi_a(x)$ leaving an *even* element in the subalgebra generated by the dual objects $\eta^a(x)$. So to get numbers the second and seemingly trivial step is to expand that quantity with respect to the "source" η.

There is a combinatorial aspect of recursion relation (10.2.5) which is most easily expressed by stating that the free Fermi field is characterized by a Gaussian generating functional:

$$E\{\eta\} := \exp\left(-S\{\eta\}\right) = \exp\left\{\tfrac{1}{2}\langle \Psi(\eta)^2 \rangle\right\} = 1 + \sum_{n=1}^{\infty} \frac{1}{n!}\langle \Psi(\eta)^n \rangle \tag{10.2.12}$$
$$S\{\eta\} := \tfrac{1}{2} \int dx \int dy \, S_2(x,y)_{ab} \, \eta^a(x)\eta^b(y).$$

This also parallels the Bose case, where $S_2(x,y)$ is the integral kernel of the inverse of a differential operator. Consider, therefore, the operator

$$\mathcal{F} := \begin{pmatrix} 0 & C(\partial\!\!\!/ + m) \\ C(\partial\!\!\!/ + m) & 0 \end{pmatrix} \tag{10.2.13}$$

whose inverse is[47]

$$\mathcal{F}^{-1} = \begin{pmatrix} 0 & (\partial\!\!\!/ - m)C \\ (\partial\!\!\!/ - m)C & 0 \end{pmatrix} (-\Delta + m^2)^{-1}. \tag{10.2.14}$$

Now, $\langle x|(-\Delta+m^2)^{-1}|y\rangle = S(x-y;m)$, and comparison with (10.2.3) yields

$$\langle x|(\mathcal{F}^{-1})_{ab}|y\rangle = S_2(x,y)_{ab} \tag{10.2.15}$$

so that

$$S\{\eta\} = \tfrac{1}{2}(\eta, \mathcal{F}^{-1}\eta) \equiv \tfrac{1}{2} \int dx \, \eta^a(x)[\mathcal{F}^{-1}\eta]_a(x). \tag{10.2.16}$$

---

[47]Recall that Δ stands for the four-dimensional Laplace operator.

The operators $\mathcal{F}$ and $\mathcal{F}^{-1}$ are *antisymmetric* in the following sense. Suppose $u(x)$ and $u'(x)$ happen to be ordinary functions taking values in $\mathbb{C}^8$. Then $(u, \mathcal{F}u')$ and $(u, \mathcal{F}^{-1}u')$ are antisymmetric bilinear forms in u, u'. In particular, they would vanish if $u = u'$. Not so for Grassmann variables: $(\eta, \mathcal{F}^{-1}\eta')$ is *symmetric* bilinear, and thus $(\eta, \mathcal{F}^{-1}\eta) \neq 0$ in general.

The principal achievement is that properly smeared bosonic and fermionic field variables can now be treated almost on the same footing. There is Gaussian functional for every free field and there is also a covariance operator that goes with it. What remains to be answered is whether we can connect the generating functional $S\{\eta\}$ to the Euclidean action. Very soon (in Sect. 10.5.4) we are going to demonstrate that there exists the analogue of the Fourier–Laplace transformation for Grassmann functionals. For now, we shall merely state that it makes sense to write the Gaussian functional as a path integral:

$$E\{\eta\} = \exp\{-\tfrac{1}{2}(\eta, \mathcal{F}^{-1}\eta)\} = \frac{\int \mathcal{D}\Psi \, \exp(\Psi(\eta) - W\{\Psi\})}{\int \mathcal{D}\Psi \, \exp(-W\{\Psi\})} \qquad (10.2.17)$$

with W the Euclidean action of the Dirac theory:

$$W\{\Psi\} = \tfrac{1}{2}(\Psi, \mathcal{F}\Psi) \equiv \tfrac{1}{2} \int dx \, \Psi_a(x)[\mathcal{F}\Psi]^a(x), \qquad (10.2.18)$$

where one treats all eight variables Ψ_a , $a = 1, \ldots, 8$, as *independent*. This unfamiliar-looking expression changes to the traditional form that we are better acquainted with once we rewrite (10.2.18) in terms of four-spinors:

$$W\{\psi, \bar{\psi}\} =$$

$$\tfrac{1}{2} \int dx \, \{\psi(x)^T C(\slashed{\partial} + m)\psi^c(x) + \psi^c(x)^T C(\slashed{\partial} + m)\psi(x)\}$$

$$= \int dx \, \bar{\psi}(x)(\slashed{\partial} + m)\psi(x). \qquad (10.2.19)$$

In doing so we have been careful not to use anything besides the relations

$$\Psi = \begin{pmatrix} \psi \\ \psi^c \end{pmatrix}, \qquad \psi^c(x) = C\bar{\psi}(x)^T. \qquad (10.2.20)$$

We have, in particular, given up the idea that there should be a relation connecting ψ and $\bar{\psi}$. Such a relation is characteristic of the Minkowskian formalism but is unreasonable in the Euclidean context. In fact, all useful relations among the various field components are those of (10.2.20). They suggest three different ways of introducing independent variables:

- choose a bispinor Ψ, or
- choose two unrelated Dirac spinors ψ and ψ^c, or
- choose one Dirac spinor ψ and one conjugate spinor $\bar{\psi}$ (both unrelated).

Since the third of these alternatives is more widespread as far we can judge from the existing literature on the subject, we would like to make contact with the established notation and to add a few remarks about the passage from the compact notation using bispinors to the more common but less convenient notation using ψ and $\bar\psi$. First, we will decompose the bispinor η suitably:

$$\eta = \begin{pmatrix} 0 & C \\ C & 0 \end{pmatrix} \begin{pmatrix} \zeta \\ \zeta^c \end{pmatrix}, \qquad \zeta^c(x) = C\bar\zeta(x)^T. \tag{10.2.21}$$

This introduces ζ and $\bar\zeta$ as independent variables in place of η. Second, changing variables in (10.2.17) leads to

$$E\{\zeta,\bar\zeta\} = Z^{-1} \int \mathcal{D}\psi\mathcal{D}\bar\psi \exp\left(\bar\zeta(\psi) + \bar\psi(\zeta) - W\{\psi,\bar\psi\}\right)$$

$$Z = \int \mathcal{D}\psi\mathcal{D}\bar\psi \exp\left(-W\{\psi,\bar\psi\}\right) \tag{10.2.22}$$

$$\bar\zeta(\psi) + \bar\psi(\zeta) = \int dx \left(\bar\zeta(x)\psi(x) + \bar\psi(x)\zeta(x)\right).$$

For the free field of mass m,

$$E\{\zeta,\bar\zeta\} = \exp \int dx\, \bar\zeta(x)(\partial\!\!\!/ + m)^{-1}\zeta(x). \tag{10.2.23}$$

10.2.1 External Vector Potentials

There is an immediate nontrivial extension of the above formulas which deals with a Dirac particle coupled to an external vector potential. The most familiar expressions of this sort come from studying electromagnetism. Consider therefore a fixed (Euclidean) electromagnetic vector potential $A_k(x)$, define $A\!\!\!/ = \gamma^k A_k$, and write the action as

$$W_A\{\Psi\} = \int dx\, \bar\psi(\partial\!\!\!/ + m + ieA\!\!\!/)\psi = \tfrac{1}{2}(\Psi, \mathcal{F}_A\Psi), \tag{10.2.24}$$

where we have introduced the antisymmetric operator

$$\mathcal{F}_A = \begin{pmatrix} 0 & C(\partial\!\!\!/ + m - ieA\!\!\!/) \\ C(\partial\!\!\!/ + m + ieA\!\!\!/) & 0 \end{pmatrix}. \tag{10.2.25}$$

Our example demonstrates a remarkable fact true for most fermionic models: the action is *bilinear* in the Fermi fields. It is thus easy to extend the previous reasoning and do a Gaussian integral (at least formally). As for the above situation, we would get an immediate answer for the generating functional:

$$E_A\{\eta\} = \exp\{-\tfrac{1}{2}(\eta, \mathcal{F}_A^{-1}\eta)\}$$

$$= \exp \int dx\, \bar\zeta(x)(\partial\!\!\!/ + m + ieA\!\!\!/)^{-1}\zeta(x), \tag{10.2.26}$$

where

$$\mathcal{F}_A^{-1} = \begin{pmatrix} 0 & -(\partial\!\!\!/ + m + ieA\!\!\!/)^{-1}C \\ -(\partial\!\!\!/ + m - ieA\!\!\!/)^{-1}C & 0 \end{pmatrix}. \tag{10.2.27}$$

By expanding (10.2.26) to second order we obtain the two-point function

$$\langle \psi(x)\bar{\psi}(y) \rangle = \langle x | (\partial\!\!\!/ + m + ieA\!\!\!/)^{-1} | y \rangle. \tag{10.2.28}$$

It should be noted, however, that the inversion of the operators involved is purely formal as it stands but can be made rigorous under suitable technical assumptions about the vector potential, a subject of its own which we shall not go into. An even more formal but challenging object needed later is the so-called *fermion determinant*. This determinant arises as the normalizing constant of the path integral for a fixed external vector potential:

$$\begin{aligned} Z_A &= \int \mathcal{D}\psi \mathcal{D}\bar{\psi} \, \exp\left\{ -\int dx \, \bar{\psi}(\partial\!\!\!/ + m + ieA\!\!\!/)\psi \right\} \\ &= \det(\partial\!\!\!/ + m + ieA\!\!\!/). \end{aligned} \tag{10.2.29}$$

It makes no sense to write down expressions like (10.2.29) unless one specifies an approximation procedure (replacing operators by matrices).

The following sections are devoted to the underlying algebraic structure. Here we shall treat mainly those parts of multilinear algebra which are closely related to the above circle of ideas. In particular we are aiming at the development of an integral calculus on Grassmann algebras. As it turns out, the technique of path integration can be adapted to this situation. The resulting expressions are well defined provided the Grassmann algebra is finite-dimensional but can be very dubious if the algebra is infinite-dimensional.

Exercise 1. Take another look at (10.2.29). Explain why the quotient

$$Z_A/Z_0 = \det\left(1 + ieA\!\!\!/(\partial\!\!\!/ + m)^{-1}\right)$$

is better behaved, i.e., expand $\log(Z_A/Z_0)$ into a power series around $A = 0$ using the formal identity $\log \det = \operatorname{tr} \log$ and find conditions that guarantee convergence. Argue why the above quotient is an even function of the coupling e.

Exercise 2. Demonstrate that recursion relation (10.2.5) is equivalent to

$$\langle \Psi(\eta)^n \rangle = (n-1) \langle \Psi(\eta)^{n-2} \rangle \langle \Psi(\eta)^2 \rangle$$

$(n = 4, 6, 8, \ldots)$. Use this to verify that

$$\langle \Psi(\eta)^{2n} \rangle = \frac{(2n)!}{n!2^n} \langle \Psi(\eta)^2 \rangle^n$$

$(n = 1, 2, 3, \ldots)$. Since odd moments vanish, this yields

$$\exp\left\{\tfrac{1}{2}\langle\Psi(\eta)^2\rangle\right\} = 1 + \sum_{n=1}^{\infty} \frac{1}{n!}\langle\Psi(\eta)^n\rangle.$$

10.3 Grassmann Algebras

Let E be an n-dimensional complex vector space. A map

$$S : \underbrace{E \times \cdots \times E}_{p} \longrightarrow \mathbb{C} \tag{10.3.1}$$

is said to be *p-linear* if $S(u_1, \ldots, u_p)$ is separately linear in each argument $u_k \in E$. It is said to be *antisymmetric* if, for any permutation π of $\{1, \ldots, p\}$, we have

$$S(u_{\pi(1)}, \ldots, u_{\pi(p)}) = \mathrm{sgn}\,\pi\, S(u_1, \ldots, u_p), \tag{10.3.2}$$

where $\mathrm{sgn}\,\pi$ denotes the signature of the permutation π. For each $p \geq 1$ we consider the space $A^p(E)$ of p-linear antisymmetric functions on E. Since functions in $A^p(E)$ may be multiplied by complex numbers and added, it is clear that $A^p(E)$ has the structure of a complex vector space. By convention, $A^0(E) = \mathbb{C}$. It it then fairly easy to prove that

$$\dim A^p(E) = \binom{n}{p} \qquad 0 \leq p \leq n$$
$$A^p(E) = 0 \qquad p > n. \tag{10.3.3}$$

The Grassmann product map $A^p(E) \times A^q(E) \rightarrow A^{p+q}(E)$ assigns to any two vectors $S \in A^p$ and $T \in A^q$ the vector $ST \in A^{p+q}$, where[48]

$$ST(u_1, \ldots, u_{p+q}) =$$
$$\frac{1}{p!q!} \sum_{\pi} \mathrm{sgn}\,\pi\, S(u_{\pi(1)}, \ldots, u_{\pi(p)}) T(u_{\pi(p+1)}, \ldots, u_{\pi(p+q)}) \tag{10.3.4}$$

(the sum extends over all permutations of $\{1, \ldots, p+q\}$), and ST is said to be the *Grassmann product* of S and T. One checks that the associative law $R(ST) = (RS)T$ is satisfied; notice that the product is not commutative. But there is the rule

$$ST = (-1)^{pq} TS \qquad \text{if } S \in A^p,\ T \in A^q. \tag{10.3.5}$$

---

[48]Mathematicians prefer to write $S \wedge T$ in place of ST.

This is so since the permutation π that takes $(1,\ldots,p,p+1,\ldots,p+q)$ into $(p+1,\ldots,p+q,1,\ldots,p)$ obeys $\text{sgn}\,\pi = (-1)^{pq}$.

The Grassmann product (10.3.4) makes the direct sum of vector spaces,

$$A(E) = \bigoplus_{p=0}^{n} A^p(E), \tag{10.3.6}$$

a graded algebra, called the *Grassmann algebra over E*. Accordingly, an element of $A(E)$ can always be written as a sum $S_0 + S_1 + \cdots + S_n$ such that $S_p \in A^p(E)$. It follows from $\sum_p \binom{n}{p} = 2^n$ that

$$\dim A(E) = 2^{\dim E}. \tag{10.3.7}$$

As is typical of a graded algebra, A may be decomposed into an even and an odd part:

$$A = A_+ \oplus A_-$$
$$A_+ = A^0 \oplus A^2 \oplus \cdots \qquad \text{(even subspace)}$$
$$A_- = A^1 \oplus A^3 \oplus \cdots \qquad \text{(odd subspace)}.$$

The reason this decomposition helps to grasp the essentials of the multiplicative structure can be seen from the following rule, a consequence of (10.3.5):

$$ST = \begin{cases} TS & \text{if } S \in A_+ \text{ or } T \in A_+ \\ -TS & \text{if } S \in A_- \text{ and } T \in A_-. \end{cases} \tag{10.3.8}$$

In particular:

The even part A_+ of a Grassmann algebra A is a commutative subalgebra.

This is essential when it comes to studying functions on A_+. Without commutativity there would be no functional calculus. We may, for instance, take polynomials, logarithms, and exponentials, add and multiply functions, form compositions $f \circ g$, and do a lot more, just as if we were dealing with ordinary numbers. Even functional equations remain valid such as

$$e^S e^T = e^{S+T} \qquad S, T \in A_+. \tag{10.3.9}$$

However, we should not expect $S \in A_+$ to always have an inverse S^{-1}. To illustrate this point we take some element $S \in A^2$ and consider arbitrary powers:

$$S^m = \underbrace{SS \cdots S}_{m} \in A^{2m} \tag{10.3.10}$$

If $m > 0$, there does not seem to be any problem and, moreover, $S^m = 0$ for $2m > n$. If $m = 0$, the result is $S^0 = 1 \in A^0$ by convention. Inverse powers however cannot be defined. By contrast, the exponential function

$$e^{-S} = \sum_{m \geq 0} \frac{(-1)^m}{m!} S^m \in A_+,$$

(10.3.11)

though it truncates when $2m > n$, has an inverse element which of course is e^S. The reason: $e^{-S}e^S = 1$ by the functional calculus.

Let $(e_i)_{i=1,\ldots,n}$ be a basis for E. So any vector $u \in E$ may be written $\sum u^i e_i$ with coordinates $u^i \in \mathbb{C}$. We define special elements $\eta^i \in A^1$:

$$\eta^i(u) = u^i, \qquad i = 1,\ldots,n, \qquad u \in E,$$

(10.3.12)

that is to say, η^i assigns the ith coordinate to any vector $u \in E$. The following elementary properties express the fact that the η^i *generate the whole Grassmann algebra*:

1. *The η^i anticommute:*

$$\eta^i \eta^k + \eta^k \eta^i = 0 \qquad (i, k = 1,\ldots,n).$$

(10.3.13)

2. *Each vector $S \in A^p$ may be represented as*

$$S = \frac{1}{p!} \sum_{i_1 \cdots i_p} s_{i_1 \cdots i_p} \eta^{i_1} \cdots \eta^{i_p}$$

(10.3.14)

with $s_{i_1 \cdots i_p}$ certain complex coefficients which are antisymmetric with respect to permutations of the indices.

Returning to the above example, we infer that, with respect to a basis in E, any element $S \in A^2(E)$ assumes the form

$$S = \tfrac{1}{2} \sum_{ik} s_{ik} \eta^i \eta^k, \qquad s_{ik} = -s_{ki} \in \mathbb{C}.$$

(10.3.15)

Conversely, given some antisymmetric $n \times n$ matrix (s_{ik}), one may assign some element $S \in A^2(E)$ to it. Moreover,

$$e^{-S} = \sum_{m \geq 0} \frac{(-1)^m}{m!} S^m,$$

(10.3.16)

where

$$\frac{1}{m!} S^m = \frac{1}{m!} \left(\tfrac{1}{2} \sum_{ik} s_{ik} \eta^i \eta^k \right)^m = \frac{1}{(2m)!} \sum_{i_1 \cdots i_{2m}} s_{i_1 \cdots i_{2m}} \eta^{i_1} \cdots \eta^{i_{2m}}.$$

(10.3.17)

This serves to define antisymmetric complex coefficients $s_{i_1 \cdots i_{2m}}$. After some labor one finds that these coefficients satisfy a recursion relation:

$$s_{i_1 \cdots i_n} = \sum_{p=1}^{n-1} (-1)^{p+1} s_{i_1 \cdots \hat{i}_p \cdots i_{n-1}} s_{i_p i_n} \qquad (n = 2m).$$

(10.3.18)

This relation has been encountered in (10.2.4) and is characteristic of any free Euclidean Fermi field. The lesson to be drawn is this: Grassmann algebras help to express complicated combinatorial relationships in a compact form.

10.3.1 When E Is a Function Space

The discussion so far has assumed, needlessly, that E is finite-dimensional. However, it is pertinent to observe that, for any field on the continuum, the underlying vector space E is unavoidably infinite-dimensional. Fortunately, almost all the statements of the preceding section remain true without change, while some constructions to be discussed in the subsequent sections become obscure. Here we would like to point out that the infinite-dimensional situation deserves to be considered in its own right.

Take E to be the Schwartz space of test functions $f \colon E_4 \to \mathbb{C}^8$. If f is evaluated at some spacetime point x, we get numbers $f^a(x)$, where $a = 1, \ldots, 8$. By definition, elements of $A^p(E)$ are antisymmetric distributions of p arguments. It makes sense to consider the Grassmann algebra[49]

$$A(E) = \prod_{p=0}^{\infty} A^p(E) = A_+(E) \oplus A_-(E) \tag{10.3.19}$$

and to decompose it into an even and an odd subspace. Exponentials like e^{-S}, where $S \in A^2(E)$, continue to make sense as elements of $A_+(E)$, but their Tailor series do not truncate in the infinite-dimensional situation. Also, the definition of the generators $\eta^a(x)$ poses no problem since their role is to evaluate test functions:

$$\eta^a(x)(f) = f^a(x), \qquad f \in E. \tag{10.3.20}$$

Comparison with (10.3.12) shows that the pair (a, x) now stands in place of the index i. Again, we find the relations

$$\eta^a(x)\eta^b(y) + \eta^b(y)\eta^a(x) = 0. \tag{10.3.21}$$

In Sect. 10.2 we have considered the two-point function $S_2(x, y)_{ab}$ of a free Dirac field writing

$$S = \tfrac{1}{2} \int dx \int dy \, S_2(x, y)_{ab} \, \eta^a(x)\eta^b(y). \tag{10.3.22}$$

---

[49]This defines A as the *algebraic direct product* of the linear spaces A^p: a technical desideratum. For infinitely many spaces, there is a profound difference between their *direct product* and their *direct sum*, while for a finite set of spaces both definitions coincide. As for our case, S is said to an element of the direct product $\prod_p A^p$ if there exists a component $S_p \in A^p$ for any $p \geq 0$. It is said to be an element of the direct sum $\bigoplus_p A^p$ if $S_p = 0$ except for finitely many p's.

This defines S as an element in $A^2(E)$. What does this mean? Well, the integral in (10.3.22) involves a product of two generators and, by definition,

$$\eta^a(x)\eta^b(y)(f,g) = f^a(x)g^b(y) - g^a(x)f^b(y) \tag{10.3.23}$$

for any pair of test functions (f,g). Therefore, S is a map taking such a pair into the number

$$S(f,g) = \int dx \int dy\, S_2(x,y)_{ab}\, f^a(x)g^b(y). \tag{10.3.24}$$

The exponential of S has significance in field theory. Writing e^{-S} as an infinite sum works perfectly well in the present situation:

$$e^{-S} = 1 + \sum_{m=1}^{\infty} \frac{(-1)^m}{m!} S^m$$

$$\frac{1}{m!}S^m = \frac{1}{(2m)!} \int dx_1 \cdots \int dx_{2m}\, S_{2m}(x_1,\ldots,x_{2m})_{a_1\cdots a_{2m}} \times$$
$$\eta^{a_1}(x_1) \cdots \eta^{a_{2m}}(x_{2m}). \tag{10.3.24}$$

There is no problem with regard to convergence: it does not even make sense to ask whether the sum over m converges. The point is that e^{-S} has components $(-1)^m S^m/m!$ in every subspace A^n where $n = 2m$. Each component involves an n-fold product of generators. By definition,

$$\eta^{a_1}(x_1) \cdots \eta^{a_n}(x_n)(f_1,\ldots,f_n) =$$
$$\sum_{\pi} \mathrm{sgn}\pi\, f^{a_1}_{\pi(1)}(x_1) \cdots f^{a_n}_{\pi(n)}(x_n) \tag{10.3.25}$$

for any n-tuple $f_1,\ldots,f_n \in E$. This clearly states that

$$\frac{1}{m!}S^m(f_1,\ldots,f_{2m}) =$$
$$\frac{1}{(2m)!} \int dx_1 \cdots \int dx_{2m}\, S_{2m}(x_1,\ldots,x_{2m})_{a_1\cdots a_{2m}} \times$$
$$f^{a_1}_1(x_1) \cdots f^{a_{2m}}_{2m}(x_{2m}). \tag{10.3.26}$$

The n-point Schwinger functions, though distributions, may thus be treated as if they were ordinary coefficients of a Taylor expansion of e^{-S} with respect to the generators $\eta^a(x)$.

10.4 Formal Derivatives

Classical analysis centers on the notion of the *derivative of a real function*. Analytic function theory then extends and exploits *differentiability* as a basic concept. In classical mechanics, one encounters *Poisson brackets with the Hamiltonian* having properties quite analogous to those of a derivative. Finally, in Heisenberg's way of treating quantum mechanics, the *commutator with the Hamiltonian* plays a similar role. In a sense these examples illustrate the notion of a *derivation in an algebra*. The key feature of a derivation then is that it acts linearly and that there is a *product rule*.

We will now explain how to introduce derivations in a Grassmann algebra $A(E)$ over some vector space E. Namely, to any $u \in E$ we associate the map

$$d_u : A^p(E) \to A^{p-1}(E) \qquad (p > 0) \tag{10.4.1}$$

by setting

$$(d_u S)(u_2, \ldots, u_p) = S(u, u_2, \ldots, u_p), \qquad S \in A^p(E). \tag{10.4.2}$$

It is agreed that $d_u S = 0$ for $S \in A^0$. This declares d_u to be a linear map from A to A. If (u, v) is a pair of vectors, $d_u d_v = -d_v d_u$. Given some basis in E, we may write

$$\begin{aligned}
S(u_1, \ldots, u_p) &= \sum_{i_1 \cdots i_p} s_{i_1 \ldots i_p} u_1^{i_1} \cdots u_p^{i_p} \\
(d_u S)(u_2, \ldots, u_p) &= \sum_{i_2 \cdots i_p} s'_{i_2 \ldots i_p} u_2^{i_2} \cdots u_p^{i_p},
\end{aligned} \tag{10.4.3}$$

where $s_{i_1 \ldots i_p} \in \mathbb{C}$ and

$$s'_{i_2 \ldots i_p} = \sum_{i_1} s_{i_1 \ldots i_p} u^{i_1}. \tag{10.4.4}$$

Still another way to describe the action of d_u is to use the generators η^i:

$$\begin{aligned}
S &= \frac{1}{p!} \sum_{i_1 \cdots i_p} s_{i_1 \ldots i_p} \eta^{i_1} \cdots \eta^{i_p} \\
d_u S &= \frac{1}{(p-1)!} \sum_{i_1 \cdots i_p} s_{i_1 \ldots i_p} u^{i_1} \eta^{i_2} \cdots \eta^{i_p}.
\end{aligned} \tag{10.4.5}$$

In the theory of second quantization for fermions, one encounters an algebraic structure usually referred to as CAR (canonical anticommutation relations). This deals with creation operators a_i^* and annihilation operators a_i such that $\{a_i, a_k^*\} = \delta_{ik}$, where $i, k = 1, \ldots, n$ and n is said to be the number of degrees of freedom. The operators are supposed to act on a Fock space whose dimension is 2^n. There is a distinguished vector Ω called the vacuum which is characterized by $a_i \Omega = 0$. Translated into the the present language this means that η^i acts as a creation operator on $A(E)$, the Fock space, while

d_{e_i} plays the role of an annihilation operator, and to the vacuum there corresponds the unit element $1 \in A(E)$.

What is the *product rule*? A moment's reflection tells us that

$$d_u(ST) = (d_u S)T + (-1)^p S(d_u T) \qquad S \in A^p,\ T \in A. \tag{10.4.6}$$

Owing to the extra sign $(-1)^p$ in the product rule, the derivative d_u is oddly behaved and for this reason is frequently said to be an *antiderivative* of the algebra A. We shall not, however, follow that terminology. Let us focus on the fact that d_u depends *linearly* on $u \in E$. So with respect to a basis we may write

$$d_u = \sum_{i=1}^{n} u^i \frac{\partial}{\partial \eta^i}\ . \tag{10.4.7}$$

This method of notation is very suggestive since the fundamental message of (10.4.5) may now be phrased as

$$\boxed{\ \frac{\partial}{\partial \eta^i} \sum_{i_1 \cdots i_p} s_{i_1 \ldots i_p} \eta^{i_1} \cdots \eta^{i_p} = p \sum_{i_2 \cdots i_p} s_{i\, i_2 \ldots i_p} \eta^{i_2} \cdots \eta^{i_p}.\ } \tag{10.4.8}$$

The result is best summarized by stating three rules which characterize the formal derivative:

Rule 1: $\quad \frac{\partial}{\partial \eta^i}(\alpha S + \beta T) = \alpha \frac{\partial}{\partial \eta^i} S + \beta \frac{\partial}{\partial \eta^i} T \qquad (\alpha, \beta \in \mathbb{C})$

Rule 2: $\quad \frac{\partial}{\partial \eta^i} 1 = 0$

Rule 3: $\quad \frac{\partial}{\partial \eta^i}(\eta^k S) = \delta_i^k S - \eta^k \frac{\partial}{\partial \eta^i} S.$

Both S and T are arbitrary elements of the algebra A. As opposed to classical analysis, we find, not surprisingly, that in a Grassmann algebra

$$\frac{\partial^2}{\partial \eta^i \partial \eta^k} = -\frac{\partial^2}{\partial \eta^k \partial \eta^i}, \qquad \text{and in particular} \qquad \left(\frac{\partial}{\partial \eta^i}\right)^2 = 0. \tag{10.4.9}$$

Our main interest lies in *even* elements of the algebra. So let us take $S \in A_+$ and examine the way in which powers are differentiated:

$$\frac{\partial}{\partial \eta^i} S^m = \frac{\partial}{\partial \eta^i} (\underbrace{SS \cdots S}_{m}) = \sum_{k=1}^{m} S^{k-1} \left(\frac{\partial}{\partial \eta^i} S\right) S^{m-k} = m S^{m-1} \frac{\partial}{\partial \eta^i} S.$$

$$\tag{10.4.10}$$

The reason for getting this simple result is that $S \in A_+$ and $\frac{\partial}{\partial \eta^i} S \in A_-$ commute. What is true for powers continues to hold for power series $f(S) = \sum a_m S^m$ $(a_m \in \mathbb{C})$, i.e., if $S \in A_+$,

$$\frac{\partial}{\partial \eta^i} f(S) = f'(S) \frac{\partial}{\partial \eta^i} S. \tag{10.4.11}$$

Notice that on the right-hand side there appears a product where the first factor is in A_+, while the second factor is in A_-. So the order of the factors does not matter.

We have space for a small example. It concerns the exponential function

$$\frac{\partial}{\partial \eta^i} e^{-S} = -e^{-S} \frac{\partial}{\partial \eta^i} S \qquad (S \in A_+). \tag{10.4.12}$$

Specifically, if

$$S = \tfrac{1}{2}(\eta, \mathcal{F}_A^{-1}\eta) = \tfrac{1}{2} \int dx \int dy \, \langle x|(\mathcal{F}_A^{-1})_{ab}|y\rangle \eta^a(x)\eta^b(y) \tag{10.4.13}$$

for some vector potential $A_k(x)$, then

$$\left[\frac{\partial^2}{\partial \eta^a(x)\partial \eta^b(y)} e^{-S} \right]_0 = \langle x|(\mathcal{F}_A^{-1})_{ab}|y\rangle, \tag{10.4.14}$$

where $[.]_0$ designates the canonical projection $A \to A^0 = \mathbb{C}$. Rule (10.4.9) for dealing with partial derivatives guarantees the antisymmetry of the two-point function with respect to an interchange $(x, a) \leftrightarrow (y, b)$.

Exercise 1. Prove the rule

$$\frac{\partial}{\partial \eta^i}(ST) = \left(\frac{\partial}{\partial \eta^i}S\right)T + S\frac{\partial}{\partial \eta^i}T, \qquad S \in A_+, \ T \in A.$$

Exercise 2. Find the solutions $T \in A(E)$ of the differential equation

$$\frac{\partial}{\partial \eta^i}T = (a\eta)_i T,$$

where $(a\eta)_i = \sum_k a_{ik}\eta^k$, with $a = (a_{ik})$ a complex antisymmetric matrix.

Exercise 3. Solve $d_u T = S$ for $T \in A^{p+1}(E)$, where $S \in A^p(E)$, $p < n = \dim E$, and $u \in E$.

Exercise 4. With E the Schwartz space of test functions $f\colon E_4 \to \mathbb{C}^8$, write

$$d_f = \int dx \sum_a f^a(x) \frac{\partial}{\partial \eta^a(x)}$$

for the derivation of $A(E)$ associated with f so that $d_f d_g = -d_g d_f$ for any pair of test functions. With S given by (10.4.13), determine

$$e^S d_f d_g e^{-S}$$

and confirm that this element has components in $A^0(E)$ and $A^2(E)$ only.

10.5 Formal Integration

This section is devoted to the *algebraic* concept of an integral. The discussion of such a notion is by no means complete, but is limited to those results essential to an exposition of *path integrals in a Grassmann algebra* with applications to the Dirac field. The results presented here constitute one of the main pillars of present-day field theory. We do not suggest regarding the integral as some sort of process which reverses differentiation, but think of it as generalizing the concept of 'volume' integration.

10.5.1 Integrals in $A(E)$

As before, E is some complex vector space with $\dim E = n < \infty$, and let there be given a distinguished basis for E. Assume that $S \in A(E)$ is arbitrary. We require:

Rule 1:   The integral $\int d\eta\, S$ is a complex number and, moreover, the map $A(E) \to \mathbb{C}$, $S \mapsto \int d\eta\, S$ is linear.

Rule 2: $\int d\eta\, \frac{\partial}{\partial \eta^i} S = 0$ for $i = 1, \dots, n$.

Rule 3: $\int d\eta\, \eta^1 \eta^2 \cdots \eta^n = 1$ (normalization).

This essentially says that the integral is a linear functional on the algebra with special properties. We demonstrate now the existence and uniqueness of the integral in a constructive fashion.

To prove existence we recall that any element $S \in A(E)$ may be decomposed in a canonical manner:

$$S = S_0 + S_1 + \cdots + S_n \quad , \qquad S_p \in A^p(E). \tag{10.5.1}$$

The component of highest degree can be written

$$S_n = I\eta^1 \eta^2 \cdots \eta^n \quad , \qquad I \in \mathbb{C}.$$

The complex number I characterizes the component S_n uniquely since $A^n(E)$ is one-dimensional. We take I to be the integral $\int d\eta\, S$. It is obviously linear (Rule 1) and normalized (Rule 3). It follows from (10.4.1) that $d_u S$ has components in A^p, $p = 0, \dots, n-1$, but not in A^n and thus $\int d\eta\, d_u S = 0$ (Rule 2).

There is an alternative way to obtain the number I which is to take as many derivatives as possible:

$$\frac{\partial}{\partial \eta^n} \cdots \frac{\partial}{\partial \eta^2} \frac{\partial}{\partial \eta^1} S_p = \begin{cases} 0 & \text{if } p < n \\ I & p = n. \end{cases} \tag{10.5.2}$$

This way we would get the following peculiar-looking definition relating integration to differentiation:

$$\boxed{\int d\eta\, S = \frac{\partial}{\partial\eta^n} \cdots \frac{\partial}{\partial\eta^2}\frac{\partial}{\partial\eta^1} S.} \qquad (10.5.3)$$

To prove uniqueness we argue as follows. In the decomposition $S = S_0 + \cdots + S_n$, each component S_p may be written as a derivative $d_u T$ (see Exercise 3 of the preceding section) provided $p < n$. By the second rule, $\int d\eta\, S_p = 0$, and by the first rule, $\int d\eta\, S = \int d\eta\, S_n$. Thus

$$\int d\eta\, S = I \int d\eta\, \eta^1\eta^2\cdots\eta^n. \qquad (10.5.4)$$

Finally, by the third rule, $\int d\eta\, S = I$ which completes the proof.

We now turn to the properties of the integral.

Formula 1 (*integration by parts*):

$$\int d\eta\, \left(\frac{\partial}{\partial\eta^i}S\right) T = (-1)^{p+1} \int d\eta\, S\frac{\partial}{\partial\eta^i}T \qquad S \in A^p,\ T \in A. \qquad (10.5.5)$$

This is a direct consequence of $\int d\eta\, \frac{\partial}{\partial\eta^i}(ST) = 0$ and the product rule of differentiation.

The next formula concerns the behavior of the integral with respect to a linear transformation $a\colon E \to E$ of the integrand. For $S \in A^p$ we define

$$S^a(u_1,\ldots,u_p) = S(au_1,\ldots,au_p) \qquad (10.5.6)$$

which extends to a linear map $A \to A$, $S \mapsto S^a$. With respect to a given basis, we would write

$$(au)^i = \sum_k a^i{}_k u^k \qquad (a\eta)^i = \sum_k a^i{}_k \eta^k \qquad (10.5.7)$$

and thus get

$$S^a = \frac{1}{p!}\sum s_{i_1\cdots i_p}(a\eta)^{i_1}\cdots(a\eta)^{i_p} \qquad (S \in A^p). \qquad (10.5.8)$$

These formulas suggest introducing the notation[50] $S = S\{\eta\}$ for any $S \in A$ and writing $S^a = S\{a\eta\}$. The case $p = n = \dim E$ is special: if $S \in A^n$, it is straightforwardly seen that $S\{a\eta\} = (\det a)\, S\{\eta\}$ taking (10.5.3) into account. The result may be stated as

---

[50]The notation serves to remind ourselves that the Grassmann algebra has generators η^i and that the element S is given in terms of an expansion with respect the generators and their products. It does by no means imply that the η^i are 'variables' in the traditional sense of the word. Nor does it make sense to assign any particular numerical value to the η^i. The widespread diction 'Grassmann variable' therefore ought to be used with care, since it may lead to severe confusion.

Formula 2: *For all $S \in A$ and $a\colon E \to E$ a linear transformation,*

$$\int d\eta \, S\{a\eta\} = \det a \int d\eta \, S\{\eta\}. \tag{10.5.9}$$

This parallels a similar result in classical analysis,

$$\int dx \, f(ax) = |\det a|^{-1} \int dx \, f(x) \qquad (x \in \mathbb{R}^n), \tag{10.5.10}$$

which holds provided a is nonsingular. There is no problem with a singular transformation a (so that $\det a = 0$) in (10.5.9).

10.5.2 Integrals in $A(E \oplus F)$

Many of the more profound applications of the Grassmann integral are best explained by noting that the structure of the underlying vector space is that of a direct sum $E \oplus F$ (frequently F is the dual of E). Vectors in $E \oplus F$ could be expressed as pairs (u, v) with $u \in E$ and $v \in F$. Let be (e_i) a basis in E and (f_k) a basis in F. We define the generators of the Grassmann algebra $A(E \oplus F)$ to be

$$\begin{aligned}
\eta^i(u, v) &= u^i & \left(u = \textstyle\sum u^i e_i\right) \\
\zeta^k(u, v) &= v^k & \left(v = \textstyle\sum v^k f_k\right).
\end{aligned} \tag{10.5.11}$$

All anticommutators of the generators vanish by definition. We regard $A(E)$ and $A(F)$ as two subalgebras of $A(E \oplus F)$, generated by the η^i and the ζ^k respectively. Note that, in order to specify an element $S \in A(E \oplus F)$, it must be expanded with respect to *all* generators, η^i and ζ^k. For convenience we express this fact by writing $S = S\{\eta, \zeta\}$ and define the partial integral by

$$\int d\eta \, S\{\eta, \zeta\} = \frac{\partial}{\partial \eta^n} \cdots \frac{\partial}{\partial \eta^2} \frac{\partial}{\partial \eta^1} S\{\eta, \zeta\} \qquad (n = \dim E) \tag{10.5.12}$$

and

$$\int d\zeta \, S\{\eta, \zeta\} = \frac{\partial}{\partial \zeta^m} \cdots \frac{\partial}{\partial \zeta^2} \frac{\partial}{\partial \zeta^1} S\{\eta, \zeta\} \qquad (m = \dim F). \tag{10.5.13}$$

The first integral defines an element in $A(F)$, while the second integral defines an element in $A(E)$. The notion of partial integration offers the opportunity to give meaning to the integral $\int d\eta \int d\zeta \, S\{\eta, \zeta\}$ as a composition of two integrals. Observe that the order of integration is important, i.e., we have, as a substitute of Fubini's theorem of classical analysis,

$$\int d\eta \int d\zeta \, S\{\eta, \zeta\} = (-1)^{nm} \int d\zeta \int d\eta \, S\{\eta, \zeta\}. \tag{10.5.14}$$

Partial integrals behave like ordinary Grassmann integrals in one respect:

Formula 3: *For $a\colon E \to E$ a linear transformation,*

$$\int d\eta\, S\{a\eta, \zeta\} = \det a \int d\eta\, S\{\eta, \zeta\}. \tag{10.5.15}$$

We shall have occasion in what follows to apply the foregoing results to the situation where $\dim E =\dim F = n$, in particular when some element $S \in A(E \oplus F)$ is such that it posseses an expansion in the generators $\xi^i = \eta^i + \zeta^i$. In such a case we write $S\{\eta + \zeta\}$ and get

Formula 4 (*translational invariance*):

$$\int d\eta\, S\{\eta + \zeta\} = \int d\eta\, S\{\eta\}. \tag{10.5.16}$$

The proof proceeds algebraically. Expand $S\{\eta\}$ with respect to the generators η^i to get the term of highest order, $I\eta^1\eta^2\cdots\eta^n$, so that $I = \int d\eta\, S\{\eta\}$. Next, expand $S\{\eta + \zeta\}$ in the η^i. The term of highest order is contained in $I(\eta^1 + \zeta^1)\cdots(\eta^n + \zeta^n)$ and coincides with the former expression.

10.5.3 Integrals of the Exponential Type

As before, the spaces E an F are assumed to have the same dimension n. Thus, the underlying space $E \oplus F$ has dimension $2n$, while the Grassmann algebra has dimension 2^{2n}. We assert that, for any complex $n \times n$ matrix $a = (a_{ik})$,

$$\int d\eta \int d\zeta \, \exp \sum_{ik} a_{ik}\zeta^i\eta^k = (-1)^{\binom{n}{2}} \det a. \tag{10.5.17}$$

In order to prove this statement, simply set $S\{\eta, \zeta\} = \exp \sum_i \zeta^i\eta^i$ and apply (10.5.15). It remains to evaluate $\int d\eta \int d\zeta\, S\{\eta, \zeta\}$. The canonical decomposition $S = S_0 + \cdots + S_{2n}$ yields

$$\begin{aligned}
S_{2n} &= \frac{1}{n!}\left(\sum_i \zeta^i\eta^i\right)^n \\
&= (\zeta^1\eta^1)(\zeta^2\eta^2)\cdots(\zeta^n\eta^n) \\
&= (-1)^{\binom{n}{2}}(\zeta^1\cdots\zeta^n)(\eta^1\cdots\eta^n).
\end{aligned} \tag{10.5.18}$$

This establishes the relation $\int d\eta \int d\zeta\, S\{\eta, \zeta\} = (-1)^{\binom{n}{2}}$ and hence formula (10.5.17).

Definition *Let $n = 2m$ be even and $A = (A_{ki})$ be some antisymmetric $n \times n$ matrix. The complex number $\operatorname{Pf} A$ defined by*

$$\frac{1}{m!}\left(\tfrac{1}{2}\sum_{ik} A_{ik}\eta^i\eta^k\right)^m = \mathrm{Pf}\,A\;\eta^1\eta^2\cdots\eta^n \tag{10.5.19}$$

is called the Pfaffian[51] *of A.*

The Pfaffian is a polynomial of the matrix elements, homogeneous of degree m. For low dimensions ($n = 2, 4, \ldots$), we find that

$$\begin{aligned} m = 1: &\qquad \mathrm{Pf}\,A = A_{12}\\ m = 2: &\qquad \mathrm{Pf}\,A = A_{12}A_{34} - A_{13}A_{24} + A_{14}A_{23}. \end{aligned} \tag{10.5.20}$$

If $C = B^T A B$ for B some complex $n \times n$ matrix, then C is antisymmetric as well, and it follows from (10.5.19) that $\mathrm{Pf}\,C = \det B\,\mathrm{Pf}\,A$. Hence, $\mathrm{Pf}\,C = \mathrm{Pf}\,A$ if $\det B = 1$. This essentially says that, although the Pfaffian of a bilinear functional A on E has been defined with reference to a particular basis in E, it is invariant under unimodular transformations of that basis.

More importantly, relation (10.5.19) may be rewritten as

$$\int d\eta\,\exp\left(\tfrac{1}{2}\sum_{ik} A_{ik}\eta^i\eta^k\right) = \mathrm{Pf}\,A. \tag{10.5.21}$$

The reason: the highest-order term of the Taylor expansion of the exponential function is precisely what appears on the left-hand side of (10.5.19). Notice that the integral in (10.5.21) exists and is in fact zero if A is a matrix in odd dimensions.

The next assertion throws some light on the relationship between the notion of a *determinant* and that of a *Pfaffian*. The special block structure of the antisymmetric matrix A (in even dimensions) assumed below mimics a characteristic feature of the covariance operator for Dirac fields when the bispinor formalism is applied (see Sect. 10.2).

Let a be some complex $m \times m$ matrix and consider the block matrix in dimension $n = 2m$,

$$A = \begin{pmatrix} 0 & a \\ -a^T & 0 \end{pmatrix}. \tag{10.5.22}$$

Then A is antisymmetric and we have the relation

$$\mathrm{Pf}\,A = (-1)^{\binom{m}{2}} \det a. \tag{10.5.23}$$

To prove this assertion, we first write

$$\sum_{ik} a_{ik}\zeta^i\eta^k = \tfrac{1}{2}\xi^T A\xi, \qquad \xi = \begin{pmatrix} \zeta \\ \eta \end{pmatrix} \tag{10.5.24}$$

---

[51]Johann Friedrich Pfaff (1765–1825) was a German mathematician.

and then

$$\int d\xi \exp \tfrac{1}{2}\xi^T A\xi = \int d\eta \int d\zeta \exp \zeta^T a\eta. \tag{10.5.25}$$

Both sides may now be evaluated separately utilizing (10.5.17) and (10.5.21) which establishes (10.5.23).

10.5.4 The Fourier–Laplace Transformation

The classical Fourier transformation is initially a linear map taking functions $f(x)$ defined on $x \in E$ (E is some n-dimensional real vector space) into functions $\tilde{f}(k)$ defined on $k \in E^*$ (E^* is the dual space of E). The same characterization applies to the two-sided Laplace transform. When the concept is extended to complex vector spaces, the two notions merge and it is no longer reasonable to distinguish Fourier from Laplace transforms. We shall refer to them as *FL transforms* as a unifying term.

Not unexpectedly, in a Grassmann algebra the FL transformation sends elements $S \in A(E)$ to elements $\tilde{S} \in A(E^*)$ and vice versa. This is reminiscent of the Hodge map familiar from linear algebra,

$$\star : \; A^p(E) \to A^{n-p}(E^*) \qquad (p = 0, \ldots, n = \dim E), \tag{10.5.26}$$

which we now turn to. Choose a basis (e_i) in E and let (e_*^i) be the induced basis in E^* so that $\langle e_*^i, e_k \rangle = \delta_k^i$. Every vector $u \in E$ is written $\sum u^i e_i$ and, correspondingly, every vector $v \in E^*$ is represented as $\sum v_i e_*^i$. We take η^i and η_i^* to be the generators of $A(E)$ and $A(E^*)$ respectively, with the defining equations

$$\eta^i(u) = u^i, \qquad \eta_i^*(v) = v_i. \tag{10.5.27}$$

Each element $S \in A^p(E)$ is of the form

$$S = \frac{1}{p!} \sum_{i_1 \cdots i_p} s_{i_1 \cdots i_p} \eta^{i_1} \cdots \eta^{i_p} \tag{10.5.28}$$

and the Hodge map takes S into

$$\star S = \frac{1}{(n-p)!} \sum_{i_{p+1} \cdots i_n} s_*^{i_{p+1} \cdots i_n} \eta_{i_{p+1}}^* \cdots \eta_{i_n}^* \tag{10.5.29}$$

with coefficients

$$s_*^{i_{p+1} \cdots i_n} = \frac{1}{p!} \sum_{i_1 \cdots i_p} s_{i_1 \cdots i_p} \epsilon^{i_1 \cdots i_n} \tag{10.5.30}$$

(ϵ is the Levi-Civita symbol). The Hodge map is almost an involution:

$$\star \star S = (-1)^{p(n-p)} S \; , \qquad S \in A^p(E). \tag{10.5.31}$$

As p runs from 0 to n, the individual maps on A^p cannot be fitted into one map on A as an involution up to a sign because the sign in (10.5.31) is p-dependent. The defect, however, can be cured by adopting a different convention concerning the signs involved. We propose to introduce

$$\tilde{S} = (-1)^\sigma \star S \quad , \quad \sigma = \binom{n-p+1}{2}, \qquad S \in A^p(E) \tag{10.5.32}$$

and find

$$\tilde{\tilde{S}} = (-1)^{\binom{n+1}{2}} S. \tag{10.5.33}$$

Since the sign relating S and $\tilde{\tilde{S}}$ comes out independent of p, the map $S \to \tilde{S}$ can be extended by linearity to some map $A \to A$ obeying (10.5.33). It will be called the *FL transformation*.

Actually, we could have discovered the 'twisted' Hodge map (which we choose to call the FL transformation) by examining certain integrals that come up in connection with exponential functions. To explain this connection, let us canonically embed $A(E)$ and $A(E^*)$ into the larger algebra $A(E \oplus E^*)$ and then write $\langle \eta^*, \eta \rangle := \sum_i \eta_i^* \eta^i = -\sum_i \eta^i \eta_i^* = -\langle \eta, \eta^* \rangle$. Consider the (partial) integral

$$\tilde{S}\{\eta^*\} = \int d\eta \, S\{\eta\} \, \exp\langle \eta^*, \eta \rangle. \tag{10.5.34}$$

The claim that this map is in fact the same as the one described earlier in (10.5.32) is substantiated by observing that

$$\int d\eta \, \eta^1 \eta^2 \cdots \eta^p \, \exp\langle \eta^*, \eta \rangle = (-1)^{\binom{n-p+1}{2}} \eta_{p+1}^* \cdots \eta_n^*. \tag{10.5.35}$$

Clearly, the inversion of the transformation (10.5.34) reads:

$$(-1)^{\binom{n+1}{2}} S\{\eta\} = \int d\eta^* \, \tilde{S}\{\eta^*\} \, \exp\langle \eta, \eta^* \rangle. \tag{10.5.36}$$

In ordinary Fourier transformation, one shows that Gaussian functions are transformed into Gaussian functions. There is an analogue in Grassmann algebras:

Let $A = (A_{ik})$ be some antisymmetric nonsingular $n \times n$ matrix (n is necessarily even: $n = 2m$) and

$$S\{\eta\} = \exp\left[\tfrac{1}{2} \sum A_{ik} \eta^i \eta^k\right]. \tag{10.5.37}$$

Then the the FL transform of S is given by

$$\tilde{S}\{\eta^*\} = \mathrm{Pf}\, A \, \exp\left[\tfrac{1}{2} \sum (A^{-1})^{ik} \eta_i^* \eta_k^*\right] \tag{10.5.38}$$

with $\mathrm{Pf}\, A$ the Pfaffian of A.

Again the proof proceeds algebraically and along the same lines as in the classical situation: we write

$$\tfrac{1}{2}\eta^T A\eta + \langle \eta^*, \eta \rangle = \tfrac{1}{2}(\eta + \zeta)^T A(\eta + \zeta) + \tfrac{1}{2}\eta^{*T} A^{-1}\eta^* \qquad (10.5.39)$$

noticing that all terms are in $A_+(E \oplus E^*)$, where $\zeta = -A^{-1}\eta^*$. Using translational invariance, $\int d\eta\, S\{\eta + \zeta\} = \int d\eta\, S\{\eta\} = \mathrm{Pf}\,A$, and the proof is complete.

The twofold application of the FL transformation to the Gauss element (10.5.37) reproduces this element except for the factor $\mathrm{Pf}\,A\,\mathrm{Pf}\,A^{-1}$. Comparison with (10.5.33) establishes the result

$$\mathrm{Pf}\,A\,\mathrm{Pf}\,A^{-1} = (-1)^{\binom{n+1}{2}} = (-1)^m \qquad (2m = n). \qquad (10.5.40)$$

The discussion of Gaussian elements in a Grassmann algebra and their Fourier transforms has assumed that the underlying vector space E is finite-dimensional making the Pfaffian a well-defined quantity. It is pretty clear that field theory calls for an extension of the above formulas to infinite dimensions. However, unless one is willing to invoke topology so that there is some concept of convergence, such an extension will have no immediate meaning. For now, it seems thus preferable to adhere to finite dimensions using *algebra without topology*. This requires that we discretize spacetime, i.e., replace E_4 by a finite lattice.

Let us investigate what kind of path integrals would emerge from the formula (10.5.38) if we were to mimic Dirac field theory postulating that the action $W\{\eta^*\}$ be bilinear in the field η^* on the lattice. We therefore write

$$W\{\eta^*\} = \tfrac{1}{2} \sum_{i,k=1}^{n} (A^{-1})^{ik}\eta_i^* \eta_k^* \qquad (10.5.41)$$

and

$$d\mu(\eta^*) = Z^{-1}\, d\eta^* \exp\left(-W\{\eta^*\}\right) \qquad (10.5.42)$$

with A some (very large) antisymmetric matrix depending on field variables other than the η_i^* (e.g. scalar and vector fields). The normalizing constant is

$$Z^{-1} = (-1)^{\binom{n+1}{2}}\mathrm{Pf}(-A) = \mathrm{Pf}\,A. \qquad (10.5.43)$$

It follows from (10.5.36) that the characteristic functional of μ is

$$\int d\mu(\eta^*)\, \exp\langle \eta, \eta^* \rangle = \exp[-\tfrac{1}{2}\textstyle\sum A_{ik}\eta^i\eta^k]. \qquad (10.5.44)$$

By expanding the exponentials on both sides, we get the following formulas for the lowest-order moments of the η_i^*:

$$\int d\mu(\eta^*) = 1$$

$$\int d\mu(\eta^*)\, \eta_i^* = 0$$

$$\int d\mu(\eta^*)\, \eta_i^* \eta_j^* = A_{ij} \qquad\qquad (10.5.45)$$

$$\int d\mu(\eta^*)\, \eta_i^* \eta_j^* \eta_k^* = 0$$

$$\int d\mu(\eta^*)\, \eta_i^* \eta_j^* \eta_k^* \eta_\ell^* = A_{ij} A_{k\ell} - A_{ik} A_{j\ell} + A_{i\ell} A_{jk}.$$

It is characteristic of most (if not all) matrices A and A^{-1} modelling interacting Euclidean Dirac fields on a lattice that they are structured as follows:

$$A = \begin{pmatrix} 0 & a \\ -a^T & 0 \end{pmatrix}, \qquad A^{-1} = \begin{pmatrix} 0 & -a^{-1T} \\ a^{-1} & 0 \end{pmatrix}. \qquad (10.5.46)$$

where a is a complex $m \times m$ matrix and m is even so that

$$(-1)^{\binom{n+1}{2}} = (-1)^m = 1, \qquad \mathrm{Pf}\, A = \det a \qquad (n = 2m). \qquad (10.5.47)$$

One beneficial effect of this structure is that it avoids Pfaffians altogether since $\mathrm{Pf}\, A$ may be replaced by $\det a$ wherever it occurs.

Exercise 1.   Find a way to prove the relation

$$\det a = \begin{cases} (\mathrm{Pf}\, a)^2 & \text{if } m = \text{even} \\ 0 & \text{if } m = \text{odd} \end{cases}$$

which is valid for any antisymmetric complex $m \times m$ matrix a.

Exercise 2.   Show that $S \in A(E)$ is *even* iff $S\{\eta\} = S\{-\eta\}$. Prove also that the FL transform of an even (odd) element of $A(E)$ yields an even (odd) element of $A(E^*)$.

10.6 Functional Integrals of QED

As yet, no direct coupling of gauge fields to dynamical fermions ("matter") has been considered. Nor have we offered path integrals even for the simplest of all realistic models, i.e., for quantum electrodynamics (QED). However, armed with the sophisticated technique of Grassmann integration, we may now resume our previous analysis of $U(1)$ gauge theory to formulate a continuum theory that incorporates the photon and a Dirac particle of (bare) mass m and charge e (think of an electron or a quark). As we have already stressed, there is no question that, in order to give meaning to various expressions like path integrals, fermion determinants or Pfaffians in a continuum theory, some kind of regularization procedure has to be devised and carefully applied beforehand. But, for now, we will only discuss the formal expressions, not their regularized counterparts. This way we are emphasizing the *structural* rather than the *computational* aspects of QED.

After gauge fixing, the Euclidean action takes the form

$$W(A, \psi, \bar{\psi}) = \int dx \left(\tfrac{1}{4} F^{k\ell} F_{k\ell} + \tfrac{1}{2}\lambda(\partial_k A^k)^2 + \bar{\psi}(\partial\!\!\!/ + m + ie A\!\!\!/)\psi \right).$$

$$(10.6.1)$$

As before, we propose to replace the pair $(\psi, \bar{\psi})$ by the bispinor Ψ, which is thought of as including the components of the Dirac spinor ψ and ψ^c, where $\psi^c = C\bar{\psi}^T$. After this change of variables, the action splits naturally into two bilinear terms:

$$W(A, \Psi) = \tfrac{1}{2}(\Psi, \mathcal{F}_A \Psi) + \tfrac{1}{2}(A, \mathcal{C}_\lambda A). \tag{10.6.2}$$

The operators $\mathcal{F}_A$ and $\mathcal{C}_\lambda$ are matrices of differential operators defined by (10.2.25) and (9.1.13). Complete information about QED is believed to be encoded in the generating functional associated with the action W,

$$S\{\eta, j\} = Z^{-1} \int \mathcal{D}A\, \mathcal{D}\Psi\, \exp\big(\Psi(\eta) + A(j) - W(A, \Psi)\big), \tag{10.6.3}$$

where η and j are external sources of different character: $\eta^a(x)$ is of Grassmann type while $j^k(x)$ is of ordinary type. The normalizing condition that determines Z is $S\{0, 0\} = 1$. The key observation is this: the path integral with respect to the fermionic degrees of freedom can be performed since the action is bilinear in the field Ψ. The FL transformation of a Gaussian element of the Grassmann algebra yields another Gaussian element. So

$$S\{\eta, j\} = \int d\mu(A)\, \exp\big(A(j) - \tfrac{1}{2}(\eta, \mathcal{F}_A^{-1}\eta)\big) \tag{10.6.4}$$

with μ a non-Gaussian probability measure on the gauge field given by

$$d\mu(A) = Z^{-1}\mathcal{D}A\, \det(\partial\!\!\!/ + m + ie A\!\!\!/)\, \exp\big(-\tfrac{1}{2}(A, \mathcal{C}_\lambda A)\big). \tag{10.6.5}$$

The condition that determines Z is $\int d\mu(A) = 1$. The result (10.6.5) follows from $\det(-C) = \det C = 1$ and

$$\begin{aligned} \mathrm{Pf}(-\mathcal{F}_A) &= \det\big(-C(\partial\!\!\!/ + m + ie A\!\!\!/)\big) = \det(\partial\!\!\!/ + m + ie A\!\!\!/) \\ &= \det\big(-C(\partial\!\!\!/ + m - ie A\!\!\!/)\big) = \det(\partial\!\!\!/ + m - ie A\!\!\!/). \end{aligned} \tag{10.6.6}$$

Since the fermion determinant is an even function of e, it is certainly an even functional of the gauge field A. Hence, we have $d\mu(-A) = d\mu(A)$. This symmetry of the measure reflects charge conjugation invariance.

Needless to say, the "measure" μ is a formal object: existence in any rigorous mathematical sense has not yet been established. The difficulty lies with the fermion determinant. Positivity of the measure requires that the determinant be positive, which can formally be shown as follows. Consider the covariant derivative $D_k \equiv \partial_k + ieA_k$. It is an anti-self-adjoint operator[52]: $D_k = -D_k^*$. Taking representation (10.2.1) of the γ matrices into account, we may write

$$\slashed{\partial} + m + ie\slashed{A} = \begin{pmatrix} m & \mathcal{D} \\ -\mathcal{D}^* & m \end{pmatrix},$$
(10.6.7)

where

$$\begin{aligned} \mathcal{D} &= i\sigma_1 D_1 + i\sigma_2 D_2 + i\sigma_3 D_3 + \mathbb{1}D_4 \\ \mathcal{D}^* &= i\sigma_1 D_1 + i\sigma_2 D_2 + i\sigma_3 D_3 - \mathbb{1}D_4. \end{aligned}$$
(10.6.8)

We argue as if all entries in (10.6.7) were matrices[53], i.e.,

$$\det \begin{pmatrix} m & \mathcal{D} \\ -\mathcal{D}^* & m \end{pmatrix} = \det(m^2 + \mathcal{D}^*\mathcal{D}) = \det(m^2 + \mathcal{D}\mathcal{D}^*),$$
(10.6.9)

which is obviously positive. It also shows that the fermion determinant is an even function of the bare mass m.

Normalization of the measure μ allows division by its worst part, which is $\det(\slashed{\partial}+m)$, with the effect that fermion determinant changes to

$$\frac{\det(\slashed{\partial} + m + ie\slashed{A})}{\det(\slashed{\partial} + m)} = \det\left(1 + ie\slashed{A}(\slashed{\partial} + m)^{-1}\right).$$
(10.6.10)

One interesting aspect of the formula (10.6.4) for the generating functional is that the coupling constant e occurs at two places: (1) in the measure μ, and hence in (10.6.10), and (2) in the operator $\mathcal{F}_A^{-1}$. This suggests applying the perturbation method (power series expansion in e) to the fermion determinant only. The technique is also called the *loop expansion*, since, in the Feynman diagrammar, it amounts to sorting all graphs according to the number of closed fermion loops encountered in them. In zero order, i.e., when $e \to 0$ or, equivalently, $m \to \infty$, one simply replaces the fermion determinant by 1 (often called the *quenched approximation*). The resulting measure μ_0 is Gaussian with characteristic functional

$$\int d\mu_0(A) \exp A(j) = \exp\{\tfrac{1}{2}(j, \mathcal{C}_\lambda^{-1} j)\}.$$
(10.6.11)

When μ_0 replaces μ, the gauge field becomes Gaussian, i.e., there is no vacuum polarization and the photon has lost its structure.

For most purposes we would be satisfied with the first few terms of an expansion of the generating functional (10.6.4) with respect to the sources η and j. These are needed to get the low-order Schwinger functions of QED. Consider, for instance, the two-point function of the fermion, which is an 8×8 matrix (we are suppressing all indices):

$$\langle \Psi(x)\bar{\Psi}(y) \rangle = \int d\mu(A) \, \langle x|\mathcal{F}_A^{-1}|y \rangle.$$
(10.6.12)

---

[52]This property is indispensible and has to be preserved under all circumstances by the regularization procedure.
[53]See Exercise 1 at the end of this section.

This formula alone provides complete information about four different two-point functions among Dirac spinor fields. However, the relevant information is already coded in

$$\langle \psi(x)\bar{\psi}(y)\rangle = \int d\mu(A)\,\langle x|(\slashed{\partial} + m + ie\slashed{A})^{-1}|y\rangle. \tag{10.6.13}$$

In principle, the evaluation of $\langle \psi(x)\bar{\psi}(y)\rangle$ would enable us to extract the physical (renormalized) mass of the fermion from the exponential fall-off as $|x - y| \to \infty$.

The two-point function of the gauge field may also be written as a path integral with respect to the measure μ:

$$\langle A_k(x)A_\ell(y)\rangle = \int d\mu(A)\,A_k(x)A_\ell(y). \tag{10.6.14}$$

This function contains information about the vacuum polarization. It is positive definite in the sense that $\langle A(j)^2\rangle \geq 0$ for all currents $j^k(x)$.

The next object in the hierarchy of expectation values is the three-point function of the three basic fields,

$$\langle \psi(x)A_k(y)\bar{\psi}(x')\rangle = \int d\mu(A)\,\langle x|(\slashed{\partial} + m + ie\slashed{A})^{-1}|x'\rangle\,A_k(y), \tag{10.6.15}$$

which reveals details about the coupling of the photon to matter. As concerns the electromagnetic structure of the fermion, another three-point function would be more appropriate,

$$\langle \psi(x)J^k(y)\bar{\psi}(x')\rangle =$$

$$\int d\mu(A)\,\langle x|(\slashed{\partial} + m + ie\slashed{A})^{-1}|y\rangle\gamma^k\langle y|(\slashed{\partial} + m + ie\slashed{A})^{-1}|x'\rangle, \tag{10.6.16}$$

where $J^k(x)$ is the electromagnetic current[54]. Actually, (10.6.16) arises from the four-point function of the Ψ field with two coinciding arguments.

Exercise 1.   Let A be some $m \times n$ matrix so that A^* is an $n \times m$ matrix. Prove that

$$\det\begin{pmatrix} \mathbb{1}_m & A \\ -A^* & \mathbb{1}_n \end{pmatrix} = \det(\mathbb{1}_m + AA^*) = \det(\mathbb{1}_n + A^*A) \geq 1.$$

Suppose next that ∂_k acts on a function space of finite dimension so that $\partial_k^* = -\partial_k$ with respect to some scalar product. Prove that

$$\det(\slashed{\partial} + m) = \big(\det(-\Delta + m^2)\big)^2,$$

where $\slashed{\partial}^2 = \mathbb{1}_4\Delta$.

---

[54]Formally, $J^k(x) = \bar{\psi}(x)\gamma^k\psi(x) - \langle\bar{\psi}(x)\gamma^k\psi(x)\rangle$, or better, $J^k = \frac{1}{2}(\bar{\psi}\gamma^k\psi - \bar{\psi}_c\gamma^k\psi_c)$.

Exercise 2. From

$$(\not{\partial} + m + ie\not{A})^{-1} - (\not{\partial} + m)^{-1} = -ie(\not{\partial} + m)^{-1}\not{A}(\not{\partial} + m + ie\not{A})^{-1}$$

deduce that

$$\langle \psi(x)\bar{\psi}(y)\rangle - \langle x|(\not{\partial} + m)^{-1}|y\rangle =$$

$$-ie\int dx'\, \langle x|(\not{\partial} + m)^{-1}|x'\rangle\langle \not{A}(x')\psi(x')\bar{\psi}(y)\rangle$$

thereby showing that two- and three-point functions are related.

10.7 The $SU(n)$ Gauge Theory with Fermions

We now turn to fermions interacting with a $SU(n)$ gauge field on a lattice, mainly in order to illustrate the most common regularization procedure. Let us thus consider a multiplet of n Dirac fields:

$$\psi(x) = \begin{pmatrix} \psi_1(x) \\ \vdots \\ \psi_n(x) \end{pmatrix}, \qquad \bar{\psi}(x) = (\bar{\psi}_1(x), \ldots, \bar{\psi}_n(x)). \tag{10.7.1}$$

The group $SU(n)$ acts naturally on such a multiplet.

More concretely, if $n = 3$, the above multiplet models a quark field having three colors but one flavor. The aim would be then to formulate quantum chromodynamics (QCD) on the lattice. The particles usually associated with the gauge field are called *gluons*.

We expect the action to be of the general (gauge invariant) form

$$W(\psi, \bar{\psi}, U) = W_2 + W_1 + W_0, \tag{10.7.2}$$

where W_2, W_1, and W_0 represent sums over plaquettes, links, and sites respectively[55]. The most straightforward ansatz would be

$$W_2 = \frac{1}{4g^2}\sum_p \mathrm{tr}\,(1 - U_{\partial p})^*(1 - U_{\partial p})$$

$$W_1 = \tfrac{1}{2}\sum_{xk}(\bar{\psi}(x)\gamma^k U_{xk}^*\psi(x + e_k) - \bar{\psi}(x + e_k)\gamma^k U_{xk}\psi(x)) \tag{10.7.3}$$

$$W_0 = m\sum_x \bar{\psi}(x)\psi(x)$$

---

[55] As always, the lattice is $(\mathbb{Z}_N)^4$. So it is four-dimensional, hypercubic, and periodic with period N. The lattice spacing is 1.

with variable unitary matrices $U_{\partial p}, U_{xk} \in SU(n)$ attached to each plaquette p and link xk. As before, we are led to take $U_{\partial p}$ as the product of the four link variables around a plaquette. The choice is clearly dictated by the requirement of local $SU(n)$ gauge invariance, i.e., for $u(x) \in SU(n)$ arbitrary, the action ought to be invariant under the replacements

$$
\begin{aligned}
U_{xk} &\longrightarrow u(x + e_k) U_{xk} u(x)^{-1} \\
\psi(x) &\longrightarrow u(x)\psi(x) \\
\bar{\psi}(x) &\longrightarrow \bar{\psi}(x)u(x)^*.
\end{aligned}
\tag{10.7.4}
$$

Whereas the expressions for W_2 and W_0 look familiar, the particular form chosen for W_1 calls for a comment. Notice that, if x and y are neighboring sites, either $y = x + e_k$ or $x = y + e_k$ for some k. Now write W_1 more suggestively:

$$
W_1 = \sum_{xy} \bar{\psi}(x) \not{D}_{xy} \psi(y), \qquad \not{D} \equiv \gamma_k D^k,
\tag{10.7.5}
$$

where D^k stands for the lattice version of the covariant derivative:

$$
D^k_{xy} = \begin{cases} \frac{1}{2} U^*_{xk} & \text{if } y = x + e_k \\ -\frac{1}{2} U_{yk} & \text{if } x = y + e_k \\ 0 & \text{otherwise.} \end{cases}
\tag{10.7.6}
$$

In the special case, where $U_{xk} = \mathbb{1}$ (all x, k), this becomes the lattice gradient satisfying $\partial^* = -\partial$:

$$
\partial^k_{xy} = \begin{cases} \frac{1}{2} & \text{if } y = x + e_k \\ -\frac{1}{2} & \text{if } x = y + e_k \\ 0 & \text{otherwise.} \end{cases}
\tag{10.7.7}
$$

The form (10.7.7) of the lattice gradient is plagued with the so-called "fermion doubling problem". This problem arises since the free fermion propagator $(\not{\partial} + m)^{-1}_{xy}$, in momentum space, has the denominator $m^2 + \sum_k \sin^2(p_k)$. On a scaled lattice with spacing a, this expression changes to $m^2 + \sum_k a^{-2} \sin^2(ap_k)$. What strikes the eye is that, as we approach the continuum (a becomes small), we get contributions not only from $p_k = 0$ (all k) but also from the corners of the Brillouin zone, where $p_k = \pm\pi/a$ (some k), owing to the zeros of the sine function. This way we get 16 contributions (taking periodicity into account), each contribution giving rise to a (Euclidean) momentum distribution characteristic of a free particle propagator. Therefore, the continuum limit gives weird results unless one invents a scheme to eliminate the extra fermions. There are a number of proposals that have already been made in this direction. We must, however, refrain from discussing them here.

Following our general strategy, which is to pass to the bispinor formalism, we are led to introduce the following expressions:

$$
\Psi = \begin{pmatrix} \psi \\ \psi^c \end{pmatrix}, \qquad \mathcal{F}_U = \begin{pmatrix} 0 & -[C(\not{D} + m)]^T \\ C(\not{D} + m) & 0 \end{pmatrix} \otimes \mathbb{1}_n.
\tag{10.7.8}
$$

This enables us to write part of the action as a bilinear form of the fermion field Ψ:

$$W_1 + W_0 = \tfrac{1}{2}(\Psi, \mathcal{F}_U \Psi). \qquad (10.7.9)$$

As before, the path integral with respect to the fermionic degrees of freedom can be performed to yield a closed-form expression for the generating functional:

$$\int \mathcal{D}\Psi \, \exp\big(\Psi(\eta) - W_1 - W_0\big) = \det(\slashed{D} + m)^n \, \exp\big(-\tfrac{1}{2}(\eta, \mathcal{F}_U^{-1}\eta)\big).$$

$$(10.7.10)$$

It should be clear that, for any x, the Grassmann variable $\eta(x)$ has $8n$ components (indices are suppressed).

The set of link variables is collectively denoted $U = \{U_{xk}\}$. The fermion determinant $\det(\slashed{D}+m)$ has to be evaluated as a function of U. It occurs in (10.7.10) raised to the power n. Because, on a lattice, $\slashed{D}+m$ is an ordinary matrix (no longer a differential operator), the fermion determinant is well defined, even positive[56]. As most entries of the matrix $\slashed{D}+m$ are zero, it is said to be a "sparse matrix". Experience with matrices in general shows that the more sparse they are, the easier they are to compute with. Sparseness, though welcome, is not sufficient to make the computation of the fermion determinant a tractable problem even on moderately sized lattices.

Having performed the integration with respect to the site variables, we are left with a positive normalized measure with respect to the link variables:

$$d\mu(U) = Z^{-1}\mathcal{D}U \, \det(\slashed{D} + m)^n \, \exp(-W_2). \qquad (10.7.11)$$

Again, the definition of this measure poses no problem because the domain of integration is compact, as has been emphasized previously (in Sect. 9.4). Notice that the fermion determinant disappears from the measure μ in the limit $m \to \infty$. To see the influence of a finite fermion mass m on expectation values of loop variables, one may try to compute

$$\langle U_{\partial G} \rangle = \int d\mu(U) \, U_{\partial G}, \qquad (10.7.12)$$

where ∂G is a Wegner–Wilson loop on the lattice. This then provides a method of determining the "average curvature" present in the system and, physically speaking, a way to study the long-range quark–antiquark potential.

---

[56]Positivity follows from the properties of the γ matrices and the fact that $D_k = -D_k^*$. The proof proceeds along the lines of Sect. 10.6. By the same token, $\det(\slashed{D} + m) \geq (m^2)^\alpha$, where $\alpha = 2N^4$.

Appendix A
List of Symbols and Glossary

1	the number *one*, the constant function 1, the unit matrix, the neutral element of a group, or the unit operator in a vector space. Often, the unit matrix is denoted by $\mathbb{1}$. If $c1$ (c real or complex) is defined, we do not distinguish between $c1$ and c.
$\mathbb{N}, \mathbb{Z}, \mathbb{R}, \mathbb{C}$	the set of natural, integer, real, and complex numbers respectively
$\mathbb{R}_+$	the set of nonnegative real numbers
$a\mathbb{Z}$	the set of integer multiples of a number a, also viewed as an infinitely extended one-dimensional lattice; a is called the *lattice constant* or the *lattice spacing*.
$A \times B$	the Cartesian product of the sets A and B. Elements of $A \times B$ are ordered pairs (a, b), where $a \in A$ and $b \in B$.
A^d	the d-fold Cartesian product of the set A: $A \times \cdots \times A$ (n factors)
$\mathbb{R}^d$	the "d-dimensional space". As a rule, it is equipped with the standard Euclidean structure. Special situations: $\mathbb{R}^3$ = position space of a particle, $\mathbb{R}^{3n}$ = configuration space of n particles, $\mathbb{R}^4 = E_4 =$ Euclidean spacetime. The scalar product $\sum_k x_k x'_k$ of two vectors $x, x' \in \mathbb{R}^d$ with components x_k or x'_k is written xx' or $x \cdot x'$ or $x^T x'$. We call $x^2 = x \cdot x$ the *square* and $|x| = \sqrt{x^2}$ the *length* of x.

$(a\mathbb{Z})^d$	the infinitely extended hypercubic lattice embedded into the space $\mathbb{R}^d$, where a is the lattice constant.
$(\mathbb{Z}_N)^d$	the finite lattice of period N and lattice spacing 1. Formally, $\mathbb{Z}_N = \mathbb{Z}/N\mathbb{Z}$.
$O(h^n)$	some error of the order h^n when $h \to 0$
$\mathrm{Re}z$, $\mathrm{Im}z$	real and imaginary part of a complex number z
A^T	transposed matrix: $(A^T)_{ik} = A_{ki}$
A^*	adjoint matrix: $(A^*)_{ik} = \bar{A}_{ki}$
$\det A$, $\mathrm{tr}\,A$	determinant and trace respectively of a square matrix A
$O(n)$, $SO(n)$	the group of (real) orthogonal and special orthogonal $n \times n$ matrices A: $A^T A = 1$. "Special" means $\det A = 1$.
$U(n)$, $SU(n)$	the group of (complex) unitary and special unitary $n \times n$ matrices U: $U^*U = 1$. "Special" means $\det U = 1$.
$\mathbf{su}(n)$	the Lie algebra of the group $SU(n)$
$\tilde{f}(p)$	the Fourier transform of a function $f(x)$. Generally, x as well as p is a vector with d components.
∇f	the gradient of a function $f(x)$ ($x \in \mathbb{R}^d$). The meaning of the symbols grad, rot, and div is reserved for the situation $d = 3$.
Δ, $\boldsymbol{\Delta}$	the Laplace operators in three and four dimensions respectively: $\sum \partial^2/\partial x_k^2 \equiv \sum \partial_k^2$
dx	the volume element of $\mathbb{R}^d$, also called the *Lebesgue measure*: $dx = dx_1 dx_2 \cdots dx_d$
$\delta(x)$	Dirac's delta function for $x \in \mathbb{R}^d$: $\int dx\, \delta(x)f(x) = f(0)$

$\delta_*(\tau)$ — the periodic delta function for $\tau \in \mathbb{R}$: $\delta_*(\tau) = \sum_{n=-\infty}^{\infty} \delta(\tau + n)$

$K(x,z)$ — the Gauss function $(2\pi z)^{-d/2} \exp(-x^2/2z)$, where $x \in \mathbb{R}^d$, $z = s + it \in \mathbb{C}$ (complex time). In the theory of diffusion, we have $z = s$ (time is *real*). In quantum theory, $z = it$ (time is *imaginary*).

$\exp$, $\log$ — exponential function and natural logarithm

$I_\nu(z)$, $K_\nu(z)$ — modified Bessel function of order ν (see [2.19]). The ordinary Bessel functions are denoted by $J_\nu(z)$, the spherical Bessel functions by $j_n(z)$.

$\|f\|$ — the norm of $f(x)$. Often $\|f\|^2 = \int_G dx\, |f(x)|^2$ where $G \subset \mathbb{R}^d$

$L^2(G)$ — the Hilbert space of functions $f\colon G \to \mathbb{C}$ (or $\mathbb{R}$) satisfying $\|f\| < \infty$. Often identified with the *state space* of some quantum mechanical system. We write $L^2(a,b)$ if G is the real interval $a \leq x \leq b$.

$V(x)$, V — the scalar potential and the associated multiplication operator: $[V\psi](x) = V(x)\psi(x)$, $\psi \in L^2(\mathbb{R}^d)$

H — the *Hamiltonian* of a quantum mechanical system, frequently of the form $H = -\frac{1}{2}\Delta + V$, sometimes $H = \frac{1}{2}(i\nabla + A)^2 + V$ with $A(x)$ some vector potential.

$\operatorname{spec} P$ — the spectrum of the operator P consisting of the eigenvalues and the continuous spectrum

Ω — the ground state of a Hamiltonian satisfying $\Omega(x) > 0$ and $\|\Omega\| = 1$, the "vacuum" in field theory. Occasionally, some set of Brownian paths with specific properties, the phase space or the space of field configurations

e^{-sH} $(s \in \mathbb{R}_+)$ — the Schrödinger semigroup associated with the Hamiltonian H, which is assumed to be bounded from below

e^{-itH} $(t \in \mathbb{R})$ the unitary Schrödinger group describing the time evolution in quantum theory

$\langle x'|T|x\rangle$ Dirac's notation for the integral kernel (the "kernel" for short) of an operator T, i.e., $[Tf](x') = \int dx\, \langle x'|T|x\rangle f(x)$ with f in the domain of T

D frequently a differential operator given in terms of some *differential expression* and *boundary conditions*. In a different context, D designates the diffusion constant.

$\langle x'|D^{-1}|x\rangle$ the *Green's function* of the differential operator D, i.e., the kernel of D^{-1}

$\langle x', s'|x, s\rangle$ $(s' \geq s)$ the *transition amplitude*. It coincides with the kernel $\langle x'|e^{-(s'-s)H}|x\rangle$ of the Schrödinger semigroup if H is independent of the time t. More generally, if $[H(t), H(t')] = 0$, it is the kernel of the operator $\exp\{-\int_s^{s'} dt\, H(t)\}$, and in the most general case, it is $\langle x'|T \exp\{-\int_s^{s'} dt\, H(t)\}|x\rangle$, where T exp denotes the time-ordered exponential.

$\hbar = h/2\pi$ Planck's constant, frequently put to one by choosing appropriate physical units

$\beta^{-1} = k_\mathrm{B}T$ $T = $ absolute temperature, $k_\mathrm{B} = $ Boltzmann's constant

ℓ a characteristic length. Either $\ell^2 = \hbar^2/(mk_\mathrm{B}T)$ or $\ell^2 = \hbar(s'-s)/m$ ($s'-s = $ time interval, $m = $ mass).

τ a scaled time variable carrying no physical dimension such that $0 \leq \tau \leq 1$. If the time t is confined to the interval $s \leq t \leq s'$, we have $t = s + (s'-s)\tau$.

$Z = \exp(-\beta F)$ the partition function in the sense of statistical mechanics; F is the *free energy*. In the thermodynamic limit, $f = \lim_{V \to \infty} F/V$, where f is the free energy per unit volume.

$X,\ Y,$ etc.

random variables taking values either in $\mathbb{R}$ or, more generally, in $\mathbb{R}^d$. In the latter case, d is said to be the *dimension* of X. The components of X are denoted by X_k $(k = 1, \ldots, d)$.

$P(X \in A)$

the *distribution* of a random variable X, a probability measure (positive and normed); $A \subset \mathbb{R}^d$

$E(X)$

the *expectation value* of the random variable X, also called its *mean value*. If X is one-dimensional, $E(X^n)$ is said to be the nth moment of the distribution.

$\mathrm{Var}(X) := E(X^2)$

the *variance* of a one-dimensional random variable provided $E(X) = 0$ (X is *centered*). When X is d-dimensional, one considers the covariances $c_{kk'} := E(X_k X_{k'})$. Then the real $d \times d$ matrix $C = (c_{kk'})$ is said to be the *covariance matrix*.

$f(t) = E(\exp(itX))$

the *characteristic function* of a random variable X, at the same time the *generating function* of the moments $E(X^n)$

$C_n(X)$

the nth cumulant of a random variable X. Cumulants are generated by the function $\log E(\exp(itX))$.

$X_t,\ Y_t$ etc.

stochastic processes, where time is continuous, taking values in $\mathbb{R}$ or $\mathbb{R}^d$. In most cases, $t \in I$ where $I = \mathbb{R},\ \mathbb{R}_+,$ or $[0,1]$. As a rule, X_t is the *Wiener process*, Q_t denotes the *oscillator process*, and $\bar{X}_\tau$ the *Brownian bridge* (see Appendix B for details).

dX_t

the *differential* of the Wiener process. It is defined for *all* $t \in \mathbb{R}$. Differential increments at different times are independent. Mean value $= E(dX_t) = 0$, variance $= E(dX_t^2) = dt$. Formally, $dX_t = W_t dt$, where W_t is white noise.

$P(X_{s'} \in A | X_s = x)$

the transition function of a Markov process, a conditional probability. It is assumed that $s' > s$.

$E(X_s X_{s'})$

the *covariance* of a stochastic process. If X_s is d-dimensional with components X_{sk}, the covariance (for fixed s and s') should be regarded as a matrix with elements $c_{kk'} := E(X_{sk} X_{s'k'})$.

$\omega(t)$ $(t \in \mathbb{R}_+)$

some path of the Wiener process X_t, very often a special path with fixed endpoints: $s \leq t \leq s'$, $\omega(s) = x$, $\omega(s') = x'$. We then write $\omega : (x, s) \rightsquigarrow (x', s')$.

$d\mu(\omega)$

the *conditional Wiener measure* normed such that $\int d\mu(\omega) = K(x' - x, s' - s)$. It serves to define the path integral $I(f) := \int d\mu(\omega)\, f(\omega)$, where the integration is over paths $\omega : (x, s) \rightsquigarrow (x', s')$. The conditional Wiener measure and hence $I(f)$ depend on the parameters x, s, x', s'.

$\bar{\omega}(\tau)$ $(\tau \in [0, 1])$

some path of the Brownian bridge. The endpoints are fixed: $\bar{\omega}(0) = \bar{\omega}(1) = 0$.

$d\bar{\omega}$

the probability measure on the set of all paths of the Brownian bridge such that $\int d\bar{\omega}\, f(\bar{\omega}) = E(f(\bar{X}))$

W_t $(t \in \mathbb{R})$

so-called *white noise*. W_t is a generalized Markov process. It is a Gaussian process with mean zero and covariance $\delta(t - t')$.

$W(f)$

the *stochastic integral* (not a path integral!) of a real function $f \in L^2(\mathbb{R})$, a Gaussian random variable with mean zero and variance $\|f\|^2$, representable as $\int f(t) dX_t$ ($X_t =$ Wiener process) or as $\int dt\, f(t) W_t$ ($W_t =$ white noise). Analogously, one defines $\bar{W}(f) = \int f(\tau) d\bar{X}_\tau$ with $f \in L^2(0, 1)$ and $\bar{X}_\tau$ the Brownian bridge.

$\int A(X_t) \cdot dX_t$

the *Itô integral* (not a path integral!) of a vector function $A : \mathbb{R}^d \rightarrow \mathbb{R}^d$ (X_t is the d-dimensional Wiener process). It generalizes the notion of a line integral. The Itô integral extends over some interval $s \leq t \leq s'$.

$\text{var}(f)$	abbreviation for $\int_0^1 d\tau\, f(\tau)^2 - \left(\int_0^1 d\tau\, f(\tau)\right)^2$ where $f \in L^2(0,1)$
M_4	the four-dimensional Minkowski space modeling spacetime with the indefinite metric $(x,x) = (x^0)^2 - (x^1)^2 - (x^2)^2 - (x^3)^2$, where $x \in M_4$.
E_4	the four-dimensional Euclidean space with the metric $x^2 = (x^1)^2 + (x^2)^2 + (x^3)^2 + (x^4)^2$, where $x \in E_4$.
$\Phi(x)$	a second-quantized field on M_4, frequently a one-component scalar field, also called a *Minkowskian field*
$\boldsymbol{\Phi}(x)$	a second-quantized field on E_4, frequently a one-component scalar field, also called a *Euclidean field*
$\psi(x)$	the four-component Dirac field either on M_4 or on E_4
$\Psi(x)$	a bispinor combining $\psi(x)$ and $\psi^c(x)$
$\omega \equiv \omega(\boldsymbol{p})$	abbreviation for $\sqrt{m^2 + \boldsymbol{p}^2}$ in the context of relativistic field theory
$\mathcal{S}(E_4)$	the Schwartz space of functions $f : E_4 \to \mathbb{R}$. The extended space $\mathcal{S}^c$ consists of complex functions $f_1 + if_2$ where $f_i \in \mathcal{S}$. The subspace $\mathcal{S}_+$ consists of functions $f \in \mathcal{S}$ where $f(\boldsymbol{x}, x^4) = 0$ for $x^4 < 0$.
$\mathcal{S}'$	the space of distributions in the sense of L. Schwartz, the dual space of $\mathcal{S}$
$\phi(x)$	the random value or the "path" of a Euclidean field $\boldsymbol{\Phi}(x)$. It may be viewed as an element in $\mathcal{S}'$.
$W(\cdots)$	the Euclidean action, a functional of the fields. It serves to specify a field-theoretical model.
$S\{\phi\}$	the Schwinger functional of a scalar field, the generating functional of the Schwinger functions

$d\mu(\phi)$ — a probability measure, often the Gibbs measure obtained from the Euclidean action

$S(\mu)$ — the entropy of the probability measure μ

D_k — the covariant derivative in (Euclidean) gauge theory: $D_k = \partial_k - A_k$, $(k = 1, \ldots 4)$, $A_k = $ gauge field

$\mathcal{D}\phi, \; \mathcal{D}A$ — formal differentials of Euclidean fields used in path integration. After some regularization by introducing a finite lattice, such differentials assume a definite meaning.

γ^μ — the Dirac matrices with respect to the Minkowskian metric $(\mu = 0, \ldots, 3)$. The Dirac matrices with respect to the Euclidean metric are denoted by γ^k $(k = 1, \ldots, 4)$.

$A(E)$ — the Grassmann algebra over some vector space E

$A^p(E)$ — the space of all p-linear antisymmetric functions over E

A_+ — the even part of a Grassmann algebra, always a commutative subalgebra

$E \oplus F$ — the direct sum of two linear spaces

Appendix B
Frequently Used Gaussian Processes

Most Gaussian Markov processes can be treated on the same footing when the concept of the stochastic integral $W(f)$ is being applied. In what follows we restrict ourselves to the one-dimensional case for simplicity so that $W(f)$ becomes a random variable for each real $f \in L^2(\mathbb{R})$ taking values in $\mathbb{R}$. Moreover, the map $f \mapsto W(f)$ is linear and satisfies

$$\boldsymbol{E}\Big(\exp\{iW(f)\} \Big) = \exp\{-\tfrac{1}{2}\|f\|^2\}, \tag{B.1}$$

where $\|f\|^2 = \int_{-\infty}^{\infty} |f(s)|^2 ds$. Formally,

$$W(f) = \int_{-\infty}^{\infty} f(s) W_s ds \tag{B.2}$$

with W_s Gaussian *white noise*. For a suitably chosen family $f_t \in L^2$ with $t \in I \subset \mathbb{R}$ setting $X_t = W(f_t)$, X_t is a Markov process on the time interval I. The known examples include:

name	time interval I	$f_t(s)$
Wiener process	$[0, \infty)$	$\begin{cases} 1 & \text{if } 0 \le s \le t \\ 0 & \text{otherwise} \end{cases}$
Ornstein–Uhlenbeck process	$[0, \infty)$	$\begin{cases} e^{-\gamma(t-s)} & \text{if } 0 \le s \le t \\ 0 & \text{otherwise} \end{cases}$
oscillator process	$(-\infty, \infty)$	$\begin{cases} e^{-k(s-t)} & \text{if } s \ge t \\ 0 & \text{otherwise} \end{cases}$
Brownian bridge	$[0, 1]$	$\begin{cases} 1 - t & \text{if } 0 \le s \le t \\ -t & \text{if } t < s \le 1 \\ 0 & \text{otherwise} \end{cases}$

Both γ and k are assumed to be positive. As for the Ornstein–Uhlenbeck process, γ is interpreted as the *friction constant*, while, for the oscillator process, k is called its *frequency*. When k and γ are identified, both processes possess the same transition function:

$$P(X_{t'} \in dx' | X_t = x) = dx' \left(\frac{k/\pi}{1 - e^{-2\nu}} \right)^{1/2} \exp\left\{ -k \frac{(x' - e^{-\nu}x)^2}{1 - e^{-2\nu}} \right\},$$

$$(B.3)$$

where $\nu = k(t' - t) > 0$. The two processes differ however by their initial distribution:

$$P(X_0 \in dx) = \begin{cases} dx \, (k/\pi)^{1/2} \exp(-kx^2) & \text{oscillator} \\ dx \, \delta(x) & \text{Ornstein–Uhlenbeck.} \end{cases} \qquad (B.4)$$

The initial distribution of the oscillator process is stationary, i.e.,

$$P(X_t \in dx) = P(X_0 \in dx)$$

for all $t \in \mathbb{R}$. Since it coincides with the asymptotic distribution $\lim_{t \to \infty} P(X_t \in dx)$ of the Ornstein–Uhlenbeck process, it may well be said that both processes become indistinguishable at large times.

The covariance of the process $X_t = W(f_t)$ is quite generally given by the following formula:

$$G(t, t') := E(X_t X_{t'}) = (f_t, f_{t'}) := \int_{-\infty}^{\infty} f_t(s) f_{t'}(s) ds. \qquad (B.5)$$

In each case considered above, $G(t, t')$ is the Green's function of some second-order self-adjoint differential operator D on $L^2(I)$, where the time interval I relates to the process, i.e., $G(t, t')$ is the integral kernel of the operator D^{-1} which is therefore called the *covariance operator*:

$$G(t, t') = \langle t | D^{-1} | t' \rangle \qquad (t, t' \in I). \qquad (B.6)$$

To specify the operator D, the underlying differential expression has to be supplemented by suitable boundary conditions at the endpoints of the interval I (provided there are endpoints). Different boundary conditions generally lead to different operators and hence to different covariances. In the following list, $u \in L^2(I)$ stands for any function in the domain of D.

name of the process	covariance $G(t, t') =$ Green's function	operator $D =$ differential expression + boundary conditions
Wiener process	$\min(t, t')$	$-d^2/dt^2$, $u(0) = 0$
Ornstein–Uhlenbeck process	$\frac{1}{2\gamma}\left(e^{-\gamma|t-t'|} - e^{-\gamma(t+t')}\right)$	$-d^2/dt^2 + \gamma^2$, $u(0) = 0$
oscillator process	$\frac{1}{2k}e^{-k|t-t'|}$	$-d^2/dt^2 + k^2$
Brownian bridge	$\min(t, t') - tt'$	$-d^2/dt^2$, $u(0) = u(1) = 0$

A Gaussian process with covariance $G(t,t')$ is Markovian iff

$$G(t,t')G(t',t'') = G(t,t'')G(t',t') \tag{B.7}$$

for all $t \leq t' \leq t''$ in the domain of definition. It is easily checked that the condition (B.7) is indeed satisfied for the processes considered above. A Gaussian process with covariance $G(t,t')$ is *invariant* (temporally homogeneous) iff the covariance is a function of the difference $t - t'$ alone. White noise and the oscillator process are the only invariant Gaussian Markov processes as may be inferred from (B.7).

Nonlinear Transformations of Time. Different processes are usually related in more than one way. For instance, a nonlinear transformation of the time variable may convert one process into another, or better, into a *version* of another process. When two processes are defined on the same time interval and have identical n-distributions, we shall write $X_t \doteq Y_t$. If $X_t = W(f_t)$ and $Y_t = W(g_t)$, we have $X_t \doteq Y_t$ iff both processes have the same covariance: $(f_t, f_{t'}) = (g_t, g_{t'})$.

Let X_t be the Wiener process, $\bar{X}_\tau$ the Brownian bridge, Y_t the Ornstein–Uhlenbeck process with friction constant γ, and Q_t the oscillator process with frequency k. Then

$$
\begin{aligned}
X_t &\doteq s^{-1} X_s & t &= s^{-1} \\
X_t &\doteq e^{-ks} Q_s & t &= (2k)^{-1} e^{-2ks} \\
\bar{X}_\tau &\doteq \sqrt{2\gamma}\, e^{-\gamma s} Y_s & \tau &= e^{-2\gamma s} \\
\bar{X}_\tau &\doteq (s+1)^{-1} X_s & \tau &= s(s+1)^{-1}.
\end{aligned}
\tag{B.8}
$$

Appendix C
Jensen's Inequality

Many inequalities in physics and mathematics have their origin in the notion of convexity and the theorem of Jensen as applied to convex functions [C.1-2]. We want to discuss Jensen's inequality from the point of view of probability theory.

Recall that a real function f is said to be *convex* on some open interval $I \subset \mathbb{R}$ if

$$f\big(\alpha u_1 + (1-\alpha)u_2\big) \leq \alpha f(u_1) + (1-\alpha)f(u_2) \tag{C.1}$$

for all $u_1, u_2 \in I$ and $0 \leq \alpha \leq 1$. From this property one infers that

$$f(\textstyle\sum \alpha_i u_i) \leq \sum \alpha_i f(u_i) \tag{C.2}$$

for all $u_i \in I$ and $0 \leq \alpha_i \leq 1$, $i = 1, \ldots, n$ by induction, where $\sum \alpha_i = 1$ and $n \in \mathbb{N}$.

Think now of the numbers α_i as probabilities of certain "events" i. Then the sums occurring in (C.2) may be interpreted as expectations with respect to the discrete distribution given by the α_i. With the concept of expectation in mind, we wish to pass from sums to integrals.

A set Ω is said to be a probability space if it carries a probability measure μ. The points $\omega \in \Omega$ are then called *elementary events*. In general, an arbitrary event is modeled by some (measurable) subset of Ω. Let the system of subsets $\{\Omega_i | i = 1, \ldots, n\}$ define a disjoint decomposition of Ω and suppose that $\alpha_i = \mu(\Omega_i)$, that is to say, α_i describes the probability of the event $\omega \in \Omega_i$. For some function $g : \Omega \to I$ and certain $\omega_i \in \Omega_i$, we put $u_i = g(\omega_i)$ so that the inequality (C.2) assumes the form

$$f\big(\textstyle\sum_i \mu(\Omega_i)g(\omega_i)\big) \leq \sum_i \mu(\Omega_i)f\big(g(\omega_i)\big). \tag{C.3}$$

The sums that arise this way appear as discrete approximations of integrals with respect to the measure μ. By refining the partition of the space Ω and passing to a suitable limit,

$$f\big(\textstyle\int d\mu(\omega)\, g(\omega)\big) \leq \int d\mu(\omega)\, f\big(g(\omega)\big). \tag{C.4}$$

This is the inequality of Jensen valid under three conditions: (1) $f : I \to \mathbb{R}$ is convex, (2) μ is a probability measure on Ω, and (3) $g : \Omega \to I$ is absolutely integrable with respect to μ. Next we shall mention special situations that come up in connection with the topic of this book, where the inequality may be applied.

Let X_t be some stochastic process and $g(X)$ a real random variable depending on the section $\{X_t | s \le t \le s'\}$. For f a convex function on $\mathbb{R}$,

$$f\Big(\boldsymbol{E}\big(g(X)\big)\Big) \le \boldsymbol{E}\Big(f\big(g(X)\big)\Big). \tag{C.5}$$

In particular, the function $f(t) = \exp t$ is convex on $I = \mathbb{R}$ and so we get

$$\exp\Big(\textstyle\int d\mu(\omega)\, g(\omega)\Big) \le \textstyle\int d\mu(\omega)\, \exp\big(g(\omega)\big). \tag{C.6}$$

More specifically, if $d\mu(\omega)$ is a normalized measure on a path space, we obtain a comparison theorem for path integrals.

There is another special version of Jensen's inequality,

$$\exp\Big(\frac{1}{s}\int_0^s dt\, g(t)\Big) \le \frac{1}{s}\int_0^s dt\, \exp\big(g(t)\big), \tag{C.7}$$

where the uniform distribution on the space $\Omega = [0, s]$ has been chosen as the underlying probability measure .

Appendix D
A Table of Path Integrals

The purpose of this appendix is to list those path integrals (Wiener integrals) that can be evaluated in terms of elementary functions. We let the variable ω stand for a Brownian path with fixed endpoints x and x' at times s and s' such that $s' > s$. The diffusion constant is taken to be $D = \frac{1}{2}$ throughout. The conditional Wiener measure is denoted by $d\mu(\omega)$ without indicating explicitly its dependence on x, x', s, s'. In what follows we shall always assume the time variables t and t' to be confined to the interval $[s, s']$. For simplicity, all formulas will be written for the one-dimensional case only. Hence, x und x' are treated as real numbers, not as vectors. Most formulas, however, admit extensions to higher dimensions without further ado. In [D.11–19], we have put

$$\nu = k(s' - s)$$

for real k. In [D.20–24] we have adopted the notation

$$\inf \omega = \inf\{\omega(t) \mid s \le t \le s'\}$$
$$\sup \omega = \sup\{\omega(t) \mid s \le t \le s'\}$$
$$\langle \omega \rangle = \frac{1}{s' - s} \int_s^{s'} dt\, \omega(t).$$

In some of the subsequent formulas, we make use of functions $f(a)$ and $f(a, b)$. They are arbitrary in principle and restricted solely by the requirement that the integrals occurring on the right-hand side be well-defined.

One basic formula is provided by [D.19]. Its validity has been shown in Section 4.4. All previous formulas arise as special cases. Another basic formula is [D.20]. It rests on the reflection principle and has been derived in Sect. 2.9.

[D.01]

$$\int d\mu(\omega) = \left(2\pi(s' - s)\right)^{-1/2} \exp\left\{-\frac{(x' - x)^2}{2(s' - s)}\right\}$$

[D.02]

$$\int d\mu(\omega)\,\omega(t) = \frac{(s'-t)x + (t-s)x'}{s'-s}\,(2\pi(s'-s))^{-1/2}\exp\left\{-\frac{(x'-x)^2}{2(s'-s)}\right\}$$

[D.03]

$$\int d\mu(\omega)\,\omega(t)^n = (2\pi(s'-s))^{-1/2}\exp\left\{-\frac{(x'-x)^2}{2(s'-s)}\right\} \times$$

$$\sum_k \frac{n!}{k!(n-2k)!}\left(\frac{(s'-t)(t-s)}{2(s'-s)}\right)^k\left(\frac{(s'-t)x+(t-s)x'}{s'-s}\right)^{n-2k}$$

[D.04]

$$\int d\mu(\omega)\,\omega(t)\omega(t') = (2\pi(s'-s))^{-1/2}\exp\left\{-\frac{(x'-x)^2}{2(s'-s)}\right\} \times$$

$$\left\{\frac{(s'-t)x+(t-s)x'}{s'-s}\cdot\frac{(s'-t')x+(t'-s)x'}{s'-s}\right.$$

$$\left.+\frac{(s'-\max(t,t'))(\min(t,t')-s)}{s'-s}\right\}$$

[D.05]

$$\int d\mu(\omega)\int_s^{s'} dt\,\omega(t)^n = (2\pi(s'-s))^{-1/2}\exp\left\{-\frac{(x'-x)^2}{2(s'-s)}\right\} \times$$

$$\sum_{k,p}\frac{(s'-s)^{k+1}}{2^k(n+1)}\binom{p}{k}\frac{(n-p)!}{(n-p-k)!}x'^{p-k}x^{n-p-k}$$

[D.06]

$$\int d\mu(\omega)\int_s^{s'} dt\int_s^{s'} dt'\,[\omega(t)-\omega(t')]^2 =$$

$$(2\pi(s'-s))^{-1/2}\exp\left\{-\frac{(x'-x)^2}{2(s'-s)}\right\}\cdot\frac{1}{6}(s'-s)^2(s'-s+(x'-x)^2)$$

[D.07]

$$\int d\mu(\omega)\,\exp(a\omega(t)) = (2\pi(s'-s))^{-1/2}\exp\left\{-\frac{(x'-x)^2}{2(s'-s)}\right.$$

$$\left.+a\frac{(s'-t)x+(t-s)x'}{s'-s}+\frac{a^2}{2}\frac{(s'-t)(t-s)}{s'-s}\right\}$$

[D.08]

$$\int d\mu(\omega) \, \exp\left\{ a \int_s^{s'} dt\, \omega(t) \right\} =$$

$$(2\pi(s'-s))^{-1/2} \exp\left\{ -\frac{(x'-x)^2}{2(s'-s)} + \frac{a}{2}(s'-s)(x'+x) + \frac{a^2}{24}(s'-s)^3 \right\}$$

[D.09]

$$\int d\mu(\omega) \, \exp\left\{ \int_s^{s'} dt\, f(t)\omega(t) \right\} =$$

$$(2\pi(s'-s))^{-1/2} \exp\left\{ -\frac{(x'-x)^2}{2(s'-s)} + \int_s^{s'} dt\, f(t)\frac{(s'-t)x + (t-s)x'}{s'-s} \right.$$

$$\left. + \int_s^{s'} dt \int_s^t dt'\, f(t)f(t')\frac{(s'-t)(t'-s)}{s'-s} \right\}$$

[D.10]

$$\int d\mu(\omega) \, \exp\left\{ -\frac{k^2}{2}\omega(t)^2 \right\} = \left[2\pi\big(s'-s+k^2(s'-t)(t-s)\big) \right]^{-1/2} \times$$

$$\exp\left\{ -\frac{1}{2}\frac{(x'-x)^2 + k^2\big((s'-t)x^2 + (t-s)x'^2\big)}{s'-s+k^2(s'-t)(t-s)} \right\}$$

[D.11]

$$\int d\mu(\omega) \, \exp\left\{ -\frac{k^2}{2} \int_s^{s'} dt\, \omega(t)^2 \right\} =$$

$$\left[\frac{k}{2\pi\sinh\nu} \right]^{1/2} \exp\left\{ -\frac{k(x'^2+x^2)}{2\tanh\nu} + \frac{kxx'}{\sinh\nu} \right\}$$

[D.12]

$$\int d\mu(\omega) \, \exp\left\{ -\frac{k^2}{2} \int_s^{s'} dt'\, \big[\omega(t') - \omega(t)\big]^2 \right\} =$$

$$\left[\frac{k}{2\pi\sinh\nu} \right]^{1/2} \exp\left\{ -\frac{k(x'-x)^2}{2\sinh\nu} \cosh[k(t-s)] \cosh[k(s'-t)] \right\}$$

[D.13]

$$\int d\mu(\omega) \, \exp\left\{-\frac{k^2}{2} \int_s^t dt' \, \left[\omega(t') - \omega(t)\right]^2\right\} =$$

$$\left[\frac{k/(2\pi)}{\sinh[k(t-s)] + k(s'-t)\cosh[k(t-s)]}\right]^{1/2} \times$$

$$\exp\left\{-\frac{k(x'-x)^2/2}{\tanh[k(t-s)] + k(s'-t)}\right\}$$

[D.14]

$$\int d\mu(\omega) \, \exp\left\{-\frac{k^2}{s'-s} \int_s^{s'} dt \int_s^{s'} dt' \, \left[\omega(t') - \omega(t)\right]^2\right\} =$$

$$\frac{\nu}{\sinh\nu} \left(2\pi(s'-s)\right)^{-1/2} \exp\left\{-\frac{k(x'-x)^2}{2\tanh\nu}\right\}$$

[D.15]

$$\int d\mu(\omega) \, \exp\left\{-\frac{k^2}{2} \int_s^{s'} dt \, \omega(t)^2 + a \int_s^{s'} dt \, \omega(t)\right\} = \left[\frac{k}{2\pi\sinh\nu}\right]^{1/2} \times$$

$$\exp\left\{-\frac{k(x'^2 + x^2)}{2\tanh\nu} + \frac{kxx'}{\sinh\nu} + \frac{a}{k}(x'+x)\tanh\frac{\nu}{2} + \frac{a^2}{k^3}\left(\frac{\nu}{2} - \tanh\frac{\nu}{2}\right)\right\}$$

[D.16]

$$\int d\mu(\omega) \, \omega(t) \exp\left\{-\frac{k^2}{2} \int_s^{s'} dt \, \omega(t)^2\right\} = \left[\frac{k}{2\pi\sinh\nu}\right]^{1/2} \times$$

$$\frac{x\sinh[k(s'-t)] + x'\sinh[k(t-s)]}{\sinh\nu} \exp\left\{-\frac{k(x'^2 + x^2)}{2\tanh\nu} + \frac{kxx'}{\sinh\nu}\right\}$$

[D.17]

$$\int d\mu(\omega) \, \omega(t)\omega(t') \exp\left\{-\frac{k^2}{2} \int_s^{s'} dt \, \omega(t)^2\right\} =$$

$$\left\{\frac{x\sinh[k(s'-t)] + x'\sinh[k(t-s)]}{\sinh\nu} \cdot \frac{x\sinh[k(s'-t')] + x'\sinh[k(t'-s)]}{\sinh\nu}\right.$$

$$\left. + \frac{\sinh[k(s'-\max(t,t'))]\sinh[k(\min(t,t')-s)]}{k\sinh\nu}\right\} \times$$

$$\left[\frac{k}{2\pi\sinh\nu}\right]^{1/2} \exp\left\{-\frac{k(x'^2 + x^2)}{2\tanh\nu} + \frac{kxx'}{\sinh\nu}\right\}$$

[D.18]

$$\int d\mu(\omega)\,\exp\left\{a\omega(t) - \frac{k^2}{2}\int_s^{s'} dt\,\omega(t)^2\right\} = \left[\frac{k}{2\pi\sinh\nu}\right]^{1/2} \times$$

$$\exp\left\{-\frac{k(x'^2 + x^2)}{2\tanh\nu} + \frac{kxx'}{\sinh\nu} + a\frac{x\sinh[k(s'-t)] + x'\sinh[k(t-s)]}{\sinh\nu}\right.$$

$$\left.+a^2\frac{\sinh[k(s'-t)]\sinh[k(t-s)]}{2k\sinh\nu}\right\}$$

[D.19]

$$\int d\mu(\omega)\,\exp\left\{-\frac{k^2}{2}\int_s^{s'} dt\,\omega(t)^2 + \int_s^{s'} dt\,f(t)\omega(t)\right\} =$$

$$\left[\frac{k}{2\pi\sinh\nu}\right]^{1/2}\exp\left\{-\frac{k(x'^2 + x^2)}{2\tanh\nu} + \frac{kxx'}{\sinh\nu}\right.$$

$$+\int_s^{s'} dt\,f(t)\frac{x\sinh[k(s'-t)] + x'\sinh[k(t-s)]}{\sinh\nu}$$

$$\left.+\int_s^{s'} dt\int_s^{t} dt'\,f(t)f(t')\frac{\sinh[k(s'-t)]\sinh[k(t'-s)]}{k\sinh\nu}\right\}$$

[D.20]

$$\int d\mu(\omega)\,f(\inf\omega) =$$

$$(2\pi(s'-s))^{-1/2}\int_{-\infty}^{\min(x,x')} da\,f(a)\frac{d}{da}\exp\left\{-\frac{(x'+x-2a)^2}{2(s'-s)}\right\}$$

[D.21]

$$\int d\mu(\omega)\,f(\sup\omega) =$$

$$-(2\pi(s'-s))^{-1/2}\int_{\max(x',x)}^{\infty} da\,f(a)\frac{d}{da}\exp\left\{-\frac{(x'+x-2a)^2}{2(s'-s)}\right\}$$

[D.22]

$$\int d\mu(\omega)\,f(\sup\omega - \inf\omega) = (2\pi(s'-s))^{-1/2}\times$$

$$\int_r^{\infty} da\,f(a)\frac{d}{da}\sum_{n=-\infty}^{\infty}\left((a-r)\frac{d}{da} + 1 - 2n\right)\exp\left\{-\frac{(r+2na)^2}{2(s'-s)}\right\}$$

$$(r = |x' - x|)$$

[D.23]

$$\int d\mu(\omega)\, f(\inf \omega, \sup \omega) \;=\; -\big(2\pi(s'-s)\big)^{-1/2}$$

$$\int_{-\infty}^{\min(x',x)} da \int_{\max(x',x)}^{\infty} db\, f(a,b) \frac{\partial^2}{\partial a \partial b} \sum_{n=-\infty}^{\infty} (-1)^n \exp\left\{ -\frac{(x'-x_n)^2}{2(s'-s)} \right\}$$

$$x_n := \tfrac{1}{2}(a+b) + (-1)^n\big(x - \tfrac{1}{2}(a+b)\big) + n(b-a)$$

[D.24]

$$\int d\mu(\omega)\, f(\langle \omega \rangle) \;=\;$$

$$\frac{\sqrt{3}}{\pi(s'-s)} \int_{-\infty}^{\infty} da\, f(a) \exp\left\{ -\frac{(x'-x)^2 + 3(x'+x-2a)^2}{2(s'-s)} \right\}$$

[D.25]

$$\int d\mu(\omega) \left[\exp\left\{ \lambda \int_s^{s'} dt\, \delta(\omega(t)) \right\} - 1 \right] \;=\;$$

$$\lambda\big(2\pi(s'-s)\big)^{-1/2} \int_0^{\infty} da\, \exp\left\{ \lambda a - \frac{(r+r'+a)^2}{2(s'-s)} \right\}$$

$$(r = |x|,\ r' = |x'|)$$

[D.26]

$$\int d\mu(\omega)\, \exp\left\{ -a \int_s^{s'} dt\, \Theta(\omega(t)) \right\} = \int_{\min(0,a)}^{\infty} dE\, e^{-E(s'-s)} \frac{1}{\pi} \mathrm{Img}(-E - i0, a)$$

$$\Theta(x) = \begin{cases} 0 & \text{if } x < 0 \\ 1 & \text{if } x \geq 0 \end{cases} \qquad g(z,a) = h\big(\sqrt{2z}, \sqrt{2(z+a)}\big)$$

$$h(u,v) = \begin{cases} \dfrac{2}{u+v} e^{xu - x'v} & \text{if } x \leq 0 \leq x' \\[2ex] \dfrac{2}{u+v} e^{x'u - xv} & \text{if } x' \leq 0 \leq x \\[2ex] \dfrac{1}{u}\left\{ e^{-|x-x'|u} + \dfrac{u-v}{u+v} e^{-|x+x'|u} \right\} & \text{if } x < 0,\, x' < 0 \\[2ex] \dfrac{1}{v}\left\{ e^{-|x-x'|v} + \dfrac{v-u}{v+u} e^{-|x+x'|v} \right\} & \text{if } x > 0,\, x' > 0. \end{cases}$$

In particular,

$$\int_{-\infty}^{\infty} dx'\, h(u,v) \;=\; \begin{cases} \dfrac{2}{v^2}\left\{1 + \left(\dfrac{v}{u} - 1\right)e^{-xv}\right\} & \text{if } x > 0 \\[3ex] \dfrac{2}{u^2}\left\{1 + \left(\dfrac{u}{v} - 1\right)e^{xu}\right\} & \text{if } x < 0 \end{cases}$$

$$\int_{-\infty}^{\infty} dx'\, h(u,v)\Big|_{x=0} \;=\; \frac{2}{uv}\,, \qquad \lim_{y\to\infty} \frac{1}{2y}\int_{-y}^{y} dx\, h(u,v)\Big|_{x'=x} \;=\; \frac{1}{2}\left(\frac{1}{u} + \frac{1}{v}\right)$$

References

Chapter 1

1.1   N. Wax (ed.), *Selected Papers on Noise and Stochastic Processses*, Dover, New York 1954

1.2   A. Einstein, *Investigations on the Theory of the Brownian Movement*, ed. by R. Frth, Dover, New York 1956

1.3   M. Kac, Random walk and the theory of Brownian motion, *Amer. Math. Monthly* **54**, No. 7 (reprinted in [1.1])

1.4   K. L. Chung, *Markov chains with stationary transition probabilities*, 2nd edn., Springer, New York 1967

1.5   K. L. Chung, *Lectures from Markov Processes to Brownian Motion*, Springer, New York 1982

1.6   B. B. Mandelbrot, *The Fractal Geometry of Nature*, Freeman, New York 1982

1.7   Lie Hou-Qiang and Zhao Hua-Ming, Multifractal characteristics of Brownian motion, *Chin. Sci. Bull.* **36**, 1142 (1991)

1.8   F. Spitzer, *Principles of Random Walk*, Van Nostrand-Reinhold, New York 1964

1.9   J. K. Percus, *Combinatorial Methods*, Appl. Math. Sciences **4**, Springer, New York 1971

1.10   J. Glimm and A. Jaffe, *Quantum Physics: A Functional Integral Point of View*, 2nd edn., Springer, New York 1987

1.11   G. Polya, ber eine Aufgabe der Wahrscheinlichkeitsrechnung betreffend die Irrfahrt im Straennetz, *Math. Ann.* **84**, 149 (1921)

1.12   J. Frhlich, B. Simon, and T. Spencer, Infrared bounds, phase transitions, and continuous symmetry breaking, *Commun. Math. Phys.* **50**, 79 (1976)

1.13   G. C. Wick, A. S. Wightman, and E. P. Wigner, The intrinsic parity of elementary particles, *Phys. Rev.* **88**, 101 (1952)

1.14   J. E. Roberts and G. Roepstorff, Some basic concepts of algebraic quantum theory, *Commun. Math. Phys.* **11**, 321 (1968)

1.15   R. P. Feynman and A. Hibbs, *Quantum Mechanics and Path Integrals*, McGraw-Hill, New York 1965

1.16   P. Billingsley, *Probability and Measure*, Wiley, New York 1979

1.17   L. Breiman, *Probability*, Addison-Wesley, Reading, Mass. 1968

1.18   P. Halmos, *Measure Theory*, Van Nostrand, Princeton, N.J. 1950

1.19   J. L. Doob, *Stochastic Processes*, Wiley, New York 1953

1.20   N. Wiener, Differential space, *J. Math. and Phys. Sci.* **2**, 132 (1923)

1.21   E. Nelson, Feynman integrals and the Schrdinger equation, *J. Math. Phys.* **5**, 332 (1964)

1.22   B. Simon, *Functional Integration and Quantum Physics*, Academic, New York 1979

1.23   D. Freedman, *Brownian Motion and Diffusion*, Holden-Day, San Francisco 1971

1.24   I. M. Gel'fand and N. Y. Vilenkin, *Generalized Functions, Vol. 4, Applications of Harmonic Analysis*, Academic, New York 1964

1.25 R. A. Minlos, Generalized random processes and their extension to a measure, *Trudy Moskov. Mat. Obsc.* **8**, 497 (1959). English translation: *Selected Translations in Mathematical Statistics and Probability*, No. 3, Amer. Math. Soc., Providence, R.I. 1963

1.26 J.-P. Caubet, Relativistic Brownian motion, in: *Probabilistic Methods in Differential Equations*, Lect. Notes. Math. **451**, Springer, New York, 1975

1.27 B. Gaveau, T. Jacobson, M. Kac, and L.S. Schulman, Relativistic extension of the analogy between quantum mechanics and Brownian motion, *Phys. Rev. Letters* **53**, 419 (1984)

1.28 T. Hida, *Brownian Motion*, Springer, New York 1980

1.29 H. McKean, *Stochastic Integrals*, Academic, New York 1969

1.30 K. L. Chung and R. J. Williams, *Introduction to Stochastic Integration*, Birkhuser, Boston 1983

1.31 B. B. Mandelbrot and J. W. Ness, Fractional Brownian Motion, Fractional Noises and Applications, *SIAM Rev.* **10**, 422 (1968)

1.32 E. Nelson, *Dynamical Theories of Brownian Motion*, Princeton Univ. Press, Princeton, N.J. 1967

1.33 G. E. Uhlenbeck and L. S. Ornstein, On the theory of Brownian motion, I. *Phys. Rev.* **36**, 823 (1930)

1.34 I. I. Gihman and A. W. Skorohod, *Stochastic Differential Equations*, Springer, Berlin 1972

1.35 N. Ikeda and S. Watanabe, *Stochastic Differential Equations and Diffusion Processes*, North Holland, Amsterdam 1981

1.36 N. G. van Kampen, *Stochastic Processes in Physics and Chemistry*, North Holland, Amsterdam 1981

1.37 K. Itô and H. McKean, *Diffusion Processes and Their Sample Paths*, Springer, New York 1965

1.38 A. M. Yaglom, *An Introduction to the Theory of Stationary Random Functions*, Prentice-Hall, Englewood Cliffs, N.J. 1962

1.39 R. Courant and D. Hilbert, *Methods of Mathematical Physics*, Interscience, New York 1953 (Vol. 1) and 1962 (Vol. 2)

1.40 S. N. Ethier and T. G. Kurtz, *Markov Processes*, Wiley, New York 1986

Chapter 2

2.1 E. Nelson, Feynman integrals and the Schrödinger equation, *J. Math. Phys.* **5**, 332 (1964)

2.2 K.L. Chung and S.R.S. Varadhan, Kac functional and Schrödinger equation, *Studia Math.* **68**, 249 (1979)

2.3 H. Aizenman and B. Simon, Brownian motion and Harnack inequality for Schrödinger operators, *Comm. Pure Appl. Math.* **35**, 209 (1982)

2.4 M. Demuth and J. A. Van Casteren, On spectral theory of selfadjoint Feller generators, *Rev. Math. Phys.* **1**, 325 (1989)

2.5 M. Demuth, On topics in spectral and stochastic analysis for Schrödinger operators, in: *Recent Developments in Quantum Mechanics*, ed. by A. Boutet de Monvel et al., Kluwer, Netherlands 1991

2.6 M. Reed and B. Simon, *Methods of Modern Mathematical Physics IV: Analysis of Operators*, Academic, New York 1978

2.7 N. I. Portenko, Diffusion processes with unbounded drift coefficient, *Theor. ProbaB. Appl.* **20**, 27 (1976)

2.8 M. Berthier and B. Gaveau, Critére de convergence des fonctionelles de Kac et application en mécanique quantique et en géométric, *J. Funct. Anal.* **29**, 416 (1978)

2.9   H. Trotter, On the product of semigroups of operators. *Proc. Amer. Math. Soc.* **10**, 545 (1959)

2.10   T. Kato, Trotter's product formula for an arbitrary pair of self-adjoint contraction semigroups, in: *Topics in Functional Analysis*, ed. by I. Gohberg and M. Kac, Academic, New York 1978

2.11   J. Glimm and A. Jaffe, *Quantum physics: A Functional Integral Point of View*, 2nd edn., Springer, New York 1987

2.12   L. S. Schulman, *Techniques and applications of path integration*, Wiley, New York, 1981

2.13   M. D. Donsker and S. R. S. Varadhan, Asymptotic evaluation of certain Markov process expectations for large time, I–IV. *Comm. Pure Appl. Math.* **28**, 1 (1975); **28**, 279 (1975); **29**, 389 (1976); **36**, 183 (1983)

2.14   B. Simon, Brownian motion, L^p-properties of Schrödinger operators, and the localization of binding, *J. Funct. Anal.* **35**, 215 (1980)

2.15   B. Simon, Large time behavior of the L^p-norm of Schrödinger semigroups, *J. Funct. Anal.* **40**, 66 (1981)

2.16   B. Simon, Schrödinger semigroups, *Bull. Amer. Math. Soc.* **7**, 447 (1982)

2.17   H. L. Cycon, R. G. Froese, W. Kirsch, and B. Simon, *Schrödinger Operators with Applications to Quantum Mechanics and Global Geometry*, Textbooks in Mathematical Physics, Springer, New York 1986

2.18   H. Weyl, Das asymptotische Verteilungsgesetz der Eigenwerte linearer partieller Differentialgleichungen, *Math. Ann.* **71**, 441 (1911)

2.19   M. Abramowitz and I. A. Stegun, *Handbook of Mathematical Functions*, Dover, New York 1965

2.20   S. Golden, Lower bounds for the Helmholtz function, *Phys. Rev.* B **137**, 1127 (1965)

2.21   C. T. Thompson, Inequality with applications in statistical mechanics, *J. Math. Phys.* **6**, 1812 (1965)

2.22   K. Symanzik, Proof and refinements of an inequality of Feynman, *J. Math. Phys.* **6**, 1155 (1965)

2.23   L. D. Landau and E. M. Lifschitz, *Statistical Physics*, Vol. 5 of *Course of Theoretical Physics*, Pergamon, Oxford, 1968

2.24   E. Lieb, Calculation of exchange second virial coefficient of a hard-sphere gas by path integrals, *J. Math. Phys.* **8**, 43 (1967)

2.25   R. F. Streater and A. S. Wightman, *PCT, Spin & Statistics, and all That*, Benjamin, New York, 1964

2.26   V. de Alfaro, S. Fubini and G. Furlan, Conformal invariance in quantum mechanics, *Nuovo Cim.* **34A**, 569 (1976)

2.27   R. Jackiw, Dynamical symmetry of the magnetic monopole. *Ann. Phys. (N. Y.)* **129**, 183 (1980)

2.28   P. Y. Cai, A. Inomata, and P. Wang, Jackiw transformation in path integrals, *Phys. Lett.* **91A**, 331 (1982)

2.29   I. H. Duru and H. Kleinert, Solution of the path integral for the H-Atom, *Phys. Lett.* **84B**, 185 (1979)

2.30   I. H. Duru and H. Kleinert, Quantum mechanics of H-Atom from path integrals, *Fortschr. Phys.* **30**, 401 (1982)

2.31   H. Kleinert, *Path Integrals in Quantum Mechanics, Statistics, and Polymer Physics*, World Scientific, Singapore 1990

2.32   W. M. Frank: Class of nonanalytic perturbations in quantum mechanics, *J. Math. Phys.* **8**, 1121 (1967)

2.33   I. Davies and A. Truman, On the Laplace asymptotic expansion of conditional Wiener integrals and the Bender–Wu formula for x^{2n}-anharmonic oscillators, *J. Math. Phys.* **24**, 255 (1983)

2.34   P. A. M. Dirac, The Lagrangian in quantum mechanics, *Physikalische Zeitschrift der Sowjetunion*, Band 3, Heft 1 (1933)

2.35 R. P. Feynman, Space-time approach to non-relativistic quantum mechanics, *Rev. Mod. Phys.* **20**, 267 (1948)

2.36 S. Albeverio and R. Hoegh-Krohn, *Mathematical Theory of Feynman Path Integrals*, Lecture Notes in Mathematics **523**, Springer, New York, 1976

2.37 S. Albeverio, R. Hoegh-Krohn, and L. Streit, Energy forms, Hamiltonians, and distorted Brownian path, *J. Math. Phys.* **18**, 907 (1977)

2.38 S. Albeverio, Ph. Combe, R. Hoegh-Krohn, G. Rideau, M. Sirugue-Collin, M. Sirugue, and R. Stora (eds.), *Feynman Path Integrals, Proc. Int. Coll. Marseille 1978*, Lecture Notes in Physics **106**, Springer, New York 1979

2.39 S. Albeverio and M. Röckner, Dirichlet forms, quantum fields, and stochastic quantization, in: *Stochastic Analysis, Path Integration and Dynamics*, ed. by Elworthy, K. D., and J. C. Zambrini, Wiley, New York 1989

2.40 S. Albeverio, S. Paycha, and S. Scarlatti, A short overview of mathematical approaches to functional integration, in: *Functional Integration Geometry and Strings*, ed. by Z. Haba and J. Sobczyk, Proceedings of the XXV Karpacz Winter School of Theoretical Physics, Birkhuser, Basel, 1989

2.41 M. C. Gutzwiller, A. Inomata, J. R. Klauder, and L. Streit (eds.), *Path Integrals from meV to MeV*, World Scientific, Singapore 1986

2.42 T. Hida and L. Streit, Generalized Brownian functionals and the Feynman integral, *Stoch. Proc. Appl.* **16**, 55 (1983)

2.43 L. Streit, Energy forms: Schrödinger theory, processe, in: *New Stochastic Methods in Physics*, ed. by DeWitt-Morette, C. and K. D. Elworthy, *Phys. Rep.* **77**, 1981

2.44 C. DeWitt-Morette, Feynman's path integral. Definition without limiting procedure, *Comm. Math. Phys.* **28**, 47 (1972)

2.45 C. DeWitt-Morette, Feynman path integrals, I. Linear and affine transformations, II. The Feynman Green's function. *Comm. Math. Phys.* **37**, 63 (1974)

2.46 R. H. Cameron and D. A. Storvick, A simple definition of the Feynman integral, with applications, *Mem. Amer. Math. Soc.* **46**, No. 288 (1983)

2.47 R. P. Feynman, An operator calculus having applications in quantum electrodynamics, *Phys. Rev.* **84**, 108 (1951)

2.48 W. Tobocman, Transition amplitudes as sums over histories, *Nuov. Cim.* **3**, 1213 (1956)

2.49 C. DeWitt-Morette, A. Maheshwari, and B. Nelson, Path integration in phase space, *Gener. Rel. and Gravit.* **8**, 581 (1977)

2.50 K. Gawedzki, Construction of quantum-mechanical dynamics by means of path integrals in phase space, *Rep. Math. Phys.* **6**, 327 (1974)

2.51 R. Fanelli, Canonical transformations and phase space path integrals, *J. Math. Phys.* **17**, 490 (1976)

2.52 I. Daubechies and J. R. Klauder, Constructing measures for path integrals, *J. Math. Phys.* **23**, 1806 (1982)

2.53 I. Daubechies and J. R. Klauder, Quantum-mechanical path integrals with Wiener measure for all polynomial Hamiltonians, *J. Math. Phys.* **26**, 2239 (1985)

2.54 I. Daubechies, J. R. Klauder, and Th. Paul, Wiener measures for path integrals with affine kinematic variables, *J. Math. Phys.* **28**, 85 (1987)

2.55 J. R. Klauder, Quantization is geometry, after all, *Ann. Phys.* **188**, 120 (1988)

2.56 L. van Hove, Sur le problème des relations entre les transforations unitaires de la mécanique quantique et les transformations canonique de la mécanique classique, *Acad. Roy. Belgique, Bull. Classe Sci. Mém. (5)* **37**, 610 (1951)

2.57 M. Böhm and G. Junker, Path integration over compact and noncompact rotation groups. *J. Math. Phys.* **28**, 1978 (1987)

2.58 M. Böhm and G. Junker, Path integration over the n-dimensional Euclidean group. *J. Math. Phys.* **30**, 1195 (1989)

2.59 M. Böhm and G. Junker, Path integrals and spacetime symmetries. *Nucl. Phys.* **B 6**, 259 (1989)

2.60   G. Junker, Explicit path integration on homogeneous spaces. *J. Phys. A: Math. Gen.* **22**, L587 (1989)

2.61   N. K. Pak and I. Sokman, General new-time formalism in path integral. *Phys. Rev.* **A 30**, 1629 (1984)

2.62   Ph. Blanchard and M. Sirugue, Treatment of some singular potentials by change of variables in Wiener integrals, *J. Math. Phys.* **22**, 1372 (1981)

2.63   F. Steiner, Exact path integral treatment of the hydrogen atom, *Phys. Lett.* **106 A**, 363 (1984)

2.64   A. Young and C. DeWitt-Morette, Time substitutions in stochastic processes as a tool in path integration, *Ann. Phys.* **169**, 140 (1986)

2.65   D. Castrigiano and F. Staerk, New aspects of the path integration treatment of the Coulomb potential, *J. Math. Phys.* **30**, 2785 (1989)

2.66   D. Castrigiano and F. Staerk, General aspects of the space and time transformations for the Coulomb propagator, in: *Quantum Probability VI*, ed. by L. Accardi, World Scientific, Singapore 1991

2.67   I. Karatzas, and S. E. Shreve, *Brownian Motion and Stochastic Calculus*, Springer, New York 1988

2.68   S. Albeverio, Ph. Blanchard,and R. Høegh-Krohn, Feynman path integrals and the trace formula for Schrödinger operators, *Commun. Math. Phys.* **83**, 49 (1982)

2.69   S. Albeverio, R. Høegh-Krohn, Oscillatory integrals and the method of stationary phase in infinitely many dimensions, with applications to the classical limit of quantum mechanics I. *Inventions Math.* **40**, 59 (1977)

2.70   C. DeWitt-Morette, A. Maheshwari, and B.Nelson, Path integration in non-relativistic quantum mechanics, *Phys.Rep.* **50**, 255 (1979)

2.71   Z. Brzeźniak, Some remarks on the Feynman-Kac formula, *J. Math. Phys.* **31**, 105 (1990)

2.72   C. Dewitt-Morette, Feynman path integrals, *Acta Phys. Austr. Suppl.* **26**, 101 (1984)

2.73   B. Gaveau, L. S. Schulman, Explicit time-dependent Schrdinger operators, *J. Phys. A: Math. Gen.* **19**, 1833 (1986)

2.74   T. Hida and L. Streit, Generalized Brownian functionals and the Feynman integral, *Stoch. Proc. Appl.* **16**, 55 (1983)

2.75   M. de Faria, J. Potthoff, and L. Streit, The Feynman integrand as a Hida distribution, *J. Math. Phys.* **32**, 2123 (1991)

2.76   D. C. Khandekar and L. Streit, Constructing the Feynman integrand, *Ann. Physik* **1**, 49 (1992)

2.77   T. Hida, H.-H. Kuo, J. Potthoff, and L. Streit, *White Noise – an Infinite Dimensional Calculus*, Kluwer–Academic, New York 1993

2.78   D. C. Khandekar, S. V. Lawande, and K. V. Bhagwat, *Path-Integral Methods and their Applications*, World Scientific, Singapore 1993

Chapter 3

3.1   E. P. Wigner, On the quantum correction for thermodynamic equilibrium, *Phys. Rev.* **40**, 749 (1932)

3.2   A. Goovaerts and J. T. Devreese, Analytic treatment of the Coulomb potential in the path integral formalism by exact summation of a perturbation expansion, *J. Math. Phys.* **13**, 1070 (1972)

3.3   A. Goovaerts, F. Broeckx, and P. van Camp, Evaluation of the even wave functions of the hydrogen atom in a path-integral formalism, *Physica* **64**, 47 (1973)

3.4   H. Leschke, Some bounds on the quantum partition functions by path-integral methods, in: *Path Summation: Achievements and Goals*, S. Lundqvist, A. Ranfagni, V. Sa-yakanit, and L. S. Schulman, World Scientific, Singapore 1988

3.5   W. Fischer, H. Leschke, and P. Müller, Changing dimension and time: two well-founded and practical techniques for path integration in physics, *J. Phys. A: Math. Gen.* **25**, 3835 (1992)

3.6   R. P. Feynman and A. Hibbs, *Quantum Mechanics and Path Integrals*, McGraw-Hill, New York 1965

3.7   R. P. Feynman and H. Kleinert, Effective classical partition function, *Phys. Rev.* A**34**, 5080 (1986)

3.8   R. Giachetti and V. Tognetti, Variational approach to quantum statistical mechanics of nonlinear systems with applications to sine-Gordon chains, *Phys. Rev. Lett.* **55**, 912 (1985)

3.9   R. Giachetti and V. Tognetti, Quantum corrections to the thermodynamics of nonlinear systems, *Phys. Rev.* B **33**, 7647 (1986)

3.10 R. Giachetti and V. Tognetti, Variational approach to the thermodynamics of a quantum sine-Gordon field with out-of-plane fluctuations, *Phys. Rev.* B **36**, 5512 (1987)

3.11 R. Courant and D. Hilbert, *Methods of Mathematical Physics*, Springer, Berlin 1968

3.12 E. Lieb, Bounds on the eigenvalues of Laplace and Schrdinger operators. *Bull. Amer. Math. Soc.* **82**, 751 (1976)

3.13 E. Lieb and W. Thirring, Inequalities for moments of the eigenvalues of the Schrdinger Hamiltonians and their relation to Sobolev inequalities, in: *Studies in Mathematical Physics, Essays in Honor of Valentine Bargmann*, ed. by Lieb, E., B. Simon, and A. S. Wightman, Princeton University Press, Princeton, N. J. 1976

3.14 M. Cwickel, Weak type estimates and the number of bound states of Schrdinger operators, *Ann. Math.* **106**, 93 (1977)

3.15 G. Rosenbljum, The distribution of the discrete spectrum for singular differential operators, *Soviet Math. Dokl.* **13**, 245 (1972)

3.16 P. Lie and S. T. Yau, On the Schrödinger equation and the eigenvalue problem, *Commun. Math. Phys.* **88**, 309 (1983)

3.17 Ph. Blanchard, J. Stubbe, and J. Rezende, New estimates on the number of bound states of Schrödinger operators, *Lett. Math. Phys.* **14**, 215 (1987)

3.18 V. Glaser, H. Grosse, and A. Martin, Bounds on the number of bound states of the Schrödinger operator, *Commun. Math. Phys.* **59**, 197 (1978)

3.19 M. Aizenman and E. Lieb, On the semiclassical bounds for eigenvalues of the Schrödinger operator, *Phys. Lett.* **66** A, 427 (1978)

3.20 A. Martin, New results on the moments of eigenvalues of the Schrödinger Hamiltonian and applications, *Commun. Math. Phys.* **129**, 161 (1990)

3.21 H. Grosse, Quasiclassical estimates on moments of the energy levels. *Acta Phys. Austr.* **52**, 89 (1980)

3.22 M. Abramowitz and I. Stegun (eds.), *Handbook of Mathematical Functions*, Dover, New York 1970

3.23 S. Coleman, *Aspects of Symmetry*, Cambridge Univ. Press, Cambridge 1985

3.24 J. Zinn-Justin, *Quantum Field Theory and Critical Phenomena*, Clarendon, Oxford 1989

3.25 E. Lieb, On characteristic exponents in turbulence, *Commun. Math. Phys.* **92**, 473 (1984)

3.26 F. Dyson, and A. Lenard, Stability of matter I, II, *J. Math. Phys.* **8**, 423 (1967); **9**, 698 (1968)

3.27 E. Lieb, and W. Thirring, Bounds for the kinetic energy of fermions which proves the stability of matters, *Phys. Rev. Lett.* **35**, 687 (1975)

3.28 D. Ruelle, Large volume limit of the distribution of characteristic exponents in turbulence, *Commun. Math. Phys.* **87**, 287 (1982)

3.29 P. Lévy, *Le Mouvement Brownien*, Sci. Math. Fasc. **126**, Gauthier-Villars, Paris 1954

3.30   L. D. Fosdick and H. F. Jordan, Path-integral calculation of the two-particle Slater sum for He^4, *Phys. Rev.* **143**, 58 (1966)

3.31   W. H. Press, B. P. Flannery, S. A. Teukolsky, and W. T. Vetterling, *Numerical Recipes in Pascal*, Cambridge Univ. Press, Cambridge 1989

3.32   J. Schauder, Zur Theorie stetiger Abbildungen in Funktionalräumen, *Math. Zeitschr.* **26**, 47 (1927)

3.33   E. Lieb, Calculation of exchange second virial coefficient of a hard-sphere gas by path integrals, *J. Math. Phys.* **8**, 43 (1967)

Chapter 4

4.1   R. D. Coalson, D. L. Freeman, and J. D. Doll, Partial averaging approach to Fourier coefficient path integration. *J. Chem. Phys.* **85**, 4567 (1986)

4.2   E. P. Wigner, On the quantum correction for thermodynamic equilibrium. *Phys. Rev.* **40**, 749 (1932)

4.3   J. G. Kirkwood, Quantum statistics of almost classical assemblies, *Phys. Rev.* **44**, 31 (1933)

Chapter 5

5.1   J. Ginibre, Some applications of functional integration in statistical mechanics, in: *Statistical Mechanics and Quantum Field Theory*, ed. by C. DeWitt-Morette und R. Stora, (Les Houches 1970), Gordon & Breach, New York 1971

5.2   H. Frhlich, Electrons in lattice fields, *Advan. Phys.* **3**, 325 (1954)

5.3   E. Nelson, Interaction of nonrelativistic particles with a quantized scalar field, *J. Math. Phys.* **5**, 1190 (1964)

5.4   J. Fröhlich, Existence of dressed one-particle states in a class of persistent models, *Fortschr. Phys.* **22**, 159 (1974)

5.5   A. J. Leggett, S. Chakravarty, A. T. Dorsey, M. P. A. Fisher, A. Garg, and W. Zwerger, Dynamics of the dissipative two-state system. *Rev. Mod. Phys.* **59**, 1 (1987)

5.6   M. Fannes, B. Nachtergaele, and A. Verbeure, The equilibrium states of the spin-boson model, *Commun. Math. Phys.* **114**, 537 (1988)

5.7   H. Spohn, Ground state(s) of the spin-boson Hamiltonian, *Commun. Math. Phys.* **123**, 277 (1989)

5.8   C. G. Kuper and G. D. Whitfield (eds.), *Polarons and Exitons*, Oliver and Boyd, Edinburgh, 1963

5.9   R. P. Feynman, Slow electrons in a polar crystal, *Phys. Rev.* **97**, 660 (1955)

5.10   R. P. Feynman, *Statistical Mechanics*, Benjamin, Reading, Mass. 1972

5.11   J. T. Devreese, *Polarons and Exitons in Polar Semiconductors and Ionic Crystals*, ed. by J. T. Devreese and F. M. Peeters, Plenum, New York 1984

5.12   B. Gerlach and H. Löwen, Analytical properties of polaron systems or: Do polaronic phase transitions exist or not?, *Rev. Mod. Phys.* **63**, 63 (1991)

5.13   A. Dvoretsky, P. Erdös, and S. Kakutani, Double Points of Brownian motion in n-space, *Acta Sci. Math. (Szeged)* **12**, 75 (1950)

5.14   H. Spohn, Roughening and pinning transition for the polaron. *J. Phys.: Math. Gen.* **19**, 533 (1986)

5.15   H. Spohn, Effective mass of the polaron: a functional integral approach. *Ann. Phys.* **175**, 278 (1987)

5.16   E. H. Lieb and K. Yamazaki, Ground-state energy and effective mass of the polaron, *Phys. Rev.* **111**, 728 (1958)

5.17   M. D. Donsker and S. R. S. Varadhan, Asymptotic evaluation of certain Markov process expectations for large time, I–IV. *Comm. Pure Appl. Math.* **28**, 1 (1975); **28**, 279 (1975); **29**, 389 (1976); **36**, 183 (1983)

5.18   M. D. Donsker and S. R. S. Varadhan, Asymptotics for the polaron. *Comm. Pure Appl. Math.* **36**, 505 (1983)

5.19   S. I. Pekar, Theory of polarons, *Zh. Experim. i Theor. Fiz.* **19**, (1949)

5.20   E. H. Lieb, Existence and uniqueness of the minimizing solution of Choquard's nonlinear equation, *Studies in Applied Math.* **57**, 93 (1977)

5.21   S. J. Miyake, Strong-coupling limit of the polaron ground state. *J. Phys. Soc. Japan* **38**, 181 (1975)

5.22   E. P. Gross, Strong Coupling polaron theory and translational invariance. *Ann. Phys.* **99**, 1 (1976)

5.23   T. D. Schultz, Feynman's path-integral method applied to the equilibrium properties of polarons and related problems. In: [5.8].

5.24   J. Adamowski, B. Gerlach, and H. Leschke, Strong coupling limit of polaron energy, revisited, *Phys. Letters* **79A**, 249 (1980)

5.25   J. Adamowski, B. Gerlach and H. Leschke, Explicit evaluation of certain Gaussian functionals arising in problems of statistical physics, *J. Math. Phys.* **23**, 243 (1982)

5.26   L. Accardi and W. von Waldenfels (eds.), *Quantum Probability and Applications II*, Proc. Heidelberg 1984, Springer, New York 1985

5.27   L. Accardi and S. Olla, On the polaron asymptotics at finite coupling constant, *In* [5.26].

5.28   E. P. Gross, Path integrals and lower bounds for density matrices, *J. Stat. Phys.* **21**, 215 (1979)

5.29   E. P. Gross, Upper bounds for density matrices using path integrals, *J. Stat. Phys.* **30**, 45 (1983)

5.30   E. P. Gross, Lower bounds for Wiener integrals, *J. Stat. Phys.* **31**, 115 (1983)

5.31   H. Leschke, Functional integral representations and inequalities for Bose partition functions, in: Lecture Notes in Physics **106**, Springer, New York 1979

5.32   D. M. Larsen, Upper and lower bounds for the intermediate-coupling polaron ground state energy, *Phys. Rev.* **172**, 967 (1968)

5.33   J. M. Luttinger and C.-Y. Lu, Generalized path integral formalism of the polaron problem, *Phys. Rev.* B **21**, 4251 (1980)

5.34   W. Thirring, *Lehrbuch der Mathematischen Physik, Bd. 4: Quantenmechanik grosser Systeme*, Springer, Wien 1980

5.35   H. Falk and L. W. Bruch, Susceptibility and fluctuation. *Phys. Rev.* **180**, 442 (1969)

5.36   G. Roepstorff, Correlation inequalities in quantum statistical mechanics and their application in the Kondo problem. *Commun. Math. Phys.* **46**, 253 (1976)

5.37   G. Roepstorff, A stronger version of Bogoliubov's inequality and the Heisenberg model, *Commun. Math. Phys.* **53**, 143 (1977)

5.38   M. Fannes and A. Verbeure, Correlation inequalities and equilibrium states. *Commun. Math. Phys.* **55**, 125 (1977)

5.39   G. Roepstorff, Bounds for a Bose condensate in dimensions $\nu \geq 3$. *Commun. Math. Phys.* **18**, 191 (1978)

5.40   F. J. Dyson, E. H. Lieb, and B. Simon, Phase transitions in quantum spin systems with isotropic and nonisotropic interactions. *J. Stat. Phys.* **18**, 335 (1978)

5.41   G. Roepstorff, On inequalities of Falk, Bruch, Fortune, and Berman, in: *Colloquia Mathematica Societatis János Bolyai. 27, Random Fields*, ed. by J. Fritz, J. Lebowitz, and D. Szasz, North Holland, Amsterdam 1981

Chapter 6

6.1   M. Combes, R. Schrader, and R. Seiler, Classical bounds and limits for energy distributions of Hamiltonian operators in electromagnetic fields, *Ann. Phys.* **111**, 1 (1978)

6.2   J. Avron, I. Herbst, and B. Simon, Schrödinger operators with magnetic fields, I. General interactions, *Duke Math. J.* **45**, 847 (1978)

6.3   J. Avron, I. Herbst, and B. Simon, Schrödinger operators with magnetic fields, II. Separation of the center of mass in homogeneous magnetic fields. *Ann. Phys.* **114**, 431 (1978)

6.4   J. Avron, I. Herbst, and B. Simon, Strongly bound states of hydrogen in intense magnetic fields. *Phys. Rev. A* **20**, 2287 (1979) magnetic fields, III. Atoms in constant magnetic fields, *Commun. Math. Phys.*

6.5   W. Pauli, Über Gasentartung und Paramagnetismus, *Zeits. Phys.* **41**, 81 (1927)

6.6   L. D. Landau, Diamagnetismus der Metalle, *Zeits. Phys.* **64**, 629 (1930)

6.7   L. D. Landau and E. M. Lifshitz, *Statistical Physics*, Pergamon, London 1958

6.8   Y. Aharonov and D. Bohm, Significance of electromagnetic potentials in quantum theory, *Phys. Rev.* **115**, 485 (1959)

6.9   S. Olarin and I. I. Popescu, The quantum effects of electromagnetic fluxes, *Rev. Mod. Phys.* **57**, 339 (1985)

6.10   L. S. Schulman, Approximate topologies, *J. Math. Phys.* **12**, 304 (1971)

6.11   M. Abramowitz and I.A. Stegun, *Handbook of Mathematical Functions*, Dover, New York 1970

6.12   J. Preskill, Magnetic Monopoles, *Ann. Rev. Nucl. Part. Sci.* **34**, 461 (1984)

6.13   G. Giacomelli, in: N. S. Craigie, P. Goddard, and W. Nahm (eds.): *Monopoles in Quantum Field Theory*, World Scientific, Singapore 1982

6.14   R. P. Feynman, R. B. Leighton, and M. Sands, *The Feynman Lectures on Physics*, Vol. II. Sect. 15–5, Addison-Wesley, Reading, Mass. 1965

6.15   M. Peshkin and A. Tonomura, *The Aharonov–Bohm Effect*, Lecture Notes in Physics, **340**, Springer, Berlin 1989

6.16   H. Kleinert, *Path Integrals in Quantum Mechanics, Statistics, and Polymer Physics*, Sects. 16.2–3, World Scientific, Singapore 1990

6.17   R. P. Feynman, R. B. Leighton, and M. Sands, *The Feynman Lectures on Physics*, Vol. III, Sect. 21-7, Addison-Wesley, Reading, Mass. 1965

6.18   P. G. De Gennes, *Superconductivity of Metal and Alloys*, Benjamin, New York 1966

6.19   D. Saint-James, G. Sarma, and E. J. Thomas, *Type II Superconductivity*, Pergamon, Oxford 1969

6.20   A. Inomata and V. A. Singh, Path integrals with periodic constraint: entangled strings. *J. Math. Phys.* **19**, 2318 (1978)

6.21   K. von Klitzing, G. Droda, and M. Pepper, New method for high-accuracy determination of the fine-structure constant based on quantized Hall resistance. *Phys. Rev. Lett.* **45**, 494 (1980)

6.22   D. Thouless, M. Kohmoto, M. Nightingale, and M. den Nijs, Quantized Hall conductance in two-dimensional periodic potential, *Phys. Rev. Lett.* **49**, 405 (1982)

6.23   J. E. Avron, R. Seiler, and L. Yaffe, Adiabatic theorems and applications to the quantum Hall effect, *Commun. Math. Phys.* **110**, 33 (1987)

Chapter 7

7.1   A. Wightman, Quantum field theory in terms of vacuum expectation values. *Phys. Rev.* **101**, 860 (1956)

7.2   R. Streater and A. Wightman, *PCT, Spin & Statistics, and All That*, Benjamin, New York 1964

7.3   R. Jost, *The General Theory of Quantized Fields*, American Mathematical Society, Providence 1965

7.4   O. Steinman, Über den Zusammenhang zwischen Wightmanfunktionen and den retardierten Kommutatoren, *Helv. Phys. Acta* **33**, 257 (1960)

7.5   H. Epstein and V. Glaser, Le role de la localité dans la renormalisation perturbative en théorie quantique des champs, in: R. Stora and DeWitt, *Statistical Mechanics and Quantum Field Theory*, ed. by R. Stora and B. DeWitt, (Les Houches 1970), Gordon and Breach, London 1971

7.6   H. Epstein and V. Glaser, The role of locality in perturbation theory. *Ann. l'Institut Poincaré*, **19**, 211 (1973)

7.7   G. Scharf, *Finite Quantum Electrodynamics*, Texts and Monographs in Physics, Springer, Berlin 1989

7.8   I. M. Gel'fand and N. Ya. Vilenkin, *Generalized Functions*, Vol. 4, Academic, New York 1964

7.9   L. Schwartz, *Theory of Distributions I, II*, Hermann, Paris 1950/51

7.10 J. Glimm and A. Jaffe, *Quantum Physics: A Functional Integral Point of View*, 2nd ed., Springer, New York 1987

7.11 K. Osterwalder and R. Schrader, Axioms for Euclidean Green's functions. I, II, *Commun. Math. Phys.* **31**, 83 (1973); **42**, 281 (1975)

7.12 B. Simon, *The $P(\Phi)_2$ Euclidean (Quantum) Field Theory*, Princeton Univ. Press, Princeton, NJ, 1974

7.13 J. R. Klauder and E. C. G. Sudarshan, *Fundamentals of Quantum Optics*, Benjamin, New York 1968

7.14 G. Roepstorff, Coherent photon states and spectral condition, *Commun. Math. - Phys.* **19**, 301 (1970)

7.15 K. Symanzik, Euclidean quantum field theory, in: *Local Quantum Theory*, ed. by R. Jost, Academic, New York 1969

7.16 E. Nelson, Probability theory and Euclidean field theory, in: G. Velo and A. Wightman *Constructive Quantum Field Theory*, ed. by G. Velo and A. Wightman, Springer, New York 1973

7.17 J. Iliopoulos, C. Itzykson, and A. Martin, Functional methods and perturbation theory, *Rev. Mod. Phys.* **47**, 165 (1975)

7.18 C. Itzykson and J.-B. Zuber, *Quantum Field Theory*, McGraw-Hill, New York 1985

7.19 J. Frhlich, Schwinger functions and their generating functionals. I, *Helv. Phys. Acta* **47**, 265 (1974)

7.20 J. Frhlich, Schwinger functions and their generating functionals. II, *Adv. in Math.* **23**, 119 (1977)

7.21 J. Frhlich, *Nonperturbative Field Theory*, World Scientific, Singapore 1992

7.22 R. Haag, *Algebras and Quantum Fields*, Texts and Monographs in Physics, Springer, Berlin 1992

7.23 R. Fernández, J. Frhlich, and A. D. Sokal, *Random Walks, Critical Phenomena, and Triviality in Quantum Field Theory*, Texts and Monographs in Physics, Springer, Berlin 1992

Chapter 8

8.1   K. Symanzik, Euclidean quantum field theory, in: *Local Quantum Field Theory*, ed. by R. Jost, Academic, New York 1969

8.2   J. Glimm and A. Jaffe, *Quantum Physics: A Functional Integral Point of View*, 2nd ed., Springer, New York 1987

8.3   D. C. Brydges, J. Frhlich, and T. Spencer, The random walk representation of classical spin systems and correlation inequalities, *Commun. Math. Phys.* **83**, 123 (1983)

8.4   Saint-James, G. Sarma, and E. J. Thomas, *Type II Superconductivity*, Pergamon, Oxford 1969

8.5   L. D. Landau and E. M. Lifschitz, *Statistical Physics*, Vol. 5 of *Course of Theoretical Physics*, Pergamon Press, Oxford, 1968

8.6   A. A. Abrikosov, *Dokl. Akad. Nauk SSSR*, **86**, 489 (1952)

8.7   B. M. MacCoy and T. T. Wu, *The Two-Dimensional Ising Model*, Harvard Univ. Press, Cambridge, Mass. 1973

8.8   G. De Angelis, D. de Falco, and F. Guerra, Note on the abelian Higgs model on a lattice: absence of spontaneous magnetization. *Phys. Rev.* D **17**, 1624 (1978)

8.9   J. Frhlich, G. Morchio, and F. Strocci, Higgs phenomenon without symmetry breaking order parameter, *Nucl. Phys.* B **190**, 353 (1981)

8.10 K. Osterwalder and E. Seiler, Gauge field theories on the lattice, *Ann. Phys.* **110**, 440 (1978)

8.11 H. J. Rothe, *Lattice Gauge Theories*, World Scientific, Singapore 1992

8.12 K. Symanzik, Continuum limit and improved action in lattice theories. I. Principles and ϕ^4-theory. *Nucl. Phys.* **B 226**, 187 (1983)

8.13 G. Parisi, *Statistical Field Theory*, Addison-Wesley, Redwood City, CA 1988

8.14 D. W. Duke and J. F. Owens, *Advances in Lattice Gauge Theory*, World Scientific, Singapore 1985

8.15 C. Itzykson, Gauge fields on a lattice, *Phys. Rep.* **23**, 368 (1976)

8.16 J. M. Drouffe and C. Itzykson, Lattice gauge fields, *Phys. Rep.* **38**, 133 (1978)

Chapter 9

9.1   C. N. Yang and R. L. Mills, Conservation of isotopic spin and isotopic gauge invariance, *Phys. Rev.* **96**, 191 (1954)

9.2   E. S. Abers and B. W. Lee, Gauge theories, *Phys. Rep.* **9C**, 1 (1973)

9.3   R. Jackiw, Introduction to the Yang–Mills quantum theory, *Rev. Mod. Phys.* **52**, 661 (1980)

9.4   M. Göckeler and T. Schücker, *Differential Geometry, Gauge Theories, and Gravity*, Cambridge Univ. Press, Cambridge (England) 1987

9.5   D. Bleecker, *Gauge Theory and Variational Principles*, Addison-Wesley, Reading, Mass. 1981

9.6   A. M. Polyakov, *Gauge Fields and Strings*, Harwood, Chur 1987

9.7   A. A. Belavin, A. M. Polyakov, A. S. Schwartz and Y. S. Tyupkin, Pseudoparticle solutions of the Yang–Mills equations, *Phys. Lett.* B **59**, 85 (1975)

9.8   S. Coleman, The uses of instantons, in: A. Zichichi, *The Whys of Subnuclear Physics*, ed. by A. Zichichi, Proceedings of the 1977 International School of Subnuclear Physics, Erice 1977

9.9   L. D. Faddeev and V. N. Popov, Feynman diagrams for the Yang–Mills field, *Phys. Lett.* B **25**, 29 (1967)

9.10 R. L. Arnowitt and S. I. Fickler, Quantization of the Yang-Mills field, *Phys. Rev.* **127**, 1821 (1962)

9.11 C. Becchi, A, Rouet, and R. Stora, Renormalization of gauge theories, *Ann. Phys.* **98**, 287 (1976)

9.12 I. V. Tyutin, Gauge invariance in field theory and statistical mechanics, Lebedev preprint FIAN No. 39 (1975) (unpublished)

9.13 I. V. Tyutin and Sh. M. Schwartsman, BRST operator and an open gauge problem, *Phys. Lett.* **169B**, 225 (1986)

9.14 M. Henneaux and C. Teitelboim, BRST cohomology in classical mechanics, *Commun. Math. Phys.* **115**, 213 (1988)

9.15   J. Fisch, M. Henneaux, J. Stasheff, and C. Teitelboim, Existence, uniqueness and cohomology of the classical BRST charge with ghosts of ghosts, *Commun. Math. Phys.* **120**, 379 (1988)

9.16   V. N. Gribov, Quantization of non-abelian gauge theories. *Nucl. Phys.* B **139**,1 (1978)

9.17   I. M. Singer, Some remarks on the Gribov ambiguity. *Commun. Math. Phys.* **60**, 7 (1978)

9.18   M. Ringe, Über das Problem von Gribov in den Eichtheorien. Thesis, RWTH Aachen 1991

9.19   K. G. Wilson, Confinement of quarks, *Phys. Rev.* D **10**, 2445 (1974)

9.20   S. Elitzur, Impossibility of spontaneously breaking local symmetries, *Phys. Rev.* **D12**, 3978 (1975)

9.21   D. C. Brydges, J. Frhlich, and E. Seiler, Construction of quantized gauge fields. I. General results, *Ann. Phys.* **121**, 227 (1979), II. Convergence of lattice approximations, *Commun. Math. Phys.* **71**, 159 (1980), III. The two-dimensional abelian Higgs model without cutoffs. *Commun. Math. Phys.* **79**, 353 (1981)

9.22   E. Seiler, Gauge theories as a problem of constructive field theory and statistical mechanics, in: Lecture Notes in Physics **159**, Springer, New York 1982

9.23   J. Challifour, A path space formula for non-abelian gauge theories. *Ann. Phys.* **136**, 317 (1981)

9.24   F. J. Wegner, Duality in generalized Ising models, *J. Math. Phys.* **12**, 2259 (1971)

9.25   D. S. Fine, Quantum Yang-Mills on the two-sphere, *Commun. Math. Phys.* **134**, 273 (1990)

9.26   S. K. Blau, The Feynman path integral for constrained systems. *Ann. Phys.* **205**, 392 (1991)

9.27   K. K. Uhlenbeck, Removable singularities in Yang–Mills fields. *Commun. Math. Phys.* **83**, 31 (1982)

9.28   M. A. Solov'ev, Global gauge in a non-abelian theory. *JETP Lett.* **38**, 505 (1983)

9.29   A. J. Niemi, Pedagogical introduction to BRST, *Phys. Rep.* **184**, 147 (1989)

9.30   M. Daniel and C.-M. Viallet. The geometrical setting of gauge theories of the Yang-Mills type, *Rev. Mod. Phys.* **52**, 175 (1980)

9.31   T. Banks, J. Kogut, and R. Myerson, Phase transitions in abelian lattice gauge theories, *Nucl. Phys.* **B 129**, 493 (1977)

9.32   C. Borgs and E. Seiler, Lattice Yang–Mills theory at non-zero temperature and the confinement problem. *Commun. Math. Phys.* **91**, 329 (1983)

9.33   E. Witten, Gauge theories and integrable lattice models, *Nucl. Phys. B 322*, 629 (1989)

9.34   W. Marciano and H. Pagels, Quantum chromodynamics, *Phys. Rep.* **36**, 137 (1978)

9.35   F. Englert and P. Windey, Dynamical and topological considerations on quark confinement, *Phys. Rep.* **49**, 173 (1979)

9.36   S. Sciuto, Difficulties in fixing the gauge in non-abelian gauge theories, *Phys. Rep.* **49**, 181 (1979)

9.37   C. Rebbi, Monte Carlo simulations of lattice gauge theories, *Phys. Rep.* **67**, 55 (1980)

9.38   J. B. Kogut, Progress in lattice gauge theory, *Phys. Rep.* **67**, 67 (1980)

9.39   J. V. Narlikar and T. Padmanabhan, Quantum cosmology via path integrals, *Phys. Rep.* **100**, 151 (1983)

9.40   M. Henneaux, Hamiltonian form of the path integral for theories with a gauge freedom, *Phys. Rep.* **126**, 1 (1985)

Chapter 10

10.1   J. M. Jauch and F. Rohrlich, *The Theory of Photons and Electrons*, Addison–Wesley, New York 1955

10.2   F. A. Berezin, *The Method of Second Quantization*, Academic, New York 1966

10.3   W. Greub, *Multilinear Algebra*, Springer, New York 1978

10.4   K. Osterwalder and R. Schrader, Feynman–Kac formula for Euclidean Fermi and Bose fields. *Phys. Rev. Lett.* **29**, 1423 (1972)

10.5   K. Osterwalder and R. Schrader, Euclidean Fermi fields and a Feynman–Kac formula for boson–fermion models. *Helv. Phys. Acta* **46**, 746 (1973)

10.6   K. G. Wilson, Quarks and strings on a lattice, in: *New Phenomena in Subnuclear Physics,* ed. by A. Zichichi, Plenum, New York 1975

10.7   J. B. Kogut and L. Susskind, Hamiltonian formulation of Wilson's lattice gauge theories, *Phys. Rev.* **D 11**, 395 (1975)

10.8   R. Balian, J. Drouffe, and C. Itzykson, Gauge fields on a lattice, I. General outlook, *Phys. Rev.* **D 10**, 3376 (1974)

10.9   L. Susskind, Lattice fermions, *Phys. Rev.* **D 16**, 3031 (1977)

10.10 K. Fujikawa, Path integral for gauge theories with fermions, *Phys. Rev.* **D 21**, 2848 (1980)

10.11 J. Maalampi and M. Roos, Physics of mirror fermions, *Phys. Rep.* **186**, 53 (1990)

10.12 H. B. Nilson and M. Ninomiya, Absence of neutrinos on a lattice, *Nucl. Phys.* **B 185**, 20 (1981)

10.13 J. Challifour and D. H. Weingarten, Cluster expansion for lattice gauge theories with fermions. *J. Math. Phys.* **19**, 1134 (1978)

Appendix C

C.1   J. L. Jensen, Sur les fonctions convexes et les inégalités entre les valeurs moyennes. *Acta Math.* **30**, 175 (1906)

C.2   W. Rudin, *Real and Complex Analysis*, McGraw-Hill, New York 1974

Index

action at a distance, 208
action integral, 83
affine connection, 303
Aharonov–Bohm effect, 208
area law, 310
Avogadro's constant, 4

Bernoullian random walk, 6
Birman–Schwinger operator, 127, 207
    trace formula, 130, 207
Birman–Schwinger principle, 127
bispinor formalism, 318–319
Bogoliubov inequality, 191
Bogoliubov scalar product, 191
Boltzmann weights, 171
Boson field theory, 217-224
    relation to quantum mechanics, 231
bosons (in solid states), 170
bound states, 126
Box–Muller algorithm, 145
Brillouin function, 254
Brillouin zone, 9, 13, 179
Brownian bridge, 110–112, 180, 357–358
    Fourier decomposition, 150
    polygonal approximation, 144
Brownian motion, 7, 21
    dynamical theory, 33
    fractional, 32
    law of iterated logarithm, 182
    most probable path, 109
Brownian particle, 20, 62
    life time, 65
Brownian path, 7, 27
    decomposition, 111
    Hölder continuity, 181
    mean position, 125
    trail of a –, 8
Brownian tube, 62
BRST symmetry, 298

canonical commutation relations, 170,
    178, 221
canonical ensemble, 71

Chapman–Kolmogorov equation, 5
characteristic function of a measure, 26
charge conjugation, 318
Choquard equation, 186
classical field, 282
    connected with a quantum state, 260
classical filed, 186
classical limit, 257
coherent state, 240
Compton wave length, 249
condensate, 223
conditional expectation, 29
conjugate variables, 262
continuum limit, 13, 179, 244
convexity, 120, 122, 132, 286
Cooper pairs, 269
correlation, 75
Coulomb potential, 50, 119, 121
coupled systems, 157
covariance operator, 227, 358
covariant derivative, 289
critical point, 85
cumulants, 188
Curie point, 268
current (density), 283
curvature, 303
Cwickel–Lieb–Rosenbljum bound, 132

derivative (formal), 331
diffusion constant, 4, 19
diffusion equation, 4, 15
Dirac field, 315
    Euclidean – , 320–324
Dirac string, 213
Dirichlet boundary condition, 59, 62, 112
double-slit experiment, 17
drift velocity, 6
dual pairs of spaces, 229
Duhamel two-point function, 189, 191

effective action, 260–265
effective potential, 68, 69, 73, 154,
    265–268
    mean-field – , 273

eigenvalues, 62, 66, 127
   asymptotic distribution, 62
   moment inequalities, 135
electric field, 48, 161–169
entropy, 252–255
ergodicity, 32
Euclidean action, 230
Euclidean Bose field
   path of the – , 228
   statistical interpretation, 218
   stochastic interpretation, 218, 225
Euclidean momentum, 220
Euclidean spacetime, 216
Euler–Lagrange equations, 89, 106

Faddeev–Popov quantization, 191
Faddeev–Popov action, 298
Faddeev–Popov determinant, 294
Faddeev–Popov ghosts, 296
fermion determinant, 325
fermion doubling problem, 347
ferromagnetic behavior, 268
   Curie–Weiss theory, 274
Feynman function, 220
Feynman integral, 101
   radial, 107
   sequential definition, 102
Feynman–Kac formula, 48–50
Feynman–Kac–Itô formula, 198, 210
finite temperature fields, 221
fluctuation-dissipation connection, 4
flux quantization, 213
Fourier expansion, 161
Fourier transformation, 9
Fourier–Laplace transform, 339
fractal set, 7
free energy, 71, 73
   effective –, 186
   of a quantum field, 259
free field, 76, 238, 264
functional integrals, 329–332
   in QED, 342

Galilean transformation, 6, 205
gamma distribution, 96
gamma matrices, 317
gauge condition, 285
gauge fixing, 283, 286
   axial gauge, 292
   Backgrond gauge, 292
   Coulomb gauge, 292
   Feynman gauge, 288
   Landau gauge, 288, 292
gauge invariance, 292, 295

gauge orbits, 292
gauge theory
   abelian, 288
   nonabelian, 289
   quantum chromodynamics, 346
   Wilson approach, 300
   with fermions, 346
gauge transformation, 48, 192, 269
Gaussian density, 14
Gaussian process, 24, 357
   generalized - , 226
generating function, 10
ghost fields, 297
Gibbs measure, 230–231, 242, 252,
   256–258, 305
Ginzburg–Landau equations, 268–272
Golden–Thompson–Symanzik bound, 67,
   175
   extension of the –, 160
Grassmann algebra, 297, 326
Green's function, 30, 53, 112, 174, 358
Gribov ambiguity, 298
Gribov copies, 298
Gribov horizon, 299
ground state, 37, 58, 60, 74, 77, 88, 100,
   146
   stochastic representation, 58

Hamiltonian, 48, 50, 171, 178, 190, 192,
   207
   in field theory, 236–238
   time-dependent, 47, 159, 160, 175
hard sphere gas, 149
harmonic oscillator, 37, 67, 174
   driven, 124, 161
   inverted, 89
   Mehler's formula, 37, 57, 60, 77, 88,
   89, 124
   partition function, 37
   perturbed, 79
   trace formula, 90
   transition amplitude, 37
harmonic spin chain, 85, 86
heat equation, 4
Heisenberg picture, 75, 174
Hida distribution, 107
   S-transform of the – , 108
   T-transform of the – , 108
Higgs model, 231, 311–315
homotopy, 291

increments of a process, 27
infrared divergence, 12
instantons, 232, 291

integral kernel of an operator, 9
integration (formal), 334
Ising model, 274–278
Itô integral, 195
Itô's lemma, 197

Jensen's inequality, 67, 80, 118, 126, 360–360

Klein–Gordon equation, 217
Kustaanheimo–Stiefel transformation, 78

Lévy algorithm, 143, 148
Landau diamagnetism, 207
Landau free energy, 269
Langevin equation, 33
Laplace–DeMoivre theorem, 4
large deviations, 185
Larmor frequency, 206
lattice, 242, 300
    dual – , 245
    Laplacian on a –, 245
    oriented links, 300
    oriented plaquettes, 304
lattice spacing, 6, 13, 177
Lebesgue theorem of bounded
        convergence, 44, 50
Lie algebra, 289, 300
Lie–Trotter product formula, 56, 103
Lieb's formula, 130–131
loop variables, 308
Lorentz manifold, 285
Lyapunov exponents, 141

magnetic field, 192, 208, 271
    constant – , 201–205
    Landau levels, 270
magnetic flux, 208
magnetic moment, 206
magnetic phase, 209
Markov chain, 1
Markov process, 23, 92
    associated with a Hamiltonian, 75
    transition probabilities, 23
mass matrix, 61, 69, 81
Maxwell distribution, 96
Maxwell field, 282
Maxwell theory, 281–286
    Euclidean version, 281
mean field approximation, 272–274
mean square displaement, 14
measure space, 20
measure theory, 20

method of stationary phase, 104
min–max principle, 127, 137
Minkowski space, 216
Minlos's theorem, 41
Monte Carlo method, 142

O(4)-invariance, 221, 235
observable, 75
open systems, 159
operator norm, 51
    norm convergence, 56
    stochastic representation, 51
order parameter, 267
Ornstein–Uhlenbeck process, 33, 357–358
    covariance operator, 35
oscillator bridge, 165, 175
oscillator process, 36, 38, 77, 357–358
    covariance operator, 36
Osterwalder–Schrader postulates,
        234–237

parity transformation, 317
particle statistics, 68, 70, 72
partition function, 71
path integral
    approximation, 42
    basic concept, 41
    integrable functions, 43
    list of – , 362–367
    several particles, 61
    stochastic formulation, 45
    relation to Brownian motion, 39
path-ordered exponential, 302
Pauli paramagnetism, 207
Peierl's argument, 11
perimeter law, 310
periodic boundary conditions, 85
perturbation expansion, 79
Pfaffian, 337
phase space, 68
    sum over histories, 105
phase transition, 244, 266
phonons, 170, 177
Planck's constant, 15, 68, 110, 255
Poisson statistics, 167
Poisson summation formula, 64
polarization (dielectric), 189
polarons, 177
    effective mass, 183
    field theory, 186
    Fröhlich parameter, 184
    free energy, 182
    Pecar's result, 185
Portenko's lemma, 52, 54

positivity preserving map, 19
probability measure, 20

quantized vortex, 271
quantum fluctuations, 111, 258, 267
quantum Hall effect, 214
quark fields, 346
quenched approximation, 344

random field, 217
random Fourier coefficients, 150
random potential, 175, 180
random variable, 20
    cumulants, 79
    distribution, 21, 27
      joint –, 22
    expectation values, 29
    Gaussian –, 21, 34
      Wick power, 38
    vector–valued –, 20
random walk, 1
    on a lattice, 6, 249
recurrence (random walk), 7, 12
reflection groups, 91–94
    of infinite order, 94
reflection positivity, 234
reflection principle, 91
relativistic invariance, 28
representations of groups, 282
Rollnik class, 128
rotational symmetry, 14

scalar field, 187
    Gaussian approximation, 278–279
    Ising spin limit, 277
    Minkowskian – , 217
scale invariance, 24
scaling, 109–113, 245
Schauder basis, 148
Schrödinger equation, 15
Schrödinger picture, 75
Schrödinger semigroup, 39, 47
Schwartz space, 228, 329
Schwinger functional, 224
Schwinger functions, 220, 224, 284
section map, 285
self-similarity, 24
semiclassical approximation, 117, 152, 199
separation of coordinates, 95
simple events, 20
simple functions, 41
sine-Gordon model, 231

spatial homogeneity, 24
spectral decomposition, 63, 210
spin systems, 84
    ferromagnetic –, 85
spinor field, 316
spontaneous breakdown of symmetry, 266
spreading of the wave packet, 16
stability of matter, 141
statistical independence, 28
statistical operator, 71, 171
stochastic integral, 31, 114, 357
stochastic process, 20, 21
    vector-valued
      covariance matrix, 29
      Gaussian –, 24, 357
      generalized – , 31, 225
      increments of a –, 28
      vector-valued –, 20, 21
      versions of a –, 22
Stokes's law, 4
Stokes's theorem, 302
stopping times, 78
Stratonovitch's midpoint rule, 196
sum over histories, 102, 105
superconductivity, 268
superposition principle, 16
susceptibility, 276

tau functions, 222
Tauberian theorem, 62
temporal homogeneity, 6
thermal wavelength, 111
thermodynamic limit, 82, 179, 244
thermodynamic potential, 267
thermodynamical formalism, 81
thermodynamics, 70
time
    complex – , 216
    first entrance –, 91
    imaginary – , 15, 102, 160, 212
    mean return – , 13
    nonlinear transformation, 8, 369
    reversal of – , 51, 235, 239
    shift of – , 235
time-ordering, 159–160, 222
topological charge, 291
transfer operator, 90
transience (random walk), 12
transition amplitude, 57
    bounds, 115
    essential singularity, 57
    path-dependent –, 173
transition function, 15

transition matrix, 1
triangle problem, 98, 146
two-point function
    Euclidean – , 220
    Minkowskian – , 219

unitary group (time evolution), 16
    $U(1)$, $SU(n)$, 282, 289, 312

vacuum state, 219, 236
Van Vleck determinant, 89
variational principle, 122, 252–256, 283
    constrained – , 261, 269
    deterministic limit, 254
vector potential, 47, 192
    external –, 323
virial coefficient (second), 69, 74, 148
virial expansion, 68

Wegner–Wilson loops, 306–311, 348
Weyl spinor, 317

white noise, 31
white noise calculus, 107
Wick ordering, 38, 108, 241
Wick rotation, 220
Wiener process, 23, 30, 357–358
    anisotropic, 61
    conditional measure, 39, 41, 44, 49,
      66
    covariance operator, 30
Wightman functions, 222, 319
Wightman's reconstruction theorem, 76
Wigner–Kirkwood expansion, 154, 156
winding number, 209, 212–213
WKB approximation, 89

Yang–Mills equation, 291
Yang–Mills field theory, 289
Yang–Mills theory, 310
Yukawa potential, 54

Springer-Verlag and the Environment

We at Springer-Verlag firmly believe that an international science publisher has a special obligation to the environment, and our corporate policies consistently reflect this conviction.

We also expect our business partners – paper mills, printers, packaging manufacturers, etc. – to commit themselves to using environmentally friendly materials and production processes.

The paper in this book is made from low- or no-chlorine pulp and is acid free, in conformance with international standards for paper permanency.

Printing: Mercedesdruck, Berlin
Binding: Buchbinderei Lüderitz & Bauer, Berlin